# Fixed-Income Portfolio Analytics

David Jamieson Bolder

# Fixed-Income Portfolio Analytics

## A Practical Guide to Implementing, Monitoring and Understanding Fixed-Income Portfolios

Springer

David Jamieson Bolder
Bank for International Settlements
Basel
Switzerland

ISBN 978-3-319-12666-1    ISBN 978-3-319-12667-8 (eBook)
DOI 10.1007/978-3-319-12667-8
Springer Cham Heidelberg New York Dordrecht London

Library of Congress Control Number: 2015930328



Printed on acid-free paper

Springer International Publishing Switzerland is part of Springer Science+Business Media (www.springer.com)

*À Nancy, la plus belle saison de ma vie*

# Foreword

Global fixed-income markets are enormous and growing. Trillions of dollars in numerous currencies are invested in these markets by a broad range of investors including pension funds, insurance companies, commercial banks, corporate treasuries, endowments, sovereign-wealth funds, and central-bank reserve managers. Each of these investors allocates significant resources towards evaluating, monitoring, and understanding the day-to-day exposures, performance, and risks associated with the underlying market risk factors found in their portfolios. It is precisely this type of careful attention by individual investors that leads to liquid, well-functioning, and efficient markets.

Quantitative tools are an important aspect of this ongoing oversight and implementation of fixed-income portfolio strategies. Use of quantitative models, however, requires both expertise and caution. Models can be powerful tools for dealing with the uncertainty in fixed-income markets, but since they are only mathematical simplifications of a complex reality, they can also go wrong. This book, written by David Bolder, a member of the staff in the Banking Department at the Bank for International Settlements (BIS), is written with this point firmly in mind. It does not provide a prescriptive solution to fixed-income analytics, but rather takes a suggestive approach. In other words, instead of telling you what you should do, it indicates techniques that you might want to consider, examine further, or possibly implement. Moreover, it consistently tries to offer alternatives—which may perform better or worse under different sets of circumstances—for any given element of analysis. True quantitative expertise is based on the mastery of a wide range of techniques, where the underlying approaches are developed from alternative perspectives. This is an important element of the BIS philosophy and one of the central tenets of this book.

This book simultaneously places a high value on the ongoing validation of these quantitative models. Chapters 8 and 12 are particularly useful in this respect, because they investigate a range of techniques for gauging the accuracy and robustness of one's fixed-income performance and risk measures, respectively. Considering alternative models is always essential, but outlining, in advance, a framework for evaluating the robustness of these models is also critical.

Why does this matter? Quantitative-based approaches, such as those presented in this book, necessarily involve assumptions and approximations. These assumptions will not always hold, nor will the approximations always be good. Realizing this fact about quantitative modelling is important. Prudent fixed-income management, therefore, also requires regular evaluation of one's model results with realized market outcomes—often this is termed back-testing or model validation. This book wisely builds this concept into its overall quantitative framework.

Putting these ideas into the public domain, an approach that is also consistent with the mandate of the BIS, is another way to assess the usefulness and validity of the quantitative approaches suggested in this book. Books are written to share knowledge, suggest ideas, and create discussion. Such a discussion can only propel the fixed-income investment community forward to better, more complete, and more robust quantitative methods.

In summary, no quantitative model is perfect. Faced with the complexity of investing sizeable amounts of money in fixed-income markets, quantitative techniques are nonetheless necessary and useful for a broad spectrum of institutions. Focusing on a range of alternative techniques, making model validation a central part of one's framework, and actively seeking feedback from the academic and practitioner community foster both a better understanding of one's models and the underlying markets. It also ensures that these models continue to be useful to fixed-income investors. This book's commitment to these principles makes me confident that it will be a valuable contribution to the literature.

Basel, Switzerland
September 2014

Jaime Caruana

# Preface

*A journey of a thousand miles must begin with a single step.*

Lao Tsu

The first step in the creation of this book was the in-house development of a software application. The objective of this application was to provide decision-support analytics to a group of portfolio managers. Decision support means, in this context, the comparison of portfolios along a wide range of dimensions, the simulation of portfolio trades, the computation and attribution of risk and performance, and the provision of some optimization tools. We came, over the course of the project, to describe this collection of methods and techniques as *portfolio analytics*.

The portfolios in question were, and still are, comprised of a relatively wide range of high-credit fixed-income instruments including principally sovereign, supranational, agency, and highly rated corporate bonds. The portfolios also included a range of ancillary instruments such as foreign-exchange swaps, bond and rate futures, inflation-linked bonds, and interest-rate swaps. In other words, these portfolios hold the typical fixed-income instruments found in the tool-kit of a reserve-portfolio manager in a central bank, a sovereign-wealth fund, a pension fund, an insurance company, an endowment, or an international institution.

When one sets out to build such a system, the first step involves the establishment of a consistent framework for the classification, comparison, and analysis of different portfolios relative to their benchmark. Such a framework necessarily involves taking explicit decisions and making assumptions about the treatment of a wide range of instruments. I found this task to be particularly challenging given the relative dearth of detailed reference books describing methods for fixed-income portfolio analytics—there were references on fixed-income risk, performance, and exposure, but relatively little combining them in a single setting. The learning curve was steep, but the reward was an in-depth and practical understanding of a number of related ideas that are reasonably well described by the term, fixed-income portfolio analytics. Given the nature and mandate of my employer—the Bank of International Settlements (BIS), which is an international institution serving global central banks—it was naturally decided to share this knowledge with our customers. I consequently began to design presentations for various knowledge-sharing seminars, with central-bank reserve managers, hosted by the BIS.

Presentations can, however, be dangerous. With presentations, participants tend to forget exactly what the speaker said and later, in the comfort of their office, tend to re-interpret the meaning of a slide or a comment in a manner that the speaker did not actually intend. Moreover, an oral presentation rarely has the time—nor do listeners typically have the patience—to go sufficiently deep into the mathematical details. The appreciation of these facts was the genesis of this book.

What the reader might find appealing about this work is that it is *not* an academic work. Instead, it is written for practitioners by a practitioner. The techniques in the following pages are not theoretical—they are used daily in a living, working fixed-income portfolio analytic system. The ideas in this text are inputs to internal and external reports used to take decisions on large fixed-income portfolios. This does not mean that this book has no academic value. On the contrary, many academic concepts and references are employed. What it does mean, however, is that is a practical document intended to help solve practical problems.

Having made this point, the development in the following chapters does not represent the only, nor even the best, approach for the analysis of fixed-income portfolios. Our philosophy in the construction of the application—and the preparation of this book—was the development of a relatively simple, robust, and transparent framework. The advantage of such an approach is that one's computations and analysis are subsequently easier to explain to managers, senior management, and one's analyst colleagues. A clear disadvantage is that the system is always open to criticism that the techniques used are not sufficiently complex and that some of the approximations lack accuracy—we accept this critique and, moreover, encourage others to both challenge and improve upon the methods presented in this text.

Basel, Switzerland
September 2014

David Jamieson Bolder

# Acknowledgements

Writing a book is no small undertaking and it is rarely the work of a single person. This work is no exception. Many people were involved, directly or indirectly, in the preparation of this book. I would first of all like to sincerely thank Jean-Pierre Matt for making this project possible and for consistently supporting the necessary effort involved in the production of this text.

Understanding the needs of fixed-income investors—people who trade in, follow, and basically live in actual markets on a daily basis—is critical to building a practical approach. I would thus like to particularly thank my colleagues Danilo Maino, Alex Joia, Mark Vincent, Peter Van Der Meulen, Jacob Nelson, Mattias Will, and Miklos Endreffy for their constructive criticism and ideas for improvement in the internal software program underlying this analytical framework.

This book took form over the course of a number of years and, during this time, has formed the base material in a series of week-long workshops for central-bank reserve managers. The participants in these seminars deserve my sincere gratitude. Their constant attentiveness, ongoing interest, and active questioning of the ideas and techniques—both during and often long after the workshops—were an invaluable source of motivation and improvement.

I would like to thank Christophe Laforge for many valuable conversations during the initial stages of this project. His calm, open, and logical approach to discussing problems helped me to build a better basic framework.

Finally, and perhaps most importantly, I would like to thank my wife and son for their support, understanding, and patience with the long hours involved in preparing this work.

It should, of course, go without saying that all of my thanks and acknowledgement are entirely without implication. All errors, inconsistencies, shortcomings, or faults in logic remain entirely my responsibility.

# Contents

# List of Figures

# List of Tables

# What Is Portfolio Analytics? 1

*Strategy without tactics is the slowest route to victory. Tactics without strategy is the noise before defeat.*

Sun Tzu

When picking up a book on the topic of portfolio analytics, it is quite natural to ask: what exactly does the author mean? While it is tautologically clear that it will involve the analysis of portfolios, one would hope for a bit more clarity. More helpful, for example, would be to understand exactly what type of analysis is implied. One can analyse portfolios tactically or strategically, over long or short-term horizons, at the asset class or risk factor or instrument level, or from a forward- or backward-looking perspective. This book, so not as to comprise 1,000 pages, will only consider a *subset* of these various perspectives. It is nonetheless important to understand how and why we have selected our specific focus and why our choice might be interesting. To help the reader with these very understandable questions and to hopefully convince him or her to read further, we will attempt, in this brief introductory chapter, to answer this question.

## 1.1 Fixed-Income Portfolio Management

The first step towards answering this question starts with posing yet another question: what are the main elements needed to manage a large fixed-income portfolio? This is important because it permits us to categorize the different perspectives from which a portfolio may be analysed. In the underlying list, we break down these requirements into *five* separate components:

1. **Resources**: First, and most importantly, one requires funds to invest. With this key requirement resolved, then one requires qualified staff, adequate information



D.J. Bolder, *Fixed-Income Portfolio Analytics*,
DOI 10.1007/978-3-319-12667-8_1

systems, and suitable premises for one's investment management activities. These critical logistic elements, while obviously of great significance, are completely ignored in this book quite simply because, while they are the necessary preconditions for managing a portfolio, they do not lend themselves readily to systematic quantitative analysis.

2. **Risk Control and Governance**: Governance simply describes a framework for how an organization makes decisions and how various responsibilities are allocated. Risk control is invariably linked with governance, because risk controllers essentially have the responsibility for ensuring that decisions are taken in accordance with the governance framework and that these responsibilities are properly discharged. Despite its importance, we will not treat this component either. We will nonetheless discuss numerous elements typically computed and controlled by the risk manager.
3. **Strategic Planning**: This involves the important task of establishing one's risk preferences, determining the set of permissible assets for inclusion in one's portfolio, selecting one's desired strategic portfolio, and taking a decision on one's stance on active versus passive positioning. The entire process of strategic planning is often termed *strategic asset allocation* or SAA.
4. **Tactical Planning**: This is basically one's investment process. Broadly speaking, it may involve passively replicating the strategic choices made in the previous step, taking active deviations from the strategic benchmark to implement one's market views, or some combination of these two alternatives.
5. **Implementation**: This essentially represents everything not covered by the preceding elements including actually investing the funds, monitoring these investments and comparing them to one's benchmark.

While one could certainly provide a deeper and more extensive description of the activities involved in fixed-income investment management—with a bit of reflection, for example, one could probably find a sixth or even seventh activity—this is nonetheless a good start. Particularly useful is the distinction between strategic elements, tactical activities, and implementation. While this text will focus predominately on the two latter activities of fixed-income portfolio management, we cannot ignore the importance of strategy.

## 1.2 Strategy

Strategic thinking and planning must always be primordial in portfolio management. As previously mentioned, this is the process of strategic asset allocation.[1] While not explicitly treated in this book, it nonetheless merits some attention. In one of the

[1]Indeed, much research suggests that upwards of 90 % of the variance of the return of one's portfolio is determined by the choice of benchmark: see Brinson et al. [4] and Hood [10]. Although subsequent works, see Ibbotson and Kaplan [11] for example, debate exactly what percentage of return variance is explained, there is general agreement that it is sizable.

strongest and most complete recent treatments of this topic, Meucci [15], provides a succinct, but very useful, definition of SAA:

> An investor seeks a combination of securities that best suit their needs in an uncertain environment.

Although this definition only spans a single sentence, it incorporates, at least, *four* key points:

- "*An investor's needs*"—This hints at the reality that there are many different types of investors with varying objectives. Management of central-bank reserves will certainly involve different objectives as compared with a corporate pension fund, an insurance company, or an endowment. These objectives will in turn shape the investors risk preferences.
- "*combination of securities*"—The set of permissible securities in one's portfolio, not at all independent from one's needs and risk preferences in the previous point, is critical. A reserve manager, for example, is less likely to consider equity investments, whereas a pension fund manager would certainly consider equities to be an essential asset class.
- "*best suit*"—While perhaps not immediately obvious, when one refers to *best*, it generally implies some kind of optimization. SAA is no exception and almost invariably, one employs optimization techniques to find the best combination of securities consistent with one's needs.
- "*uncertain environment*"—This final point is the source of virtually all of the complexity of SAA. If one knew with certainty the future returns of all asset classes, then selecting assets would be a trivial exercise. Unfortunately, this is not the case. An analyst attempting to perform strategic analysis about a portfolio has only historical data, subjective opinions about the future, and market-implied expectations to assist in dealing with this uncertainty. Statistical theory, based on these inputs, is really the only available tool that may be employed to organize one's thinking in this area.

Combining these four key elements of Meucci's definition, we may conclude that SAA is essentially (1) a *prediction* problem and (2) an *optimization* problem. One tries to *predict* the distribution of return outcomes for a set of securities and find the correspondingly *optimal* portfolio conditional on one's predicted distribution and risk preferences. The optimal portfolio stemming from this analysis is technically termed one's strategic portfolio, but is more often simply termed one's benchmark or, to avoid confusion, one's strategic benchmark.[2]

While strategic portfolio analytics are challenging, interesting and essential, such analysis is only performed infrequently. The reason is relatively simple: such analysis is inherently forward-looking and focuses on the medium- to long-term.

[2]One practical work on performing strategic asset allocation analysis is found in Bernadell et al. [3].

One should not expect to revise long-term analysis on a weekly or even monthly basis. One's views on the medium- to long-term evolution of market conditions, one's preferences, and one's set of permissible instruments are highly unlikely to change at such a short frequency. In brief, benchmarks are revised every few years or so and consequently, strategic analysis is also performed at roughly this frequency.

## 1.3 Tactics

Once a strategic benchmark has been decided, it becomes the reference point for all of one's daily, weekly, and monthly decisions. The portfolio manager lives and breathes his or her benchmark and, as such, it is always there in the foreground. This brings us to the activity of tactical planning or what is often termed, tactical asset allocation (TAA). TAA is your investment process, or how you position your *actual* investments relative to your strategic benchmark. There are, in fact, really only *two* possible tactical approaches:

1. *Passive* positioning where one essentially tries to replicate the strategic benchmark; and[3]
2. *Active* positioning, where one takes views on market outcomes relative to one's strategic benchmark.

One might imagine that detailed analysis of one's portfolio is only required when one takes active positions. This is not actually true. In both cases, one needs to be very aware of how the key characteristics of one's portfolio compare to the strategic benchmark. If not, one is basically flying blind. Without a clear understanding of a portfolio's characteristics relative to the benchmark, one cannot actually know if one has taken, or continues to take, a passive stance or is unknowingly undertaking active positions.

There is an exception to this rule. The only time one is uninterested in relative portfolio characteristics is when one *perfectly* replicates the strategic benchmark. Perfect replication implies holding precisely the same set of instruments in precisely the same weights as the benchmark. While this is an extremely effective approach to passive portfolio management—one's portfolio has, by construction, zero deviation to the benchmark—it is rarely employed. The reason is simple. Strategic benchmarks, generally constructed using commercially developed external indices, are typically comprised of a large number of securities; the transaction costs associated with acquiring and regularly rebalancing each of these individual holdings in concert with the benchmark provider are significant. For most strategic benchmarks, therefore, it is simply too expensive. Instead, one typically purchases a smaller number of securities that replicate the key characteristics of one's benchmark. In

[3]A passive, *do-nothing*, tactic is still a *conscious* decision and is not necessarily easy nor inexpensive to implement.

this text, we will assume that the vast majority of passive managers do *not* perfectly replicate, but rather attempt to approximate their strategic benchmark with a smaller number of securities.

### 1.3.1 Asset Classes vs. Risk Factors

Active management involves deviating from the market exposures embedded in the strategic benchmark, whereas passive management seeks to replicate these exposures. Figure 1.1 illustrates graphically the difference between passive and active portfolio management. The idea of replicating one's market exposures is not very precise. We need to introduce a new concept: the risk factor. This notion is of such importance, that it is worth briefly returning to first principles to ensure that we have a common understanding of what precisely it means. To a finance professional, it is perhaps a bit daring to make such an obvious point, but we make it nonetheless (for the sake of completeness):

> Investors are, in expectation, compensated for taking risk.

The point becomes rather more interesting when one asks: which risks are compensated? One often talks about risks being priced—in the equity world these risks are often summarized by the beta exposure to systematic risk factors. In the fixed-income world, this concept is not as common. Investors are still compensated, in expectation of course, for these priced risks. We define these *priced* risks as fundamental risk factors. In other words, a risk factor is any variable that can influence the value of a security.

Investors are *not* uniquely compensated for taking exposure to asset classes, but instead they are compensated for taking exposure to risk factors. There is a subtle, but critical, difference between risk factors and asset classes. Two different

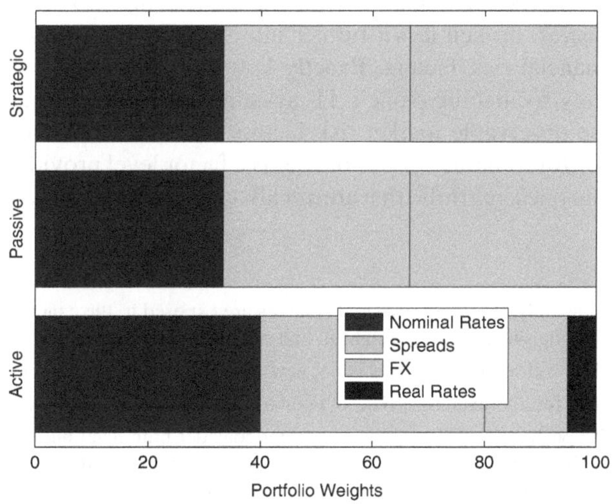


**Fig. 1.1** Active vs. passive positioning. This figure illustrates a key distinction between active and passive tactical positioning. Active management involves deviating from the risk-factor exposures embedded in the strategic benchmark, whereas passive management seeks to replicate a selected set of strategic exposures

asset classes may be exposed to the same risk factor, whereas we can think of risk factors as being separable although, generally speaking, they will be dependent.[4] The important point is that one's portfolio analytics should be performed, to the extent possible, at the risk-factor level.

Ang [1] provides a powerful analogy for understanding the idea of risk factors.[5] He compares asset classes to meals, whereas asset classes are the underlying nutrients. While most individuals plan their eating habits around favourite meals, the nutritionist selects particular foods to target the body's required nutrients. The finance professional should act like the nutritionist. Instead of focusing on asset classes, it is essential for the analyst to consider the underlying risk factors that drive the returns of all securities, albeit in different ways, in his or her portfolio.

Consider *four* different fixed-income asset classes: US Treasuries, US Agency bonds, US corporate bonds, and US mortgage-backed securities. Each of these asset classes are exposed, in varying degrees, to the underlying US Treasury curve. Each of these asset classes, with the notable exception of the US Treasury curve, is also exposed to other their own unique risk factors. Failure to appreciate this point may lead to over- or underestimating one's risk to a given market factor.

Risks may also depend on the nature of the investor. If, as a European investor, one purchases a US Agency bond, one would expect to be compensated for the exposure to:

- to the underlying US Treasury curve;
- to the residual credit of a specific US Agency;
- the relative illiquidity of the instrument purchased; and
- uncertainty associated with the USD/EUR exchange rate.

While US Agencies are a distinct asset class, it certainly shares risk factors with other asset classes.[6] In the US market, for example, all fixed-income investments have some element of risk stemming from the underlying US Treasury market.

The risk associated with the underlying US Treasury curve can be, if one desires, broken down further into separate monetary policy, macroeconomic, and financial risk factors. Exactly how deep one goes is a matter of taste, situation, daily availability, one's IT systems, and one's requirements.[7] We will tend to use observable market risk factors in this text. Whatever your choice, examining portfolio characteristics on the risk-factor level provides a number of useful insights into one's portfolio that are *not* always evident when using an asset class perspective.

[4]Logically, therefore, two asset classes are exposed to the exact same set of underlying risk factors with the same relative weight if, and only if, they are the same asset class.

[5]The notion of risk factors is not new. See Fama [7] and Fama and French [8].

[6]The foreign-exchange risk, of course, would not be present for an American investor.

[7]To push Ang's [1] analogy, we might consider high level nutrients such as protein, carbohydrates or fats. One could, however, go deeper and consider the amino acids forming the underlying proteins, or specific vitamins, or types of fats. In short, choices need to be made.

## 1.4 Strategy and Tactics

Where do all of these ideas bring us? We still have not clearly defined what precisely is meant by portfolio analytics, although we now have the necessary background to do so. Portfolio analytics in this text focuses on the technical aspects of day-to-day management of one's portfolios. Thus, our perspective is concerned about implementation and tactics and is not concerned with strategic questions.[8] An institution's strategic choice, encapsulated by its strategic benchmark, is always the point of reference or the yardstick for all of our subsequent analysis.

Figure 1.2 describes the relative differences between strategic and tactical analysis in terms of resource allocation and relative contribution to the portfolio's risk. Strategic analysis is of tremendous importance and accounts for much of one's overall risk, but it consumes a rather modest amount of an organization's overall resources. This is a simple consequence of the fact that, by construction, strategic analysis cannot be performed frequently. Tactical analysis, conversely, accounts for much less of one's total risk, but consumes substantial time and effort.

One may legitimately ask why, if strategic decisions dominate their tactical counterparts, should one focus on tactical decisions at all? While this is a good question, it underestimates the importance of tactical management as it was previously defined. Tactical management is not only active positioning of the portfolio relative to the benchmark, it is more importantly the implementation of the strategic benchmark. To this end, day-to-day tactical management is about ensuring that the strategic vision of the organization becomes a reality. While it also seeks to modestly add value by deviating slightly from the benchmark, this is not the principal task.

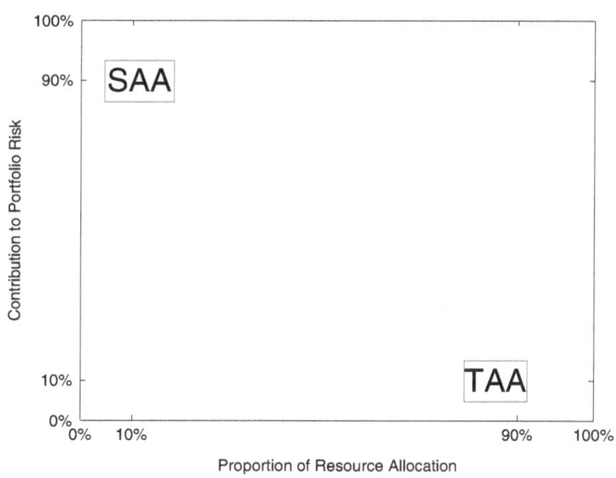


**Fig. 1.2** Strategic vs. tactical analysis. The following figure describes the relative differences between strategic and tactical analysis in terms of resource allocation and relative contribution to the portfolio's risk. It is clear that strategic analysis is of tremendous importance, but cannot be performed frequently. Tactical analysis, while less important in terms of overall risk, is a daily activity

[8] Strategy—and this cannot really be stressed enough—remains a key input into our entire framework.

Failure to expose one's portfolio's to the set of desired risk factors identified in the SAA process would undermine one's overall objectives.

Only with careful and continuous monitoring of one's daily exposure, risk, and return figures can one hope to accurately and effectively implement one's strategic vision. In short, therefore, tactical planning is the practical side of one's strategy and, as such, takes on significant importance. This work seeks to explore different techniques to assist with these day-to-day tasks.

Coming finally to the point, we define, for our purposes, portfolio analytics as:

> Understanding and comparing—along a number of quantitative dimensions—a fixed-income portfolio relative to its benchmark.

This definition implies that portfolio analytics is a collection of tools that can be used to understand, oversee, and control one's day-to-day fixed-income investment operations. Finally and importantly, our approach to this form of portfolio analytics seeks to operate not in terms of asset classes, but instead upon underlying risk factors.

## 1.5 Key Characteristics

As hinted on a number of occasions, an understanding of one's principal portfolio characteristics relative to the benchmark is primordial. There are *three* main dimensions that we will consider under the heading of portfolio analytics: exposure, risk, and return. All three of these elements operate best from a risk-factor perspective, although one can also examine them from a variety of alternative perspectives. In this section, we will attempt to expand somewhat on these three elements.

In keeping with the risk-factor perspective, exposures are essentially the sensitivity of one's portfolio return to movements in the underlying risk factors. These are snapshot measures that describe, for a given point in time, the sign and magnitude of one's position with respect to a given risk factor.[9] Exposures evolve over time with changes in market conditions. Active exposures are the differences between the portfolio sensitivities to a given risk factor and those in the underlying strategic benchmark.

Depending on the state of the market, risk factors exhibit different levels of volatility and dependence with other risk factors. Exposures do *not* capture this perspective. Capturing this time-varying element is the role of risk analysis. Risk essentially describes the uncertainty of future portfolio and strategic benchmark returns. Monitoring of one's active risk positions and attributing them to the various risk factors is thus a useful addition to the snapshot perspective afforded by analysis

[9]Transforming exposures into risk, as we will see, requires a statistical view of the joint evolution of the risk factors over time. Such a transformation is essential because the actual risk associated with a given exposure is not actually static.

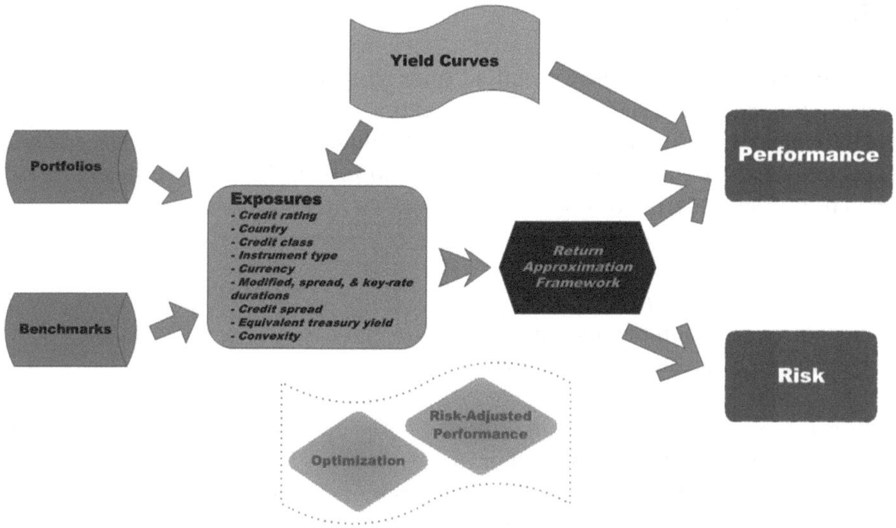


**Fig. 1.3** Portfolio analytics schematic. The underlying figure seeks to describe the key elements of portfolio analytics addressed in the following chapters

of one's exposures.[10] The risk perspective essentially transforms the snapshot exposure measures into an aggregate picture of uncertainty associated with each risk factor.

The return element is the final reckoning. It is the compensation that one has received for assuming risk. Sadly, sometimes this compensation is negative. Understanding one's return is critical to understanding the success, or failure, of one's exposure and active-risk positions. Again, such computations are not only of interest to active managers. Given that passive management typically does not involve perfect replication, comparing one's portfolio and benchmark returns is a very effective approach to assessing the efficiency of one's replication strategy.

To perform such a detailed analysis, we need to develop a framework for mapping risk-factor exposures into returns. With such a mapping, one can compute and attribute the risk and return of a portfolio to the individual risk factors. Along the way, one requires a few other tools including a yield-curve model and some optimization techniques. Each of these elements, and the associated tools, are summarized in Fig. 1.3 and treated in substantial detail in the coming chapters.

[10]Active risk is formally defined as the uncertainty of the difference between future portfolio and strategic benchmark returns.

### 1.5.1 Principles

To complete our discussion, we outline some ground rules, or principles, for the forthcoming chapters. Such principles are extremely helpful in guiding the depth, breadth, and composition of the discussion. It should also help to set the reader's expectations about the nature of the treatment to follow. To be more specific, *four* main principles underpin this book:

- no black boxes;
- a common framework;
- no unnecessary complexity; and
- *usefulness* as the ultimate criterion.

Black boxes are, in many walks of life, fairly inevitable. I would suspect that for most readers, including the author, the exact inner workings of the carburator found in one's car is something of a black box.[11] Without such knowledge, life goes on and one's capacity to drive his or her car is in no way hampered. A lack of understanding of a carburator would nevertheless be a major shortcoming for an automotive mechanic. The same applies to portfolio analysis. This book is written for professional financial analysts and, as such, it would be unfair to skip over important details necessary for each analyst to fully understand the presented techniques. The price of this principle is detail. When we have felt that the details are potentially overwhelming, they are relegated to separate shaded box, which may be read or skipped by the reader without impacting the flow of the text.

There is a plethora of books and articles in the finance literature that consider either risk or performance, but relatively few that jointly consider both. At the same time, there are numerous commercial systems that provide either risk or performance analysis, but not both. This implies that there is rarely a common foundation for performance and risk computations and analysis. This is, in our view, a shortcoming. This text offers a detailed, robust and consistent framework for the joint consideration of each dimension—exposure, risk, and performance—across a wide range of fixed-income instruments and risk factors. This common framework is important, because risk and performance are best examined jointly and consistently.

Markets are complex and fixed-income markets are no exception. There exist a wide array of different models of varying degrees of complexity that seek to describe markets. It is always tempting to add complexity to one's models in an effort to better describe the real world. Financial markets are *not* governed by anything as reliable and robust as the laws of physics, but instead, at their core, are driven by human behaviour. This makes markets subject to structural changes

[11]We probably understand, albeit very vaguely, that it combines air and fuel in some way and sends it to your engine.

in behaviour, unpredictably unpredictable and therefore risky.[12] Our approach is to employ relatively simple models that are easily understood and explained, with the knowledge that they may be at times oversimplified and fallible.[13]

The use of complexity for complexity's sake is something we seek to avoid in this text. We make ample use of simplifying assumptions—such as normality and linearity—and regularly employ mathematical approximations. Complexity is naturally unavoidable, but we argue and attempt to make use of it only when it is necessary. Finding the appropriate balance is difficult. We leave it to our readers to decide if we have succeeded or not.

Somewhat different is the notion of usefulness.[14] In seeking to find the appropriate tool or determining the appropriate balance between simplicity and complexity, we have principally employed a single criterion: usefulness. Ultimately, this is the only criterion that matters for choosing a model, a technique or a tool. If it is functional, if it can be applied robustly across many situations, and if it can be reliably interpreted and explained: then it is useful. If it useful, then it should be included.

These guiding principles describe how we have attempted to write the following chapters. Although we have done our best to provide a complete and fair description of portfolio analytics for tactical day-to-day portfolio management, the reader should not come away with the impression that this is the *only* way to perform this analysis. Our perspective on the aforementioned principles may deviate dramatically from the reader's view. This may lead to different approaches, different levels of complexity, and ultimately, different analytics. What we present in the following chapters is suggestive and we therefore encourage the reader to challenge our thinking, look for mistakes or weaknesses in logic, simultaneously seek improvements, and communicate these points to us. In this way, hopefully this book itself will prove useful.

## 1.6 An Appetizer

We have dedicated a substantial amount of ink to defining our (somewhat abstract) version of portfolio analytics. Perhaps disappointingly, we have not yet actually seen any portfolio analytics. In this penultimate section, we seek to rectify this shortcoming. In the following few pages, we will examine a simple single-currency,

[12]Unpredictable in itself is not so problematic. Card games are unpredictable, but the probabilities of given events are fixed over time. Sudden changes in the behaviour of market participants leads to this unpredictable unpredictability.

[13]Simple models often prove to be more robust to change than their more complex counterparts.

[14]This applies to all walks of life. Computer applications are a good example. As well stated by the founder of the Linux operating system, Linus Torvalds, "any program is only as good as it is useful." We wholeheartedly agree and feel that this sentiment applies, this work notwithstanding, equally to finance literature.

single-credit portfolio. We will examine this portfolio, which has a US Treasury benchmark, over the course of a single month: February 2012.

The exposition is intended to provide a flavour of what can be done and what will be covered in the following chapters. To be more specific, we will examine exposure, risk and return from a risk-factor perspective. We can, and will, drill down much deeper into the computation and interpretation of the results later in this book. This section should be treated as something of an appetizer for the subsequent discussion.

### 1.6.1 Exposure

The first step in understanding one's position relative to a given benchmark involves examining one's exposure, or sensitivity, to the underlying risk factors. The portfolio under examination in this section is composed uniquely of US Treasury bonds, bills, and notes as is its underlying benchmark. It has neither currency nor credit risk.[15] The principal remaining risk factor—and the only assumed risk factor for this portfolio—is the US Treasury yield curve. One could easily imagine that the UST curve is, in turn, itself driven by a set of macroeconomic risk factors.[16] Our approach, however, remains at the market level.

As we will see in the subsequent chapters, a simple, but useful, notion of the exposure to movements in yields is summarized by the modified duration. Figure 1.4 outlines the evolution of the active modified duration of the portfolio over the course

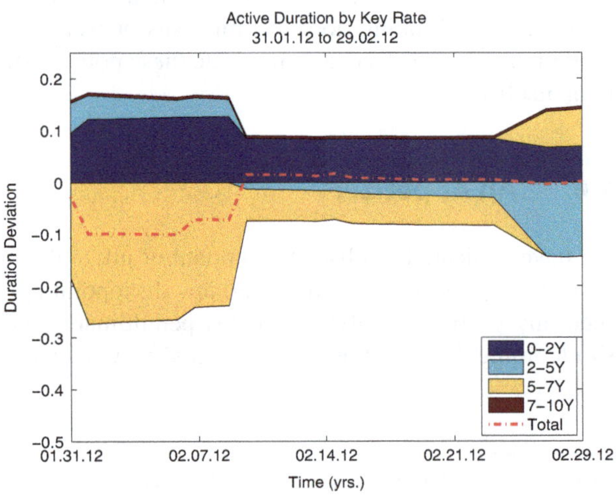


**Fig. 1.4** UST portfolio exposures by key rate. This graphic outlines the evolution, over the course of a single month, of the daily active duration exposure for a US Treasury portfolio

[15]There is, of course, some credit risk associated with the US Treasury, but it is sufficiently small that we do not model it explicitly.

[16]Excellent examples of this relationship are provided by Ang and Piazessi [2] and Diebold et al. [5, 6].

of the month—active modified duration is described as the difference between the portfolio and benchmark's modified duration.

The dashed red line in Fig. 1.4 shows the total active modified duration, which moved from a negative position at the beginning of the month to basically flat about one third into the month. The examination of active modified duration is taken a step further by examining the sectors contributing to the overall position. These sectors are organized by tenor, but could also be broken down by currency, country, instrument, or credit class. Given our simple portfolio structure, a detailed breakdown beyond tenor is not meaningful and, thus, not provided.

For the first part of the month, the portfolio was short the 5–7 year sector and long the 0–2 and 2–5 year sectors. When the short duration position was closed, it nevertheless still involved a long position in the 0–2 year sector offset by shorts in both the 2–5 and 5–7 year sectors. At the end of the month, a change in the curve positioning of the portfolio was again performed, despite no modification to the overall active duration stance. The portfolio ended the month with a short 2–5 year position offset by long 0–2 year and 5–7 year positions.

A surprisingly rich variety of positions can be taken in a relatively simple single-currency, single credit fixed-income portfolio. Active duration has *not* been constant over the month. In this example, the tenor sectors are the only risk factors—in a more complex example, this would be significantly expanded. Examining the portfolio's exposure at a single point in time—either the beginning or end of the month—would lead to erroneous conclusions about the portfolio's strategy. The computation and intertemporal examination of portfolio exposure relative to its benchmark is thus the first dimension of portfolio analytics and forms an important element of this book. Moreover, it is the foundation for the subsequent risk and performance analysis.

### 1.6.2 Risk

The second key characteristic of portfolio analytics is risk. This may be slightly confusing since often one considers the exposure as an approach to summarizing the risk position of the portfolio. Exposures are snapshots. Measuring risk is an attempt to enrich the exposure with the market dynamics of the risk factors.[17] An active modified duration of 0.5, for example, is always the same irrespective of the current market conditions. When one considers the current level of volatility and correlation present in the marketplace, it is natural to expect that an active duration position of 0.5 may vary in terms of overall risk.[18] Risk is dynamic—it combines portfolio

[17]Our principal focus in this book is market risk. Credit risk is only considered through the lens of the market. Excellent sources on credit risk include Jarrow and Turnbull [12], Schönbucher [16], Gordy [9], Jeanblanc [13] and Jeanblanc and Rutkowski [14].

[18]Of course, constructing a sensible and reliable estimate of the unobservable volatility and correlation of our portfolio's risk factors is not always obvious. This will be treated in much more detail in later chapters.

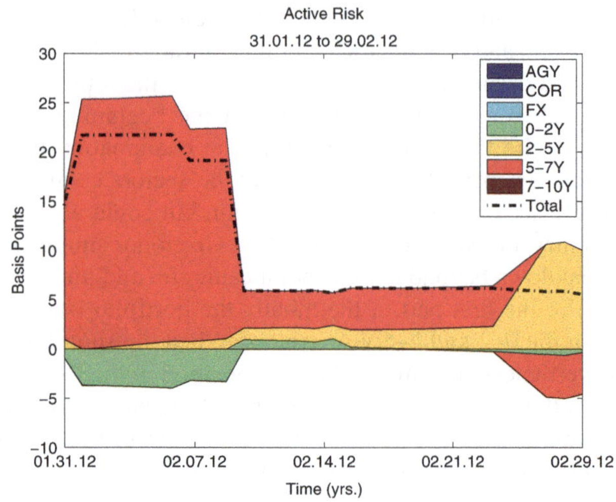


**Fig. 1.5** UST portfolio risk by key rate. This graphic outlines the evolution, over the course of a single month, of the daily ex-ante tracking error of a US Treasury portfolio

composition with market conditions. It maps portfolio exposures, volatilities, and correlations for each risk factor into a dynamic assessment of the uncertainty regarding the portfolio's return.

Figure 1.5 provides, for the same US Treasury portfolio over the same time horizon, an assessment of the portfolio's ex-ante tracking error.[19] The black dotted line indicates the evolution of the overall ex-ante tracking error, while the other colours indicate the source and relative contribution of the various risk factors to the overall total risk. While exposure to agency and corporate bonds and foreign-exchange risk are included as risk factors, they do *not* contribute to the overall tracking error. This is consistent with the point that this is a single-currency, single-credit mandate.[20]

Starting at around 20 basis points, the tracking error falls to slightly more than five basis points at precisely the same time as the short active duration position is closed. The lion's share of the initial tracking error stems from the short position in the 5–7 year sector. By the end of the month, the principal contributor to active risk switches to the 2–5 year sector, with the 5–7 sector actually acting to reduce overall ex-ante tracking error.

We have seen the evolution of active duration, curve exposures and risk of this portfolio over the course of February 2012. What can we conclude? Overall exposure and risk was reduced substantially about one third into the month. After a

[19]Ex-ante tracking error is essentially the forward-looking estimated volatility of the portfolio's active return, where active return is defined as the difference between the portfolio and benchmark return.

[20]It is useful to verify that the portfolio is indeed consistent with its mandate and has neither currency nor credit risk. If not, this would certainly lead to an urgent, and perhaps unpleasant, conversation with the portfolio manager.

few weeks of this relatively neutral position, a modest curve position was added; it involved shorting the intermediate (i.e., 2–5 year) sector with offsetting positions in the short (i.e., 0–2 year) and longer (i.e., 5–7 year) portions of the curve.

### 1.6.3 Return

The final step is to examine the relative success of these activities. To do this correctly, we need to do more than merely compute the return over the month. We need to compute the return for each day in the month and decompose this return into the contribution from each risk factor. We then need to examine how each of these risk factors contributed, over the course of the month, to the final relative performance of the portfolio. This entire process is termed performance attribution.

Figure 1.6 performs these computations for our simple illustrative portfolio. The dashed black line describes the cumulative active return (i.e., portfolio less benchmark) over the course of the month. The results are generally positive with an approximately three basis-point out-performance relative to the benchmark over the month.

The majority of the risk factors appear to be contributing positively to the overall return, with the main contributors being the 2–5 and 5–7 year sectors. The 0–2 year position was the sole negative contributor to the active performance reducing the overall active return by roughly two basis points. Interestingly, this was the only consistently long duration position in the portfolio. Given that the portfolio was short the 2–5 and 5–7 year sectors over the course of the month and that these two sectors contributed most of the out-performance, one can conclude that the curve positioning was generally successful. One might, however, be inclined to discuss with the portfolio manager the reason for closing of the overall short position, since the evidence suggests that this would have likely been successful. This type of

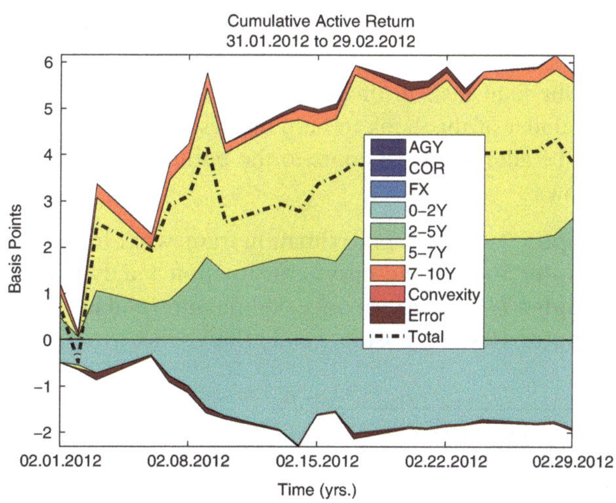


**Fig. 1.6** UST portfolio return by key rate. This graphic outlines the evolution, over the course of a single month, of the cumulative daily return of a US Treasury portfolio organized by risk factor

discussion can improve understanding of the active investment process and, when performed on a regular basis, help improve it.

We also observe, to our satisfaction, that neither credit (i.e., agency or corporate) nor foreign-exchange factors contributed to the portfolio's return. A fraction of a basis point stemmed from exposure to yields in the 7–10 year sector, but not very much as there was essentially no active exposure to this part of the curve over the month.

Two additional factors also arise in this analysis: convexity and error.[21] In general, one hopes to keep the errors, which is also termed the unexplained or residual component in a performance attribution, as small as possible. These additional elements, among others, will be derived, illustrated and discussed in significant detail throughout the course of the subsequent chapters.

This single-currency, single-credit portfolio may be interesting, but it may also be somewhat frustrating. We have glossed over many details. This is unavoidable as these details are, in fact, the subject of this book. Hopefully, we have nevertheless been successful in providing you as the reader, with a taste of what is to come and piqued your interest. Much more can be accomplished, but the overall idea is to gain high-dimensional insight into the inner workings of one's fixed-income portfolio. We also hope we've been reasonably clear about what we mean by portfolio analytics and, equally importantly, what is possible with these techniques.

## 1.7 The Coming Chapters

The goal of this chapter was to define what we mean by the idea of portfolio analytics. Our final definition was a collection of quantitative tools applied to understand, oversee, and control one's day-to-day fixed-income portfolio operations with a particular focus on risk factors. We also examined a practical example to demonstrate, at least in a very stylized manner, what might be accomplished. As must be the case in an introductory chapter, our illustrative example skipped over many important and interesting details. The subsequent chapters will address these points.

Our final task in this chapter is to provide the reader with a roadmap to the remainder of this book to help understand where, when, and how important details will be addressed. To come to the point, the subsequent chapters are organized as follows:

**Chapter 2–4** an approximation framework for our analysis;
**Chapter 5–6** yield-curve construction and dynamics;
**Chapter 7–9** performance computation and attribution;
**Chapter 10–12** absolute and relative risk computation and attribution;

[21]Convexity is a correction for the necessary approximations made in this return decomposition. The error or residual term, quite simply, is the aspect of the return that our model is incapable of explaining.

**Chapter 13–14** risk and return in combination;
**Appendix A** a technical refresher; and
**Appendix B** a few thoughts on the use of optimization.

The objective of this book is to demonstrate how one might compute the exposure, risk, and return of a fixed-income portfolio associated with its set of fundamental risk factors over time. To this end, Chaps. 2–4 are essentially concerned with how one computes exposures for a wide range of risk factors and fixed-income instruments and maps these exposures into returns. This is the foundation of our portfolio-analytic framework. Chapters 7–9, working from the ideas in Chaps. 2–4, deal with the fundamental details involved in decomposing the portfolio return to these risk factors, while Chaps. 10–12 are concerned with, across a number of different perspectives, the computation of risk. In both settings, we build upon the basic framework and extend it to the computation and attribution of return and risk associated with each of the underlying fixed-income risk factors. Risk and return attribution and analysis are both the heart of this book and our perspective of fixed-income portfolio analytics.

Chapters 5–6 and Appendices A and B are concerned with the technical tools—such as curve construction, simple optimization techniques, statistics, probability, and matrix algebra—required to accomplish our principal tasks. Many of these ideas may be known to some readers, but for the sake of completeness, and to provide an overview of the range of possible tools at one's disposal, we found it useful and important to include them. Many of these elements of the book may be skipped by the experienced analyst. We nonetheless hope that, even for the initiated, our yield-curve treatment will prove interesting and illuminating. We have attempted to present this material in a new light and tried, to the extent possible, to make it both accessible and offer simpler alternatives to the standard, and often rather complex, approaches to yield-curve modelling.

The final two chapters (Chaps. 13 and 14) deviate from the basic structure and seek to *jointly* examine the risk and return of the portfolio. This is intended to be an interesting and useful complement to principal ideas forwarded in the other chapters. Moreover, it takes a few tentative steps towards simultaneously examining the two key dimensions—risk and return—involved in the selection of one's strategic benchmark. As we have clearly indicated, this text takes the strategic benchmark as given. It does not, however, imply that we may not use the experience and insights gained through the consistent and careful application of portfolio analytics to the strategic-benchmark decision.

## References

1. A. Ang, *Asset Management: A Sytematic Approach to Factor Investing* (Oxford University Press, New York, 2013)
2. A. Ang, M. Piazzesi, A no-arbitrage vector autoregression of term structure dynamics with macroeconomic and latent variables. J. Monet. Econ. **50**, 745–787 (2003)

3. C. Bernadell, J. Coche, K. Nyholm, *Yield curve prediction for strategic investors*. European Central Bank Working Paper No. 472, April 2005
4. G.P. Brinson, B.D. Singer, G.L. Beehower, Determinants of portfolio performance ii: An update. Financ. Anal. J. **47**(3), 40–48 (1991)
5. F.X. Diebold, M. Piazzesi, G.D. Rudebusch, *Modeling bond yields in finance and macroeconomics*. Federal Reserve Bank of San Fransisco Working Paper 2005-04, January 2005
6. F.X. Diebold, G.D. Rudebusch, S.B. Aruoba, *The macroeconomy and the yield curve: a nonstructural analysis*. University of Pennsylvania Working Paper, 2004
7. E.F. Fama, Components of investment performance. J. Finance **17**, 551–567 (1972)
8. E.F. Fama, K.R. French, Common risk factors in the returns on stocks and bonds. J. Financ. Econ. **33**, 3–56 (1993)
9. M.B. Gordy, *A comparative anatomy of credit risk models*. Board of Governors of the Federal Reserve System, December 1998
10. L.R. Hood, Determinants of portfolio performance: 20 year later. Financ. Anal. J. **61**, 6–8 (1986)
11. R.G. Ibbotson, P.D. Kaplan, Does asset allocation policy explain 40, 90 or 100 percent of performance. Financ. Anal. J. **56**, 26–33 (2000)
12. R.A. Jarrow, S.M. Turnbull, The intersection of market and credit risk. J. Bank. Finance **24**, 271–299 (2000)
13. M. Jeanblanc, *Credit risk*. Université d'Evry, April 2002
14. M. Jeanblanc, M. Rutkowski, *Modelling of default risk: an overview*. Université d'Evry and Warsaw University of Technology, September 1999
15. A. Meucci, *Strategic Asset Allocation*, 1st edn. (Springer, Berlin, 2007)
16. P.J. Schönbucher, *Credit risk modelling and credit derivatives*. Rheinischen Freidrich-Wilhelms-Universität Bonn—Dissertation, January 2000

# Part I

# From Risk Factors to Returns

The following chapters set the stage by identifying the risk factors associated with a collection of generic fixed-income securities. Employing these exposures, or sensitivities, to these risk factors, we then proceed to construct an additive linear relationship between the security's return and the return associated with each individual risk factor. We also demonstrate how it may be extended to incorporate a relatively wide range of fixed-income securities found in modern fixed-income portfolios. This relationship represents the foundation of our portfolio-analytic framework.

# Computing Exposures

# 2

*An idea which can be used once is a trick. If it can be used more than once, it becomes a method.*

George Pólya

What factors affect the value of a security? That is a common, and important, question asked by investors across all asset classes. Investors of fixed-income securities are no exception. Security holders take risk. They must be, at least in expectation, compensated for taking this risk. Factors that affect the value of a security, therefore, are those factors that generate risk or, in other words, return or loss. We typically call these *risk factors*. In this chapter, we will review a number of risk factors that drive fixed-income security returns. We will particularly focus on the exposure, or sensitivity, of a fixed-income security to these risk factors. These sensitivities form the backbone of portfolio analytics since they permit us to quickly understand the nature of a security's or, more generally, a portfolio's risk both on an absolute or relative basis. The base unit of examination, however, is the security. Portfolios then are merely collections of securities.

While there are many excellent sources that describe the sensitivities of fixed-income instruments, we take the time to derive a number of key measures in order to ensure consistent notation, a common understanding, and to permit a self-contained discussion. Moreover, we also hope that the reader will be exposed to a few new ideas.

## 2.1 A Starting Point

Terminology is useful in any profession as it permits the succinct and precise description of complicated ideas. It can, on occasion, lead to confusion when the underlying ideas are not fully understood or multiple definitions for a given term exist. The concept of duration in the field of finance appears to fall into this category.



D.J. Bolder, *Fixed-Income Portfolio Analytics*,
DOI 10.1007/978-3-319-12667-8_2

There are many different notions of duration; MacCauley, modified, effective, spread, and key-rate duration are a few commonly used examples. Moreover, sometimes duration is quoted as a sensitivity and sometimes it is described as a cash-flow weighted time to maturity of a fixed-income security. It is fair to say, therefore, that when one evokes the term *duration*, not everyone immediately shares the same understanding of what it means. Given the potential for confusion arising from a loose use of terminology, we will work correspondingly hard to thoroughly define all terms used in this text.

Our true starting point is the bond-price equation. The value, at time $t$ of a generic fixed-income security is described as,

$$V(t, y) = \sum_{i=1}^{I} c_{t_i} \delta(t, t_i), \tag{2.1}$$

where the discount factor, $\delta(t, t_i)$, may be modelled as

$$\delta(t, t_i) = \frac{1}{(1 + y)^{t_i - t}}. \tag{2.2}$$

This is a relatively simple, but powerful, identity. It holds that the value of a generic fixed-income security is merely the sum of its discounted future cash-flows, $\{c_{t_i} : i = 1, \ldots, I\}$. In this case, we use the security's yield, $y$, to discount each cash-flow; as we will see in later development, this need not always be the case.[1]

## 2.2 Simple Yield Exposure

It is clear from Eq. (2.1) that the yield of the security plays an important role in the security's value. Should one increase $y$, then the present value of each cash flow becomes smaller, leading to a reduction in the current value, $V(t, y)$.[2] Conversely, decreasing the yield, $y$, has the effect of increasing the present value of each cash-flow and thereby increasing the security's value. In short, there is an inverse relationship between the value of a fixed-income security and its yield. With a bit

[1]In reality, market practice is almost always a bit more complicated than it appears in (2.1). There are day-count conventions that describe each of the individual $(t_i - t)$'s, compounding frequencies impacting the yield and the coupon, and settlement dates. These elements are market convention. For the purposes of this discussion, however, we will skip over many of these details unless, of course, they become important.

[2]We employ the terms *discounted cash flow* and *present value of a cash flow* interchangeably.

of calculus, we can formalize this relationship through the computation of the first derivative of the security's value with respect to a change in its yield,

$$\begin{aligned}\frac{\partial V(t,y)}{\partial y} &= \frac{\partial}{\partial y}\left(\sum_{i=1}^{I}\frac{c_{t_i}}{(1+y)^{t_i-t}}\right), \qquad (2.3)\\ &= \sum_{i=1}^{I}\frac{-(t_i-t)c_{t_i}}{(1+y)^{(t_i-t)+1}},\\ &= -\frac{1}{(1+y)}\sum_{i=1}^{I}\frac{(t_i-t)c_{t_i}}{(1+y)^{t_i-t}}.\end{aligned}$$

This expression describes the sensitivity of the value of a fixed-income security to an infinitesimal change in its yield. The form of the expression is fairly enlightening. It holds that the sensitivity depends on the sum of the time-weighted discounted cash-flows, $\{(t_i-t)c_{t_i} : i = 1,\ldots,I\}$. If one divides both sides by the security's value, one arrives at,

$$\frac{1}{V(t,y)}\frac{\partial V(t,y)}{\partial y} = -\frac{1}{V(t,y)(1+y)}\sum_{i=1}^{I}\frac{(t_i-t)c_{t_i}}{(1+y)^{(t_i-t)}}. \qquad (2.4)$$

What have we done? Recall that for a small change in $y$ (i.e., $\Delta y = y_1 - y_0$) that

$$\left.\frac{\partial V(t,y)}{\partial y}\right|_{y=y_0} \approx \frac{V(t,y_1)-V(t,y_0)}{\underbrace{y_1-y_0}_{\Delta y}}. \qquad (2.5)$$

Thus, we have that

$$\underbrace{\left.\frac{1}{V(t,y)}\frac{\partial V(t,y)}{\partial y}\right|_{y=y_0}}_{\text{Eq. (2.4)}} \approx \left(\frac{1}{\Delta y}\right)\underbrace{\frac{V(t,y_1)-V(t,y_0)}{V(t,y)}}_{\text{Percentage change in } V}, \qquad (2.6)$$

which implies that Eq. (2.4) is, approximately at least, a function of the percentage change in the value of our security for a small change in its yield.

The negative of this quantity has another, much more frequently used, name. It is called the *modified duration*, which we will denote as $D_M$. It is formerly defined as,

$$D_M = \frac{1}{V(t,y)}\frac{\partial V(t,y)}{\partial y}. \qquad (2.7)$$

Equation (2.7) provides, in short, the analytic representation of a security's exposure to its yield. There is good reason that this is such a well-known and often used measure of the risk of a fixed-income security. It summarizes, in a single number, the percentage gain or loss for a fixed-income security associated with a small change in its yield. Given a 25 basis-point decrease in yields, a fixed-income security with a duration of 5 would expect to gain about 125 basis points. Conversely, a security with a duration of 0.5 would only expect to earn a profit of 12.5 basis points. This capacity to succinctly describe one's risk is extremely useful.

Another, more explicit mathematical way to understand this fact is to return to the linear approximation of the derivative in Eq. (2.6) (i.e., $\frac{\partial V}{\partial y} \approx \frac{\Delta V}{\Delta y}$) and re-arrange the terms as follows,

$$
\begin{aligned}
D_M(t,y) &= -\frac{1}{V(t,y)}\frac{\partial V(t,y)}{\partial y}, && (2.8)\\
D_M(t,y) &\approx -\frac{1}{V(t,y)}\frac{\overbrace{V(t,y+\Delta y)-V(t,y)}^{\Delta V(t,y)}}{\Delta y},\\
-D_M(t,y)\Delta y &\approx \frac{\Delta V(t,y)}{V(t,y)} = r(t,\Delta y).
\end{aligned}
$$

This is another way to see the result from Eq. (2.6). In words, the product of one's modified duration and expected (or realized) yield change directly approximates the percentage change in the bond's value for a given yield movement, $r(t, \Delta y)$. This is immensely useful.

We can, of course, perform the same exercise using continuously compounded interest rates—this is merely a different model of the discount factor, $\delta(t,t_i) = e^{-y_c(t_i-t)}$. The partial derivative of the bond value with respect to the continuously compounded yield, $y_c$, is given as,

$$
\begin{aligned}
\frac{\partial V(t,y_c)}{\partial y_c} &= \frac{\partial}{\partial y_c}\left(\sum_{i=1}^{I} c_{t_i} e^{-y_c(t_i-t)}\right), && (2.9)\\
&= -\sum_{i=1}^{I} c_{t_i}\,(t_i-t)\,e^{-y_c(t_i-t)},
\end{aligned}
$$

(continued)

whereas the modified duration has the following, relatively simple, form,

$$D_M(t, y_c) = \frac{1}{V(t, y_c)} \frac{\partial V(t, y_c)}{\partial y_c}, \tag{2.10}$$
$$= -\frac{1}{V(t, y_c)} \sum_{i=1}^{I} c_{t_i} (t_i - t) e^{-y_c(t_i - t)}.$$

One may also represent the modified duration analytically as,

$$\frac{1}{V(t, y)} \frac{\partial V(y)}{\partial y} = \lim_{\epsilon \to 0} \frac{1}{V(t, y)} \left( \frac{V(t, y + \epsilon) - V(t, y - \epsilon)}{2\epsilon} \right). \tag{2.11}$$

This comes directly from the formal definition of a partial derivative. It also suggests the following possible numerical approximation,

$$\frac{1}{V(t, y)} \frac{\partial V(y)}{\partial y} \approx \frac{1}{V(t, y)} \left( \frac{V(t, y + \epsilon) - V(t, y - \epsilon)}{2\epsilon} \right), \tag{2.12}$$

for a sufficiently small and judicious choice of $\epsilon$.[3] Often, when the duration is numerically computed using something like Eq. (2.12), it is termed the *effective duration*.[4] This is particularly useful when the security value cannot be so easily represented as indicated in Eq. (2.1); a good example would be a security with embedded optionality such as a callable bond or a mortgage-backed security. In such a case, a numerical computation of the sensitivity may prove more convenient. We will, however, also see that such a numerical computation can be useful, even for straightforward fixed-income securities, for a complex model of the discount factor.[5]

It is always easier to understand an idea in the context of a concrete example. Consider, therefore, the US Treasury bond described in Table 2.1. Our plan is to demonstrate the application of Eq. (2.11) using this specific bond. At this point, it is useful to indicate where market conventions become important. When discounting cash-flows using Eq. (2.1), we arrive at what is called the dirty price. This is the value obtained by discounting all of one's cash-flows back to the settlement date, without accounting for accrued interest. The clean price, of course, is the dirty price

[3]This is formally termed a central finite-difference approximation. See Press et al. [1] for much more information on the numerical computation of derivatives.

[4]Caution should nevertheless be exercised as there is not, to the author's knowledge, a clear consensus in the finance universe on the definition of effective duration.

[5]In general, given that the numerical computation requires three full function valuations, it will only be employed in the absence of an analytical solution.

**Table 2.1** An example bond

| Characteristic | Data value |
|---|---|
| Issuer | US Treasury |
| ISIN | US912828NP10 |
| Position | \$100 |
| Coupon | 1.75 % |
| Issue date | 31 July 2010 |
| Maturity date | 31 July 2015 |
| Settle date | 10 August 2010 |
| Next coupon date | 31 January 2011 |
| Tenor | 4.980 years |
| Yield | 1.524 % |
| Clean price | \$101.078 |
| Accrued interest | \$0.048 |
| Dirty price | *\$101.126* |
| Modified duration | 4.747 years |

This table outlines the key data values for a 5-year on-the-run US Treasury bond on 9 August 2010. This information is used to practically demonstrate the analytic and numeric computation of modified duration.

**Table 2.2** The analytic computation

| Date | Days | $t_i - t$ | $c_{t_i}$ | $\delta(t, t_i)$ | $c_{t_i}\delta(t, t_i)$ | $(t_i - t)c_{t_i}\delta(t, t_i)$ |
|---|---|---|---|---|---|---|
| 31 Jan 2011 | 173 | 0.473 | 0.875 | 0.9928 | 0.87 | 0.41 |
| 31 Jul 2011 | 355 | 0.973 | 0.875 | 0.9853 | 0.86 | 0.84 |
| 31 Jan 2012 | 539 | 1.473 | 0.875 | 0.9779 | 0.86 | 1.26 |
| 31 Jul 2012 | 721 | 1.973 | 0.875 | 0.9705 | 0.85 | 1.68 |
| 31 Jan 2013 | 905 | 2.473 | 0.875 | 0.9632 | 0.84 | 2.08 |
| 31 Jul 2013 | 1,086 | 2.973 | 0.875 | 0.9559 | 0.84 | 2.49 |
| 31 Jan 2014 | 1,270 | 3.473 | 0.875 | 0.9486 | 0.83 | 2.88 |
| 31 Jul 2014 | 1,451 | 3.973 | 0.875 | 0.9415 | 0.82 | 3.27 |
| 31 Jan 2015 | 1,635 | 4.473 | 0.875 | 0.9343 | 0.82 | 3.66 |
| 31 Jul 2015 | 1,816 | 4.973 | 100.875 | 0.9273 | 93.54 | 465.15 |
| Total | n/a | n/a | n/a | n/a | *101.126* | 483.720 |

This table outlines the computation of the bond price in Table 2.1 using its yield and then the further computation of the modified duration.

adjusted for accrued interest. One must be careful, however, when taking prices from different screens or data sources not to mix clean and dirty prices.[6]

[6]It is often the case, for example, that clean prices are returned from various software functions and, as a consequence, a bit of caution is advised.

All of the necessary computations are outlined in Table 2.2. We begin, in the first column, with the actual cash-flow dates. As this bond pays a semi-annual coupon, as of August 2010 there are ten remaining cash-flows culminating with its maturity payment on 31 July 2015. Each of these calendar dates are transformed into the number of days from the settlement date, 10 August 2010. Using these days, we proceed to divide them by 365 to generate a sequence of cash-flow times. This permits us to easily compute the sequence of discount factors,

$$\delta(t, t_i) = \frac{1}{\left(1 + \frac{y}{2}\right)^{2\cdot(t_i - t)}}, \tag{2.13}$$

to account for the semi-annual compounding associated with the bond's cash-flows. Observe that each of the cash-flows is discounted using the bond's yield as the discount rate. Taking the product of the cash-flows with the discount factor, we obtain a sequence of discounted cash-flows, $\{\delta(t, t_i)c_{t_i}, i = 1, \ldots, 10\}$. The sum of this sequence is \$101.126, which coincides with the clean price outlined in Table 2.1.

The next step is the computation of the modified duration. Here we need to compute the sum of the time-weighted discounted cash-flows. This sequence, $\{(t_i - t)\delta(t, t_i)c_{t_i}, i = 1, \ldots, 10\}$, is the final column in Table 2.2. The sum of this sequence is \$483.720. Thus using the analytic expression in Eq. (2.11), we should arrive at the modified duration,

$$\begin{aligned} D_M &= -\underbrace{\frac{1}{V(t,y)}}_{\substack{\text{Dirty}\\ \text{price}}} \underbrace{\frac{1}{\left(1 + \frac{y}{2}\right)}}_{\substack{\text{Adjust for}\\ \text{compounding}}} \sum_{i=1}^{I} (t_i - t)c_{t_i} \underbrace{\frac{1}{(1 + \frac{y}{2})^{2(t_i - t)}}}_{\substack{\text{Eq. (2.13):}\\ \delta(t, t_i)}}, \\ &= -\left(\frac{1}{101.126\left(1 + \frac{1.524\,\%}{2}\right)}\right) 483.720, \\ &= -4.747. \end{aligned} \tag{2.14}$$

This is exactly the value of 4.747 that we expected and which is provided in Table 2.1.

We may now turn to the numerical computation of modified duration introduced in Eq. (2.11). This is a relatively straightforward computation; the tricky part, however, is the appropriate selection of the parameter, $\epsilon$. If it is too large, then the approximation will be poor. Conversely, if it is too small, then it can lead to instability in the computation. To make this clearer, let us work through the computation with $\epsilon$ set to 0.01, which is equivalent to 100 basis points. We, therefore, start with Eq. (2.11) and follow through with the computation as,

$$D_M = \frac{1}{V(t,y)} \left( \frac{V(t, y + \epsilon) - V(t, y - \epsilon)}{2\epsilon} \right), \tag{2.15}$$

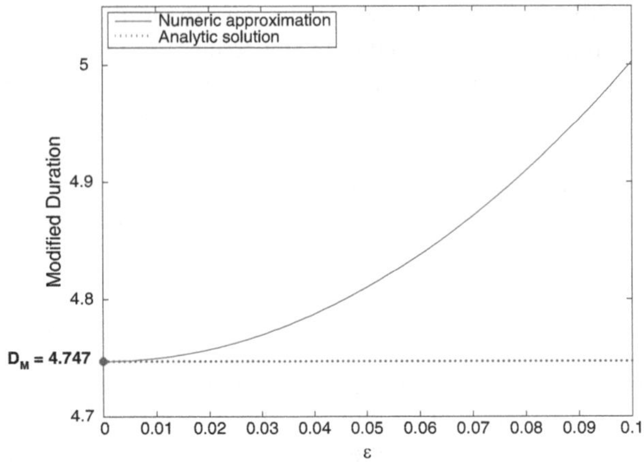


**Fig. 2.1** Numerical computation of duration. This figure demonstrates the convergence of the central finite-difference approximation introduced in Eq. (2.11) to the bond described in Table 2.1. Observe that for relatively large values of $\epsilon$, the numerical approximation is a poor estimate for the analytic value of 4.747. As $\epsilon$ tends to zero, however, it converges to the analytically computed value

$$\begin{aligned}
&= \frac{1}{101.126}\left(\frac{V(t, y+0.01) - V(t, y-0.01)}{2 \cdot 0.01}\right), \\
&= \frac{1}{101.126}\left(\frac{96.4513 - 106.0577}{0.02}\right), \\
&= -4.750.
\end{aligned}$$

Observe that, for a choice of $\epsilon = 0.01$, we do *not* reproduce the modified duration value of 4.747 in the analytic computation. The reason is that 0.01 represents a 100 basis-point movement in the bond yield, which is actually quite large. It should be stressed, however, that even with a 100 basis-point movement, the numerical approximation is fairly acceptable.[7]

Figure 2.1 takes the demonstration one step further and performs the numerical computation of the modified duration using a sequence of $\epsilon$'s from 0.1 (i.e., 100 basis points) to 0.0000001 (i.e., 1000th of a basis point). Observe that for relatively large values of $\epsilon$, the numerical estimate is a poor approximation for the analytic value of 4.747. As $\epsilon$ tends to zero, however, it converges to the analytically computed value. Indeed, it appears that for values of $\epsilon$ slightly less than 0.01 (ten basis points), the numerical computation basically converges to the analytic value.

In summary, modified duration is a key fixed-income exposure to the interest-rate risk factor and may be computed analytically or numerically. Given the simplicity of

[7]Note also that an increase of 100 basis points generates a \$4.67 decrease in the price, while a 50 basis-point decrease leads to a \$4.93 basis-point rise in the price. There are two points that should be taken from this fact. First, a 100 basis-point movement in yields leads to relatively large price changes. Second, the price movement is not symmetric for an equivalent upward and downward movement in bond yields. This is due to the fact that the relationship between bond prices and yields is *not* linear. More on this point will be discussed in the next section.

the analytic approach, a numerical approximation is generally only employed when the security has complex features that preclude the use of Eq. (2.8) on page 24.

## 2.3 Correcting for Our Linear Approximation

Up to this point, we have only examined the concept of modified duration—one dimension of the security's exposure to its yield. As we've established, this is the percentage change in the value of a fixed-income security associated with a small change in its yield. What happens, however, when the change in yield is *not* so small? Indeed, the daily changes in market yields can occasionally be quite sizeable. In Eq. (2.1), we note that the relationship between the bond's value, $V(y)$, and its yield, $y$, is not linear. Indeed, the bond-price equation has a polynomial form in terms of the discount factors. Thus, we would expect some degree of non-linearity. To understand the nature of this relationship, we have selected a specific US Treasury bond and plotted its value across yields ranging from one basis point to 10 %. The results, summarized in Fig. 2.2, clearly indicate a non-linear relationship. The bond examined is an on-the-run 10-year US Treasury bond—it has a 3.5 % coupon and a 15 May 2020 maturity date giving it, as of 9 August 2010, a tenor of about 9.8 years. The straight line passing through the current clean price represents the predicted price movement stemming from the modified duration. For relatively small yield changes, this linear approximation is quite reasonable; as the yield change increases, however, the accuracy of this approximation deteriorates.

Given this non-linearity, it would seem sensible to construct a measure that attempts to capture it. Modified duration is clearly insufficient to capture the full exposure of a fixed-income security to the yield factor. Since locally the linear approximation is quite good, one reasonable approach would be to try to capture the rate of change in the linear approximation. This leads us naturally to the second

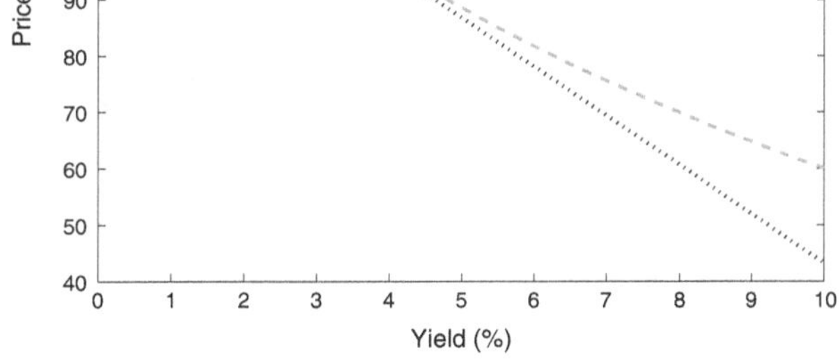


**Fig. 2.2** Relationship between price and yield. This figure outlines the relationship between a bond's yield and its price for yields ranging from 0 to 10 %. The bond examined is 10-year US Treasury as of 9 August 2010—it has a 3.5 % coupon, a 15 May 2020 maturity date, and an ISIN number of US912828ND89. Observe that the relationship between these two variables is *not* linear; the modified duration is included to demonstrate the deviation from linearity

derivative of the security's value function. The second derivative of the bond-price function with respect to yield has the form,

$$\begin{aligned}\frac{\partial^2 V(t,y)}{\partial y^2} &= \frac{\partial^2}{\partial y^2}\left(\sum_{i=1}^{I}\frac{c_{t_i}}{(1+y)^{t_i-t}}\right), \qquad (2.16)\\ &= \frac{\partial}{\partial y}\underbrace{\left(\sum_{i=1}^{I}\frac{-(t_i-t)c_{t_i}}{(1+y)^{(t_i-t)+1}}\right)}_{\text{First derivative with respect to } y},\\ &= \sum_{i=1}^{I}\frac{(t_i-t)(t_i-t+1)c_{t_i}}{(1+y)^{(t_i-t)+2}},\\ &= \frac{1}{(1+y)^2}\sum_{i=1}^{I}\frac{(t_i-t)(t_i-t+1)c_{t_i}}{(1+y)^{t_i-t}}.\end{aligned}$$

Here we see the rate of change of the first derivative, which is also somewhat more difficult to interpret. If one normalizes this quantity by its current value, $V(t,y)$, as follows

$$\frac{1}{V(t,y)}\frac{\partial^2 V(t,y)}{\partial y^2} = \frac{1}{V(t,y)(1+y)^2}\sum_{i=1}^{I}\frac{(t_i-t)(t_i-t+1)c_{t_i}}{(1+y)^{t_i-t}}, \qquad (2.17)$$

then one arrives at a second well-known quantity: the bond convexity, which we will denote as $C$. The convexity measure is not as simple to apply as the modified duration and, generally speaking, acts as a correction factor for approximations made using modified duration. Interestingly, we have seen that for a simple risk factor—the market yield—we may employ multiple exposures. We will see how one applies the convexity measure in rather more detail in the next chapter.

Again, we can perform the same exercise using continuously compounded interest rates. The second partial derivative of the bond price with respect to $y$ is given as,

$$\begin{aligned}\frac{\partial V^2(t,y)}{\partial y^2} &= \frac{\partial^2}{\partial y^2}\left(\sum_{i=1}^{I}c_{t_i}e^{-y(t_i-t)}\right), \qquad (2.18)\\ &= \sum_{i=1}^{I}c_{t_i}\,(t_i-t)^2\,e^{-y(t_i-t)}.\end{aligned}$$

(continued)

Convexity, in a continuously compounded setting, is thus defined as,

$$\frac{1}{V(t,y)}\frac{\partial^2 V(t,y)}{\partial y^2} = \frac{1}{V(t,y)}\sum_{i=1}^{I} c_{t_i}\,(t_i - t)^2\, e^{-y(t_i - t)}. \tag{2.19}$$

Once again, we observe that continuous compounding gives rise to more convenient mathematics.

## 2.4 Time Exposure

We have been careful to indicate that the value of a bond is a function of two arguments: time and yield. Modified duration and convexity provide a basis for understanding the exposure of our bond to changes in the yield, but these measures are silent on the implications associated with changes in the first argument: time. This is easily corrected. To better understand the sensitivity, or exposure, of a bond to changes in time, we need only compute the partial derivative of our bond-value function with respect to time. In principle, this is a straightforward exercise, although we need to recall a simple property of logarithms to isolate the $t$ term. In particular, we need to remember that a function of the form $(1 + y)^x$ can be alternatively written as $e^{x\ln(1+y)}$. With that in mind, the partial derivative of our bond-value function with respect to time is,

$$\begin{aligned}
\frac{\partial V(t,y)}{\partial t} &= \frac{\partial}{\partial t}\left(\sum_{i=1}^{I}\frac{c_{t_i}}{(1+y)^{t_i-t}}\right), \qquad (2.20)\\
&= \frac{\partial}{\partial t}\left(\sum_{i=1}^{I}\frac{c_{t_i}}{e^{(t_i-t)\ln(1+y)}}\right),\\
&= \frac{\partial}{\partial t}\left(\sum_{i=1}^{I}c_{t_i}e^{-(t_i-t)\ln(1+y)}\right),\\
&= \sum_{i=1}^{I}c_{t_i}\ln(1+y)e^{-(t_i-t)\ln(1+y)},\\
&= \ln(1+y)\underbrace{\left(\sum_{i=1}^{I}\frac{c_{t_i}}{(1+y)^{t_i-t}}\right)}_{V(t,y)},\\
&= \ln(1+y)V(t,y).
\end{aligned}$$

If we recall that $\ln(1+y) \approx y$ for small values of $y$, then

$$\frac{\partial V(t,y)}{\partial t} \approx yV(t,y). \tag{2.21}$$

To compute, therefore, a kind of time duration for our bond, we need only to divide both sides by $V(t,y)$ arriving at,

$$\begin{aligned} D_t &= \frac{1}{V(t,y)} \frac{\partial V(t,y)}{\partial t}, \\ &= \frac{1}{\cancel{V(t,y)}} \underbrace{\ln(1+y)\cancel{V(t,y)}}_{\text{Eq. (2.20)}}, \\ &\approx y. \end{aligned} \tag{2.22}$$

In simple words, therefore, the exposure of a fixed-income security to the passage of time—also a risk factor—is well approximated by its yield. This is related to the notion of carry.

Using the same heuristic notion of a derivative as in the previous discussion (i.e., $\frac{\partial V}{\partial t} \approx \frac{\Delta V}{\Delta t}$) and re-arrange the terms as follows,[8]

$$\begin{aligned} D_t &= \frac{1}{V(t,y)} \frac{\partial V(t,y)}{\partial t}, \\ \underbrace{D_t}_{\approx y} &\approx \frac{1}{V(t,y)} \frac{\overbrace{V(t+\Delta t,y) - V(t,y)}^{\Delta V(t,y)}}{\Delta t}, \\ y\Delta t &\approx \frac{\Delta V(t,y)}{V(t,y)}. \end{aligned} \tag{2.23}$$

This is not a terribly surprising result, but it is nonetheless encouraging that this basic framework provides a consistent and sensible answer. Moreover, the notion of time sensitivity is an important aspect of performance analysis and, as such, we will making extensive use of Eq. (2.23) in the coming chapters.

[8]The product of the security's yield and the time interval approximates the return associated with the movement of time.

## 2.5 Key-Rate Exposures

The notion of modified duration is quite useful, but it has some limitations. The principal limitation is that, in a typical fixed-income portfolio, one generally holds a collection of bonds with varying tenors. If one holds $n$ bonds in one's portfolio with durations $\{D_{M_i}, i = 1, \ldots, n\}$ and market-value weights $\{\omega_i, i = 1, \ldots, n\}$, we straightforwardly define the duration of the portfolio as,

$$D_{M_p} = \sum_{i=1}^{n} \omega_i D_{M_i}. \tag{2.24}$$

Quite simply, the modified duration of the portfolio is the weighted-average modified duration of the instruments in the portfolio.[9] The duration of the portfolio gives one an insight into the sensitivity of that portfolio to a constant change across all yields in the portfolio. It is, however, not always the case that all yields move in an identical manner across the entire yield curve. Quite often, 2-year yields move in a different way compared to, say, 5-year yields or 10-year yields. Figure 2.3 provides a graphical example of the change in the UST yield curve from 31 July 2010 to 31 August 2010. It shows an example of relatively modest yield movement in the short end of the curve and coincident reduction and flattening of the curve beyond about 4 years.

Briefly put, yields at different tenors move in different ways. To understand the sensitivity of our portfolio to changes in yields at different parts of the yield curve, therefore, we require a more *local* measure of yield changes. Local in this context

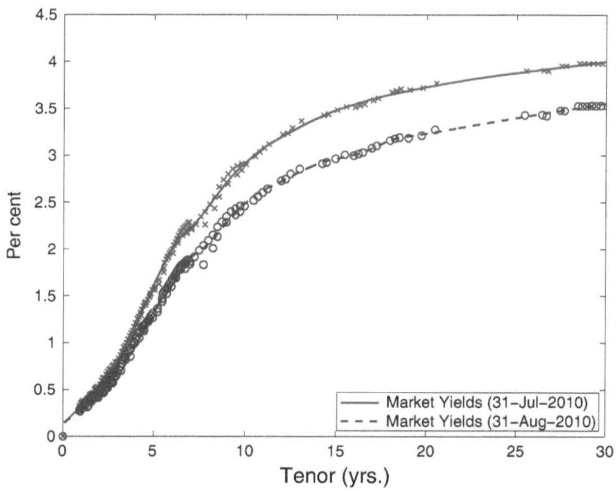


**Fig. 2.3** Typical yield-curve movements. The underlying graph describes the change in the UST yield curve from 31 July 2010 to 31 August 2010. This is a clear example of situation where, over a given period of time, not all yields change in a parallel fashion

[9]This computation is based on the fact that modified duration—essentially a derivative—is a linear quantity and, as such, can be averaged.

means restricted to a smaller sector of the yield curve. The obvious solution to determining the exposure of one's portfolio to changes in 2-year yields is to compute something like the 2-year duration of the portfolio. In other words, to compute the portfolio's mathematical exposure to movements in the 2-year yield. Determination of such a sensitivity should, in principle, be possible.

Exactly for which yields one should compute such an exposure is a very natural question. Should we stop at the 2-year sector? Clearly not. There are a range of possible points along the sovereign yield curve that may be of interest. One's exact choice of *key* yield points will depend on the needs of the user, although a reasonable, and defensible, approach is to select key rates to coincide with areas of market liquidity (i.e., on-the-run sovereign bonds) to permit hedging of one's exposure to these sectors. Very specific tenors—such as the 1-year and 9-month yield or the 2-year and 1-month yield—are probably to be avoided.

For a more concrete perspective on the computation of bond (and portfolio) exposures to specific tenors along the yield curve, we consider an example. Figure 2.4 outlines the US Treasury curve as of 11 October 2010 and highlights eight different yield points across the curve: the 6-month, 2-year, 5-year, 7-year, 10-year, 15-year, 20-year, and 30-year yields. It is important to repeat that there is nothing magic about this specific choice of eight yields and that each analyst—hopefully after consultation with his or her portfolio-manager colleagues—must decide on the set of *key* yields across the yield curve for which exposures are required.

Having defined a set of key yield tenors, the next step is the determination of the sensitivity of one's fixed-income instrument to a change in this key rate. To understand how such a sensitivity might be computed, imagine that only *one* of these key yields, or rates, moves by, say, 50 basis points? Figure 2.5 highlights just such a 50 basis-point movement in the 10-year UST rate. Observe that all of the other key rates remain unchanged, but that the intermediate yield points are

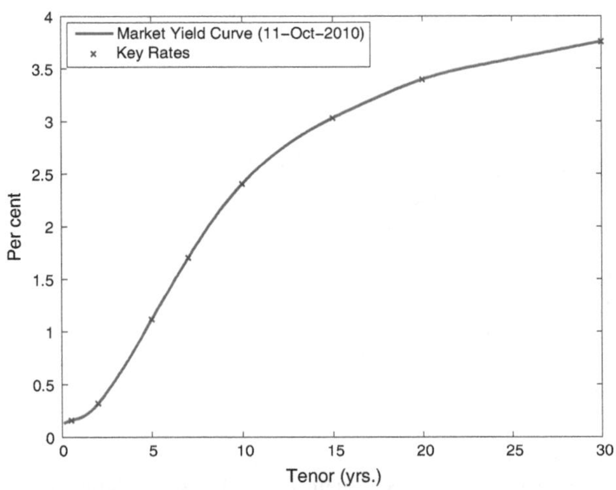


**Fig. 2.4** Selection of key rates. This figure highlights eight different yield points across the US Treasury curve on 11 October 2010

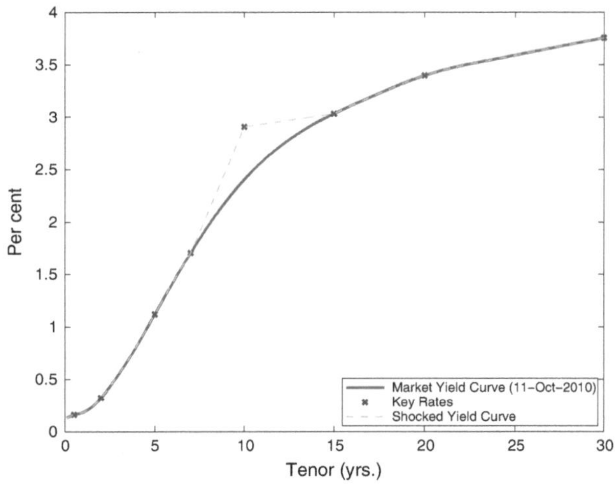


**Fig. 2.5** Perturbing a key-rate. This figure demonstrates the impact of a 50 basis-point movement in the 10-year UST rate. All of the other rates remain unchanged, except for the intermediate yield points between the nearest adjacent non-perturbed key rates and the new perturbed value. These values are linearly interpolated to create a tent-like shape around the perturbed rate

linearly interpreted between the nearest adjacent non-perturbed key rates and the new perturbed value for the 10-year rate. The end result looks something like a tent.

The next task is to transform the image in Fig. 2.5 into a concrete mathematical expression. Naturally, this requires us to return to the bond-price equation. Once again, we perform a slight modification of the original expression in Eq. (2.1). In this case, instead of discounting all cash-flows at a single yield, we discount each cash-flow with its distinct yield. This leads us to the following expression:

$$V(t, y_{t_1}, \ldots, y_{t_n}) = \sum_{i=1}^{n} \frac{c_{t_i}}{(1 + y_{t_i})^{t_i - t}}. \tag{2.25}$$

Formally, therefore, $\{y_{t_i}, i = 1, .., n\}$ denotes the set of individual yields associated with each individual cash-flow. To compute the sensitivity of our bond price to a change in the $k$th yield, we start, as usual, with the partial derivative with respect to $y_{t_k}$ for $k \in \{1, \ldots, n\}$:

$$\begin{aligned} \frac{\partial V(t, y_{t_1}, \ldots, y_{t_n})}{\partial y_k} &= \frac{\partial}{\partial y_{t_k}} \left( \sum_{i=1}^{n} \frac{c_{t_i}}{(1 + y_{t_i})^{t_i - t}} \right), \\ &= \frac{-(t_k - t)\, c_{t_k}}{(1 + y_{t_k})^{(t_k - t) + 1}} \end{aligned} \tag{2.26}$$

In this case, the resulting derivative is *not* a sum, but rather a single term.[10] This is the equivalent of modified duration, but only for a given area of the curve.

[10]This is because all terms where $i \neq k$ are, by definition, zero. That is, they do not contribute to the derivative.

Dividing both sides of Eq. (2.26) by $V(t, y_{t_1}, \ldots, y_{t_n})$ transforms this derivative into something resembling the duration concepts seen earlier in this chapter. Indeed, the resulting expression is generally termed the *key-rate* duration,

$$\begin{aligned} D_{t_k} &= \frac{1}{V(t, y_{t_1}, \ldots, y_{t_n})} \frac{\partial V(t, y_{t_1}, \ldots, y_{t_n})}{\partial y_{t_k}}, \\ &= \frac{-(t_k - t)\, c_{t_k}}{V(t, y_{t_1}, \ldots, y_{t_n})(1 + y_{t_k})^{(t_k - t)+1}}. \end{aligned} \tag{2.27}$$

We denote the key-rate duration, therefore, as $D_{t_k}$ representing the sensitivity of one's fixed-income instrument, or portfolio, to a change in $y_{t_k}$.

The use of the key-rate duration is conceptually identical to the use of the modified duration.[11] The following approximation, similar in spirit to Eq. (2.8), demonstrates this point in mathematical terms,

$$\begin{aligned} D_{t_k} &\approx -\frac{1}{V(t, y_{t_1}, \ldots, y_{t_n})} \\ &\quad \times \frac{\overbrace{V(t, y_{t_1}, \cdots, y_{t_k} + \Delta y_{t_k}, \cdots, y_{t_n}) - V(t, y_{t_1}, \ldots, y_{t_n})}^{\Delta V(t, y_{t_1}, \ldots, y_{t_n})}}{\Delta y_{t_k}}, \\ -D_{t_k} \Delta y_{t_k} &\approx \frac{\Delta V(t, y_{t_1}, \ldots, y_{t_n})}{V(t, y_{t_1}, \ldots, y_{t_n})}. \end{aligned} \tag{2.28}$$

The key-rate duration is basically the *exposure* of the bond to a small change in the $k$th yield. The local nature of this measure of portfolio sensitivity makes it a very useful supplement to the modified duration, which provides a more global view of yield-curve sensitivity.

The careful reader has probably noticed that these analytic computations only make sense when the security's cash flows coincide precisely with the desired key-rate tenors. This is quite unlikely and probably impossible for a large portfolio of fixed-income securities. In reality, key-rate durations are determined numerically—a description of a sensible algorithm for their computation is found in the underlying shaded box. We will nonetheless continue to use the analytic development, because it provides useful insight into the notion of a key-rate duration and permits easy comparison with the other exposures developed in this chapter.

[11] In words, the product of the key-rate duration and an expected or realized change in the $k$th yield approximates the approximate percentage change in the value of one's fixed-income instrument.

In practice, it is generally quite difficult and inconvenient to analytically compute key-rate durations. Instead a numerical approximation is typically employed. This essentially involves a base sovereign yield curve—computed using your favourite method or a technique borrowed from Chap. 5—and a central finite-difference approximation. This is the key input, but the algorithm nonetheless requires a number of distinct steps. For a given key rate tenor, $k$, and choice of security it involves

1. determining the cash-flows of your security from its coupon and maturity date;
2. transforming your sovereign yield curve into a zero-coupon curve using a simple bootstrapping technique;[12]
3. computing the central value of the fixed-income security from the zero-coupon curve and the previously determined cash-flows—call this $V$;
4. shocking upwards the desired key-rate, at the desired key tenor $k$, on your base sovereign yield curve by a small amount, $\epsilon$;
5. transforming—again using a bootstrap—the shocked yield curve into a correspondingly shocked zero coupon curve;
6. repricing your security with the shocked zero-coupon curve–call this $V^+$; and
7. perturbing the original sovereign curve downwards, again at the desired key tenor, by $\epsilon$, determining the associated zero-coupon curve, and recomputing the new security value, $V^-$.

This provides all of the required ingredients for the final computation. The key-rate duration, for the selected key tenor, is thus approximated by the central finite-difference technique as,

$$D_{t_k} \approx \frac{1}{V}\left(\frac{V^+ - V^-}{2\epsilon}\right). \tag{2.29}$$

This is a tedious exercise and must be repeated for each security and each key-rate tenor (i.e., $k = 1, \ldots, \kappa$). Fortunately, such tedious tasks are easily organized into a computer program and happily delegated to your computer's CPU.

Simultaneous use of both modified and key-rate durations would be much easier if we could establish a link between these two measures. It turns out, in fact, that

[12]For more information on this technique, see Chap. 5.

such a link does exist. If we take the sum of all $n$ key-rate durations, we arrive at the following expression

$$\sum_{k=1}^{n} \frac{\partial V(t, y_{t_1}, \ldots, y_{t_n})}{\partial y_{t_k}} = \sum_{k=1}^{n} \frac{-(t_k - t)\, c_{t_k}}{(1 + y_{t_k})^{(t_k - t)+1}}. \tag{2.30}$$

This looks slightly familiar. If we proceed to divide both sides by $V(t, y_{t_1,\ldots,y_{t_k}})$, then

$$\begin{aligned} \frac{1}{V(t, y_{t_1,\ldots,y_{t_k}})} \sum_{k=1}^{n} \frac{\partial P(t, y_{t_1}, \ldots, y_{t_n})}{\partial y_k} &= \underbrace{\frac{1}{V(t, y_{t_1,\ldots,y_{t_k}})} \sum_{k=1}^{n} \frac{-(t_k - t)\, c_{t_k}}{\left(1 + \boxed{y_{t_k}}\right)^{(t_k - t)+1}}}_{\substack{\text{Sum of the} \\ \text{key-rate durations}}}, \\ &\approx \underbrace{\frac{1}{V(t, y_{t_1,\ldots,y_{t_k}})} \sum_{k=1}^{n} \frac{-(t_k - t)\, c_{t_k}}{\left(1 + \boxed{y}\right)^{(t_k - t)+1}}}_{\substack{\text{Modified} \\ \text{duration}}}. \end{aligned} \tag{2.31}$$

The consequence is that the sum of the key rate durations is virtually identical to the expression for modified duration derived in Eq. (2.3). The only difference is that the individual discount factors in the sum are a sequence of values $\{y_{t_k}, k = 1, \ldots, n\}$ for the key-rate duration, but only a single value, $y$, for the modified duration. These need not be equal, of course, but generally they will be quite close. Consequently, the sum of the key-rate durations is a close approximation to the modified duration.

This relationship essentially permits us to sum the exposures from the key-rate durations and equate it to the exposure arising from the modified duration. More specifically, we can try to equate Eqs. (2.4) and (2.28) as,

$$-\sum_{k=1}^{n} D_{t_k} \Delta y_{t_k} \approx -D_M \Delta y, \tag{2.32}$$

$$\sum_{k=1}^{n} \frac{\overbrace{V(t, y_{t_1}, \cdots, y_{t_k} + \Delta y_{t_k}, \cdots, y_{t_n}) - V(t, y_{t_1}, \ldots, y_{t_n})}^{\Delta V(t, y_{t_1}, \ldots, y_{t_n})}}{V(t, y_{t_1}, \ldots, y_{t_n})}$$

$$\approx \frac{\overbrace{V(t, y + \Delta y) - V(t, y)}^{\Delta V(t,y)}}{V(t, y)}.$$

The power of the key-rate duration is that it generalizes the idea of the yield risk factor. Using modified duration, each individual security yield is a risk factor. When we employ key-rate durations, however, we use a generic set of key-rate risk factors for all securities in a given portfolio and strategic benchmark. This is very powerful.

### 2.5.1 A Word of Caution

Some caution is nevertheless required in the use of Eq. (2.32) as there are basically *two* sources of approximation. The sum of the key-durations may indeed be quite close to the modified duration, but the individual yield changes at the key-rate sectors need not be consistent.

To see how this might work, let's look at another example. Imagine that we have a UST bond with a modified duration of 5.07 years and the curve moves as described by Fig. 2.6. Over this period, the yield of this bond happily increases by 42 basis points. This would lead us to approximate the percentage change in the bond's value due to this yield movement as $-5.07 \cdot 42 = -212.9$ basis points.

Figure 2.6 and Table 2.3 also handily provide us with all of the key-rate durations and yield changes over this period. Performing the computation exactly as it appears in Eq. (2.32) with these inputs, however, we arrive at a rather different answer. Specifically, we predict a loss of $-182.0$ basis points amounting to a difference of approximately 30 basis points with the value stemming from the modified duration computation.

What is going on? Essentially, the changes in the yield curve were relatively modest at the short end of the curve with an 8 and 18 basis-point widening at the 6-month and 2-year sectors, respectively. Yields widened by 33 basis points at the 5-year point and 56 basis points at the 7-year sector. If we compute a key-rate weighted average of the total yield movement, therefore, it amounts to only about

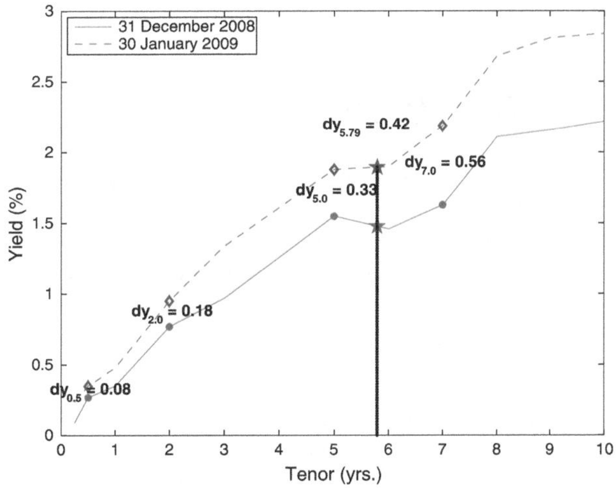


**Fig. 2.6** Key-rate and modified durations. Here we examine the change in five key rates (i.e., 6-months, 2-years, 5-years, and 7 years) for a given bond along with a 5.79 year maturity

**Table 2.3** Key-rate and modified-duration example

| Key-rate tenor | $D_{t_k}$ | $\Delta y_{t_k}$ | $-D_{t_k}\Delta y_{t_k}$ | $\Delta y$ | $-D_{t_k}\boldsymbol{\Delta y}$ |
|---|---|---|---|---|---|
| 6 months | 0.02 | 8 | −0.2 | 42 | **−0.8** |
| 2 years | 0.14 | 18 | −2.5 | 42 | **−5.9** |
| 5 years | 4.16 | 33 | −137.3 | 42 | **−174.7** |
| 7 years | 0.75 | 56 | −42.0 | 42 | **−31.5** |
| Total/weighted average | 5.07 | 36 | −182.0 | 42 | **−212.9** |

This table describes the key-rate durations and key-rate movements for a UST bond with a tenor of 5.79 years. The yield changes are also graphically represented in Fig. 2.6. All yield values are represented in basis points.

36 basis points of widening. The difference between the actual 42 basis points and the approximated 36 basis points accounts for the 30 basis points of difference in the approximation.

A more robust, albeit perhaps less satisfying, way to perform the computation, would be to use the key-rate durations, but restrict the yield changes at each key-rate point to be equal to the 42 basis-point widening experienced by the actual bond yield. This amounts to adjusting Eq. (2.32) as follows,

$$-\sum_{k=1}^{n} D_{t_k} \boxed{\Delta y} \approx -D_M \Delta y. \tag{2.33}$$

This essentially distributes the total curve return—without any under- or overstatement—across the pre-determined key tenors in the sovereign yield curve.

While this is clearly a rather extreme example, it does highlight the fact that the change in the yield of a given bond need not be equal to the key-rate weighted average of key-rate yield movements.[13] We will return to this point again in the following chapters when we discuss performance attribution and the estimation of portfolio risk.

## 2.6 Spread Exposure

Thus far, we have focused on bonds issued by so-called risk-free borrowers. Risk-free, in this context, implies that these bonds are subject to no, or at least very little, credit risk. It would be tempting to classify these risk-free borrowers as governments, but in fact, some government bonds do have a relatively high prob-

[13]For a portfolio with numerous bonds across the curve, the effect is likely to be relatively small. Nevertheless, for computations that require a high degree of accuracy, such as performance attribution, this is likely to remain an unacceptable source of error.

ability of default or losses associated with a deterioration of their credit quality.[14] A good example would be the Eurozone, where some government issuers such as Germany and France could be considered risk-free borrowers, while others—such as Greece, Portugal, Ireland, or Spain at the time this document was written—could be considered to have considerable credit risk. Simply put, there are many fixed-income instruments including bonds issued by some governments, supranational entities, government agencies, and corporations that are exposed to credit risk. Credit risk is clearly a risk factor. Moreover, it is probably a collection of risk factors that might be organized in varying degrees of granularity.[15] Our analysis thus far does not permit us to compute the exposure of a fixed-income instrument to changes in its credit quality. In this section, we will address this shortcoming.

Determination of the exposure of a fixed-income instrument to changes in its underlying credit quality requires the identification, within the context of our bond-price equation, of some element that relates to credit risk. At first glance, the price of an agency, supranational, or even a corporate bond has the same form as government bond,

$$V(t, y) = \sum_{i=1}^{n} \frac{c_{t_i}}{(1 + y)^{t_i - t}}. \tag{2.34}$$

It remains the sum of the individual cash-flows discounted by its yield. This form is unfortunately *not* very helpful.

With a bit of reflection, this drawback can be resolved. The credit element is embedded in the yield. Supranational bonds, for example, trade at a higher yield (i.e., lower price) than a treasury security with an equivalent tenor to account for this incremental credit risk. We can exploit this fact and decompose a security's yield into *two* distinct components:

1. the equivalent treasury yield; and
2. the credit spread.

What does this mean? The *equivalent treasury yield* represents the yield that a bond subject to credit risk would have if it was issued by the risk-free issuer in that currency. An example would be a US agency bond with a tenor of 4.76 years–its equivalent treasury yield would be the yield of a US Treasury bond with the same tenor of 4.76 years. To perform this computation, one requires a mathematical model of the underlying risk-free yield curve. Such a model permits one to easily determine the yield of a US Treasury bond with a 4.76 year bond, as it is unlikely that such a bond would exist in the market.

---

[14]This latter situation is also termed credit migration.

[15]One could, for example, use categories such as sovereign, agency, supranational, or corporate. It would also be possible to have subcategories with each of these groups or even, in the limit, focus on individual names.

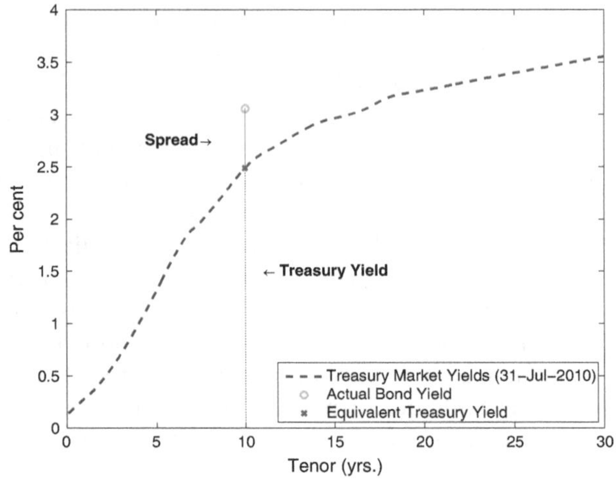


**Fig. 2.7** Decomposing the bond yield. This figure provides a graphical representation of how, armed with a risk-free yield curve, one can decompose the yield of a credit bond into the sum of its equivalent treasury yield and its credit spread

The difference between the bond's actual yield and this equivalent treasury yield is particularly interesting. It is called the *credit spread*. This credit spread, or simply spread, is essentially the additional yield demanded by the market—over and above the yield required for the risk-free borrower—to compensate for the incremental credit risk associated with that particular bond.[16] Thus, the larger the credit spread of a specific bond, the less optimistic is the market's assessment of the general credit quality of the underlying issuer. Changes in the spread over time consequently represent adjustments in the overall credit quality of the underlying issuer. Should, for example, the spread widen (narrow), then this would typically imply that the market has downgraded (upgraded) its view of the bond's credit quality.

Figure 2.7 demonstrates graphically how, with a yield-curve model for the necessary risk-free yield curve, one can decompose the yield into the equivalent treasury yield and a credit spread. Such a decomposition is quite practical as it breaks out a bond's yield into the risk-free component—common to all fixed-income instruments in that market—and an idiosyncratic credit spread component. Using this decomposition, our bond-price equation becomes,

$$V(t, \hat{y}, s) = \sum_{i=1}^{n} \frac{c_{t_i}}{(1 + \underbrace{\hat{y} + s}_{y})^{t_i - t}}, \tag{2.35}$$

[16]To be precise, the spread may also include additional yield demanded by lenders to compensate for lower liquidity of the credit bond relative to the underlying government curve. Decomposition of the credit and liquidity aspects of the credit spread is not trivial and we will, in our development ignore such effects. It is nonetheless important to be aware that part of the credit spread may be attributable to liquidity and, in some cases, its contribution can be important.

where $\hat{y}$ denotes the equivalent treasury yield and $s$ represents the associated credit spread.[17] This simple decomposition separates *two* risk factors—the treasury yield curve and the credit spread—that were previously linked through the security's yield.

We are now in familiar territory. As with modified duration for nominal and inflation-linked bonds and the notion of key-rate durations, it would seem that the derivative is a good place to start to determine the sensitivity of the bond's price to a change in the spread. The derivative of Eq. (2.35) with respect to the credit spread, $s$, is

$$\begin{aligned}\frac{\partial V(t,\hat{y},s)}{\partial s} &= \frac{\partial}{\partial s}\left(\sum_{i=1}^{n}\frac{c_{t_i}}{(1+\hat{y}+s)^{t_i-t}}\right),\\ &= -\frac{1}{(1+\hat{y}+s)}\sum_{i=1}^{n}\frac{(t_i-t)\,c_{t_i}}{(1+\hat{y}+s)^{t_i-t}}.\end{aligned} \tag{2.36}$$

If we recall, however, that $y = \hat{y} + s$ we can make a number of substitutions to simplify Eq. (2.36) as follows,

$$\begin{aligned}\frac{\partial V(t,\hat{y},s)}{\partial s} &= -\frac{1}{(1+\underbrace{\hat{y}+s}_{y})}\sum_{i=1}^{n}\frac{(t_i-t)\,c_{t_i}}{(1+\underbrace{\hat{y}+s}_{y})^{t_i-t}},\\ &= -\frac{1}{(1+y)}\sum_{i=1}^{n}\frac{(t_i-t)\,c_{t_i}}{(1+y)^{t_i-t}},\\ &= \frac{\partial V(t,y)}{\partial y}.\end{aligned} \tag{2.37}$$

Ultimately, there is nothing different about the sensitivity of the bond price whether the yield change comes from a movement in the equivalent treasury yield ($\hat{y}$), the credit spread ($s$), or the overall yield ($y$). While this is hardly a surprise, given that the terms enter additively into the bond-price equation, it is nonetheless a useful result.

If we proceed to divide both sides of Eq. (2.37) by $\frac{\partial V(\hat{y},s)}{\partial s}$, we arrive at the equivalent of modified duration, but for spread movements. It is termed, quite naturally, the spread duration and has the following form,

$$D_S = \frac{1}{V(t,\hat{y},s)}\frac{\partial V(t,\hat{y},s)}{\partial s}, \tag{2.38}$$

[17] The explicit introduction of the credit spread into the bond price equation permits us to include the credit risk factor into our mathematical framework.

$$= -\frac{1}{V(1+y)} \sum_{i=1}^{n} \frac{(t_i - t)\, c_{t_i}}{(1+y)^{t_i - t}},$$

$$= D_m.$$

We see again that spread duration and modified duration are equivalent for a plain-vanilla bond.

The spread duration is used in practice by replacing the partial derivative in the spread duration with its linear approximation $\frac{\Delta V}{\Delta s}$ as follows,

$$D_S \approx -\frac{1}{V(t,\hat{y},s)} \frac{\overbrace{V(t,\hat{y},s+\Delta s) - V(t,\hat{y},s)}^{\Delta V(t,\hat{y},s)}}{\Delta s}, \tag{2.39}$$

$$-D_s \Delta s \approx \frac{\Delta V(t,\hat{y},s)}{\Delta V(t,\hat{y},s)(t,\hat{y},s)}.$$

The spread duration, therefore, is basically the *exposure* of the bond to a small change in its credit spread—or, more generally, a small change in the credit quality of the bond issuer. It should be noted that for some fixed-income instruments, most notably floating-rate notes, the spread and modified durations do *not* coincide. In these cases, it is typical to compute the actual spread duration using an alternative approach. Floating-rate notes are addressed in Chap. 4.

In many commercial applications, the modified and spread durations, while close, do not usually perfectly coincide. For complex securities, the reason is obvious. The cash flows also depend on the spread, which leads to the following bond-price equation,

$$V(t,\hat{y},s) = \sum_{i=1}^{n} \frac{c_{t_i}(s)}{(1 + \underbrace{\hat{y} + s}_{y})^{t_i - t}}, \tag{2.40}$$

Clearly, differentiating Eq. (2.40) with respect to $s$ will not reduce to the modified duration. For plain-vanilla instruments, the bond-price equation is also often written as follows,

$$V(t,\hat{y},s) = \sum_{i=1}^{n} \frac{c_{t_i}}{(1 + \underbrace{z(t,t_i) + s_z}_{y})^{t_i - t}}, \tag{2.41}$$

(continued)

implying that each cash flow is discounted at its own individual spot rate, $z(t, t_i)$ for $i = 1, \ldots, n$. In this case, the credit spread, which is denoted as $s_z$, will not be the same as the $s$ computed using our simple decomposition. That is, $s_z \neq s$. In this case, when the spread duration is calculated—using a numerical approach—the result is not equal to the modified duration. For normal plain-vanilla instruments, however, the difference is minimal and, for portfolio analytic purposes, we simply assume that spread duration and modified duration are equal. In the case of complex securities, as described in Eq. (2.40), this may be a poor assumption. In this case, it is safer to employ a numerical approximation.

## 2.7 Foreign-Exchange Exposure

Fixed-income securities are not always denoted in one's base currency. Often, they are denominated in foreign currency. This implies that a movement in the exchange rate between the foreign currency and your base currency—all else being equal—leads to changes in the value of your investment. As such, the foreign-exchange rate is a risk factor.

Since the foreign-exchange rate is a risk factor, it would be useful to understand the exposure of a generic fixed-income security to the exchange rate. To do this, we return as usual to the bond-price equation. Denote $E_t$ as the exchange rate between the security and your base currency. Consequently, your security's value may be written as,

$$
\begin{aligned}
V(t, y, E_t) &= E_t \underbrace{\sum_{i=1}^{I} c_{t_i} \delta(t, t_i)}_{\substack{\text{Base currency} \\ \text{value: } V(t,y)}}, \\
&= E_t V(t, y).
\end{aligned}
\tag{2.42}
$$

If we differentiate $V$ with respect to $E_t$, we have

$$
\begin{aligned}
\frac{\partial V(t, y, E_t)}{\partial E_t} &= \frac{\partial E_t V(t, y)}{\partial E_t}, \\
&= V(t, y).
\end{aligned}
\tag{2.43}
$$

The result is quite intuitive. A security's exposure to a foreign-exchange rate is the entire investment.[18]

[18]A possible exception would be a dual-currency bond, where the coupon and notional amounts are denominated in different currencies. In this case, the security would need to be split into two

**Table 2.4** Summarizing exposures

| Factor | Exposure | Definition | Value |
|---|---|---|---|
| Yield | Modified duration | $\frac{1}{V}\frac{\partial V}{\partial \mathbf{y}}$ | $D_M$ |
| | Convexity | $\frac{1}{V}\frac{\partial^2 V}{\partial \mathbf{y}^2}$ | $C$ |
| | $k$th key-rate duration | $\frac{1}{V}\frac{\partial V}{\partial \mathbf{y}_{\mathbf{t_k}}}$ | $D_{t_k}$ |
| Time | Carry | $\frac{1}{V}\frac{\partial V}{\partial \mathbf{t}}$ | $y$ |
| Credit spread | Spread duration | $\frac{1}{V}\frac{\partial V}{\partial \mathbf{s}}$ | $D_s$ |
| FX | FX exposure | $\frac{\partial V}{\partial \mathbf{E_t}}$ | $V$ |

This table summarizes the exposures computed to the various risk factors identified throughout the course of this chapter

## 2.8 Concluding Thoughts

This has been a detailed chapter with numerous mathematical digressions. The common thread among these computations and derivations was the determination of the exposure of a fixed-income security to a collection of different risk factors: yield, time, spread, key rates and foreign-exchange rates. In short, this chapter has been about computing the exposure, or sensitivity, of a fixed-income security to various risk factors. Table 2.4 summarizes all of the operations that we performed on the bond-price expression in terms of factors and exposures.

These exposures provide an insight into the change in the value of a generic fixed-income security associated with a given movement in the underlying risk factor. Note, however, that they consider each risk factor in isolation. We naturally wish to examine them in a joint fashion—this requires additional effort. In the next chapter, we will make ample use of these exposures and employ them to link our set of risk factors to the return of our fixed-income security.

## Reference

1. W.H. Press, S.A. Teukolsky, W.T. Vetterling, B.P. Flannery, *Numerical Recipes in C: The Art of Scientific Computing*, 2nd edn. (Cambridge University Press, Cambridge, 1992)

separate instruments: an annuity for the coupon stream and a zero-coupon bond for the notional value. Each would then require a separate currency definition.

# A Useful Approximation

# 3

*It is better to be roughly right than precisely wrong.*

John Maynard Keynes

One of our principal interests for any portfolio is its return. The return is of interest from both backward and forward-looking perspectives. Not only are the size and magnitude of historical returns interesting, but our interest also extends to the factors contributing to this return. How, for example, did our portfolio manager generate his or her return? This is termed performance attribution. We also have an natural interest in future returns. Predicting specific future returns is extremely difficult, but understanding the uncertainty inherent in future returns is less difficult and can be very helpful. Assessing the uncertainty of future returns is essentially equivalent to measuring the risk of one's portfolio.[1] We often wish to take a step beyond the mere measurement of overall risk and break it down by the factors contributing to the portfolio's risk. Where, for example, is our portfolio manager exposing our portfolio to risk? This is termed risk attribution.

Performance and risk attribution are powerful techniques in portfolio analytic that share an important requirement: a *sensible* mathematical description of a fixed-income portfolio's return as a function of a relevant set of risk factors. Without such a representation one can neither attribute performance nor risk to an underlying set of factors. The objective this chapter is to construct such a description. We will begin with the construction of a general framework and then consider how more complex instruments—such as inflation-linked bonds and derivative contracts—can be handled in this setting.

[1]In measuring risk, we are essentially interested in the statistical distribution of our portfolio returns. With this distribution in hand, we may compute its moments, its tails, or its risk relative to a pre-defined strategic benchmark.



D.J. Bolder, *Fixed-Income Portfolio Analytics*,
DOI 10.1007/978-3-319-12667-8_3

Before we proceed to the heavy-lifting associated with building a mathematical link between return and risk factors, we should have a basic idea of what we're trying to accomplish. What are these risk factors? The following list highlights the principal drivers of return for a fixed-income security:

1. Perhaps most importantly, all fixed-income securities depend, to a lesser or a greater extent, on the movements of the underlying risk-free yield curve.[2]
2. All securities generate some return due to the passage of time.
3. Non-government securities also depend on the movements of credit spreads.
4. Securities denominated in currencies other than one's base currency will, of course, depend on movements in exchange rates.
5. Inflation-linked securities will depend on the movements in the expected and realized rate of future and past inflation.
6. Fixed-income securities with optionality, such as callable bonds or mortgage-backed securities, will depend upon movements in volatilities.

This list is a good start, but it should not be considered exhaustive. Moreover, different sets of underlying risk factors may be employed for different purposes.[3] For the time being, however, we will focus on the first *four* elements of this list. The fifth element will be addressed later in next chapter, while the final element is beyond the scope of this book.

These risk factors are consistent with the collection of risk-factor exposures developed in the previous chapter. With a bit of hard effort and reflection, we should be able to use these exposures in the construction of our desired mapping between return and risk factors.

## 3.1 What We Want

Starting from first principles, what do we know about the return of an individual fixed-income security? Denoting our security value as $V(t)$ and considering the interval, $[t_0, t_1]$ where $t_1 > t_0$, then the security return can be trivially defined as,

$$r(t_0, t_1) = \frac{V(t_1) - V(t_0)}{V(t_0)}. \tag{3.1}$$

[2]Note, of course, that any sovereign yield curve is a relatively complicated object that itself requires a few factors to adequately describe it. The underlying factors driving the yield curve likely include macroeconomic variables such as output, inflation, and monetary policy. Statistical factors such as level, slope, and curvature may also be employed.

[3]The credit spread is a classic example. Is it sufficient to consider the spread as a risk factor or does one wish to identify the underlying macroeconomic elements driving the credit spread? The answer depends on what one is trying to accomplish. Be aware, however, that the latter approach involves significantly more complexity.

The return is thus the percentage change in the value of the portfolio over a given period.[4] While technically correct, Eq. (3.1) is not terribly useful as it provides no insight into the sources of the return. The computation merely incorporates information about the security's value at the beginning and ending of the period. What we really *want* is a representation of the return of a fixed-income security that looks something like,

$$r(t_0, t_1) = A(t_0, t_1) + B(t_0, t_1) + C(t_0, t_1) + \cdots, \tag{3.3}$$

where the quantities $A(\cdot)$, $B(\cdot)$, and $C(\cdot)$ denote the return contributions from the underlying set of risk factors. This has great conceptual and practical appeal. We are generally quite good at understanding, interpreting, and using additive relationships. Multiplicative relationships, in contrast, are much less intuitive and difficult to use.

An additive decomposition, as described in Eq. (3.3), need not be unique. One could readily invent a wide variety of arbitrary decompositions of one's portfolio return. What is required is a breakdown that encompasses a meaningful description of the respective return contribution of each underlying risk factor. Such a meaningful decomposition will necessarily require the use of some mathematics.

The first step in defining a possible decomposition is to write the security's value as a mathematical function. Happily, this has already been done in the previous chapter. The value function of a fixed-income security was given the following form,

$$V(t, y) = \sum_{i=1}^{I} \frac{c_{t_i}}{(1 + y)^{t_i - t}}. \tag{3.4}$$

Equation (3.4) sadly does *not* offer a nice additive form. Try as we might to algebraically manipulate this expression, we will not succeed in successfully creating an additive form as described in Eq. (3.3).

What Eq. (3.4) does offer, however, is well-behaved function of *two* arguments, time and yield, as follows

$$V \triangleq V(t, y). \tag{3.5}$$

These two risk factors are thus explicitly represented in our bond value function. Although much work remains to obtain our ultimate goal of building an additive

[4] An alternative definition involves the log differences,

$$r(t_0, t_1) = \ln\left(\frac{V(t_1)}{V(t_0)}\right). \tag{3.2}$$

This continuously compounded return definition is commonly used by economists, but sadly not by market practitioners.

decomposition, Eq. (3.4) provides us with a natural starting point.[5] Going further will require some mathematical theory.

## 3.2 The Taylor Series

There is a wide variety of mathematical techniques that permit one to describe a complicated function as an additive sum of a set of simpler functions. This is really just a clever mathematical trick.[6] An approximation of this form represents a given complicated or inconvenient function—let's call it $f(x)$—as the infinite sum of *simpler* or more *useful* functions. There are many possible approximations and they generally have the form:

$$f(x) = \sum_{i=0}^{\infty} \underbrace{(\text{Coefficients})_i}_{\alpha_i} \cdot \underbrace{(\text{Approximating functions})_i}_{g_i(x)}. \tag{3.6}$$

The trick is to find a sensible set of coefficients, $\{\alpha_i; i = 1, \ldots\}$, and functions, $\{g_i; i = 1, \ldots\}$, that adequately and efficiently describe our function, $f(x)$. Note the additive form—this appears to be consistent with what we are seeking. There are a number of choices including:

- Fourier series;
- orthogonal polynomials; and
- the Taylor or Maclaurin series expansions.

The ultimate choice typically depends on the underlying function and the final application. Whatever one's choice, these powerful techniques can permit a dramatic simplification and understanding of complicated functions.

Using the mathematical description of the security's value in Eq. (3.4), we are free to employ one of these techniques. Our objective is to find an additive collection of simpler functions that meaningfully capture risk-factor returns. To take a step towards our additive decomposition, we therefore introduce the underlying mathematical result. This is a very simple, and not entirely complete, statement of this result. For more detailed information along with the full conditions and associated proofs, please see Apostol [2].[7]

[5]We will see in the following development that the results from the previous chapter make a startling reappearance.

[6]See Abramovitz and Stegun [1], Burges [4], Hastie et al. [5], or Bolder and Rubin [3] for much more detail on the theoretical and practical aspects of this idea.

[7]The same applies for the multivariate result provided in a few pages.

**Theorem 3.1 (Univariate Taylor Series I)** *If $f : \mathbb{R}^n \rightarrow \mathbb{R}$ is $n+1$ times continuously differentiable, and $a \in \mathbb{R}$, then*

$$f(x) = f(a) + \frac{1}{1!}\frac{\partial^1 f(a)}{\partial x^1}(x-a)^1 + \frac{1}{2!}\frac{\partial^2 f(a)}{\partial x^2}(x-a)^2 + \frac{1}{3!}\frac{\partial^3 f(a)}{\partial x^3}(x-a)^3 + \cdots + \frac{1}{k!}\frac{\partial^k f(a)}{\partial x^k}(x-a)^k + \mathbf{R_{k+1}(a)}. \tag{3.7}$$

*Moreover, if* $\lim_{\mathbf{k}\rightarrow\infty} \mathbf{R_{k+1}(a)} = \mathbf{0}$ *then,*

$$f(x) = f(a) + \sum_{k=1}^{\infty} \frac{1}{k!}\frac{\partial^k f(a)}{\partial x^k}(x-a)^k. \tag{3.8}$$

To those unaccustomed to reading theorems, this may appear somewhat intimidating.[8] The natural question is: what does it mean? We should be able to extract the form of Eq. (3.6) in the preceding theorem. Cutting through the technical details, it basically implies that for a reasonably well-behaved function, $f$, we can write its value at point $x$ as a function of its value at another point $a$,

$$f(x) = \sum_{i=0}^{\infty} \alpha_i \underbrace{(x-a)^i}_{g_i(x)}. \tag{3.9}$$

In words, the function's value at $x$ can be written as the (infinite) sum of a collection of polynomials in the distance between $x$ and $a$. The coefficients, $\{\alpha_k, k = 0, 1, 2, \ldots\}$, are, up to a constant, merely the derivatives of $f$.[9] Our, potentially complex function, $f$, is thus reduced to the sum of a collection of simpler polynomial functions.

[8] The theorem indicates that if this remainder term is essentially well-behaved, then the infinite Taylor series converges to $f$. The term $R_{k+1}(a)$ is a remainder term. A number of results, such as Lagrange's formula, provide descriptions of the magnitude of the remainder function, $R_{k+1}(a)$.

[9] More precisely, the result permits us to expand the value of a one-dimensional function around a given point an infinite sum of polynomials of increasing order. The coefficient of each term in this series depends on the function's derivatives evaluated at the known value, $a$.

At first glance, Theorem 3.1 might appear difficult to use as it includes an infinite sum. It is, however, typical to implement Theorem 3.1 as a finite number of terms plus an error as follows:

$$f(x_1) = f(x_0) + \underbrace{\left.\frac{\partial f(x)}{\partial x}\right|_{x=x_0}(x_1 - x_0)}_{\substack{\text{Linear} \\ \text{term}}} + \underbrace{\left.\frac{1}{2}\frac{\partial^2 f(x)}{\partial x^2}\right|_{x=x_0}(x_1 - x_0)^2}_{\substack{\text{Quadratic} \\ \text{term}}} + \underbrace{\left.\frac{1}{6}\frac{\partial^3 f(x)}{\partial x^3}\right|_{x=x_0}(x_1 - x_0)^3}_{\substack{\text{Cubic} \\ \text{term}}} + \underbrace{\left.\frac{1}{24}\frac{\partial^4 f(x)}{\partial x^4}\right|_{x=x_0}(x_1 - x_0)^4}_{\substack{\text{Quartic} \\ \text{term}}} + \underbrace{\left.\frac{1}{120}\frac{\partial^5 f(x)}{\partial x^5}\right|_{x=x_0}(x_1 - x_0)^5}_{\substack{\text{Quintic} \\ \text{term}}} + \epsilon. \tag{3.10}$$

where $\epsilon$ denotes the error, or remainder, term.[10]

To better see what is going on, let's apply this approximation to an actual function and see how well it performs. Figure 3.1 demonstrates a fifth-order Taylor-series expansion of the following function,

$$f(x) = 5 + 4x + 6x^2 + 7x^3 + 15\ln(x) - 20\sin(x), \tag{3.11}$$

around the point, $x_0$, for all points in the interval $[x_0, x_1]$.[11] In other words, given information about Eq. (3.11) at the point $f(x_0)$, we wish to determine its value at $f(x_1)$. Figure 3.1 illustrates both the true function and the use of the first five terms in the Taylor expansion to perform the approximation.

A first, albeit naive, approximation would be to use the original value of the function, $f(x_0)$; this is termed a zero-order Taylor series expansion. It is tantamount to assuming a constant function and, as we see in Fig. 3.1, it's *not* terribly successful. A first-order Taylor-series expansion—that is, including the first two terms in the infinite sum described in Theorem 3.1—does somewhat better. It uses the first derivative (or slope) at $f(x_0)$. For points relatively close to $x_0$, the approximation

[10] One can loosely think of the remainder term in Eq. (3.10) as the distance between the true and estimated value of the function $f(x_1)$ that could *not* be explained by the first *six* terms of the Taylor series expansion.

[11] Note that there is nothing terribly special about this function, it was really just (mostly) randomly selected to demonstrate Theorem 3.1.

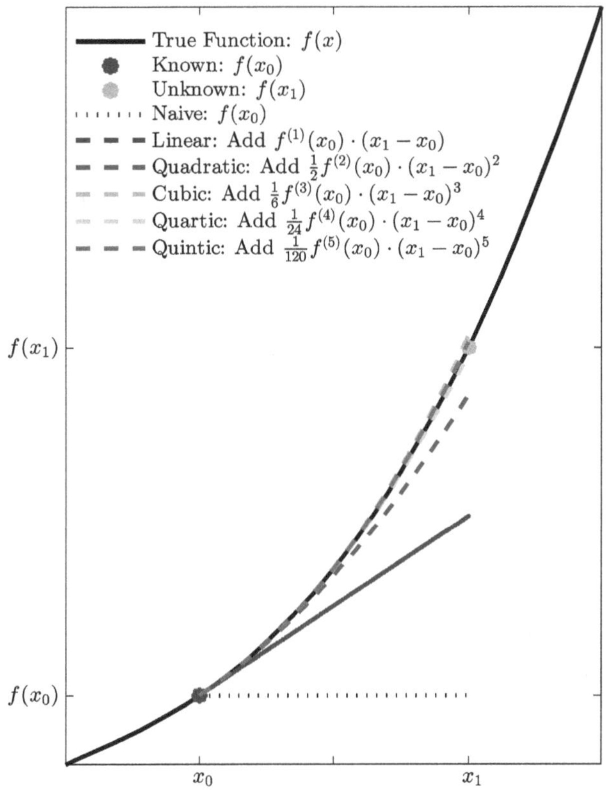


**Fig. 3.1** Approximation using Taylor polynomials. This figure describes each term in a fifth-order Taylor-series expansion around the point, $x_0$, for all points in the interval $[x_0, x_1]$. The function being approximated is $f(x) = 5 + 4x + 6x^2 + 7x^3 + 15\ln(x) - 20\sin(x)$. Here we see clearly how well a linear approximation performs over small movements in $x$

is quite good. As one moves further from $x_0$, however, the local (and linear) approximation of the function's slope deteriorates.[12]

The second-order (or quadratic) expansion—the first three terms in our infinite sum—provides a fairly dramatic improvement as it adds a non-linear element to the approximation. With only three terms from our infinite sum, we have already obtained a fairly sensible approximation. Again, note that for small to medium movements in $x$, the approximation is excellent. Addition of the cubic, quartic, and quintic terms brings us, as expected, progressively closer to the true function value, $f(x_1)$.[13] Far from requiring an infinite number of terms, the Taylor series appears to do quite well with a relatively modest number of elements in its sum. The punchline is that this mathematical result permits us to write a function as the additive sum of

[12]This is a typical result: almost all continuous functions are locally linear and, thus, small movements are generally quite well described by a linear approximation. There are, of course, some exceptions to this rule. Brownian motion is one classic example.

[13]As indicated in Theorem 3.1, adding additional terms in the Taylor-series expansion permits us to converge to the true value, $f(x_1)$.

a small number relatively simple polynomial functions plus an error term. This is a step in the desired direction.

Its promising additive form aside, how precisely does this mathematical result help us with our practical return problem? To answer this question, let's begin with a second-order Taylor-series expansion of the function $f$ around $x_0$ and perform two simple algebraic manipulations: move the $f(x_0)$ term from the right-hand-side to the left-hand-side of Eq. (3.12) and divide both sides by $f(x_0)$. The result is,

$$\begin{aligned} f(x_1) &\approx f(x_0) + \left.\frac{\partial f(x)}{\partial x}\right|_{x=x_0} \underbrace{(x_1 - x_0)}_{\Delta x} + \frac{1}{2}\left.\frac{\partial^2 f(x)}{\partial x^2}\right|_{x=x_0} \underbrace{(x_1 - x_0)}_{\Delta x}^2, \\ \underbrace{f(x_1) - f(x_0)}_{\Delta f} &\approx \left.\frac{\partial f(x)}{\partial x}\right|_{x=x_0} \Delta x + \frac{1}{2}\left.\frac{\partial^2 f(x)}{\partial x^2}\right|_{x=x_0} (\Delta x)^2, \\ \boxed{\frac{\Delta f}{f(x_0)}} &\approx \frac{1}{f(x_0)}\left.\frac{\partial f(x)}{\partial x}\right|_{x=x_0} \Delta x + \frac{1}{f(x_0)}\frac{1}{2}\left.\frac{\partial^2 f(x)}{\partial x^2}\right|_{x=x_0} (\Delta x)^2. \end{aligned} \tag{3.12}$$

With these two elementary manipulations, we have transformed the left-hand-side of Eq. (3.12) into the percentage change in the function, $f$, over the interval $[x_0, x_1]$. Recalling that the return of a security is merely the percentage change, we have approximated a return as an *additive* function of two (admittedly ugly) mathematical terms. Our remaining task is to transform these two terms into something meaningful.

Now for some bad news. The value of our fixed-income security is a function of more than one argument and Theorem 3.1 applies only to functions of a single argument. We need to generalize the result in Theorem 3.1 to multidimensional functions as follows:

**Theorem 3.2 (Multivariate Taylor Series)** *If $f : \mathbb{R}^n \rightarrow \mathbb{R}$ is $n + 1$ times continuously differentiable, $a = (a_1, \ldots, a_n) \in \mathbb{R}^n$, and $x = (x_1, \ldots, x_n) \in \mathbb{R}^n$, then*

$$\begin{aligned} f(x_1, \cdots, x_n) &= f(a_1, \cdots, a_n) + \frac{1}{1!}\sum_{i=1}^{n} \frac{\partial^1 f(a_1, \ldots, a_n)}{\partial x_i}(x_i - a_i) \\ &+ \frac{1}{2!}\sum_{i=1}^{n}\sum_{j=1}^{n} \frac{\partial^2 f(a_1, \ldots, a_n)}{\partial x_i \partial x_j}(x_i - a_i)(x_j - a_j) \\ &+ \frac{1}{3!}\sum_{i=1}^{n}\sum_{j=1}^{n}\sum_{\nu=1}^{n} \frac{\partial^3 f(a_1, \ldots, a_n)}{\partial x_i \partial x_j \partial x_\nu}(x_i - a_i)(x_j - a_j)(x_\nu - a_\nu) \end{aligned}$$

$$+\sum_{k=4}^{n}\left(\frac{1}{k!}\sum_{j_1=1}^{n}\cdots\sum_{j_k=1}^{n}\frac{\partial^k f(a_1,\ldots,a_n)}{\partial x_{j_1}\cdots\partial x_{j_k}}(x_{j_1}-a_{j_1})\cdots(x_{j_k}-a_{j_k})\right)$$
$$+\mathbf{R_{k+1}}(\mathbf{a_1},\ldots,\mathbf{a_n}). \tag{3.13}$$

*Moreover, if* $\lim_{\mathbf{k}\to\infty}\mathbf{R_{k+1}}(\mathbf{a_1},\ldots,\mathbf{a_n}) = \mathbf{0}$ *then,*

$$f(x_1,\ldots,x_n) = f(a_1,\cdots,a_n) + \sum_{k=1}^{\infty}\left(\frac{1}{k!}\sum_{j_1=1}^{n}\cdots\sum_{j_k=1}^{n}\frac{\partial^k f(a_1,\ldots,a_n)}{\partial x_{j_1}\cdots\partial x_{j_k}}\right.$$
$$\left.\times(x_{j_1}-a_{j_1})\cdots(x_{j_k}-a_{j_k})\right). \tag{3.14}$$

We now have a result that applies to functions with multiple arguments.[14] It is conceptually identical to the more intuitive version in Theorem 3.1 and essential for the treatment of multiple risk factors. For an $n$-order expansion, there are now $n$ terms associated with each of the function's arguments plus numerous cross (or interaction) terms between the individual arguments.

## 3.3 Applying the Taylor Series

We have been abstract about the function in the Taylor's series and employed the classical function, $f$. It's time to be more concrete and use the value function of our fixed-income security,

$$V \equiv V(t, y). \tag{3.15}$$

Our value function, to repeat, depends on two arguments: time and yield. The return over the period $[t_0, t_1]$ is simply written as,

$$r(t_0, t_1) = \frac{V(t_1, y_1) - V(t_0, y_0)}{V(t_0, y_0)}, \tag{3.16}$$

[14]It is unfortunately a bit ugly. In mathematics textbooks, it looks a bit better since it is typically written in vector notation.

If we use Theorem 3.2 and take a second-order expansion with respect to $t$ and $y$, we arrive at the following expression,

$$
\begin{aligned}
\underbrace{V(t_1,y_1)}_{V_1} \approx \underbrace{V(t_0,y_0)}_{V_0} &+ \overbrace{\left.\frac{\partial V(t,y)}{\partial t}\right|_{(t,y)=(t_0,y_0)} \underbrace{(t_1-t_0)}_{\Delta t}}^{\text{First-order term with respect to } t} + \cancel{\overbrace{\frac{1}{2}\left.\frac{\partial^2 V(t,y)}{\partial t^2}\right|_{(t,y)=(t_0,y_0)} \underbrace{(t_1-t_0)^2}_{\Delta t}}^{\text{Second-order term with respect to } t}} + \\
&\overbrace{\left.\frac{\partial V(t,y)}{\partial y}\right|_{(t,y)=(t_0,y_0)} \underbrace{(y_1-y_0)}_{\Delta y}}^{\text{First-order term with respect to } y} + \overbrace{\frac{1}{2}\left.\frac{\partial^2 V(t,y)}{\partial y^2}\right|_{(t,y)=(t_0,y_0)} \underbrace{(y_1-y_0)^2}_{\Delta y}}^{\text{Second-order term with respect to } y} + \\
&\cancel{\overbrace{\left.\frac{\partial^2 V(t,y)}{\partial t \partial y}\right|_{(t,y)=(t_0,y_0)} \underbrace{(y_1-y_0)}_{\Delta y}\underbrace{(t_1-t_0)}_{\Delta t}}^{\text{Cross-order term with respect to } y \text{ and } t}}.
\end{aligned}
\tag{3.17}
$$

This lengthy and formal-looking expression follows directly from the multivariate version of Taylor's theorem (i.e., Theorem 3.2). We have eliminated the second-order term with respect to time and the cross-order term.[15] We set these two terms to zero, because they are typically small and difficult to interpret. The interested reader will find a shaded box justifying the logic behind this choice.

Simplifying the notation will also help enormously. We thus propose writing the partial derivative evaluated at the point, $(t_0, y_0)$ as,

$$
\left.\frac{\partial V(t,y)}{\partial t}\right|_{(t,y)=(t_0,y_0)} \equiv \frac{\partial V(t_0,y_0)}{\partial t}. \tag{3.18}
$$

This permits us to re-write Eq. (3.17) as,

$$
V_1 \approx V_0 + \frac{\partial V(t_0,y_0)}{\partial t}\Delta t + \frac{\partial V(t_0,y_0)}{\partial y}\Delta y + \frac{1}{2}\frac{\partial^2 V(t_0,y_0)}{\partial y^2}(\Delta y)^2. \tag{3.19}
$$

This should begin to remind the reader of some of the results developed in the previous chapter. The partial derivatives of our value function with respect to time and yield are critical elements of the sensitivities used to measure the exposure of the security to different risk factors. Moreover, they enter into Eq. (3.22) in an additive manner. This is an encouraging sign.

[15]Note that there are two cross terms in the second-order expansion: $\frac{\partial^2 V(t,y)}{\partial t \partial y}$ and $\frac{\partial^2 V(t,y)}{\partial y \partial t}$. Since, under normal conditions that we assumed to be fulfilled, second partial derivatives are symmetric, we may collect them into a single term.

Employing our trick of moving $V_0$ from the right-hand to left-hand-side of Eq. (3.22) and dividing both sides by $V_0$, we arrive at the following (less intimidating) expression,

$$\begin{aligned}
V_1 - V_0 &\approx \frac{\partial V(t_0, y_0)}{\partial t}\Delta t + \frac{\partial V(t_0, y_0)}{\partial y}\Delta y + \frac{1}{2}\frac{\partial^2 V(t_0, y_0)}{\partial y^2}(\Delta y)^2, \\
\underbrace{\frac{V_1 - V_0}{V_0}}_{r(t_0,t_1)} &\approx \underbrace{\frac{1}{V_0}\frac{\partial V(t_0, y_0)}{\partial t}}_{A(t_0,t_1)}\Delta t + \underbrace{\frac{1}{V_0}\frac{\partial V(t_0, y_0)}{\partial y}}_{B(t_0,t_1)}\Delta y + \underbrace{\frac{1}{2}\frac{1}{V_0}\frac{\partial^2 V(t_0, y_0)}{\partial y^2}}_{C(t_0,t_1)}(\Delta y)^2, \\
r(t_0, t_1) &= A(t_0, t_1)\Delta t + B(t_0, t_1)\Delta y + B(t_0, t_1)(\Delta y)^2. \qquad (3.20)
\end{aligned}$$

We've succeeded in representing the return on a fixed-income security as an additive function of three separate components relating to time and yield-related risk factors. Having achieved our objective of constructing an additive link between return and risk factors, we need to identify the coefficients: $A$, $B$, and $C$.

The second-order term with respect to $t$ warrants some discussion and justification. If the return horizon is short, say one day, then $\Delta t = \frac{1}{365} \approx 0.003$. This implies that $\Delta t^2$ is zero to five decimal points. Consequently, the coefficient must be enormous to justify its inclusion. The second derivative of our bond-price function with respect to time, however, is merely,

$$\begin{aligned}
\frac{\partial^2 V(t, y)}{\partial t^2} &= \frac{\partial}{\partial t}\left(\ln(1+y)\sum_{i=1}^{I} c_{t_i} e^{-(t_i - t)\ln(1+y)}\right), \\
&= \sum_{i=1}^{I} c_{t_i} \ln(1+y) e^{-(t_i - t)\ln(1+y)}, \\
&= \ln(1+y)\ln(1+y)\underbrace{\left(\sum_{i=1}^{I}\frac{c_{t_i}}{(1+y)^{t_i - t}}\right)}_{V(t,y)}, \\
&= 2\ln(1+y)V(t, y). \qquad (3.21)
\end{aligned}$$

Recalling that $\ln(1+y) \approx y$ for small values of $y$—we arrive at what might be termed *time convexity* of approximately $2y$. The return associated with $2y\Delta t$ will be very small and can be safely ignored.

(continued)

The interaction, or cross, term is a bit more involved. Computation of the appropriate coefficient requires an application of the product rule as follows,

$$\begin{aligned}\frac{\partial^2 V(t,y)}{\partial t \partial y} &= \frac{\partial}{\partial y}\left(\underbrace{\frac{\partial V(t,y)}{\partial t}}_{\ln(1+y)V(t,y)}\right),\\ &= \frac{\partial \ln(1+y)}{\partial y}V(t,y) + \ln(1+y)\frac{\partial V(t,y)}{\partial y},\\ &= \frac{V(t,y)}{1+y} + \ln(1+y)\frac{\partial V(t,y)}{\partial y}.\end{aligned} \tag{3.22}$$

By the symmetry of second derivatives, we achieve the same result by computing $\frac{\partial V^2(t,y)}{\partial y \partial t}$, although it turns out to be rather more tedious. If we normalize by $V(t,y)$ to get the exposure, we have

$$\begin{aligned}\frac{1}{V(t,y)}\frac{\partial^2 V(t,y)}{\partial t \partial y} &= \frac{1}{V(t,y)}\left(\frac{\cancel{V(t,y)}}{1+y} + \ln(1+y)\frac{\partial V(t,y)}{\partial y}\right),\\ &= \frac{1}{1+y} + \ln(1+y)D_M,\\ &\approx 1 + yD_M.\end{aligned} \tag{3.23}$$

The final result is that the interaction exposure can be approximated as a simple function of the yield and the modified duration—that is, it depends on the yield and time exposures of the underlying security. The question is, what is the typical size of,

$$\frac{1}{V(t,y)}\frac{\partial^2 V(t,y)}{\partial t \partial y}\Delta y \Delta t \approx (1 + yD_M)\,\Delta y \Delta t. \tag{3.24}$$

Let's consider an extreme case. We have a bond with a 10 % yield, a duration of 10 years, a 100 basis-point yield movement and a 1-month return horizon. The return impact associated with the interaction between the yield and time risk factors is approximately 17 basis points. In summary, the interaction term can be different from zero, but even for extreme values, it remains relatively small. Given that it is also quite difficult to interpret, it seems reasonable to ignore it.

**Table 3.1** Identifying our coefficients

| Coefficient | Mathematical form | What is it? |
|---|---|---|
| $A(t_0, t_1)$ | $\frac{1}{V}\frac{\partial V(t_0, y_0)}{\partial t}$ | Yield |
| $B(t_0, t_1)$ | $\frac{1}{V}\frac{\partial V(t_0, y_0)}{\partial y}$ | Modified duration |
| $C(t_0, t_1)$ | $\frac{1}{2V}\frac{\partial^2 V(t_0, y_0)}{\partial y^2}$ | Convexity |

This table illustrates the coefficients from Eq. (3.24) and provides some insight into their identity.

Table 3.1 provides a closer inspection of the mathematical form of each coefficient and reveals that they each all related to the risk-factor exposures developed in the previous chapter. In particular, they are none other than the yield, the modified duration and the convexity of a fixed-income security. This permits us to dramatically simplify (3.24) as,

$$\begin{aligned}
\underbrace{\frac{V_1 - V_0}{V_0}}_{\substack{r(t_0,t_1)\\ \text{Equation 3.15}}} &\approx \underbrace{\frac{1}{V_0}\frac{\partial V(t_0, y_0)}{\partial t}}_{\substack{\text{Yield to}\\ \text{maturity: } y}} \Delta t + \underbrace{\frac{1}{V_0}\frac{\partial V(t_0, y_0)}{\partial y}}_{\substack{\text{Modified}\\ \text{duration: } -D_M}} \Delta y + \frac{1}{2}\underbrace{\frac{1}{V_0}\frac{\partial^2 V(t_0, y_0)}{\partial y^2}}_{\text{Convexity: } C} (\Delta y)^2, \\
r(t_0, t_1) &= A(t_0, t_1)\Delta t + B(t_0, t_1)\Delta y + B(t_0, t_1)(\Delta y)^2, \\
r(t_0, t_1) &\approx y\Delta t - D_M \Delta y + \frac{1}{2} C\,(\Delta y)^2.
\end{aligned} \tag{3.25}$$

This is a remarkable result. We have constructed a meaningful additive decomposition of our security's return as a function of a set of well-known, easily accessible exposures.

We can also write our approximation as,

$$r(t_0, t_1) = y\Delta t - D_M \Delta y + \frac{1}{2} C\,(\Delta y)^2 + \epsilon, \tag{3.26}$$

where $\epsilon$ denotes the amount of return unexplained by our additive risk-factor decomposition. Conceptually, we know from Theorems 3.1 and 3.2 that we can basically make $\epsilon$ as small as we like. It merely requires adding additional terms to our approximation. The problem, however, is that we have already eliminated the second-order time and yield-time interaction terms on the grounds that they are not only small, but difficult to interpret. How would we interpret $\frac{\partial^4 V(t,y)}{\partial y^4}$ or $\frac{\partial^3 V(t,y)}{\partial t^3}$? There is, therefore, a trade-off. We require an acceptably small value of $\epsilon$ to trust the decomposition, but we simultaneously desire the ability to interpret the individual additive terms. This tension will be discussed and analysed extensively in later chapters.

Some conceptual simplification may be helpful in better understanding the structure of Eq. (3.25). Translating our mathematical symbols into words, the equation has the form,

$$\text{Return} \approx \underbrace{\boxed{\begin{array}{c}\text{Exposure}\\ \text{to time}\end{array}} \times \boxed{\begin{array}{c}\text{Change}\\ \text{in time}\end{array}}}_{\substack{\text{Time}\\ \text{factor}}} + \underbrace{\boxed{\begin{array}{c}\text{Exposure}\\ \text{to yield}\end{array}} \times \boxed{\begin{array}{c}\text{Change}\\ \text{in yield}\end{array}}}_{\substack{\text{Yield}\\ \text{factor}}} + \underbrace{\boxed{\begin{array}{c}\text{Exposure}\\ \text{to duration}\end{array}} \times \boxed{\begin{array}{c}\text{Change}\\ \text{in yield}^2\end{array}}}_{\substack{\text{Duration}\\ \text{factor}}} \tag{3.27}$$

If we generalize this idea, we have that the return on a fixed-income security has the form,

$$\text{Return} \approx \sum_{k=1}^{n} \underbrace{\boxed{\begin{array}{c}\text{Exposure}\\ \text{to factor } k\end{array}} \times \boxed{\begin{array}{c}\text{Change}\\ \text{in factor } k\end{array}}}_{\substack{\text{Contribution of}\\ \text{factor } k\\ \text{to return}}}. \tag{3.28}$$

With judicious application of Theorem 3.2 to the fixed-income security's value function, therefore, we've shown that the return on a fixed-income security can be approximated as the sum of the products of its exposure to a given factor and the change in that factor.[16] This is a powerful and flexible result.

### 3.3.1 Adding Risk Factors

Our additive decomposition only incorporates the time and yield elements of return. This is problematic since our original list of risk factors was somewhat more extensive. With some algebraic sleight of hand, however, we can add some additional factors as follows:

$$\begin{aligned} r(t_0, t_1) &\approx y\Delta t - D_M \Delta y + \frac{1}{2} C (\Delta y)^2, \\ &\approx y\Delta t - D_M (\Delta y_{\text{TRE}} + \Delta s_{\text{OAS}}) + \frac{1}{2} C (\Delta y)^2 \\ &\approx y\Delta t - D_{\text{M}} \Delta y_{\text{TRE}} - \underbrace{D_{\text{S}} \Delta s_{\text{OAS}}}_{\substack{D_S \approx D_M \text{ is}\\ \text{spread duration}}} + \frac{1}{2} C (\Delta y)^2. \end{aligned} \tag{3.29}$$

[16] Note that in some cases, such as convexity, we need to raise the change in the factor to a power.

If we recall that the modified duration is approximately equal to the sum of the key-rate durations,

$$D_M \approx \sum_{i=1}^{\nu} \kappa_i, \tag{3.30}$$

where $\{\kappa_i, i = 1, \ldots, \nu\}$ denote the key-rate durations, then we may take a step further as,

$$\begin{aligned} r(t_0, t_1) &\approx y\Delta t - \underbrace{\sum_{i=1}^{\nu} \kappa_i}_{D_M} \Delta y_{\text{TRE},i} - D_S \Delta s_{\text{OAS}} + \frac{1}{2} C (\Delta y)^2, \\ &\approx \underbrace{y\Delta t}_{\text{Carry}} - \underbrace{\sum_{i=1}^{\nu} \kappa_i \Delta y_{\text{TRE},i}}_{\text{Curve}} - \underbrace{D_S \Delta s_{\text{OAS}}}_{\text{Credit}} + \underbrace{\frac{1}{2} C (\Delta y)^2}_{\text{Convexity}}. \end{aligned} \tag{3.31}$$

The credit risk factor and a perspective on different sectors of the yield-curve have been separately incorporated.

This is good news, but what precisely have we done? Two adjustments were made:

1. **Credit**: We expand the change in the security's yield ($\Delta y$) into an equivalent treasury-yield movement ($y_{\text{TRE}}$) and a credit spread movement ($s_{\text{OAS}}$).[17] A key assumption of this expansion is to assume that the modified duration and spread duration are identical. For most fixed-income securities this is the case, although for some (such as floating-rate notes), this is not always true.
2. **Key Rates**: In the basic Taylor series expansion, there is no insight regarding how the security reacts to the movement of different parts of the yield curve. Through the replacement of the modified duration with security's key-rate durations, we substantially augment our understanding of how the return reacts to different aspects of the yield curve.

Careful application of the Taylor's series expansion to our fixed-income security value function has permitted us to write its return as a additive function of the first three risk factors: time, credit, and the yield curve. In the following section, we will add yet another *important* risk factor.

[17] Such an expansion is always possible. One merely requires a mathematical description, or model, of the underlying treasury yield curve.

## 3.4 The Foreign-Exchange Dimension

Despite good progress, we are still missing the foreign-exchange element. Our development to this point permits us to compute the return of a security as long as that security is denominated in our base currency. Naturally, it is entirely possible to have a base currency of USD, a USD benchmark, and yet invest in a CAD-denominated asset. Consequently, one is not only exposed to movements in CAD interest rates and spreads, but also to the exchange rate between CAD and USD. If one's base currency was USD, one would typically compute the return of this CAD (or any other non-USD) security in USD terms. This leads to a slight, but important, adjustment to the return computation introduced in Eq. (3.1),

$$\tilde{r}(t_0, t_1) = \frac{E(t_1)V(t_1, y_1) - E(t_0)V(t_0, y_0)}{E(t_0)V(t_0, y_0)}, \tag{3.32}$$

where $E(t)$ denotes the exchange rate at time $t$.

This is an unfortunate expression, because it is impossible to algebraically isolate the independent influences of the local currency return and the exchange rate. We find ourselves with a multiplicative relationship between risk factors and the overall return. In other words, the exchange rate enters into Eq. (3.32) in an undesirable multiplicative manner, whereas we wish to maintain our additive description of time, curve, and spread effects in local currency terms (as in Eq. (3.29) and then merely add on the foreign-currency impact.

Fortunately, we may once again use Theorem 3.2 to resolve this problem. In this expanded case, our value function can be written to include foreign-currency denominated securities as,

$$V \equiv E(t)V(t, y). \tag{3.33}$$

We merely compute a first-order Taylor series expansion of Eq. (3.33). The development, as evidenced by the manipulations in Eq. (3.34), is a bit long and tedious, but the final result is very pleasant.

$$E(t_1)V(t_1, y_1) \approx E(t_0)V(t_0, y_0) + \overbrace{\underbrace{\frac{\partial E(t)V(t,y)}{\partial E(t)}\bigg|_{(t,y)=(t_0,y_0)}}_{V(t_0,y_0)}(E(t_1) - E(t_0))}^{\text{First-order term with respect to } E(t)}$$

$$+ \overbrace{\underbrace{\frac{\partial E(t)V(t,y)}{\partial V(t,y)}\bigg|_{(t,y)=(t_0,y_0)}}_{E(t_0)}(V(t_1, y_1) - V(t_0, y_0))}^{\text{First-order term with respect to } V(t,y)},$$

$$E(t_1)V(t_1, y_1) - E(t_0)V(t_0, y_0) \approx V(t_0, y_0)\left(E(t_1) - E(t_0)\right) + E(t_0)\left(V(t_1, y_1) - V(t_0, y_0)\right),$$

$$\underbrace{\frac{E(t_1)V(t_1,y_1)-E(t_0)V(t_0,y_0)}{E(t_0)V(t_0,y_0)}}_{\tilde{r}(t_0,t_1)} \approx \frac{\cancel{V(t_0,y_0)}(E(t_1)-E(t_0))}{E(t_0)\cancel{V(t_0,y_0)}} + \frac{\cancel{E(t_0)}(V(t_1,y_1)-V(t_0,y_0))}{\cancel{E(t_0)}V(t_0,y_0)},$$

$$\tilde{r}(t_0,t_1) \approx \underbrace{\frac{E(t_1)-E(t_0)}{E(t_0)}}_{\substack{\text{Foreign-}\\ \text{exchange}\\ \text{return}}} + \underbrace{\frac{V(t_1,y_1)-V(t_0,y_0)}{V(t_0,y_0)}}_{\substack{\text{Local-currency}\\ \text{return: } r(t_0,t_1)}}. \tag{3.34}$$

This second application of Theorem 3.2 reduces the multiplicative representation of the foreign-currency element into the sum of the foreign-exchange and local-currency returns. This approximation enormously simplifies our analysis.

There is also a conceptual argument supporting this mathematical analysis. Imagine that you want to understand the return of a foreign-currency denominated investment—let's call this $r_{\text{FC}}$. We know that this will be a function of the local currency return, call it $r_{\text{Local}}$, and the foreign-exchange return, denoted $r_{\text{FX}}$. The geometric sum of the two returns is given as,

$$1 + r_{\text{FC}} = (1 + r_{\text{Local}})(1 + r_{\text{FX}}), \tag{3.35}$$

$$r_{\text{FC}} = r_{\text{Local}} + r_{\text{FX}} + \underbrace{r_{\text{Local}} r_{\text{FX}}}_{\substack{\text{Interaction}\\ \text{Term}}}, \tag{3.36}$$

which, up to an interaction term, corresponds to the development from Eq. (3.34). The formal origin of this interaction term is described in the underlying shaded box.

Two additional points are required: the addition of the elements in Eq. (3.29) and the determination of an exposure term for the foreign-exchange component. The first aspect is easily rectified; one need only replace $r(t_0, t_1)$ with Eq. (3.29). The second aspect requires some reflection. We have already seen, in the previous chapter, the foreign-exchange exposure of a fixed-income security. It is, however, not completely sufficient to handle multiple currencies. Foreign-exchange exposure is a binary affair. Either you have it, or you do not. If the security is denominated in one's base currency, then it has *no* foreign-exchange exposure. Conversely, if the currency is *not* denominated in one's base currency, then one has foreign-exchange exposure. The mathematical representation of exposure should reflect this fact. Quite simply, therefore, the exposure to each of the permissible $\alpha$ foreign-exchange rates is an indicator variable,

$$\mathbb{I}_{FX_i} = \begin{cases} 0 : \text{Not exposed to currency } i \\ 1 : \text{Exposed to currency } i \end{cases}. \tag{3.37}$$

The foreign-exchange exposure simply takes a value of 1, if exposed to currency $i$ and zero, if not. This is entirely consistent with the notion of foreign-exchange exposure.

In this particular case, it is possible to compute the *full* Taylor-series expansion. The only term we have not yet described is the second-order cross term. All the other terms of the Taylor-series expansion are strictly null.

$$\begin{aligned} E(t_1)V(t_1,y_1) &= E(t_0)V(t_0,y_0) + \\ &\left.\frac{\partial E(t)V(t,y)}{\partial E(t)}\right|_{(t,y)=(t_0,y_0)} (E(t_1)-E(t_0),) + \\ &\left.\frac{\partial E(t)V(t,y)}{\partial V(t,y)}\right|_{(t,y)=(t_0,y_0)} (V(t_1,y_1)-V(t_0,y_0)) + \\ &\left.\frac{\partial E(t)V(t,y)}{\partial E(t)\partial V(t,y)}\right|_{(t,y)=(t_0,y_0)} (E(t_1)-E(t_0))\,(V(t_1,y_1)-V(t_0,y_0)). \end{aligned} \tag{3.38}$$

We can compute easily the derivatives and reorganise the terms leading us to the following result:

$$\begin{aligned} &\frac{E(t_1)V(t_1,y_1)-E(t_0)V(t_0,y_0)}{E(t_0)V(t_0,y_0)} = \\ &\underbrace{\frac{E(t_1)-E(t_0)}{E(t_0)}}_{\substack{\text{Currency}\\ \text{return}}} + \underbrace{\frac{V(t_1,y_1)-V(t_0,y_0)}{V(t_0,y_0)}}_{\substack{\text{Local-currency}\\ \text{return}}} + \\ &\underbrace{\left(\frac{E(t_1)-E(t_0)}{E(t_0)}\right)\left(\frac{V(t_1,y_1)-V(t_0,y_0)}{V(t_0,y_0)}\right)}_{\substack{\text{Extra interaction}\\ \text{term}}} \end{aligned} \tag{3.39}$$

Let's note $r_e(t_0,t_1)$ the foreign-exchange return, we find the contribution of the foreign-exchange convexity:

$$\tilde{r}(t_0,t_1) = r(t_0,t_1) + r_e(t_0,t_1) + r(t_0,t_1)r_e(t_0,t_1). \tag{3.40}$$

The importance of this result mainly lies in the fact that it is no longer an approximation, but a strict equality. Even in very volatile foreign-exchange market, this relationship will still hold. It is also exactly consistent with the conceptual development provided in Eq. (3.35).

Combining all of our previous efforts, we have the following expression,[18]

$$\tilde{r}(t_0,t_1) \approx \sum_{i=1}^{\alpha} \mathbb{I}_{FX_i} \frac{E(i,t_1)-E(i,t_0)}{E(i,t_0)} + \underbrace{r(t_0,t_1)}_{\substack{\text{Equation}\\ (3.29)}},$$

$$\begin{aligned} \tilde{r}(t_0,t_1) \approx & \underbrace{\sum_{i=1}^{\alpha} \mathbb{I}_{FX_i} \frac{E(i,t_1)-E(i,t_0)}{E(i,t_0)}}_{\text{Foreign-exchange return}} \\ & + \underbrace{\underbrace{y\Delta t}_{\text{Carry}} - \underbrace{\sum_{i=1}^{\nu} \kappa_i \Delta y_{\text{TRE}}}_{\text{Curve}} - \underbrace{D_{\text{S}} \Delta s_{\text{OAS}}}_{\text{Credit}} + \underbrace{\frac{1}{2} C (\Delta y)^2}_{\text{Convexity}}}_{\text{Local-currency return}}. \end{aligned} \tag{3.41}$$

We have thus achieved our objective of decomposing a security's return into the sum of the returns attributable to a set of different factors: time, credit, currency, and the various sectors of the yield curve. The heavy lifting is now complete, but we are not completely finished yet. In the next chapter, we will consider how different, less standard fixed-income securities such as inflation-linked bonds and futures contracts, can be included in this framework.

## 3.5 Closing Thoughts

Performance and risk attribution require a *sensible* mathematical description of a fixed-income portfolio's return as a function of a relevant set of risk factors. In this chapter, we performed the hard work necessary to construct such a description. This involved applying the Taylor series expansion to the bond-price equation to construct a sequence of approximating functions. Caution was required to find a reasonable trade-off between accuracy of the overall approximation and capacity to interpret the individual terms. Our final result is an additive risk-factor-based decomposition, which happily incorporates the exposures derived in the previous chapter. This mapping between risk-factor exposures and return represents the common foundation for the subsequent discussion of portfolio-analytic techniques. This decomposition, in other words, will be used for both performance and risk computations and their associated attributions.

[18]We have enriched our notation for an exchange rate to $E(i,t)$, where $i$ denotes the currency and $t$ continues to denote the point in time for which the exchange rate applies.

## References

1. M. Abramovitz, I.A. Stegun, *Handbook of Mathematical Functions* (Dover Publications, New York, 1965)
2. T. Apostol, *Calculus Volume II: Multi Variable Calculus and Linear Algebra, with Applications to Differential Equations and Probability* (Wiley, New York, 1969)
3. D.J. Bolder, T. Rubin, Optimization in a simulation setting: Use of function approximation in debt-strategy analysis. Bank of Canada: Working Paper 2007–13 (2007)
4. C.J.C. Burges, Dimension reduction: A guided tour. Microsoft Research Tech Report MSR-TR-2009-2013 (2013)
5. T. Hastie, R. Tibshirani, J. Friedman, *The Elements of Statistical Learning: Data Mining, Inference and Prediction*, 1st edn. (Springer, New York, 2001)

# 4 Extending Our Framework

*Give me six hours to chop down a tree and I will spend the first four sharpening the axe.*

Abraham Lincoln

Our useful approximation, developed in the previous chapter, is the foundation for our portfolio-analytic framework. It represents an additive decomposition of the return of a fixed-income security to a collection of key risk factors including time, key points along the underlying treasury curve, credit spreads, and foreign exchange movements. The reader may be wondering—with some justification—how generally these results apply to a broad range of fixed-income instruments. This framework applies to a generic fixed-income security. Our work, therefore, is not yet complete. While it is easy to see how this approach might be applied to a cash account, a treasury bill, or a plain-vanilla coupon-bearing bond, most fixed-income portfolios are nevertheless comprised of a rather more varied collection of instruments. This shortcoming is not addressed in the basic framework.

This chapter, therefore, is dedicated to tying up this rather important loose end. It may be skipped entirely without jeopardizing one's understanding of subsequent chapters. One can also read this chapter by merely focusing on the securities of interest, since each is allocated its own separate section. It may also be read in its entirety by those readers interested in the practical details of how to incorporate more complex fixed-income instruments into our general framework. It is common to find rate- or bond-futures contracts, foreign-exchange forwards, inflation-linked bonds, and floating-rate notes in modern fixed-income portfolios. We will address each of these assets in turn.



D.J. Bolder, *Fixed-Income Portfolio Analytics*,
DOI 10.1007/978-3-319-12667-8_4

## 4.1 Handling Inflation-Linked Bonds

We begin with inflation-linked securities. In principle, the treatment of these instruments is practically quite straightforward. Theoretically, however, it requires a significant amount of effort to convince ourselves that this straightforward treatment is, in fact, justified.

### 4.1.1 Revisiting Exposures

Our treatment of inflation-linked bonds requires re-examining their basic exposure computations—it is not clear, *a priori*, that the usual computations apply. This turns out to be rather involved. Up to this point, we have written the value of a bond as follows:

$$V(t,y) = \sum_{i=1}^{I} \frac{c_{t_i}}{(1+y)^{t_i - t}}. \tag{4.1}$$

For bonds with nominal future cash-flows $\{c_{t_i}, i = 1, \ldots, I\}$ this is a perfectly sufficient representation. Not all fixed-income instruments, however, have a known stream of nominal future cash-flows. Inflation-linked bonds, for example, are an excellent example of a more complicated fixed-income instrument where this property does *not* apply.[1] It does not, fortunately, mean that one cannot compute the sensitivities of an inflation-linked bond. Indeed, in this section, we will see how our framework can be extended to accommodate inflation-linked bonds. It does, however, require a bit more effort.

The first step is to slightly adjust the form of our bond-value equation as follows,

$$V(t,y) = \sum_{i=1}^{I} \frac{c \cdot N}{(1+y)^{t_i - t}} + \frac{N}{(1+y)^{t_I - t}}. \tag{4.2}$$

We have basically made two changes. First, we have explicitly written out the cash-flows as the product of a coupon and notional repayment. $c$ denotes the appropriate *fractional* coupon, while $N$ is the notional amount.[2] Second, we have broken up the cash-flows into a sum of coupon payments ranging from $\{t_i, i = 1, \ldots, I\}$

[1]Cash-flows of inflation-linked bonds are not known with certainty in advance, as with a typical nominal fixed-income instruments. Instead, an inflation-linked bond's cash-flows depend on the past and future evolution of inflation. This implies, in a strict sense, that the future cash-flows of an inflation-linked bond are unknown.

[2]We have abstracted from the coupon frequency to ease the notation. It will become sufficiently cluttered in the forthcoming pages without coupon-payment frequencies.

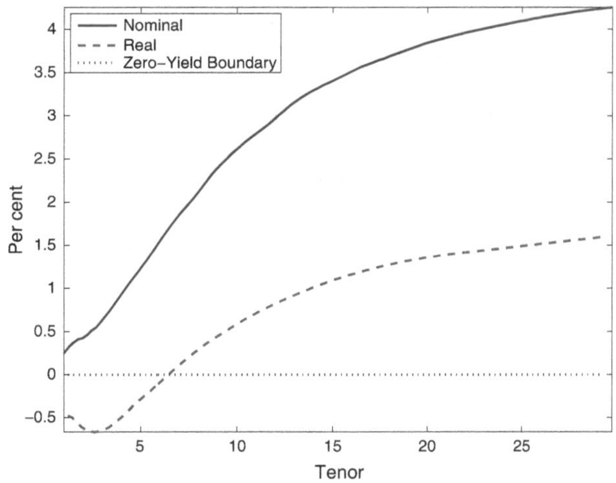


**Fig. 4.1** Real vs. nominal yields. This figure summarizes the UST real and nominal yield curves as of 9 November 2010

and the final nominal payment. These two slight changes will help us handle the complicated nature of inflation-linked bond cash-flows.

At this point, we will make a claim that essentially makes handling inflation-linked bonds relatively easy. The claim is that the value of a nominal bond is given as,

$$V(t, y_n) = \sum_{i=1}^{I} \frac{c \cdot N}{(1 + y_n)^{t_i - t}} + \frac{N}{(1 + y_n)^{t_I - t}}, \tag{4.3}$$

while the real value of an inflation-linked bond can be represented as,

$$V(t, y_r) = \sum_{i=1}^{I} \frac{c \cdot N}{(1 + y_r)^{t_i - t}} + \frac{N}{(1 + y_r)^{t_I - t}}, \tag{4.4}$$

where $y_n$ and $y_r$ denote the nominal and *real* yields, respectively. The claim, therefore, is that to compute a nominal or inflation-linked bond, one need only sum the discounted cash-flows using the appropriate discount factor: nominal cash-flows are discounted with the nominal yield, while real cash-flows are discounted with the real yield. Figure 4.1 outlines, as of 9 November 2010, the real and nominal UST yield curves.[3] The important point is that one does not mix nominal cash flows with real discount factors or vice versa.

While one might expect that we denote the nominal and real values as $V_n$ and $V_r$, respectively, we would prefer to use the generic notation, $V$. In other words, we prefer to differentiate the two values through the choice of yield: nominal $(y_n)$ or

[3]See Sun [8] for a useful source on fitting real yield curves.

real ($y_r$). The underlying rationale is that we want be able to move back and forth, between the real and nominal worlds, by correctly adjusting for inflation. The idea is that we could discount a nominal bond's cash-flows with a real yield if we adjust those cash-flows for expected future inflation. In a similar vein, we may discount an inflation-linked bond's cash-flows with a nominal yield if we adjust its cash-flows for expected future inflation.

It is quite likely that the reader is somewhat suspicious of this claim despite the fact that it significantly eases the task of handling inflation-linked bonds. The suspicion likely stems from the fact that the cash-flows of an inflation-linked bond are adjusted for past and future inflation. Why, therefore, does this realized and expected inflation not show up in the ILB bond-price equation? The simple answer is: it *cancels* out. This is nevertheless not terribly obvious at first glance. Moreover, it is only, as we will see, approximately and not precisely true. We will attempt, therefore, to demonstrate that our claim is (generally) true. To do so, however, we need to introduce *two* important ideas: the

1. index ratio; and
2. Fisher's theorem.[4]

Armed with these two ideas and some tedious algebra, we will demonstrate that the inflation terms (*mostly*) cancel out, leaving us with a straightforward and easy-to-use equation for the price of an inflation-linked bond.[5]

The mechanism relating real to nominal cash-flows is termed the *index ratio*. Let's denote the price level at time $t$ as,

$$\Lambda_t. \tag{4.5}$$

This quantity is typically known as the consumer price level or CPI. It is the price of a representative basket of goods and services in an economy at a specific point in time and is generally computed by an independent government statistical authority. There are typically many different available baskets, which are computed either including or excluding certain types of goods or services (i.e., such as energy or tobacco). The specific basket used for a given inflation-linked bond is contractually defined. For US treasury inflation protected securities (TIPS), the specific basket of goods and services is termed the seasonally unadjusted US Consumer Price Index for Urban Consumers and is computed by the US Bureau of Labour Statistics. It

[4]This is also termed Fisher's identity. The original idea was published in the 1930's by Irving Fisher. Most readers would be better served looking to a modern macroeconomics textbook. One good example is Siklos [7]. Another financial application of this theorem, from a liability perspective, can be found in Anthony et al. [2].

[5]Much of the following development can be found in Deacon et al. [3]. The interested reader will find much useful information on inflation-linked bonds in this text.

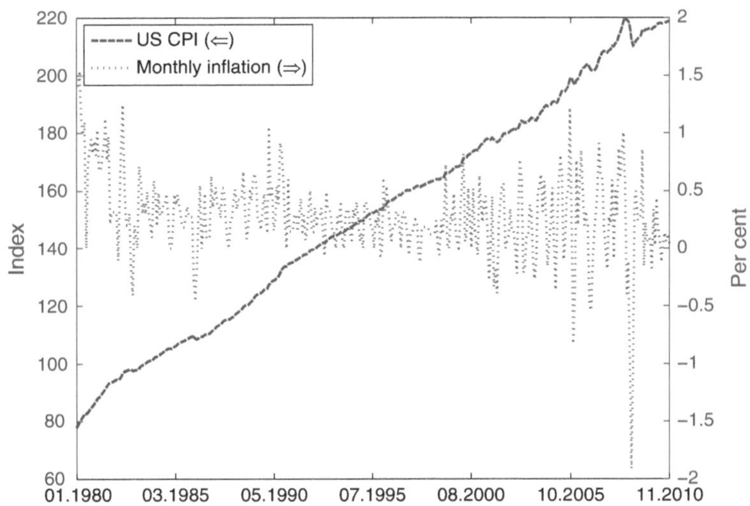


**Fig. 4.2** Historical US CPI and monthly inflation. This figure shows the evolution of the non-seasonally adjusted US Urban Consumers CPI from January 1980 to November 2010. It also demonstrates the monthly inflation rate stemming from the changes in the price index as described in Eq. (4.7)

is computed on a monthly basis and the value for the preceding month is generally published towards the end of the third week in the month.

Over the time interval from $[\tau, t]$, the ratio

$$\frac{\Lambda_t}{\Lambda_\tau}, \tag{4.6}$$

describes the movement in the price level. The percentage change in the price level is, of course, nothing other than inflation: the change in the general level of prices over a given period of time. Thus,

$$\frac{\Lambda_t}{\Lambda_\tau} = 1 + \pi(t, T), \tag{4.7}$$

basically describes the inflation over the period $[\tau, t]$. Figure 4.2 shows the evolution of US price levels and inflation over the last 30 years.

Let's now turn to look at a practical example of how inflation enters into the computation of an inflation-linked bond price. Imagine that you purchased the 3 % TIPS maturing 15 July 2012 (US912828AF74) for settlement on 15 November 2010. This bond was initially issued on 15 July 2002 when the US price level, as measured by the CPI, stood at 179.8000. From issuance to the settlement date of the purchase, 101 months or about 8.3 years has elapsed. The inflation-linked bond compensates the bond holder for the movement in inflation over this period. It does

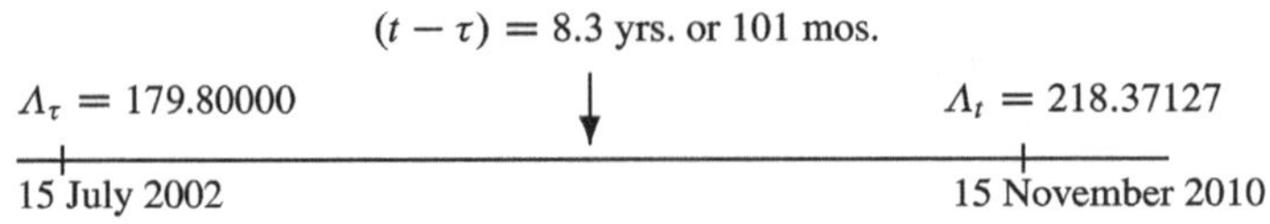


**Fig. 4.3** Movement in the US CPI. This describes how the CPI has moved from 15 July 2002 to 15 November 2010

not compensate for future inflation since it is *not* yet known. Figure 4.3 graphically describes how inflation has moved from initial issuance to the settlement of the bond purchase.

If we take the ratio of the two US price levels, as described in Fig. 4.3, we have

$$\begin{aligned}\frac{\Lambda_t}{\Lambda_\tau} &= \frac{218.37127}{179.80000},\\ &= \underbrace{1.21452}_{\substack{\text{Index}\\ \text{ratio}}}.\end{aligned} \tag{4.8}$$

This simple ratio of price levels is, in the vernacular of inflation-linked bonds, termed the *index ratio*. It represents the cumulative increase in general price level over the time interval, $[\tau, t]$. The index ratio is also, as we've seen, merely the inflation over the period. It can be written in a number of different ways. It may, for example, be considered as the 21.452 % cumulative inflation from 15 July 2002 to 15 November 2010 or,

$$\begin{aligned}\frac{\Lambda_t}{\Lambda_\tau} &= 1 + \underbrace{\pi\left(\tau_{15\text{ July }2002}, t_{15\text{ November }2010}\right)}_{\text{Cumulative inflation}}\\ &= 1 + 21.452\,\%.\end{aligned} \tag{4.9}$$

It can also be thought of as the geometric sum of the monthly inflation over the 101 month period beginning on 15 July 2002 and ending on 15 November 2010,

$$\begin{aligned}\frac{\Lambda_t}{\Lambda_\tau} &= \prod_{i=1}^{101} \underbrace{1 + \pi\left(t_{i-1}, t_i\right)}_{\text{Monthly inflation}},\\ &= \left(1 + \underbrace{\frac{19.261\text{ bps.}}{10^4}}_{\substack{\text{Average}\\ \text{monthly}\\ \text{inflation}}}\right)^{101}.\end{aligned} \tag{4.10}$$

One can easily compute that the average monthly inflation over this 101-month period was approximately 19.3 basis points. Finally, one can also compute the average *annualized* inflation over this period which, as we see below,

$$\frac{\Lambda_t}{\Lambda_\tau} = (1 + \underbrace{2.358\,\%}_{\substack{\text{Annualized}\\ \text{inflation}}})^{8.336}, \tag{4.11}$$

amounts to about 2.36 %.[6] The key point is that the index ratio is the mechanism for compensating the bondholder for the realized path of inflation.

Now we are ready to apply the index ratio to the real price of an inflation-linked bond, $V_r$, as described in Eq. (4.4). The nominal marked-to-market value (or settlement value) of an index-linked bond may be written as

$$\text{Settlement Price} = \underbrace{V(t, y_r)}_{\substack{\text{Equation}\\ (4.4)}} \frac{\Lambda_t}{\Lambda_\tau},$$

$$\begin{aligned} V\left(t, y_r, \frac{\Lambda_t}{\Lambda_\tau}\right) &= \left(\sum_{i=1}^{I} \frac{c \cdot N}{(1 + y_r)^{t_i - t}} + \frac{N}{(1 + y_r)^{t_I - t}}\right) \frac{\Lambda_t}{\Lambda_\tau}, \\ &= \sum_{i=1}^{I} \frac{c \cdot N \cdot \frac{\Lambda_t}{\Lambda_\tau}}{(1 + y_r)^{t_i - t}} + \frac{N \cdot \frac{\Lambda_t}{\Lambda_\tau}}{(1 + y_r)^{t_I - t}}. \end{aligned} \tag{4.12}$$

That is, the settlement price is merely the product of the real price and the index ratio.[7] What is striking in Eq. (4.12) is the fact that there is no forward-looking inflation term. Basically, when one purchases an inflation-linked bond, one must compensate the seller for the cumulative inflation from its issue date up until the settlement date: this is somewhat analogous to an accrued interest computation. Future inflation is *not* considered.

Imagine further that the real price, viewed on the broker screen, for our US TIPS is \$106.25. We note that there are 123 days of accrued interest implying that the real

---

[6] These are all equivalent ways of thinking about the index ratio and the inflation rate.

[7] We have added the current index ratio as an argument to the bond-value function. In general, the real value of the bond has the following form,

$$V(t, y_r) = V(t, y_r, 1), \tag{4.13}$$

where the index ratio is suppressed, but ultimately is equal to unity.

accrued interest for our bond is $\frac{123}{365} \cdot 3 = \$1.01$. Bringing this all together, we arrive at a settlement price of,

$$\begin{aligned}\text{Settlement Price} &= (\$106.25 + \$1.01)\,1.21452,\\ V\left(t, y_r, \frac{\Lambda_t}{\Lambda_\tau}\right) &= (\$106.25 + \$1.01)\,1.21452,\\ &= \$130.27.\end{aligned} \tag{4.14}$$

To purchase this bond, you must pay \$130.27. Again, to this point, there is no forward-looking notion of inflation.

To demonstrate our claim, we need to introduce exactly such a forward-looking notion of inflation. The most direct way to write out the bond-price equation for an inflation-linked bond is to adjust each of the real cash-flows for the expected index ratio. This would transform the cash-flows into nominal terms, permitting us to discount them all at the nominal yield. Mathematically, the bond-value equation would become,

$$V(t, y_n) = \sum_{i=1}^{I} \frac{c \cdot N \cdot \mathbb{E}\left(\frac{\Lambda_{t_i}}{\Lambda_t}\right)}{(1 + y_n)^{t_i - t}} + \frac{N \cdot \mathbb{E}\left(\frac{\Lambda_{t_I}}{\Lambda_t}\right)}{(1 + y_n)^{t_I - t}}. \tag{4.15}$$

Here we demonstrate the previous point that it is possible to discount the inflation-adjusted cash-flows of an inflation-linked bond with the nominal yield. This is only possible if the cash flows are also written in nominal terms. Equation (4.15) is thus the starting point for the demonstration of our claim: that is, that Eq. (4.4) holds.

The next step is to establish a link between expected inflation and current interest rates. If things are going to cancel out, we need both expected inflation in the numerator and denominator of Eq. (4.15). Fortunately, there is a well-known and well-established result from macroeconomics that we can use in this respect. The basic idea is that expected inflation, plus a risk premium, is built into the nominal bond yield according to the following identity,

$$\underbrace{(1 + y_n)}_{\substack{\text{Nominal}\\ \text{yield}}} = \underbrace{(1 + y_r)}_{\substack{\text{Real}\\ \text{yield}}} \underbrace{(1 + \mathbb{E}(\pi))}_{\substack{\text{Expected}\\ \text{inflation}}} \underbrace{(1 + \delta)}_{\substack{\text{Inflation}\\ \text{risk}\\ \text{premium}}}. \tag{4.16}$$

This identity is termed Fisher's theorem and it holds that nominal interest rates are comprised of real interest rates, expected inflation, and a risk premium demanded by bondholders for the uncertainty surrounding future expected inflation. The risk premium is relatively easy to understand when one recalls that, for a nominal bondholder, unexpected inflation erodes the value of the investment. The inflation risk premium is basically extra compensation demanded to protect the bondholder against unexpected inflation.[8] Economies with stable inflation and credible monetary authorities should, therefore, expect to observe relatively small and stable inflation risk premia.

The Fisher identity implies that if we know three of the four elements, we can determine the fourth. Imagine, for example, that nominal rates are 6 %, expected inflation is 2.5 %, and the inflation risk premium is estimated at 40 basis points. Then, following Fisher's theorem,

$$\underbrace{(1+y_n)}_{\text{(i.e., } 1+6\,\%)} = \underbrace{(1+y_r)}_{\text{(i.e., } 1+3\,\%)} \underbrace{(1+\mathbb{E}(\pi))}_{\text{(i.e., } 1+2.5\,\%)} \underbrace{(1+\delta)}_{\text{(i.e., } 1+\frac{40\text{ bps.}}{10^4})}, \tag{4.17}$$

the real yield is 3 %.

Another way to use Fisher's theorem is to recall that for small $x$:

$$1 + x \approx e^x. \tag{4.18}$$

Applying this to Eq. (4.16), we have

$$\begin{aligned} e^{y_n} &\approx e^{y_r} e^{\mathbb{E}(\pi)} e^{\delta} = e^{y_r + \mathbb{E}(\pi) + \delta}, \\ \ln\left(e^{y_n}\right) &\approx \ln\left(e^{y_r + \mathbb{E}(\pi) + \delta}\right), \\ y_n &\approx y_r + \underbrace{\mathbb{E}(\pi) + \delta}_{\substack{\text{Break-even} \\ \text{inflation}}}. \end{aligned} \tag{4.19}$$

The implication is that nominal yields are approximately equal to the sum of real yields, expected inflation, and the inflation risk premium. The latter two quantities are often termed break-even inflation and can be easily computed by taking the difference between nominal and real yields of the same tenor. It is also common in economies—such as the US—where the inflation risk premium is relatively small and stable to assume that it is equal to zero. This simplifies Fisher's theorem further to,

$$y_n \approx y_r + \mathbb{E}(\pi). \tag{4.20}$$

We now have all the necessary ingredients to establish the claim made a few pages back in Eq. (4.4). We return to Eq. (4.15) where we used the expected index

[8]The certainty around the inflationary expectations in an economy are intimately related to the confidence that markets have in the capacity of the monetary authority to meet its explicit or implicit inflation targets. There are other risk premia embedded in bond yield, but, for high quality fixed-income instruments, it is generally agreed to be the most important. These premia, of course, may also be estimated. See Fung et al. [5] for one possible approach.

ratio to transform the real cash-flows into their nominal equivalents that may be safely discounted with the set of nominal discount factors. We now employ Fisher's theorem to simplify as follows

$$
\begin{aligned}
\boxed{V(t, y_n)} &= \sum_{i=1}^{I} \frac{c \cdot N \cdot \mathbb{E}\left(\frac{\Lambda_{t_i}}{\Lambda_t}\right)}{(1+y_n)^{t_i - t}} + \frac{N \cdot \mathbb{E}\left(\frac{\Lambda_{t_I}}{\Lambda_t}\right)}{(1+y_n)^{t_I - t}}, \\
&= \sum_{i=1}^{I} \frac{c \cdot N \cdot \overbrace{\prod_{k=1}^{i} \mathbb{E}\left(\frac{\Lambda_{t_k}}{\Lambda_{t_{k-1}}}\right)}^{\text{Geometric sum of monthly index ratios}}}{(1+y_n)^{t_i - t}} + \frac{N \cdot \overbrace{\prod_{k=1}^{I} \mathbb{E}\left(\frac{\Lambda_{t_k}}{\Lambda_{t_{k-1}}}\right)}^{\text{Geometric sum of monthly index ratios}}}{(1+y_n)^{t_I - t}}, \\
&\approx \sum_{i=1}^{I} \frac{c \cdot N \cdot \overbrace{\prod_{k=1}^{i} \mathbb{E}(1+\pi_{t_{k-1},t_k})}^{\text{Geometric sum of monthly inflation}}}{\underbrace{(1+y_r)^{t_i - t}\mathbb{E}(1+\pi_{t,t_i})^{t_i - t}}_{\text{Fisher's theorem: } \delta = 0}} + \frac{N \cdot \overbrace{\prod_{k=1}^{I} \mathbb{E}(1+\pi_{t_{k-1},t_k})}^{\text{Geometric sum of monthly inflation}}}{\underbrace{(1+y_r)^{t_I - t}\mathbb{E}(1+\pi_{t,t_I})^{t_I - t}}_{\text{Fisher's theorem: } \delta = 0}}, \\
&\approx \sum_{i=1}^{I} \frac{c \cdot N \cdot \cancel{\prod_{k=1}^{i} \mathbb{E}(1+\pi_{t_{k-1},t_k})}}{(1+y_r)^{t_i - t} \cancel{\prod_{k=1}^{i} \mathbb{E}(1+\pi_{t_{k-1},t_k})}} + \frac{N \cdot \cancel{\prod_{k=1}^{I} \mathbb{E}(1+\pi_{t_{k-1},t_k})}}{(1+y_r)^{t_I - t} \cancel{\prod_{k=1}^{I} \mathbb{E}(1+\pi_{t_{k-1},t_k})}}, \\
&\approx \sum_{i=1}^{I} \frac{c \cdot N}{(1+y_r)^{t_i - t}} + \frac{N}{(1+y_r)^{t_I - t}}, \\
&= \boxed{V(t, y_r)}.
\end{aligned}
\tag{4.21}
$$

The expected inflation terms in the numerator of each term in the sum *cancel* with the expected inflation terms in the denominator—each arising from the application of Fisher's theorem. We have demonstrated our claim that one can represent the price of an inflation-linked bond as the sum of its real cash-flows discounted with the appropriate real yield.[9]

As often is the case in life, there is somewhat more to the story. First, the previous development is something of an approximation because we have assumed that the

[9]In other words, it is equivalent to think of inflation-adjusted, or nominal, cash-flows discounted at the nominal yield and real cash-flows discounted at the real yield.

inflation risk premium term in Fisher's theorem to be equal to zero. Second, and more importantly, not all of the terms actually *cancel*, because inflation, in contrast to what we have shown you in Eq. (4.21), is not perfectly indexed. The problem arises because inflation is *neither* observed

- contemporaneously; nor
- on a daily basis.

Instead, we observe inflation with a lag and only at a monthly frequency.[10] In practice, these two challenges have been cleverly solved by issuers of inflation-linked bonds through the use of lagged (known) values of the CPI level and an interpolation technique to compute a daily CPI value. It nevertheless implies that the terms in the numerator and denominator described in Eq. (4.21) do *not* entirely cancel out. In particular, the cash-flows in the numerator are adjusted using the index ratio computed with a 3-month lag. The denominator depends on interest rates that, through Fisher's theorem, embed a contemporaneous view of expected inflation. If we adjust Eq. (4.21) for the 3-month lag in the index ratio, we arrive at

$$
\begin{aligned}
V_r(t, y_r) &= \sum_{i=1}^{I} \frac{c \cdot N \cdot \prod_{k=-2}^{i-3} \mathbb{E}(1+\pi_{t_{k-1},t_k})}{(1+y_r)^{t_i-t} \prod_{k=1}^{i} \mathbb{E}(1+\pi_{t_{k-1},t_k})} + \frac{N \cdot \prod_{k=-2}^{I-3} \mathbb{E}(1+\pi_{t_{k-1},t_k})}{(1+y_r)^{t_I-t} \prod_{k=1}^{I} \mathbb{E}(1+\pi_{t_{k-1},t_k})}, \\
&= \sum_{i=1}^{I} \frac{c \cdot N \cdot \prod_{k=-2}^{0} (1+\pi_{t_{k-1},t_k}) \cancel{\prod_{k=1}^{i-3} \mathbb{E}(1+\pi_{t_{k-1},t_k})}}{(1+y_r)^{t_i-t} \cancel{\prod_{k=1}^{i-3} \mathbb{E}(1+\pi_{t_{k-1},t_k})} \prod_{k=i-2}^{i} \mathbb{E}(1+\pi_{t_{k-1},t_k})} \\
&\quad + \frac{N \cdot \prod_{k=-2}^{0} (1+\pi_{t_{k-1},t_k}) \cancel{\prod_{k=1}^{I-3} \mathbb{E}(1+\pi_{t_{k-1},t_k})}}{(1+y_r)^{t_I-t} \cancel{\prod_{k=1}^{I-3} \mathbb{E}(1+\pi_{t_{k-1},t_k})} \prod_{k=I-2}^{I} \mathbb{E}(1+\pi_{t_{k-1},t_k})}, \\
&= \sum_{i=1}^{I} \frac{c \cdot N \cdot \prod_{k=-2}^{0} (1+\pi_{t_{k-1},t_k})}{(1+y_r)^{t_i-t} \prod_{k=i-2}^{i} \mathbb{E}(1+\pi_{t_{k-1},t_k})} + \frac{N \cdot \prod_{k=-2}^{0} (1+\pi_{t_{k-1},t_k})}{(1+y_r)^{t_I-t} \prod_{k=I-2}^{I} \mathbb{E}(1+\pi_{t_{k-1},t_k})},
\end{aligned}
$$

[10] As previously mentioned, we observe last month's inflation at some point in the third week of the following month. Moreover, our estimate of inflation remains fixed for the entire month.

$$
\begin{aligned}
&\approx \sum_{i=1}^{I} \frac{c \cdot N}{(1+y_r)^{t_i - t}} + \frac{N}{(1+y_r)^{t_I - t}}, \\
&= \boxed{V(t, y_r)}. 
\end{aligned} \tag{4.22}
$$

The numerator has the known inflation for the more recent 3 months, while the denominator retains the expected inflation over the last 3 months of the contract. To the extent there is a large difference between recent realized inflation and expected future inflation across these adjacent 6 months of inflation-linked bond cash flows, there will be an error in the approximation. It is, however, common practice to ignore this lag and assume that everything cancels.[11] In this text, we opt to follow this common practice and assume all the terms cancel out.

We are finally in a position to compute the sensitivity of the value of an inflation-linked bond to a change in its yield. To do this, we merely repeat the previous analysis used for the computation of the modified duration of a nominal bond. An index-linked bond market value, in nominal terms, is,

$$
\begin{aligned}
V(t, y) &\equiv V(t, y_r) \underbrace{\frac{\Lambda_{\text{Settlement Date}}}{\Lambda_{\text{Issue Date}}}}_{\text{Index ratio: } \alpha(t)}, \\
&= V(t, y_r, \alpha(t)).
\end{aligned} \tag{4.23}
$$

From this point on, we will stop explicitly denoting the nominal or real yield (i.e., $y_n$ or $y_r$) and merely use $y$.[12] Equation (4.23) will, therefore, act as the link between the real value of an inflation-linked bond (i.e., $V(t, y_r)$) and its nominal, or current market, value denoted by $V(t, y, \alpha(t))$ in a manner consistent with a nominal bond. If one prefers, one can imagine that the $\alpha(t)$ is always present, but that it is equal to one for nominal bonds and the appropriate index ratio for inflation-linked bonds.

With this important notational point resolved, we proceed to compute the partial derivative with respect to $y \equiv y_r$ and dividing both sides by the bond price, we arrive at

$$
\begin{aligned}
\frac{\partial V(t, y, \alpha(t))}{\partial y} &= \frac{\partial \left( V(t, y)\alpha(t) \right)}{\partial y}, \\
&= \alpha(t) \frac{\partial V(t, y)}{\partial y},
\end{aligned}
$$

[11]This is (i) because there is typically relatively little difference between the two terms and (ii) it facilitates the mathematics dramatically.

[12]It will be understood that when discussing an nominal bond we use its nominal yield, whereas an inflation-linked bond uses its real yield.

$$
\begin{aligned}
\left(\frac{1}{V(t,y,\alpha(t))}\right)\frac{\partial V(t,y,\alpha(t))}{\partial y} &= \left(\frac{1}{V(t,y,\alpha(t))}\right)\alpha(t)\frac{\partial V_r(t,y)}{\partial y} \\
&= \frac{\cancel{\alpha(t)}}{V(t,y)\cancel{\alpha(t)}}\frac{\partial V(t,y)}{\partial y}, \\
\frac{1}{V(t,y,\alpha(t))}\frac{\partial V(t,y,\alpha(t))}{\partial y} &= \frac{1}{V(t,y)}\frac{\partial V(t,y)}{\partial y}, \\
\boxed{D_M} &= \frac{1}{V(t,y)(1+y)}\left(\sum_{i=1}^{I}\frac{(t_i-t)\cdot c\cdot N}{(1+y)^{t_i-t}}+\frac{(t_I-t)\cdot N}{(1+y)^{t_I-t}}\right).
\end{aligned}
\tag{4.24}
$$

The result is a modified duration of an inflation-linked bond. It has *precisely* the same form as the modified duration for a nominal bond–the only difference is that the former is computed using its real yield and price while the latter uses the nominal yield and price.[13]

The convexity of an inflation-linked bond also follows this pattern. As before we compute the second partial derivative of the inflation-linked bond price, adjusted by the index ratio, and divide both sides by the inflation-linked bond's value. The result is,

$$
\begin{aligned}
\frac{\partial^2 V(t,y,\alpha(t))}{\partial y^2} &= \frac{\partial^2\left(V(t,y)\alpha(t)\right)}{\partial y^2}, \\
&= \alpha(t)\frac{\partial^2 V(t,y)}{\partial y^2}, \\
\left(\frac{1}{V(t,y,\alpha(t))}\right)\frac{\partial^2 V(t,y,\alpha(t))}{\partial y^2} &= \left(\frac{1}{V(t,y,\alpha(t))}\right)\alpha(t)\frac{\partial^2 V(t,y)}{\partial y^2}, \\
&= \frac{\cancel{\alpha(t)}}{V(t,y)\cancel{\alpha(t)}}\frac{\partial^2 V(t,y)}{\partial y^2}, \\
\frac{1}{V(t,y,\alpha(t))}\frac{\partial^2 V(t,y,\alpha(t))}{\partial y^2} &= \frac{1}{V(t,y)}\frac{\partial^2 V(t,y)}{\partial y^2} = \boxed{C}.
\end{aligned}
\tag{4.25}
$$

Again, the form is identical to the convexity for a nominal bond. What is clear with this approach, therefore, is that the index ratio neither plays a role in the duration nor the convexity of an inflation-linked bond. This greatly simplifies the computation of inflation-linked bond sensitivities.

We have reconsidered the classical yield risk-factor exposures associated with an inflation-linked bond. In the next section, we will consider what implications this will have for our additive decomposition.

---

[13]The point is that the index ratio does *not* influence the modified duration, because it simply cancels out.

### 4.1.2 Adjusting our Useful Approximation

The complexity associated with inflation-linked bonds, as we've seen, arises from the fact that the settlement amount of the inflation-linked bond is determined using the real value and the accumulated inflation since inception. The accumulated inflation happily cancels out in the computation of the modified duration, key-rate durations, and convexity of the inflation-linked bond. It does *not*, however, cancel out when computing the contribution of time to the return of an inflation-linked bond.[14]

The time contribution—also generally known as carry return—arises from the partial derivative of the bond-price function with respect to time. We will see that, for inflation linked bonds, this has a somewhat more complex form than is the case for nominal bonds. Recall that the settlement amount of an inflation linked bond is given as,

$$V(t, y, \alpha(t)) = \alpha(t)V(t, y) \tag{4.26}$$

If we then proceed to compute the partial derivative of the nominal value of the inflation-linked bond with respect to $t$, we arrive at

$$\begin{aligned} \frac{\partial V(t, y, \alpha(t))}{\partial t} &= \frac{\partial \left(V(t, y)\alpha(t)\right)}{\partial t}, \\ &= \underbrace{\frac{\partial V(t, y)}{\partial t}\alpha(t) + V(t, y)\frac{\partial \alpha(t)}{\partial t}}_{\text{By application of the product rule}}. \end{aligned} \tag{4.27}$$

Instead of reducing to a single term, as is the case with a nominal bond, we now have two terms to handle.

We may simplify the partial derivative in Eq. (4.27) with some cancellation and the computation of a *time duration* by dividing both sides by $V(t, y, \alpha(t))$ and recalling the form of Eq. (4.27) as follows,

$$\begin{aligned} \frac{1}{V(t, y, \alpha(t))}\frac{\partial V(t, y, \alpha(t))}{\partial t} &= \frac{1}{V(t, y, \alpha(t))}\left(\frac{\partial V(t, y)}{\partial t}\alpha(t) + V(t, y)\frac{\partial \alpha(t)}{\partial t}\right), \\ &= \frac{1}{V(t, y)\alpha(t)}\left(\frac{\partial V(t, y)}{\partial t}\alpha(t) + V(t, y)\frac{\partial \alpha(t)}{\partial t}\right), \end{aligned}$$

[14]The inflation-linked bond price equation merely involves discounting real cash-flows with the real yield. The consequence is that the modified duration, key-rate duration, and convexity terms are identical to those found in Eq. (3.41)—the yield changes are, of course, those of the real yield associated with the inflation-linked bond.

$$= \underbrace{\frac{1}{V(t,y)}\frac{\partial V(t,y)}{\partial t}}_{\approx y} + \frac{1}{\alpha(t)}\frac{\partial \alpha(t)}{\partial t},$$

$$\frac{1}{V(t,y)}\frac{\partial V(t,y)}{\partial t} \approx y + \underbrace{\frac{1}{\alpha(t)}\left(\frac{\Delta\alpha(t)}{\Delta t}\right)}_{\substack{\text{Recall } \alpha(t) \\ \text{is linear}}}. \tag{4.28}$$

The last step arises by recalling that $\alpha(t)$ is linear, within a given month, and thus its derivative with respect to time is merely $\frac{\Delta\alpha(t)}{\Delta t}$. The important point is that the exposure of an index-linked bond to time has a more complex form than is the case with a nominal bond.

Equation (4.28) has provided us with all of the necessary ingredients for the addition of the inflation-linked carry into our normal Taylor-series expansion of the inflation-linked bond's return. There are, in fact, two terms: a typical carry term stemming from the real coupon and a second term, which depends on the percentage change in the bond's index ratio. We can see already that this will be different than the carry term associated with a garden-variety nominal bond.

Returning to our Taylor-expansion of the bond return with particular focus on the boxed expressions, a number of simple algebraic manipulations lead to

$$\begin{aligned}
r &= \underbrace{\boxed{\frac{1}{V(t,y)}\frac{\partial V(t_0,y_0)}{\partial t}\Delta t}}_{\substack{\text{Equation} \\ \text{(4.28)}}} + \frac{1}{V(t,y)}\frac{\partial V(t_0,y_0)}{\partial y}\Delta y + \frac{1}{2V(t,y)}\frac{\partial V(t_0,y_0)}{\partial^2 y}(\Delta y)^2, \\
&\approx \boxed{\left(y + \frac{1}{\alpha(t)}\frac{\Delta\alpha(t)}{\Delta t}\right)\Delta t} + \frac{1}{V(t,y)}\frac{\partial V(t_0,y_0)}{\partial y}\Delta y + \frac{1}{2V(t,y)}\frac{\partial V(t_0,y_0)}{\partial^2 y}(\Delta y)^2, \\
&\approx \boxed{y\Delta t + \frac{\Delta\alpha(t)\cancel{\Delta t}}{\alpha(t)\cancel{\Delta t}}} + \frac{1}{V(t,y)}\frac{\partial V(t_0,y_0)}{\partial y}\Delta y + \frac{1}{2V(t,y)}\frac{\partial V(t_0,y_0)}{\partial^2 y}(\Delta y)^2, \\
&\approx \underbrace{\boxed{y\Delta t}}_{\substack{\text{Real} \\ \text{carry}}} + \underbrace{\boxed{\frac{\Delta\alpha(t)}{\alpha(t)}}}_{\substack{\text{Inflation} \\ \text{carry}}} + \underbrace{\frac{1}{V(t,y)}\frac{\partial V(t_0,y_0)}{\partial y}}_{\substack{\text{Real modified} \\ \text{duration}}}\Delta y + \frac{1}{2}\underbrace{\frac{1}{V(t,y)}\frac{\partial^2 V(t_0,y_0)}{\partial y^2}}_{\text{Real convexity}}(\Delta y)^2.
\end{aligned} \tag{4.29}$$

The consequence is an expression that looks virtually identical to that of a nominal bond with the exception of an additional inflation carry term. This additional term basically describes the percentage change in the bond's index ratio over the return period. This is somewhat inconvenient as each inflation-linked bond has a different index ratio depending upon its original date of issuance.

It turns out, however, that some further simplification is possible. Let's look more closely at the inflation carry. First, we recall that the index ratio is defined as

$$\alpha(t) = \frac{\Lambda_t}{\Lambda_{\text{Base}}}. \tag{4.30}$$

Thus, revisiting the second term in Eq. (4.29), we have

$$\begin{aligned}
\frac{\Delta\alpha(t)}{\alpha(t)} &= \frac{\alpha(t+1) - \alpha(t)}{\alpha(t)} \\
&= \frac{\frac{\Lambda_{t+1}}{\Lambda_{\text{Base}}} - \frac{\Lambda_t}{\Lambda_{\text{Base}}}}{\frac{\Lambda_t}{\Lambda_{\text{Base}}}} \\
&= \underbrace{\frac{\Lambda_{t+1} - \Lambda_t}{\Lambda_t}}_{\substack{\text{Percentage}\\ \text{change in}\\ \text{CPI}}} \\
&= \underbrace{\pi(t, t+1)}_{\substack{\text{Inflation from}\\ t \text{ to } t+1}}.
\end{aligned} \tag{4.31}$$

The percentage change in the index ratio reduces to the inflation rate over the return period. This implies that all inflation-linked bonds can be treated in the same fashion and that one only requires the rate of inflation over the reporting period. The end result is that the return for an inflation-linked bond is approximated as,

$$r \approx \underbrace{y\Delta t}_{\substack{\text{Real}\\ \text{carry}}} + \underbrace{\pi(t, t+1)}_{\substack{\text{Inflation}\\ \text{carry}}} + \underbrace{\frac{1}{P(t,y)} \frac{\partial P(t,y)}{\partial y}}_{\substack{\text{Real modified}\\ \text{duration}}} \Delta y + \underbrace{\frac{1}{P(t,y)} \frac{\partial^2 P(t,y)}{\partial y^2}}_{\text{Real Convexity}} (\Delta y)^2. \tag{4.32}$$

This development allows us to update our general framework. Including all of the risk factors developed in this chapter, we can approximate our fixed-income return through the sum of the following factors,

$$
\begin{aligned}
r \approx \overbrace{\underbrace{y\Delta t}_{\substack{\text{Carry}\\\text{return}}} + \mathbb{I}_{\text{ILB}} \underbrace{\pi(t, t+1)}_{\substack{\text{Inflation}\\\text{carry}}} - \underbrace{\sum_{i=1}^{\nu} \kappa_i \Delta y_{\text{TRE},i}}_{\substack{\text{Treasury curve}\\\text{return}}} - \underbrace{D_{\text{S}} \Delta s_{\text{OAS}}}_{\substack{\text{Credit}\\\text{return}}} + \underbrace{\frac{1}{2} C(\Delta y)^2}_{\substack{\text{Convexity}\\\text{return}}}}^{\text{Local currency return}} \\
+ \underbrace{\sum_{i=1}^{\alpha} \mathbb{I}_{FX_i} \left( \frac{E_{i,1} - E_{i,0}}{E_{i,0}} \right)}_{\text{FX return}},
\end{aligned} \tag{4.33}
$$

where FX exposure is,

$$
\mathbb{I}_{FX_i} = \begin{cases} 0 : \text{Not exposed to currency } i \\ 1 : \text{Exposed to currency } i \end{cases}, \tag{4.34}
$$

and exposure to inflation is,

$$
\mathbb{I}_{ILB} = \begin{cases} 0 : \text{A nominal bond} \\ 1 : \text{An ILB} \end{cases}. \tag{4.35}
$$

Recalling Eq. (3.4) in the previous chapter, we have succeeded in writing our portfolio return as the sum of a number of distinct factors. The individual risk factors and their associated exposures are summarized in Table 4.1.

**Table 4.1** Return factors

| Return | Factor | Exposure |
|---|---|---|
| Carry | Time $(t)$ | Yield to maturity $(y)$ |
| **Inflation carry** | **Inflation** $(\pi(t, t+1))$ | **Indicator variable** $(\mathbb{I}_{\text{ILB}})$ |
| Curve | Sovereign yield $(y_{TRE})$ | Modified duration $(D_M)$ |
| Credit | OA spread $(s_{\text{OAS}})$ | Spread duration $(D_S)$ |
| Convexity | Yield $(\Delta y)^2$ | Convexity $(C)$ |
| FX | FX return $(r_{\text{FX}})$ | Indicator variable $(\mathbb{I}_{FX_i})$ |

This table outlines five important risk factors and their associated exposures for the return of a fixed-income security.

This approximation will apply generally to a wide range of fixed-income securities: coupon-bearing, zero-coupon, and inflation-linked bonds.[15] Derivative securities, at first glance, do not seem to readily fit into this framework. In the following sections, however, we will see how the principal derivative contracts can be represented as portfolios of more common instruments. Consequently, we will not need to update our general framework, but merely use logical reasoning to represent our derivative contracts in a simpler fashion.

## 4.2 Handling Floating-Rate Notes

A deeper examination into the pricing and sensitivity of floating-rate notes is a worthwhile exercise. It highlights a situation where the spread and modified duration are *not* the same. Moreover, it forms, when combined with a fixed-rate bond, an interest-rate swap contract, which is another important tool in fixed-income portfolios. To get started, let us define the simply compounded pure discount bond, or discount rate, as

$$P(t,T) = \frac{1}{1 + z(t,T)(T-t)}, \tag{4.36}$$

where $z(t,T)$ denotes the zero-coupon interest rate prevailing over the time interval, $[t,T]$. This is yet another model for the discount factor employed principally in money markets. A floating-rate note has future cash-flows that are determined by a specific interest-rate index. As a consequence, its cash-flows are not known in advance. In a classical setting, one projects the future interest-rate values from the zero-coupon curve.[16] These projections are none other than the implied forward rates over the future periods in question. The forward interest rate at time $t$ prevailing from time $s$ to time $T$ in the future is determined by solving the following equation for $f(t,s,T)$,

$$(1 + z(t,s)(s-t)) \underbrace{(1 + f(t,s,T)(T-s))}_{\substack{\text{Break-even}\\ \text{rate}}} = 1 + z(t,T)(T-t), \tag{4.37}$$

$$f(t,s,T) = \frac{1}{T-s}\left(\frac{1 + z(t,T)(T-t)}{1 + z(t,s)(s-t)} - 1\right).$$

[15] The notable exception *not* addressed in this work are instruments with embedded optionality such as callable bonds and mortgage-backed securities.

[16] The modern approach involves what is termed OIS discounting. This is a generic term for separating the reference curve used for the projection of future cash flows from the discounting of these same cash flows. While we remain in the classic setting, those employing large portfolios of interest-rate futures would be advised to investigate this growing literature. A good place to start is Ametrano and Bianchetti [1].

The implied-forward rate, as we can see from Eq. (4.37), is the break-even rate such that if you invested your funds from time $t$ to $s$ at $z(t,s)$ and then reinvested at the rate, $f(t,s,T)$, you would, at the end of time $T$, have realized a return equal to the original zero-coupon rate, $z(t,T)$. We can further simplify Eq. (4.37) by using the definition of the discount factor in Eq. (4.36) as,

$$f(t,s,T) = \frac{1}{T-s}\left(\frac{P(t,s)-P(t,T)}{P(t,T)}\right). \tag{4.38}$$

The consequence is that the implied forward rate is essentially a ratio of pure discount bond prices.

We now have the necessary background to determine the price of a floating-rate note. If we denote the current time as $t$ and the most recent coupon-reset date as $t_0$, the floating-rate note value function can be written as,

$$V(t) = N\sum_{i=\gamma+1}^{I}\frac{\overbrace{(t_i-t_{i-1})f(t,t_{i-1},t_i)}^{\text{Cash-flows}}}{1+z(t,t_i)(t_i-t)} + \frac{N}{1+z(t,t_I)(t_I-t)}, \tag{4.39}$$

where $N$ represents the notional amount of the contract, $\{(t_i-t_{i-1})f(t,t_{i-1},t_i), i = 1,\ldots,m\}$ are the future cash-flows approximated with the appropriate forward rates, and $\gamma$ is the next reset date. This implies that $\gamma+1$ is the next payment date. The expression can be dramatically simplified by making use of our definitions of discount factors and implied forward rates as follows,

$$\begin{aligned}
V(t) &= N\sum_{i=\gamma+1}^{I}(t_i-t_{i-1})f(t,t_{i-1},t_i)P(t,t_i) + NP(t,t_I),\\
&= N\sum_{i=\gamma+1}^{I}\underbrace{\cancel{(t_i-t_{i-1})}\frac{1}{\cancel{t_i-t_{i-1}}}\left(\frac{P(t,t_{i-1})-P(t,t_i)}{\cancel{P(t,t_i)}}\right)}_{\text{Equation (4.38)}}\cancel{P(t,t_i)} + NP(t,t_I),\\
&= N\left(\underbrace{\sum_{i=\gamma+1}^{I}P(t,t_{i-1})-P(t,t_i)}_{\text{Telescoping sum}} + P(t,t_I)\right),\\
&= N\left(P(t,t_\gamma) - \cancel{P(t,t_I)} + \cancel{P(t,t_I)}\right),\\
&= NP(t,t_\gamma).
\end{aligned} \tag{4.40}$$

Thus, we have quite simply shown that the value of a floating-rate note is its face-value, or notional amount, discounted back from the next reset date.[17]

Equation (4.40) describes the textbook definition of the value function for a floating-rate note. In reality, it is slightly more complex. The difference arises from the fact that the next coupon, which is paid at time $t_\gamma$ and was determined at time $t_{\gamma-1}$, is not included. The next coupon is, in fact, known with certainty. We denote it as $c_\gamma$ and it has the following form,

$$
\begin{aligned}
c_\gamma &= f(t_{\gamma-1}, t_{\gamma-1}, t_\gamma), \\
&= \left(\frac{1}{t_\gamma - t_{\gamma-1}}\right)\left(\frac{\overbrace{P(t_{\gamma-1}, t_{\gamma-1})}^{=1} - P(t_{\gamma-1}, t_\gamma)}{P(t_{\gamma-1}, t_\gamma)}\right), \\
&= \left(\frac{1}{t_\gamma - t_{\gamma-1}}\right)\left(\underbrace{\frac{1}{P(t_{\gamma-1}, t_\gamma)}}_{1+z(t_{\gamma-1},t_\gamma)(t_\gamma - t_{\gamma-1})} - 1\right), \\
&= z(t_{\gamma-1}, t_\gamma),
\end{aligned} \tag{4.41}
$$

which is exactly as we should have expected. The next known coupon, which pays at time $t_\gamma$, is the level of the index at time, $t_{\gamma-1}$. The actual market price from Eq. (4.40) is modified as,

$$
\begin{aligned}
V(t) &= \underbrace{N z(t_{\gamma-1}, t_\gamma)\left(t_\gamma - t_{\gamma-1}\right) P(t, t_\gamma)}_{\text{Discounted value of next coupon}} + NP(t, t_\gamma), \\
&= \left(1 + z(t_{\gamma-1}, t_\gamma)\left(t_\gamma - t_{\gamma-1}\right)\right) NP(t, t_\gamma).
\end{aligned} \tag{4.42}
$$

While technically this is the correct form for the value of a floating-rate note, it makes no difference to the forthcoming sensitivity computations. For notational simplicity, we will exclude this term. We should stress, however, that for the pricing of actual floating-rate notes, this additional term must not be ignored.

[17]Observe that if we compute the value at a reset date, then $t = t_\gamma$. Since the value of the discount factor $P(t_\gamma, t_\gamma) = 1$, the floating-rate note will trade at par.

What is the sensitivity of a floating-rate note to a change in interest rates? In other words, what is the modified duration of a floating-rate note? This is easily accomplished if we re-write Eq. (4.40) in terms of yields instead of discount factors,

$$V(t, z(t, t_\gamma)) = \frac{N}{1 + z(t, t_\gamma)(t_\gamma - t)}, \tag{4.43}$$

and differentiate with respect to $z(t, t_\gamma)$ as follows,

$$\begin{aligned}
\frac{\partial V(t, z(t, t_\gamma))}{\partial z(t, t_\gamma)} &= \frac{\partial}{\partial z(t, t_\gamma)} \left( \frac{N}{1 + z(t, t_\gamma)(t_\gamma - t)} \right), \\
&= \frac{-(t_\gamma - t)N}{\big(1 + z(t, t_\gamma)(t_\gamma - t)\big)^2}, \\
&= \frac{-(t_\gamma - t)}{1 + z(t, t_\gamma)(t_\gamma - t)} \underbrace{\frac{N}{1 + z(t, t_\gamma)(t_\gamma - t)}}_{V(t, z(t, t_\gamma))}, \\
&= \frac{-(t_\gamma - t)V(t, z(t, t_\gamma))}{1 + z(t, t_\gamma)(t_\gamma - t)}.
\end{aligned} \tag{4.44}$$

If we divide both sides by $V(t, z(t, t_\gamma))$ we will, of course, arrive at the modified duration,

$$\begin{aligned}
D_M &= \frac{1}{V(t, z(t, t_\gamma))} \underbrace{\frac{\partial V(t, z(t, t_\gamma))}{\partial z(t, t_\gamma)}}_{\text{Equation (4.44)}}, \\
&= \frac{1}{\cancel{V(t, z(t, t_\gamma))}} \frac{-(t_\gamma - t)\cancel{V(t, z(t, t_\gamma))}}{1 + z(t, t_\gamma)(t_\gamma - t)}, \\
&= \underbrace{\frac{-(t_\gamma - t)}{1 + z(t, t_\gamma)(t_\gamma - t)}}_{\approx 1}, \\
&\approx -(t_\gamma - t).
\end{aligned} \tag{4.45}$$

The final result is that the duration of a floating-rate note is very close to the time remaining until the next reset date. Although it is *not* exact, it is a reasonable approximation and is consistent with the general result. A floating-rate note, therefore, is essentially a zero-coupon bond with a maturity falling on the next reset date, $t_\gamma$.[18]

This is for a floating-rate note absent of credit risk. Often, however, a floating-rate note's coupon is not flat to the index, but instead involves a spread over and above the index level. If we denote the spread as $s$, then each of the cash-flows will no longer be the forward rate, but instead the forward rate plus $s$. If we return to the value function of the floating-rate note in Eq. (4.40), we must make the following adjustment,

$$\begin{aligned}
V(t,s) &= N\sum_{i=\gamma+1}^{I}(t_i-t_{i-1})\left(\underbrace{f(t,t_{i-1},t_i)+s}_{\substack{\text{Adjusted}\\ \text{cash-flow}}}\right)P(t,t_i)+NP(t,t_I),\\
&= N\sum_{i=\gamma+1}^{I}\cancel{(t_i-t_{i-1})}\underbrace{\frac{1}{\cancel{t_i-t_{i-1}}}\left(\frac{P(t,t_{i-1})-P(t,t_i)}{\cancel{P(t,t_i)}}+\frac{s\,(t_i-t_{i-1})\,P(t,t_i)}{\cancel{P(t,t_i)}}\right)}_{\text{Equation 4.38}}\\
&\quad\times\cancel{P(t,t_i)}+NP(t,t_I),\\
&= N\left(\underbrace{\sum_{i=\gamma+1}^{I}P(t,t_{i-1})-P(t,t_i)}_{\text{Telescoping sum}}+\sum_{i=\gamma+1}^{I}s\,(t_i-t_{i-1})\,P(t,t_i)+P(t,t_I)\right),\\
&= N\left(P(t,t_\gamma)-\cancel{P(t,t_I)}+\sum_{i=\gamma+1}^{I}s\,(t_i-t_{i-1})\,P(t,t_i)+\cancel{P(t,t_I)}\right),\\
&= N\left(P(t,t_\gamma)+\sum_{i=\gamma+1}^{I}s\,(t_i-t_{i-1})\,P(t,t_i)\right).
\end{aligned}\tag{4.47}$$

This is an interesting result. Here we see that, when one adds a spread to the interest-rate index, the value of a floating-rate note becomes more complex. It includes, as before, the discounted value of the notional to the next reset date, but it also includes the discounted stream of spread values over the life of the floating-rate

[18]One may also compute the convexity of a floating-rate note, although given the very short effective maturity of this instrument, we should expect it to be quite small. Indeed, the convexity is found by a second differentiation of Eq. (4.45) as follows,

$$C \approx (t_\gamma - t)^2, \tag{4.46}$$

which is, indeed, rather small for contracts with a reset frequency of typically less than a year.

note. This makes inherent sense, because the spread, $s$, does not cancel out of the value function.

If we compute the partial derivative of our extended floating-rate bond value function with respect to the spread, we have

$$\begin{aligned}\frac{\partial V(t,s)}{\partial s} &= \frac{\partial}{\partial s}\left(N\left(P(t,t_\gamma) + \sum_{i=\gamma+1}^{I} s\,(t_i - t_{i-1})\,P(t,t_i)\right)\right),\\ &= N\sum_{i=\gamma+1}^{I}(t_i - t_{i-1})\,P(t,t_i). \end{aligned} \tag{4.48}$$

This is essentially the value of an annuity over the life of the floating-rate note adjusted for the reset frequency, $t_i - t_{i-1}$. If we normalize by the price of the floating-rate note, we have the spread duration. It can be written analytically as follows,

$$\begin{aligned} D_S &= \frac{1}{V(t,s)}\frac{\partial V(t,s)}{\partial s},\\ &= \left(\frac{1}{V(t,s)}\right)N\sum_{i=\gamma+1}^{I}(t_i - t_{i-1})\,P(t,t_i),\\ &= \left(\frac{1}{V(t,s)}\right)N\sum_{i=\gamma+1}^{I}(t_i - t_{i-1})\,P(t,t_i),\\ &= \frac{\displaystyle\sum_{i=\gamma+1}^{I}(t_i - t_{i-1})\,P(t,t_i)}{\displaystyle P(t,t_\gamma) + s\sum_{i=\gamma+1}^{I}(t_i - t_{i-1})\,P(t,t_i)}, \end{aligned} \tag{4.49}$$

although, unfortunately, it does not reduce to a simpler expression. The spread duration is also often numerically computed.

The important, and interesting, point in this development is that a floating-rate note is an instrument where the modified and spread durations do *not* necessarily coincide. The remaining treatment of floating-rate notes within the context of our additive decomposition remains the same as for plain-vanilla fixed-income instruments.

## 4.3 Handling Fixed-Income Derivatives Contracts

Our general framework, as outlined in the previous chapter, only considered cash instruments. Fixed-income portfolio managers typically, however, have more than merely cash instruments—bills, nominal bonds, and inflation-linked bonds—in their investment universe. It is common, in fact, for fixed-income managers to also use derivative instruments. The *four* most popular fixed-income derivative instruments likely include:

1. interest-rate swaps;
2. foreign-exchange swaps;
3. interest-rate futures; and
4. bond futures.

The first two instruments, interest-rate swaps and foreign-exchange swaps, are relatively easy to handle because they are, in essence, combinations of simple securities.[19] An interest-rate swap is merely a linear combination of a fixed-rate bond and a floating rate note. A foreign-exchange swap is simply a long and short position in two zero-coupon bonds denominated in two different currencies. These instruments, therefore, can readily be incorporated in the basic framework laid out earlier in the chapter. As a consequence, we need not allocate any time to their discussion.

Bond and interest-rate futures are more involved. In both cases, these instruments are essentially forward contracts—with a daily margining mechanism—on underlying fixed-income instruments. The underlying for an interest-rate future is a deposit, while the underlying instrument in a bond future contract is a government bond. Conceptually straightforward but practically tricky, we will discuss how one might incorporate both of these instruments into our approximation framework. We start with the relative simple treatment of interest-rate futures and progress to the more complex bond futures.

### 4.3.1 Interest-Rate Futures

We will define an interest-rate future, or simply *rate future*, as a fixed-income future where the underlying instrument is a deposit.[20] The classic example is the Euro-

[19]As previously indicated, the divergence of LIBOR and OIS interest rates during the crisis has made this rather more complicated and lead to the notion of OIS discounting, although we will not have much to say about these issues in this discussion.

[20]The term interest-rate future is perhaps a bit misleading as it could, in principle, relate to any futures contract with an underlying that depends upon interest rates. In this respect, therefore, bond futures would also logically fall under this definition.

**Table 4.2** Euro-dollar rate futures

| Year | March (H) | June (M) | September (U) | December (Z) |
|---|---|---|---|---|
| 2010 | ~~EDH0~~ | ~~EDM0~~ | ~~EDU0~~ | EDZ0 |
| 2011 | EDH1 | EDM1 | EDU1 | EDZ1 |
| 2012 | EDH2 | EDM2 | EDU2 | EDZ2 |
| 2013 | EDH3 | EDM3 | EDU3 | EDZ3 |
| 2014 | EDH4 | EDM4 | EDU4 | EDZ4 |
| ⋮ | ⋮ | ⋮ | ⋮ | ⋮ |

This table describes, from the perspective of December 2010, the set of available Euro-dollar rate future contracts. The first three contracts from 2010 are crossed out given that, as of December 2010, they have already matured.

dollar future offered on the Chicago Mercantile Exchange (CME).[21] Being long a single Euro-dollar future contract, for example, is equivalent to purchasing a 90-day one million USD deposit at a given time in the future. Unlike many futures contracts, you will not actually receive a physical 90-day deposit. Instead, Euro-dollar futures are cash settled.[22]

There are numerous active rate-future contracts available and traded at any given point in time. Table 4.2 summarizes, from the perspective of December 2010, the set of available 90-day Euro-dollar futures contracts. We have described the contracts using their four-character security tickers where the third character denotes the month–March (H), June (M), September (U), and December (Z)—and the final character denotes the maturity. Specifically, 0 denotes 2010, 1 represents 2011, and so on.

There is a new contract starting every 3 months and that the number of available contracts spans several years. As of March 2011, one could have transacted in a March 2018 contract. At any given point in time, there are 40 outstanding contracts covering a 10-year period. The further one moves out the curve, however, the thinner the amount of trading. The first 3 or 4 years, and particularly the first year, are nonetheless very liquid.

A useful way to think about the range of available Euro-dollar futures is to examine how they span the yield curve. Figure 4.4 displays the US yield curve in December 2010 and superimposes the available Euro-dollar maturities—each maturity is represented by a vertical line. The individual contracts span virtually the entire yield curve. This implies that, with the correct choice of contract, one

[21]Our objective here is to get to the key elements of these contracts so as to incorporate them into our additive decomposition. As such, we gloss over numerous details. An excellent reference for a detailed discussion of the intricacies of interest-rate future contracts is Fabozzi [4]. Hull [6] is also very useful.

[22]Other rate future contracts, such as 90-day Euribor and sterling futures offered by the London International Financial Futures and Options Exchange (LIFFE), operate in virtually the same manner as the Euro-dollar future contract. As a consequence, we will focus on Euro-dollar futures.

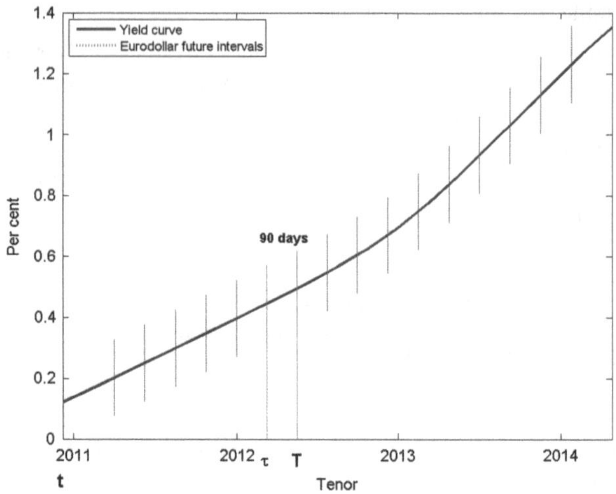


**Fig. 4.4** Range of Euro-dollar future contracts. This figure describes, from the perspective of December 2010, the range of euro-dollar future contracts across the yield curve. A forward 90-day deposit maturing in June 2012 is highlighted

may gain exposure to a 90-day deposit across a wide range of future points in time. Exposure to a 90-day deposit in the future, is basically equivalent to gaining exposure to a 90-day *interest rate* in the future. Rate futures, therefore, permit investors to gain very specific exposure to future interest rates.

To correctly handle rate futures, we will need some notation. The time dimension is particularly important. We denote, as usual, the current time as $t$. The future time point when the deposit begins is represented by $\tau$, while the maturity of the deposit is $T$. For a Euro-dollar future contract with a 90-day deposit, therefore, $T-\tau$ is 90-days or about $\frac{1}{4}$. The difference $\tau - t$, conversely, denotes the maturity of the Euro-dollar future contract. In Fig. 4.4, the inception and maturity of our arbitrary Euro-dollar future contract is represented with the bold lines descending to the horizontal axis.

Armed with this notation, we proceed to reflect on how our rate future can be incorporated into our approximation framework. The first question is: what risk factors drive the return of our rate future? Recalling Table 4.1 on page 83, we have identified six different risk factors: time, inflation, treasury curve, option-adjusted spread, foreign-exchange, and convexity. Since a rate future pays no coupon, there is no time aspect to its return and consequently no carry return. Rate futures do not receive any explicit inflation compensation and, thus, we can safely ignore the inflation factor. A rate future is also exposed to very little credit exposure.[23] Technically, the deposit rates in a rate future are determined by the LIBOR setting.[24]

[23]The margining mechanism eliminates counterparty risk. The remaining creditor is the futures clearing house itself, which is typically of very high credit quality.

[24]The unsecured nature of LIBOR rates does, however, incorporate an element of credit risk. This is the source of much of the complexity in pricing interest-rate swap contracts after the crisis. OIS discounting basically implies that one should use the swap curve for the computation of the credit spread, but the OIS curve for discounting cash flows.

The credit spread is thus equivalent to the swap spread. This effect is relatively small and ignored in this discussion.

Rate futures, like all futures contracts, do not have any explicit foreign-exchange risk. The basic reason stems from the daily margining mechanism. The daily value of any futures contract is typically very close to zero. This is because, each day based on market movements and associated daily profit-and-loss (P&L), funds are added or withdrawn from a margin account with the futures clearing-house. These cash positions, which are considered part of the portfolio, will certainly given rise to foreign-exchange risk, but this risk will be captured by computing the foreign-exchange exposure of one's cash balances. The consequence, therefore, is that one can ignore the foreign-exchange component in futures contracts even when they are not denominated in one's base currency. It is not that it does not exist, but rather that it is already handled through the cash account. Explicitly incorporating it into our computations would generally lead to double counting.

Rate futures, like bond futures, also do not have any convexity. To appreciate this statement, we need to review the pricing conventions of rate futures. A key aspect of futures market is the specification of the tick size and tick value. All financial markets have a tick size—it is defined as the minimum movement that can be observed in the change of an asset price. Futures contracts, however, have a *fixed* value associated with each tick. In the Euro-dollar futures market, for example, the tick size is 0.005. Each tick has a fixed value of USD 12.50 per contract. This implies that if the price of the rate future falls, from one day to the next by three ticks, then the margin account must be increased by USD 37.50. In other words, the P&L on this contract over this period was –USD 37.50. This is a linear function.

To make these ideas a bit more concrete, let's consider a simple example. Table 4.3 describes the computation of P&L and return for a fictitious Euro-dollar future contract. With prices of \$94.75 and \$94.65 at times $t_0$ and $t_1$, respectively, the price change is \$0.1. Given a tick size of 0.005, this amounts to a decrease of 20

**Table 4.3** Euro-dollar pricing example

| Statistic | Price $P_{t_i}$ | Rate $100 - P_{t_i}$ | Price $\frac{100}{(1+r_{t_i})^{\frac{90}{360}}}$ |
|---|---|---|---|
| Value ($t_0$) | \$94.75 | 5.25 % | **\$98.73** |
| Value ($t_1$) | \$94.65 | 5.35 % | **\$98.71** |
| Price/yield change: $\Delta$ | –\$0.10 | 0.10 % | **–\$0.02** |
| Ticks: $\frac{\Delta}{\text{Tick Size}}$ | –20.000 | 20.000 | **n/a** |
| P&L: Ticks · \$*12.50* | –\$250.00 | –\$250.00 | **–\$234.37** |
| Return (bps.): $\frac{\text{PnL}}{\text{USD 1 million}}$ | –2.50 | –2.50 | **–2.34** |

This table describes the computation of P&L and return for a fictitious Euro-dollar future contract. The typical tick and tick-value market convention is employed and supplemented with the treatment of these contracts as typical deposits. The difference between these two approaches clearly indicates the lack of convexity in the Euro-dollar futures contract.

ticks, which at \$12.50 per tick, amounts to a loss of −\$250.[25] One may also attempt, as we do in the final column of Table 4.3, to use the Euro-dollar futures rate and actually properly discount the deposits following money-market conventions. The daily loss on the true deposit is actually −\$234 instead of \$250. The difference is due to the linear P&L computation for a Euro-dollar future. It does not incorporate the convexity effect arising from the discounting of future cash-flows. Interest-rate futures thus do not have any convexity and, as such, we need not compute a convexity correction.

Having established that a rate future has no carry, inflation, foreign-exchange, or convexity risk, what is the risk of an interest-rate future? A rate future is exposed to movements in the yield curve.[26] A rate future has all of the characteristics of a deposit or zero-coupon bond. As a consequence, we could incorporate it into our framework as a 90-day zero-coupon bond.

To compute the yield return of a zero-coupon bond, we require its modified duration. What is the modified duration of a 90-day deposit or, equivalently, a 90-day zero-coupon bond? Conventional wisdom suggests that the duration of a zero-coupon is equal to its tenor.[27] This is based on an approximation that we readily adopt. For the curious reader, the underlying box provides more detail into the derivation of our deposit's modified duration.

To determine the modified duration of a deposit, let's consider a zero-coupon bond at time $t$ with a maturity of $\tau$. Its price, $P(t,\tau)$, is simply the final payment of one unit of currency discounted to time $t$ at the appropriate zero-coupon rate, $z(t,\tau)$. Mathematically, this is represented as follows,

$$P\left(t,\tau,z_{t,\tau}\right) = \frac{1}{\left(1+z_{t,\tau}\right)^{\tau-t}}. \tag{4.51}$$

(continued)

[25] An identical computation can be made with the associated rates of 5.25 % and 5.35 %. This is possible because the rate associated with a Euro-dollar contract is not, in fact, a true interest rate. Instead, the rate is determined as,

$$\underbrace{\text{Rate}}_{r_{t_i}} = 100 - \underbrace{\text{Price}}_{P_{t_i}}, \tag{4.50}$$

or merely 100 less the price. The price is thus a linear combination of the rate and vice versa. Consequently, one may compute the number of ticks and the associated P&L on either quantity; although, one needs to place a negative sign in front of the rate computation.

[26] This can be further decomposed, if desired, into a treasury-curve and credit spread components.

[27] That is, a 6-month treasury bill should have a modified duration of 0.5.

The first step in computing the modified duration is to take the derivative of Eq. (4.51) with respect to $z_{t,\tau}$,

$$\frac{\partial P(t,\tau,z_{t,\tau})}{\partial z_{t,\tau}} = \frac{\partial}{\partial z_{t,\tau}} \underbrace{\left(\frac{1}{(1+z_{t,\tau})^{\tau-t}}\right)}_{\text{Equation (4.51)}},$$

$$= \frac{1}{(1+z_{t,\tau})} \frac{-(\tau-t)}{(1+z_{t,\tau})^{\tau-t}}. \tag{4.52}$$

We arrive at the modified duration by merely dividing both sides of Eq. (4.52) by $-P(t,\tau,z_{t,\tau})$,

$$\begin{aligned}
\underbrace{-\frac{1}{P(t,\tau,z_{t,\tau})}\frac{\partial P(t,\tau,z_{t,\tau})}{\partial z_{t,\tau}}}_{\text{Modified Duration}} &= \\
&- \underbrace{\frac{1}{P(t,\tau,z_{t,\tau})}}_{\text{Equation (4.51)}} \underbrace{\frac{1}{(1+z_{t,\tau})}\frac{-(\tau-t)}{(1+z_{t,\tau})^{\tau-t}}}_{\text{Equation (4.52)}}, \\
&= \cancel{(1+z_{t,\tau})^{\tau-t}} \underbrace{\frac{1}{(1+z_{t,\tau})}}_{\substack{\text{Simple}\\ \text{discount}\\ \text{factor}}} \frac{(\tau-t)}{\cancel{(1+z_{t,\tau})^{\tau-t}}}, \\
&= \underbrace{\left(\frac{1}{1+z_{t,\tau}}\right)}_{\approx 1}(\tau-t), \\
&\approx \tau - t.
\end{aligned} \tag{4.53}$$

The modified duration of a zero-coupon bond is close, but not quite equal, to its tenor. With continuous compounding the modified duration is exactly equal to the tenor. For small interest rates, however, the tenor is a very good approximation. For very large interest rates, however, the modified duration could differ significantly from the tenor. For our purposes, we will use the tenor as a very reasonable approximation of the modified duration.

Returning to our fictitious example in Table 4.3 and using our zero-coupon bond representation of the rate future, we compute the return decomposition as,

$$
\begin{aligned}
r(t,T) &= -D_M \underbrace{(y_{t_1} - y_{t_0})}_{\Delta y}, \\
&= -\left(\frac{90}{360}\right) \cdot (5.35\,\% - 5.25\,\%), \\
&= -2.50 \text{ basis points.}
\end{aligned} \tag{4.54}
$$

This value corresponds exactly with the tick-based methodology presented in Table 4.3. The duration approximation provides the *exact* return in this case, precisely because the tick-based computation used for the pricing of futures contracts is computed in a linear fashion. Computing the return on a rate future only requires the yield change—easily inferred from the price change using Eq. (4.50)—and the modified duration. The duration for all Euro-dollar futures contracts is always 0.25, or $\frac{90}{360}$, across the entire curve because the underlying instrument is always a 90-day deposit.[28]

We have resolved almost all of the inherent problems in the handling of rate futures. One problem, however, remains. While the duration of each Euro-dollar future is fixed at 0.25, the location of each individual duration contribution is different across the entire yield curve. Although each Euro-dollar future has the *same* duration, every contract has a slightly *different* key-rate duration profile. This is the whole point of having a wide range of different rate future contracts: it permits market participants to take positions or hedge exposures to very specific elements of the yield curve.

Treating our rate future as a 90-day deposit, or zero-coupon bond, will provide the appropriate return, but the *incorrect* key-rate duration. We need the key-rate duration to occur in the right place along the curve. Our representation of a rate future will need to be a bit more complex than merely treating it as a single zero-coupon bond.

A bit more reflection is required. The rate future price depends on the 90-day forward rate at the time of the contract's maturity. Stripping away the margining aspect, a rate future is a forward contract on a forward interest rate. We could treat our rate future, for the purposes of computing key-rate duration, as a forward deposit. We need, therefore, to consider the forward rate in more detail.

Using our basic time notation, a forward deposit can be represented as a portfolio of two zero-coupon bonds with the following characteristics,[29]

- a *short* position in a zero coupon bond with maturity $\tau$, or $P\,(t,\tau,z_{t,\tau})$; and
- a *long* position in a zero-coupon bond with maturity $T$, $P\,(t,T,z_{t,T})$

[28]Note, of course, that if you are dealing with a rate future with an underlying instrument other than a 90-day deposit, you will need to make the necessary adjustment to the duration.

[29]The current time is $t$, the future maturity is $\tau$, and the maturity of the underlying deposit is $T$.

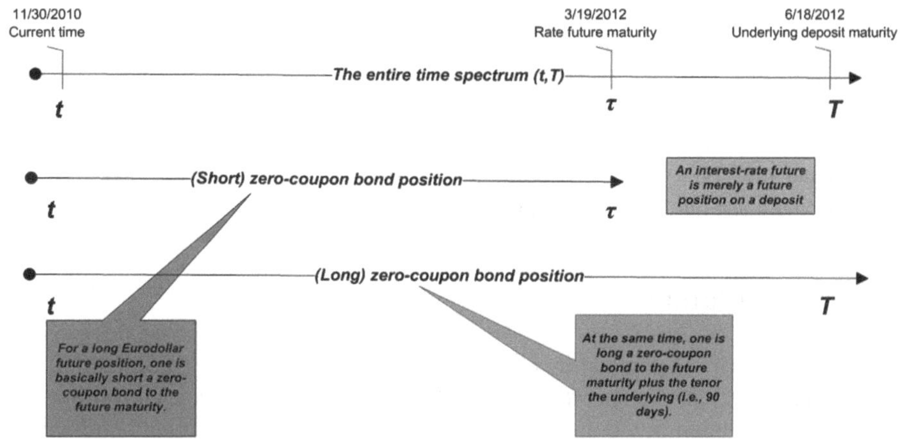


**Fig. 4.5** Rate future schematic. This figure describes how an interest-rate future can be broken into two zero-coupon-bond positions in a manner conceptually identical to the construction of an implied forward-interest rate

Figure 4.5 describes how an interest-rate future can be broken into two zero-coupon-bond positions in a manner conceptually identical to the construction of a forward interest rate.[30] This portfolio is a forward deposit, which is a reasonable approximation of a rate-future contract and suitable for the computation of key-rate durations.

Representing our interest-rate future as a portfolio of two zero-coupon bonds, we can easily show that the duration of the total portfolio remains equal to the tenor of the underlying deposit. The duration of this position is simply,

$$\begin{aligned} D_{\text{Rate Future}} &= -\underbrace{\left(\frac{1}{P(t,\tau,z_{t,\tau})}\frac{\partial P(t,\tau,z_{t,\tau})}{\partial z_{t,\tau}}\right)}_{\textit{Short}\text{ position}} + \underbrace{\left(\frac{1}{P(t,T,z_{t,T})}\frac{\partial P(t,T,z_{t,T})}{\partial z_{t,T}}\right)}_{\textit{Long}\text{ position}}, \\ &= -(\tau-t)+(T-t), \\ &= T-\tau, \end{aligned} \tag{4.56}$$

[30] If you borrow one unit of currency at time $t$ for $\tau$ periods and invest it for $T$ periods, you will have an obligation at time $\tau$ and a receipt of funds at time $T$. The annualized return associated with this portfolio of zero-coupon bonds is

$$f(t,\tau,T) = \left(\frac{(1+z(t,T))^{T-t}}{(1+z(t,\tau))^{\tau-t}}\right)^{\frac{1}{T-\tau}} - 1, \tag{4.55}$$

which is the forward-interest rate.

which is the tenor of our deposit. In summary, we obtain the appropriate modified and key-rate durations in the correct place along the curve between $\tau$ and $T$.[31]

A rate-future contract, from a portfolio-analytic perspective, is well described as a simple portfolio of a short and long zero-coupon bond with the appropriate tenors and maturity dates. No adjustment to our general framework was required, but instead merely a clever representation of the security. In the next section, we will see that a similar approach is possible for bond futures.

### 4.3.2 Bond Futures

A bond future is a contract to buy (or sell) forward a government-issued bond. If the bond future referred to a specific bond, our life would much easier. Unfortunately, the government bond in a bond future contract is a *virtual* government bond with a fixed maturity and a fixed coupon rate.

Complexity arises because this virtual bond does *not* exist and, as such, naturally cannot be delivered upon the maturity of the contract.[32] Instead, a basket of physical bonds are identified at the inception of the contract and any of these bonds may be delivered upon the maturity of the futures contract. At any given point in time, however, one of the bonds in the basket is typically less expensive to deliver than the others—this is termed the cheapest-to-deliver bond.

An enormous amount of complexity is embedded in the delivery aspect of bond futures contracts.[33] While these aspects of the bond-future contract are very important for a bond-future trader, they are not generally necessary for the purpose of portfolio analytics. For portfolio analysis, our task is to find a reasonable approximation of the interaction between the return of a bond-future contract and our set of risk factors.

---

[31]The return in Eq. (4.54) can be allocated across the curve using the key-rate durations as follows,

$$\begin{aligned} r(t,T) &= -D_M \underbrace{\left(y_{t_1} - y_{t_0}\right)}_{\Delta y}, \\ &= -\sum_{\kappa=1}^{\nu} D_\kappa \Delta y. \end{aligned} \tag{4.57}$$

[32]This would not be true if the contract were cash settled, which is the case for some bond futures contracts. The majority of bond futures contracts are physically settled.

[33]These delivery features are often modelled as financial options and, depending on market conditions, can be quite valuable.

As with interest-rate futures, there is no carry, foreign-exchange, credit-spread, or convexity risk associated with bond futures. The risk associated with a bond-future contract is treasury-curve risk. In this section, we will compare and contrast two possible approaches for approximating this treasury-curve risk. In particular, we consider modelling the risk characteristics of the bond future contract with either:

- the cheapest-to-deliver bond; or
- the virtual bond defined in the bond-future contract.

We will thus represent a more complex derivative contract as a simpler instrument—in this case a government bond—to permit its incorporation into our general framework.

Before considering these two approaches, however, let's invest some time to precisely define the return of a bond future contract. This tedious algebraic exercise nonetheless provides clarity as to how one should compute future returns and weight them in one's overall portfolio. Let us begin with the treatment of a typical bond position. The return of the $k$th bond in a portfolio over the time interval, $[t_0, t_1]$, can be written as,

$$\begin{aligned} \begin{array}{c}\text{Return of } k\text{th} \\ \text{bond in} \\ \text{portfolio}\end{array} &= \begin{pmatrix}\text{Weight of} \\ k\text{th bond} \\ \text{in portfolio} \\ \end{pmatrix} \cdot \begin{pmatrix}\text{Return of} \\ k\text{th bond}\end{pmatrix}, \\ r_{p,k} &= \omega_k \underbrace{\left(\frac{V_{t_1,k} - V_{t_0,k}}{V_{t_0,k}}\right)}_{r_k}, \end{aligned} \tag{4.58}$$

where $\omega_k$ is the market weight of the $k$th position in the portfolio at time $t_0$. Defining the value of the $k$th position at time $t_i$ as,

$$V_{t_i,k} = V_{t_i,k} \underbrace{N_{t_i,k}}_{\substack{\text{Nominal} \\ \text{value}}}, \tag{4.59}$$

then the market weight is merely,

$$\omega_k = \frac{V_{t_0,k}}{\sum_{i=1}^{n} V_{t_0,i}}. \tag{4.60}$$

If we plug Eq. (4.60) into our original return expression for $r_k$ in Eq. (4.58), we have

$$
\begin{aligned}
r_{p,k} &= \underbrace{\left(\frac{V_{t_0,k}}{\sum_{i=1}^{n} V_{t_0,i}}\right)}_{\text{Equation (4.60)}} \left(\frac{V_{t_1,k} - V_{t_0,k}}{V_{t_0,k}}\right), \\
&= \left(\frac{\cancel{V_{t_0,k}} N_{t_0,k}}{\sum_{i=1}^{n} V_{t_0,i} N_{t_0,i}}\right) \left(\frac{V_{t_1,k} - V_{t_0,k}}{\cancel{V_{t_0,k}}}\right), \qquad (4.61) \\
&= \frac{V_{t_1,k} N_{t_0,k} - V_{t_0,k} N_{t_0,k}}{\sum_{i=1}^{n} V_{t_0,i} N_{t_0,i}}, \\
&= \frac{N_{t_0,k} \left(V_{t_1,k} - V_{t_0,k}\right)}{\sum_{i=1}^{n} V_{t_0,i} N_{t_0,i}}.
\end{aligned}
$$

The consequence is that, using the typical weighting scheme, the return contribution of an arbitrary individual bond in one's portfolio is basically the bond's P&L over the period divided by the market value of the portfolio; hardly a surprising result. Using the same logic, the return contribution of the $k$th future contract is,

$$
\begin{aligned}
\underbrace{\text{Return of } k\text{th future}}_{r_k^f} &= \frac{\left(\frac{P^f_{t_1,k} - P^f_{t_0,k}}{\text{Tick Size}_k}\right) \cdot \text{Tick Value}_k}{P^f_{t_0,k} \cdot \text{Contract Size}_k}, \qquad (4.62) \\
&= \frac{\text{P\&L from one futures contract}_k}{\text{Exposure from one futures contract}_k},
\end{aligned}
$$

where $P^f_{t_i,k}$ is the price of the $k$th future contract at time $t_i$. Thus the return of futures contract is defined as the ratio of the P&L on a single contract to the exposure from a single contract. Readers interested in the derivation of this definition may review the following shaded section.

The result in Eq. (4.62), while conceptually appealing, requires an assumption and bit of effort for its formal derivation. We wish to use the same basic idea as the result in Eq. (4.61). That is, the return contribution of an arbitrary future contract to one's portfolio return should be,

$$\begin{aligned}\text{Return of } k\text{th} \\ \text{future in portfolio} &= \frac{\begin{array}{c}\text{P\&L of}\\ k\text{th future}\end{array}}{\text{Portfolio MtM}},\\ r^f_{p,k} &= \frac{\begin{array}{c}\text{P\&L of}\\ k\text{th future}\end{array}}{\sum_{i=1}^{n} V_{t_0,i}}.\end{aligned}\tag{4.63}$$

Although a futures contract has little, or no, market value, it nonetheless makes a P&L contribution to the portfolio over every period. The exact form of the P&L, as discussed in the previous section, for the $k$th future position is described as,

$$\begin{aligned}\begin{array}{c}\text{P\&L of}\\ k\text{th future}_k\end{array} &= \begin{pmatrix}\#\text{ of}\\ \text{Ticks}\end{pmatrix}_k \cdot \begin{pmatrix}\text{Tick}\\ \text{Value}\end{pmatrix}_k \cdot \begin{pmatrix}\#\text{ of}\\ \text{Contracts}\end{pmatrix}_k,\\ &= \left(\frac{\overbrace{P^f_{t_1,k} - P^f_{t_0,k}}^{\text{Change in price}}}{\text{Tick Size}_k}\right) \begin{pmatrix}\text{Tick}\\ \text{Value}\end{pmatrix}_k \cdot \begin{pmatrix}\#\text{ of}\\ \text{Contracts}\end{pmatrix}_k,\end{aligned}\tag{4.64}$$

where $P^f_{t_1,k}$ denotes the price of the $k$th future contract at time $t_1$. It is the P&L on each individual contract adjusted for the number of contracts in the portfolio.

Requiring an assumption on the form of the return weight of a futures contract in our portfolio, we employ

$$\omega_k = \frac{\overbrace{P^f_{t_0,k} \cdot \#\text{ of Contracts}_k \cdot \text{Contract Size}_k}^{\text{Bond-Equivalent Exposure}}}{\sum_{i=1}^{n} V_{t_0,i}}.\tag{4.65}$$

(continued)

The numerator is the bond-equivalent exposure of the futures contract. It is basically a representation of what the market value of the future contract would be if it was transformed into a bond position.

One cannot merely use the market-value-based weight, as in Eq. (4.60), because the market value of a futures contract, through daily margining, is approximately zero. Such a weight function would give a zero weight to the bond-future return. This weighting function also incorporates the leverage aspect of the futures contract—one may, with very little initial investment, gain exposure to the futures risk factors. Other weighting schemes can be used but, in our view, this is the most sensible.

Returning to the form of Eq. (4.58), we solve for the return of the $k$th future ($r_k^f$) as,

$$
\begin{aligned}
\underbrace{\begin{array}{c}\text{Return of } k\text{th}\\ \text{future in}\\ \text{portfolio}\end{array}}_{r^f_{p,k}} &= \underbrace{\begin{pmatrix}\text{Weight of } k\text{th}\\ \text{future in}\\ \text{portfolio}\\ \\ \end{pmatrix}}_{\omega_k} \cdot \underbrace{\begin{pmatrix}\text{Return of}\\ k\text{th future}\end{pmatrix}}_{r^f_k}, \\
\underbrace{\left(\frac{\begin{array}{c}\text{P\&L of}\\ k\text{th future}\end{array}}{\sum_{i=1}^{n} V_{t_0,i}}\right)}_{\text{Equation (4.63)}} &= \underbrace{\left(\frac{P^f_{t_0,k}\cdot \begin{array}{c}\#\text{ of}\\ \text{Contracts}_k\end{array}\cdot \begin{array}{c}\text{Contract}\\ \text{Size}\end{array}_k}{\sum_{i=1}^{n} V_{t_0,i}}\right)}_{\text{Equation (4.65)}} \cdot r^f_k, \\
r^f_k &= \frac{\overbrace{\begin{array}{c}\text{P\&L of}\\ k\text{th future}\end{array}}^{\text{Equation (4.64)}}}{\cancel{\sum_{i=1}^{n} V_{t_0,i}}}\,\frac{\cancel{\sum_{i=1}^{n} V_{t_0,i}}}{P^f_{t_0,k}\cdot \begin{array}{c}\#\text{ of}\\ \text{Contracts}_k\end{array}\cdot \begin{array}{c}\text{Contract}\\ \text{Size}\end{array}_k}, \\
&= \frac{\left(\frac{P^f_{t_1,k}-P^f_{t_0,k}}{\text{Tick Size}_k}\right)\cdot \begin{array}{c}\text{Tick}\\ \text{Value}_k\end{array}}{P^f_{t_0,k}\cdot \begin{array}{c}\text{Contract}\\ \text{Size}\end{array}_k}.
\end{aligned}
\tag{4.66}
$$

The futures return is thus the P&L of a single futures contract divided by the contract size adjusted for the time $t_0$ price. It required rather significant effort to find a reasonable definition of a futures return, but now we have a useful and consistent result that can be widely applied.

**Table 4.4** 10-year German Bund future details

| Summary statistic | Value |
|---|---|
| Identifier | RXZ0 |
| Virtual bond | 10-year, 6 % German bund |
| Contract size | EUR 100,000 |
| Tick size | 0.01 |
| Tick Value | EUR 10 |
| First trade date | 9 Mar 2010 |
| Last trade date | 8 Dec 2010 |
| First and last delivery date | 10 Dec 2010 |
| Delivery basket | *Bunds with 8.5 to 10.5 tenors* |

This table provides a summary for the key details associated with the December 2010, 10-year German Bund-future contract.

The most effective way to examine our two alternative approaches to modelling the return on a bond-future contract is to consider a practical example. Table 4.4 summarizes the key details of the December 2010 German Bund futures contract. This contract has a notional size of EUR 100,000, originated in March 2010 and matured in December 2010. It existed, therefore, for only about 9 months. Its liquidity was highest from about September 2010 to mid-December 2010.[34]

The virtual bond represented by this contract is a 6 %, 10-year German Bund. Since the bond does *not* exist, the physical bond to be delivered must be selected out of a basket of bonds. These bonds are not specifically named, but instead a set of criteria are provided for their definition. In the case of the German Bund futures, any Bund with a tenor between 8.5 and 10.5 years at the maturity of the futures contract may be delivered.

Table 4.5 provides some basic information for the basket of bonds that could have be delivered into this futures contract on 10 December 2010.[35] Although all meet the base criteria, each of the bonds in the delivery basket has slightly different characteristics relative to the virtual bond. To handle differences in maturity dates and coupons, each individual bond has a conversion factor (CF), which is used to adjust the futures price prior to comparison with an individual bond price.

To represent a bond future as the cheapest-to-deliver bond, we need an approach to identifying it. There are two measures that can help us identify which bonds might, at any given point in time, be the cheapest to deliver. The first measure is called the *basis*. It is defined as the difference between the cash bond price and

[34]This is because, unlike rate futures, most of the trading for a given bond-future contract occurs in the active contract. Active, in this context, means the contract with the nearest maturity. Thus, trading in the December contract picks up substantially a few weeks before the maturity of the September contract.

[35]Three of the delivery bonds, where the issue date is italicized, were issued after the inception of the futures contract in March 2010. This entirely possible. If a new bond is issued before the maturity of the futures contract and it meets the basic criteria, then it is included in the basket.

**Table 4.5** Details of delivery basket

| ISIN | Issue | Maturity | Tenor | C (%) | CF |
|---|---|---|---|---|---|
| DE0001135*382* | 22 May 2009 | 4 Jul 2019 | 8.6 | $3\frac{1}{2}$ | 0.836047 |
| DE0001135*390* | 13 Nov 2009 | 4 Jan 2020 | 9.1 | $3\frac{1}{4}$ | 0.811794 |
| DE0001135*408* | *30 Apr 2010* | 4 Jul 2020 | 9.6 | 3 | 0.785987 |
| DE0001135*416* | *20 Aug 2010* | 4 Jul 2020 | 9.7 | $2\frac{1}{4}$ | 0.729277 |
| DE0001135*424* | *23 Nov 2010* | 4 Jan 2021 | 10.1 | $2\frac{1}{2}$ | 0.740991 |

This table provides detailed information on the five bonds in the 10-year German Bund future contract. Note that three of the delivery bonds, with the issue date italicized, were issued after the inception of the futures contract in March 2010.

the converted futures price. One can either be long or short the basis, where each position is defined as,

Long: Sell (short) the futures contract and buy the cash bond
Short: Buy the futures contract and sell (short) the cash bond

The basis is essentially the difference between the spot and the forward price. It is analogous to the forward premium or discount associated with a forward contract and is often used to determine if the futures contract to over- or under-valued or, albeit rather loosely, to identify the cheapest to deliver.

Let's see how the basis is computed. On 12 October 2010 the closing future price was EUR 132.01, while the German Bund in the delivery basket, with an ISIN of DE0001135382, was priced at EUR 110.75. The basis is simply computed as,

$$\begin{aligned} \text{Basis} &= \text{Bond Price} - \text{CF} \cdot \text{Future Price}, \\ &= 110.785 - 0.836047 \cdot 132.01, \\ &= 0.418. \end{aligned} \tag{4.67}$$

By itself, this computation does not mean much to the casual bond-future holder. The basis is more meaningful in comparison with the other bonds in the delivery basket. Figure 4.6 provides the evolution of the bond-future basis for each of the bonds in the December 2010 German Bund future delivery basket from inception to maturity of the contract. The first bond in the delivery basket, with an ISIN of DE0001135382, has the lowest basis over the entire life of the contract.

The basis is sadly *not* the best measure to tell us which bond is its cheapest to deliver.[36] The definitive way to compute the cheapest-to-deliver bond involves computing the actually cost of delivering each bond. One basically constructs the following portfolio:

- shorting the futures contract; and
- borrowing money to purchase the cash bond .

[36]It is nevertheless popular given its ease of computation.

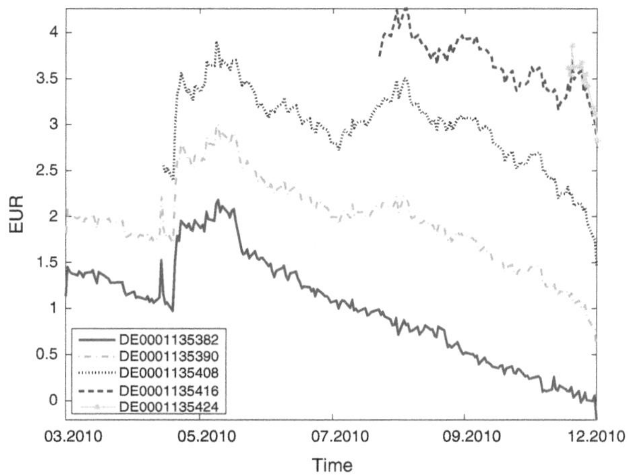


**Fig. 4.6** The basis. This figure provides the daily evolution of the basis for each of the bonds in the 10-year German Bund futures contract's delivery basket over the 3-month period from August to October 2010

By entering into this position, one is neutral since one has a physical bond to deliver into the short futures position.[37] One has essentially taken a financed long position in the basis. The implied return from this position is termed the *implied repo rate*. The bond with the *highest* implied repo rate, or rather implied return, is the *cheapest* to deliver. That is, this is the choice of bond that permits the lowest cost associated with covering your futures position with borrowed money.[38]

Again, the implied repo rate is quite interesting when all of the bonds in the delivery basket are compared. Figure 4.7 describes the daily evolution of the implied repo rate over our December 2010 contract's lifetime.[39] We can see that, up until the issuance of the third bond, the first delivery bond was the cheapest to deliver. The third bond remained cheapest to deliver for a few months until it was overtaken by the first bond, which was cheapest to deliver for the rest of the contract's tenor.

---

[37]This portfolio is termed a long basis position. The least expensive bond to deliver has the highest return associated with taking a (financed) long position in the basis.

[38]The exact formula for computing the implied repo rate is ugly and non-intuitive, but the basic idea is:

$$\begin{array}{c}\text{Implied}\\ \text{Repo}\\ \text{Rate}\end{array} = \frac{\begin{array}{c}\text{Cash}\\ \text{coming in}\end{array} - \begin{array}{c}\text{Cash}\\ \text{going out}\end{array}}{\begin{array}{c}\text{Cash}\\ \text{going out}\end{array}} \frac{360}{\begin{array}{c}\text{Trade}\\ \text{days}\end{array}}. \tag{4.68}$$

The actual computation involves taking into account the accrued interest at the settlement date and the futures delivery date as well as any intermediate coupons.

[39]Fabozzi [4, Chap. 58] is an excellent reference if one wishes to gain a more detailed understanding of the basis and the implied repo rate.

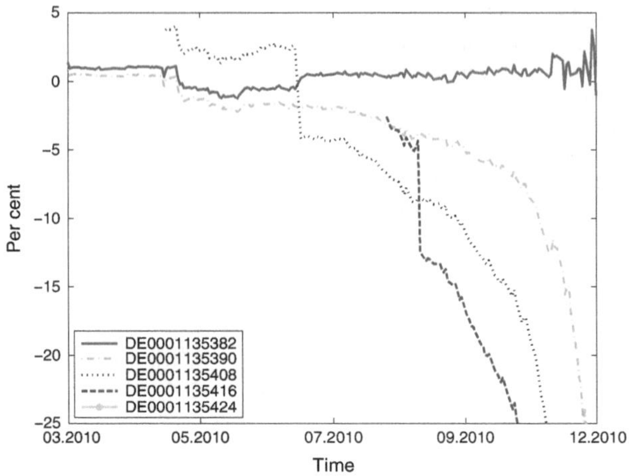


**Fig. 4.7** Implied repo rate. This figure provides the daily evolution of the implied repo rate for each of the bond in the 10-year German bund futures contract's delivery basket over the 3-month period from August to October 2010

Now that we know how to identify it, how do we use the cheapest-to-deliver bond? It merely involves use of the modified and key-rate durations of the cheapest-to-deliver (CTD) bond to approximate movements in one's futures positions. Returning to our additive risk-factor-based return decomposition and inputting the key values from our cheapest-to-deliver bond (CTD),

$$r \approx \underbrace{y_{\text{CTD}}\Delta t}_{\text{Carry}} - \underbrace{D_{M,\text{CTD}}\Delta y_{\text{CTD}}}_{\text{Curve}} - \underbrace{D_{\text{S,CTD}}\Delta s_{\text{OAS}}}_{\text{Credit}} + \underbrace{\frac{1}{2}C_{\text{CTD}}(\Delta y)^2}_{\text{Convexity}}, \tag{4.69}$$

we have a return approximation for our bond-future contract. If we eliminate the irrelevant aspects of the return, we arrive at,

$$\begin{aligned} r &\approx \cancel{\underbrace{y_{\text{CTD}}\Delta t}_{\text{Carry}}} - \underbrace{D_{M,\text{CTD}}\Delta y_{\text{CTD}}}_{\text{Curve}} - \cancel{\underbrace{D_{\text{S,CTD}}\Delta s_{\text{OAS}}}_{\text{Credit}}} + \cancel{\underbrace{\frac{1}{2}C_{\text{CTD}}(\Delta y)^2}_{\text{Convexity}}}, \\ &\approx -\left(\frac{D_{M,\text{CTD}}}{\text{Conversion Factor}}\right)\Delta y_{\text{CTD}}. \end{aligned} \tag{4.70}$$

This is the standard approximation with one exception. We've made a slight adjustment to the modified duration of the cheapest-to-deliver bond—it is merely divided by its conversion factor. The underlying shaded section works through the rationale behind this approximation.

Note from the discussion of the basis, the futures price is approximately,

$$\text{CF} \cdot \text{Futures Price} \approx \begin{array}{c}\text{Price of}\\ \text{CTD Bond},\end{array} \tag{4.71}$$

$$\text{CF} \cdot P^f(t, y^f) \approx P(t, y_{\text{CTD}}).$$

where $P^f$ and $y^f$ are the price and yield of the bond-futures contract, respectively. CF denotes the conversion factor. We proceed to take the derivative of both sides with respect to their yields and divide both sides by their respective prices and simplify as follows,

$$\frac{1}{P^f(t, y^f)} \frac{\partial\left(\text{CF} \cdot P^f(t, y^f)\right)}{\partial y^f} \approx \frac{1}{P^f(t, y_{\text{CTD}})} \frac{\partial P(t, y_{\text{CTD}})}{\partial y},$$

$$\frac{\text{CF}}{P^f(t, y^f)} \frac{\partial\left(P^f(t, y^f)\right)}{\partial y^f} \approx \underbrace{\frac{1}{P^f(t, y_{\text{CTD}})} \frac{\partial P(t, y_{\text{CTD}})}{\partial y}}_{D_{M,\text{CTD}}},$$

$$D_{M,\text{Futures Contract}} \approx \frac{D_{M,\text{CTD}}}{\text{CF}} \tag{4.72}$$

This brings us to the following approximate relationship between the duration of the futures contract and the duration of the cheapest-to-deliver bond. One need simply divides the modified duration of the cheapest-to-deliver bond by its conversion ratio. This is heuristic reasoning that leads to an approximation and not, by any stretch of the imagination, a formal proof.

How well does our cheapest-to-deliver bond approximation perform? Figure 4.8 provides a description of the accuracy of the cheapest-to-deliver approximation over the life of the futures contract. The first graphic compares the daily bond-future returns—as described in Eq. (4.62)—and the approximation using the cheapest-to-deliver bond in Eq. (4.70). The second graphic summarizes the difference between the true return and the approximation.

The approximation errors are typically relatively small with an error of around ± 10 basis points. During some periods, particularly in May 2010, the error terms are quite large and occasionally exceed 20 basis points. While not terribly encouraging, we should recall that during the first months our German Bund futures' life, it was not a terribly liquid instrument. Over the period from September to December 2010, which is the period of most liquidity in this contract, we can see that the approximation errors are relatively smaller.

The second approach to approximating the return of our bond-future contract is extremely simple. We merely create a *virtual bond* with a 6 % coupon and a maturity date of 10 years from the expiry date of our futures contract. On each date

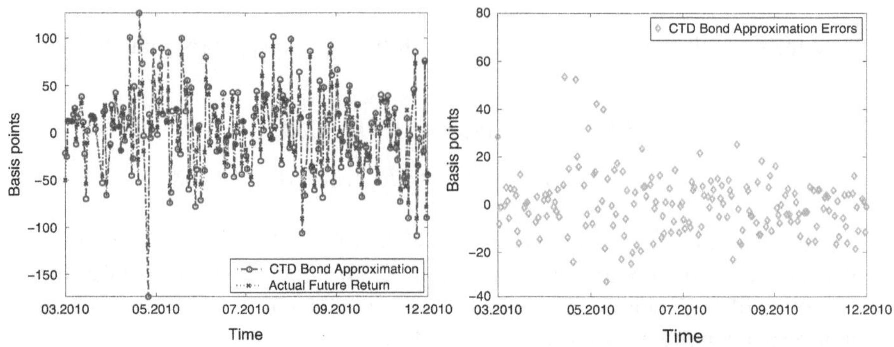


**Fig. 4.8** Cheapest-to-deliver approximation. This figure outlines the daily changes and approximation errors in the 10-year German bond future contract compared against the approximated changes in the cheapest-to-deliver bond price using modified duration

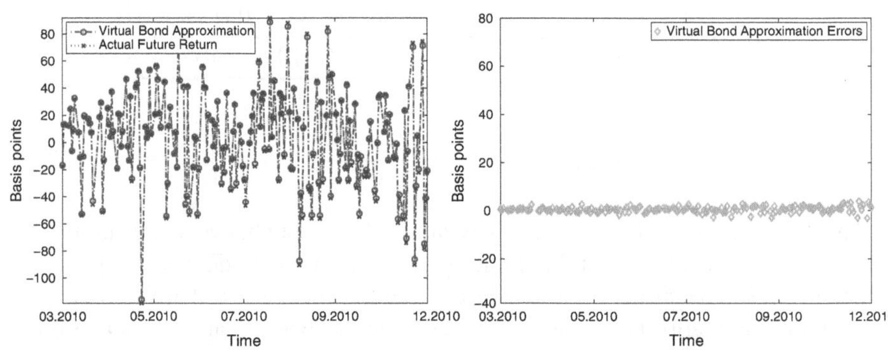


**Fig. 4.9** Virtual-bond approximation. outlines the daily changes and approximation errors in the 10-year German bond future contract compared against the approximated changes using a virtual 10-year bond

we assign the bond-future price to this virtual bond, compute the associated yield, and calculate the associated modified duration. The return approximation becomes,

$$r \approx -D_{M,\text{Virtual Bond}} \underbrace{\Delta y_{\text{Virtual Bond}}}_{\substack{\approx \text{ change in} \\ \text{future yield}}}. \tag{4.73}$$

While relatively naive, this appears to be a sensible approximation. Figure 4.9 describes this virtual bond approximation. The left-hand graphic compares the daily bond-future returns against the virtual-bond approximation in Eq. (4.73), while the right-hand graphic provides the approximation errors. Note that a yield-curve model is required to determine the yield changes of the virtual bond.

The results suggest that the virtual-bond approximation is superior to the cheapest-to-deliver approach. The reason appears to arise from the fact that one single bond from the delivery basket has difficulty in describing the price behaviour of the bond-future contract. Both approximations are, for the purposes of portfolio analytics, quite reasonable approaches for describing the return dynamics of one's bond-future positions. It is up to the individual analyst to decide on his or her preference.

## 4.4 Closing Thoughts

In the previous chapter, we performed the heavy lifting necessary to construct an additive decomposition of the security return into a broad range of risk factors. This mapping between risk-factor exposures and return represents the foundation for the subsequent discussion of portfolio-analytic techniques. It was, however, constructed for a generic fixed-income security. Since modern fixed-income portfolios contain a broad range of occasionally rather complex securities, our framework could be considered incomplete. This chapter rectified this shortcoming by demonstrating how more complex products—such as inflation-linked bonds, floating-rate notes and derivative contracts—can be incorporated into our generic framework. In summary, the principal derivative contracts—foreign-exchange swaps, interest-rate swaps, rate futures and bond futures—can be incorporated into this setting by representing these complex securities as portfolios of simpler instruments. Inflation-linked bonds, while conceptually simple to handle, require a slight adjustment to our additive decomposition to account for the inflation-related carry associated with these instruments.

## References

1. F. Ametrano, M. Bianchetti, Bootstrapping the illiquidity: Multiple yield curves construction for market coherent forward rates estimatio, in *Modeling Interest Rates: Latest Advances for Derivatives Pricing*, ed. by F. Mercurio (Risk Books, New York, 1994)
2. M.L. Anthony, A. Marès, D.J. Bolder, N. Sagnes, Liability management with inflation-linked products: Uk and canada approaches, in *Modeling Interest Rates: Latest Advances for Derivatives Pricing*, ed. by B. Benaben, S. Goldenberg (Risk Books, New York, 2008)
3. M. Deacon, A. Derry, D. Mirfendereski, *Inflation-Indexed Securities: Bonds, Swaps and Other Derivatives*, 2nd edn. (Wiley Finance Series, Chichester, 2004)
4. F.J. Fabozzi, *The Handbook of Fixed Income Securities*, 5th edn. (Irwin, Chicago 1997)
5. B.S. Fung, S. Mitnick, E. Remolona, Uncovering inflation expectations and risk premiums from internationally integrated financial markets. Bank of Canada Working Paper No. 99-6 (1999)
6. J.C. Hull, *Options, Futures, and Other Derivatives*, 4th edn. (Prentice Hall, Englewood Cliffs, 2000)
7. P. Siklos, *Money, Banking and Financial Institutions: Canada in the Global Environment* (McGraw-Hill Ryerson Limited, Toronto, 1994)
8. T.-S. Sun, Real and nominal interest rates: A discrete-time model and its continuous time limit. Rev. Financ. Stud. **5**, 581–611 (1992)

# Part II
# The Yield Curve

The most fundamental underlying driver of fixed-income securities' returns is the sovereign yield curve. In fixed-income portfolio analytics, it simply cannot be ignored. The challenge, however, is that yield curve are complex objects. The following two chapters consider a variety of techniques for the two main problems in yield-curve analysis: (i) fitting the yield curve at a given point in time and (ii) describing the intertemporal evolution of the yield curve as a dynamic system.

# 5 Fitting Yield Curves

*Hell, there are no rules here - we're trying to accomplish something.*

Thomas Edison

The yield curve is unquestionably the central concept in the fixed-income world. It represents the relationship between investment yields and tenor for a given issuer at a given point in time. A wide range of important information is embedded in this relationship ranging from fundamental issues such as the time value of money, expected monetary policy actions, and inflationary expectations to more complicated, but equally important ideas such as risk premia, assessments of creditworthiness, and relative liquidity.

There is no single *yield curve*, but rather a range of different possible curves for different points in time, different currencies, different issuers, and different levels of credit. Teasing out these key elements from different yield curves and describing their movements over time is a rich area of academic and practitioner study. Models of the yield curve find themselves at the heart of fixed-income asset pricing, strategic asset allocation, tactical positioning of fixed-income portfolios, monetary policy implementation, and, of course, portfolio analytics.

Any treatment of fixed-income portfolio analytics would be incomplete without a discussion of yield-curve analytics. Yield curves play an important role in helping us understand the exposure, risk, and performance of a fixed-income portfolio. Yield-curve analytics play such an important role that we feel this topic warrants special treatment in the form of two separate chapters.

This first chapter seeks to answer an extremely simple question: how do we actually build a yield curve? The range of possible answers may come as a surprise to some finance practitioners who have been long-time consumers of yield curves: there is *no* single accepted approach to the construction of the yield curve. Instead, there is a wide range of methods varying dramatically in complexity.



D.J. Bolder, *Fixed-Income Portfolio Analytics*,
DOI 10.1007/978-3-319-12667-8_5

In the following pages, we will introduce the fundamental concepts required for the construction of a yield curve. We will then consider the traditional methods employed for this purpose. The chapter concludes with the consideration of some less traditional, but computationally less challenging, approaches. These latter approaches may be profitably used in our portfolio-analytic applications.

## 5.1 Getting Started

A natural starting point is to examine bond yields at a given point in time. Figure 5.1 displays a collection of US Treasury bond yields as at 29 May 2009. The horizontal axis describes the tenor, or remaining time to maturity, for each of the US Treasury securities outstanding on that date. The vertical axis is the observed yield. Even a cursory examination of Fig. 5.1 reveals a strong relationship between the bond yield and its tenor. Shorter tenor bonds, on this particular date, exhibit starkly lower yields than US Treasury bonds with longer tenors. The reader will naturally recognize the typical upward-sloping yield curve.

We ask the reader to put aside their current understanding of the yield curve—which is likely to be considerable—and examine Fig. 5.1 from the perspective of a mathematician. Denoting the tenor of an arbitrary bond as $T$ and unimaginatively representing the bond yield as $y$, then

$$y = f(T). \tag{5.1}$$

Equation (5.1) suggests a mathematical relationship between a bond's yield and its tenor. The actual interaction between bond yields and tenors is, of course, fairly complex. Equation (5.1)—despite its simplicity—allows us to address this

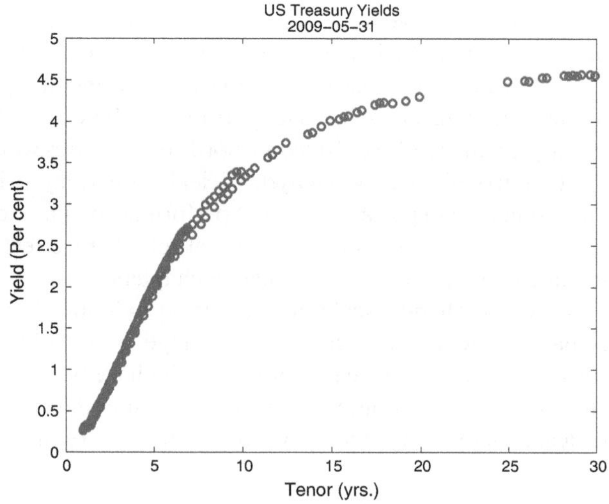


**Fig. 5.1** Bond yields. This is the collection of US Treasury bond yields from 29 May 2009. Even without a model, a strong relationship is evident between these individual yields and their associated tenors

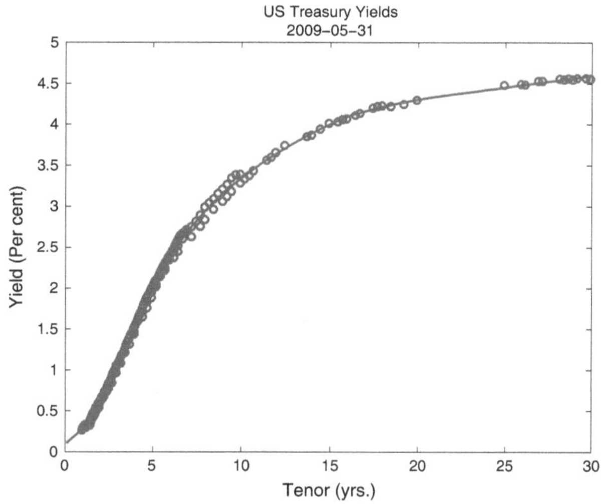


**Fig. 5.2** The yield curve at a point in time. Here we describe the yield-tenor relationship for an instant in time

complexity by employing a wide range of ideas from mathematics, statistics, and engineering. In short, it allows us to mathematically *model* the yield curve. A model seeks to simplify this interaction and help us understand it better. It is precisely these models that we wish to describe in more detail so that they can assist us in developing useful portfolio analytics.

When we talk about a yield-curve *model*, however, some caution is required. A yield-curve model can describe one of two things:

1. the form of the yield curve at a given *point in time*; or
2. the dynamics of the yield curve *through time*.

While these two types of model are not completely independent—indeed, some models simultaneously address both questions—it is conceptually cleaner and practically more convenient to think about them separately. The principal reason is that they typically require very different mathematical techniques. When discussing or considering a given yield-curve model, it is important to know exactly which of these perspectives one is talking about.

The differences between these two types of models are, in fact, quite sizeable. In Fig. 5.1, we postulated a mathematical relationship between yield and tenor. Finding such a relationship is essentially an engineering exercise. Figure 5.2, for example, highlights one possible approach to fitting a mathematical function to this collection of points.[1] Concretely, it fits a function, $f(\cdot)$, to the bond-tenor observations. The first question thus merely asks what this relationship looks like at a given point in

[1]The approach used is a multivariate adaptive regression spline. See Friedman [17], Bolder and Rubin [9], or Lewis and Stevens [21] for more detail.

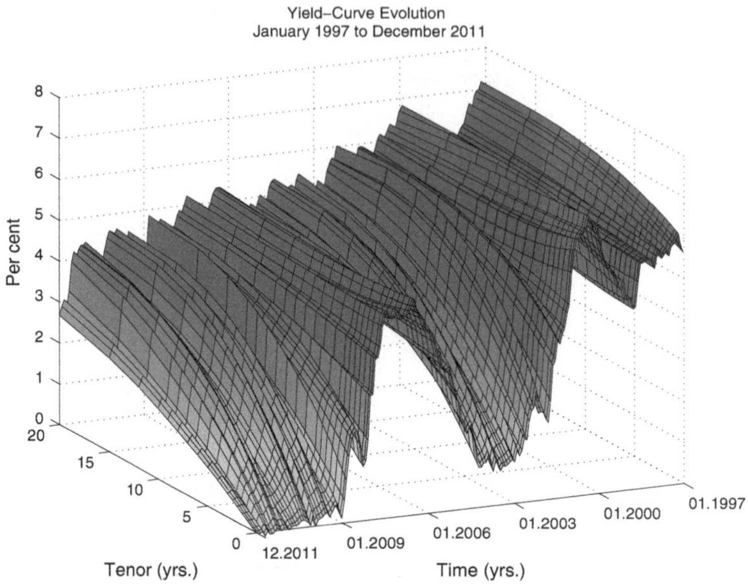


**Fig. 5.3** The yield curve through time. Here we describe the evolution of the yield-tenor relationship over a period of time. This is a sequence of Fig. 5.2's organized chronologically at many different instants in time

time. Since time is held constant, it is a *static* analysis of the yield curve. Simply put, it is a curve-fitting exercise.[2]

The second question seeks to understand how the yield curve moves across time. As time is not held constant, it is a *dynamic* analysis of the yield curve. Figure 5.3 provides a graphical description of a sequence of fitted yield curves ranging from January 1997 to June 2010.

Figure 5.3 indicates that the yield curve is not constant over time, but rather constantly evolves.[3] A dynamic yield curve model, therefore, seeks to use statistical techniques to describe the future evolution of the yield curve in a manner that is consistent with its observed behaviour. This is not merely an engineering exercise. It involves economic reasoning and is also, mathematically speaking, much more difficult.

[2]See Lancaster and Salkauskas [20] is a good introduction to the general realm of curve-fitting.

[3]During some periods the curve is flat, or even inverted, with short rates equal to or greater than long-term yields. During most periods, however, long-term yields dominate the short rates. The magnitude of the difference between short and long yields—what is often termed the slope or steepness of the yield curve—varies substantially over the 10 years displayed in Fig. 5.3.

Figure 5.3 is actually a sequence of monthly static views of the yield curve. It was constructed by computing Fig. 5.2 at the end of each month over the interval from January 1997 to June 2010. A solution to the *static* question of how yields and their tenors are related at a given point in time is thus essential to answering the *dynamic* question of how yields move across time. We will address both the static and the dynamic questions, but given its fundamental role, we begin in this chapter with the static perspective.

## 5.2 Yield Curves 101

A few words of perspective are necessary. Our objective in this chapter is not to make you an expert in fitting yield curves, but instead to make you an informed consumer. Becoming an expert requires more than a brief chapter, but instead lengthy readings and substantial practical experience.[4] All fixed-income professionals are, however, consumers of yield curves. Many have a good understanding of how these yield curves are constructed. Others do not. In our view, a moderate degree of understanding of the various techniques and their relative strengths and weaknesses is essential. It is nevertheless an admittedly dry and technical subject. We'll try to keep our discussion to a (relatively) high level with the use of examples and, wherever possible, graphics.

To fit a sensible mathematical relationship between yields and their tenors, we first need to review some basic concepts. Armed with these fundamental concepts, we will be able to properly review alternative techniques for fitting a yield curve.

There are *four* fundamental yield-curve building blocks:

1. discount factors;
2. spot (or zero-coupon) rates;
3. par yields; and
4. implied-forward rates.

Each of these four building blocks unfortunately has a number of different names, which often appear to be randomly employed. When in doubt, it is always useful to verify what exactly is meant.

Terminology aside, each building block provides an alternative perspective on the yield curve. They are *not* independent. Knowledge of any one of these building blocks permits one to solve for the other three elements. Consequently, each perspective essentially encapsulates the same information. This key insight is exploited consistently in fitting yield curves.

[4]There are many very good references on fitting yield curves. A good start would include Anderson et al. [3], Dierckyx [14], and deBoor [13].

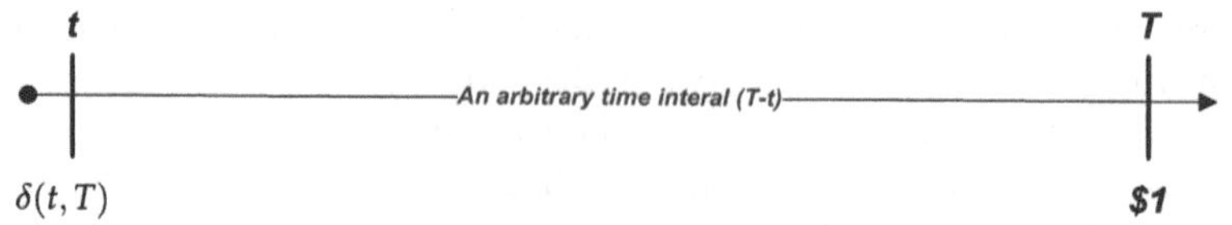


**Fig. 5.4** The very beginning. This is the simplest possible situation: you receive a single cash-flow at some point of time in the future

### 5.2.1 Pure-Discount Bond Prices

The most basic yield-curve element comes back to the time value of money. Imagine that at time $t$, you are promised \$1 at a future point in time, $T$. This is the simplest possible financial instrument that one can imagine. Figure 5.4 illustrates this extremely simple, but powerful, situation.

How much would you pay at time $t$ for this future cash-flow occurring at time $T$? Would you pay \$0.50, \$0.75, or \$0.95? If it was a freely traded instrument, the actual price would be determined like most other good and services in an economy: through the interaction of the supply and demand conditions prevailing in the economy. When determining the price, market participants nonetheless take a number of elements into consideration. In particular, the market price will depend upon:

- The distance between $t$ and $T$. You would observe a rather different price if $T - t$ was 1 week as opposed to 30 years.
- Current economic conditions. If similar contracts were selling for \$0.85, this would presumably be used as an important reference point.
- The creditworthiness of the institution or individual promising to pay the money in the future. Instruments issued by a large and financially stable entity will likely sell for a higher price, all else equal, than similar claims from, say, your brother-in-law.

Although the price is determined by the market, we can logically infer some restrictions on this price. What, for example, is the most you would ever pay for this future cash-flow? Some reflection reveals that the price should not exceed \$1. Paying more than the future value essentially implies a negative interest rate—if equal to 1, it would imply an interest rate of precisely zero. What about the lower bound? Again, anything less than zero does not make much logical sense. A negative price would imply that your counterparty is paying you money for the pleasure of paying you more money in the future! Although many of us would enjoy entering into such a contract, the prospect of finding someone to undertake the other side seems rather unlikely. Consequently, the market price must lie in the interval, $(0, 1)$.

We denote this price, whatever it may be, as $\boxed{\delta(t, T)}$. The first argument $t$ is the current time and its second argument $T$ is the instrument's tenor. Technically, $\delta(t, T)$

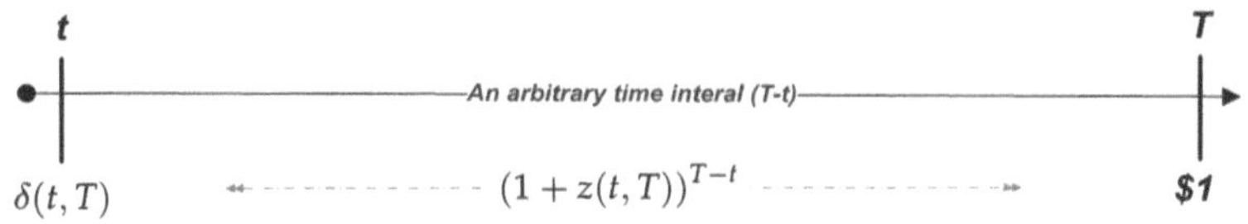


**Fig. 5.5** Rates or prices. Pure-discount prices are very convenient, from a computational perspective, but they are nonetheless hard to interpret. Here we transform prices into rates

is termed the zero-coupon or pure-discount bond price. It is also often referred to as the *discount factor*. This is because the product of this price and any future cash-flow occurring on the same date—be it \$10, \$100, or \$100,000,000—immediately provides one with its present value.

Imagine that we had a pure-discount price for every month starting from the current time ranging out for the next 30 years. These prices would provide significant insight into the cost of money across the entire maturity spectrum. If such information was freely available, we could immediately stop this chapter, because fitting the yield curve would be a trivial and uncontroversial affair. This information is not available.[5] Instead, we are forced to use fairly complicated techniques to deduce these values from a conceptually less appealing instrument—the coupon-bearing bond.

### 5.2.2 Spot Rates

Even with a extensive, uniformly spaced collection of pure-discount prices, we would not be entirely happy. This is because pure-discount prices are difficult to interpret. A pure-discount price of \$0.75 for a 3-year tenor does not immediately resonate with a finance practitioner. We are likely rather more interested in a percentage *rate* of return. Is this \$0.75 price, for example, equivalent to a 10 % or a 2 % interest rate? To find out, we must transform the pure-discount price.

Figure 5.5 provides a graphic description of how we might transform a pure-discount bond price into an equivalent interest rate.

The rate of return associated with a pure-discount bond price is termed the pure-discount rate, the zero-coupon rate or simply the spot rate. Computing the spot rate begins with the ratio of the final cash flow, one unit of currency, and the initial price: $\frac{1}{\delta(t,T)}$. This is basically the return from holding the zero-coupon bond.

[5]Sadly, a broad range of such instruments does not exist. The closest thing to a pure-discount bond in the real world is a treasury bill. Unfortunately, most sovereign issuers do not issue such instruments beyond the 1 year tenor. In some markets, interest and principal payments from coupon bonds can be separately sold, thus creating long-dated pure-discount bonds—see Bolder and Boisvert [5] for more details on this practice. These are rarely used for fitting yield curves, however, due to their relative poor liquidity and lack of transparent pricing.

A simple return is insufficient as we would prefer an annualized figure. Taking annualization into account, zero-coupon rates are related one-to-one with the pure-discount rate through the following expression:

$$\underbrace{\frac{1}{\delta(t,T)}}_{\substack{\text{Price return}\\ \text{over } T-t}} = (1+z(t,T))^{T-t}, \tag{5.2}$$

$$\boxed{z(t,T)} = \left(\frac{1}{\delta(t,T)}\right)^{\frac{1}{T-t}} - 1.$$

Let's recall our 3-year pure-discount bond with a difficult-to-interpret price of \$0.75. Using Eq. (5.2), we find that such a price is equivalent to a zero-coupon rate of 10.1 %. This representation, although it contains the same information, is more easily interpreted. For this reason, most market participants prefer the use of rates to prices.

Equation (5.2) provides the representation of the spot rate in the context of simple compounding. While simple compounding is the standard for financial-market computations, it does lead to somewhat unwieldy formulae. Academics, therefore, often employ continuous compounding to permit simpler representations that allow for easier manipulations of the basic yield-curve building blocks. With the application of continuous compounding, Eq. (5.2) becomes,

$$\underbrace{\frac{1}{\delta(t,T)}}_{\substack{\text{Price return}\\ \text{over } T-t}} = e^{z(t,T)(T-t)}, \tag{5.3}$$

$$\delta(t,T) = e^{z(t,T)(T-t)},$$

$$\boxed{z(t,T)} = -\frac{\ln \delta(t,T)}{T-t}.$$

### 5.2.3 Par Yields

Zero-coupon bonds and prices are not terribly popular, beyond tenors of about 1 year, in actual markets. Instead, the preferred fixed-income vehicle is the coupon-bearing bond. Is there a link between the bond price (or yield) and the zero-coupon bond? Yes, there is. It can be seen by carefully examining the classic bond-price equation,

$$P(t,t_I) = \underbrace{\frac{N}{(1+z(t,t_I))^{t_I-t}}}_{\text{Notional repayment}} + \underbrace{\sum_{i=1}^{I} \frac{c \cdot N}{(1+z(t,t_i))^{t_i-t}}}_{\text{Coupon payments}}, \tag{5.4}$$

$$= N \cdot \delta(t,t_I) + \sum_{i=1}^{I} c \cdot N \cdot \delta(t,t_i).$$

Each individual cash-flow can be considered as a pure-discount bond. The bond price is the sum of each of the individual cash-flows—or pure-discount bonds—discounted back to the current point in time, $t$. The discount factor for each cash-flow is naturally the pure-discount bond price associated with its cash-flow date. In short, Eq. (5.4) reveals that a coupon-bearing bond is merely a portfolio of pure-discount bonds.

Unfortunately, Eq. (5.4) is a theoretical construct. We only observe a bond's price or its yield.[6] The price or yield is silent on the level of the individual discount rates for each of the individual cash-flows. One can construct an infinite number of combinations of discount factors that are consistent with the bond yield.[7] This is an important challenge. A yield-curve model is an attempt to address and overcome this challenge. It essentially involves identifying *one* reasonable set of discount factors that is consistent with observed bond prices. Bond yields themselves are an input to the process, but not a fundamental building block.

The next building block instead involves computing the bond yields for a collection of special bonds: par bonds. Par bonds are coupon-bearing bonds that trade at par, or $1. While par yields are not the same as the observed set of bond yields at a given time, the par-yield curve is nevertheless typically quite close to the observed yield curve.

Determining the collection of par yields given zero-coupon rates or prices, while not complicated, is slightly involved. It requires exploitation of the basic characteristics of a coupon-bearing bond trading at par. Figure 5.6 provides a graphical illustration of the technique. The computation starts with a recognition that the shortest possible par bond has only a single cash-flow. As such, its price must coincide with the associated zero-coupon rate. Mathematically, the first par yield is described as,

$$\boxed{y(t,t_1)} = z(t,t_1). \tag{5.5}$$

[6]One can think of the bond yield, therefore, as a kind of complicated weighted average of the individual zero-coupon rates.

[7]Going in the other direction, however, is no problem. That is, given the set of zero-coupon rates or pure-discount bond prices, we can easily compute the corresponding unique set of bond yields.

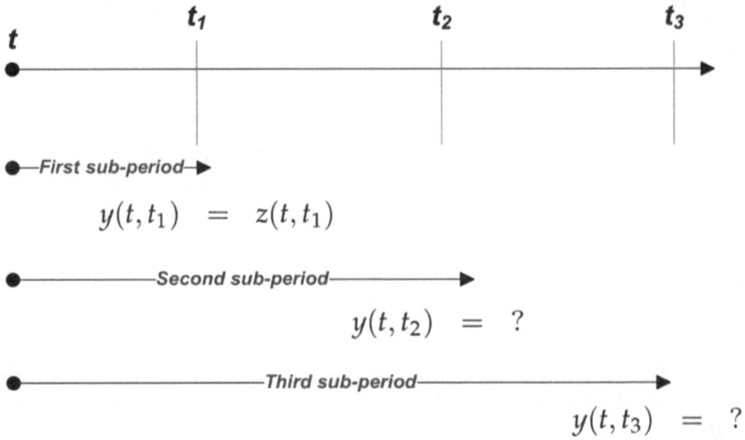


**Fig. 5.6** Back to bond yields. When we discuss bond yields, we very often refer to the theoretical concept of par-bond yields. These are the yields associated with bonds that trade at par—par yields are typically quite close to the observed yield curve

So far so good, but now we need to determine the second par-yield, $y(t, t_2)$. This requires us to recall *two* features of a bond trading at par:

1. a par bond trades with a price of \$1; and
2. a par bond's yield is equal to its coupon rate.

The trick is to write out the bond-price equation with these two facts and solve for $y(t, t_2)$ as follows,

$$1 = \underbrace{\frac{\frac{y(t,t_2)}{2}}{(1+z(t,t_1))^{t_1-t}}}_{\text{First coupon}} + \underbrace{\frac{\frac{y(t,t_2)}{2}}{(1+z(t,t_2))^{t_2-t}}}_{\text{Second coupon}} + \underbrace{\frac{1}{(1+z(t,t_2))^{t_2-t}}}_{\text{Principal}}, \tag{5.6}$$

$$1 - \underbrace{\frac{1}{(1+z(t,t_2))^{t_2-t}}}_{\delta(t,t_2)} = \frac{y(t,t_2)}{2}\left(\underbrace{\frac{1}{(1+z(t,t_1))^{t_1-t}}}_{\delta(t,t_1)} + \underbrace{\frac{1}{(1+z(t,t_2))^{t_2-t}}}_{\delta(t,t_2)}\right),$$

$$\boxed{y(t,t_2)} = 2\left(\frac{1-\delta(t,t_2)}{\sum_{i=1}^{2}\delta(t,t_i)}\right).$$

The par yield is a simple function of the pure-discount prices. Armed with the first two par yields, one may proceed to compute the third, the fourth, and so on until you have them all.

Using the previous logic, we may compute the par yield for any maturity. The general par yield for maturity $t_k$ is written as,

$$\boxed{y(t,t_k)} = 2\left(\frac{1-\delta(t,t_k)}{\sum_{i=1}^{k}\delta(t,t_i)}\right). \tag{5.7}$$

Equation (5.7) clearly shows that if one has the set of discount factors or spot rates, one can easily derive the par yield curve.

Par yields are theoretical bond yields for a collection of bonds that are trading at par. Given zero-coupon prices or rates, par yields are easily determined and can occasionally be useful. More often, however, one assumes that observed bonds yields are sufficiently close to par to extract zero-coupon prices or rates. Such a technique is only effective in limited circumstances. This process, termed bootstrapping, is described in the underlying shaded section.

Par yields are typically used in the opposite direction. Instead of using discount rates to solve for par yields, one often uses par yields to solve for the discount rates. This process is called *bootstrapping*. The classic example is the use of swap rates to solve for the associated zero-coupon rates. Interest-rate swaps are always quoted in the market at par and, as such, meet all of the characteristics of a par bond.

The key element of bootstrapping is that one must work recursively. One starts from the shortest possible par-yield tenor and works gradually out to the longest tenor. The first zero-coupon rate, with the shortest tenor, is equal to the first par yield. In the second step, one knows the coupon and first discount factor. We merely need to solve for the second discount factor. At each step, one reduces the problem into a single equation with a single unknown that can easily be resolved. The general term, comparable to Eq. (5.6) is described as,

$$\boxed{\delta(t,t_k)} = \frac{1-\frac{c}{2}\sum_{i=1}^{k-1}\delta(t,t_i)}{\frac{c}{2}+1}. \tag{5.8}$$

(continued)

To use Eq. (5.7) to find $\delta(t, t_k)$ one must have already solved for all of the rates from $1, .., k-1$. In other words, the spot rates must be determined in successive order from par yields. Par yields conversely can be computed in any order you wish—as evidenced by Eq. (5.6)—if you have the discount factors.

### 5.2.4 Implied-Forward Rates

Is it possible to transact today, at a known rate, for a contract that both begins and ends in the future? The simple answer is yes. This is basically a forward contract not on a commodity, an exchange rate or a stock, but on a pure-discount bond. Conceptually, this is like any other financial forward contract. Two parties can agree on a price for a future sale or purchase of a commodity or an equity. With interest rates it is also possible. The underlying asset that is being purchased or sold, however, is a deposit.

For an interest-rate forward contract the idea is to enter, at a rate determined today, into a deposit starting at some future point in time. Figure 5.7 outlines graphically the notion of a forward interest rate. This rate is termed the implied forward interest rate, or quite simply the forward rate. These rates can easily be determined through the use of zero-coupon rates or prices.

Practically, the interest-rate forward contract is a bit more complicated than its counterparts. An interest-rate forward contract has *three* different important time points, instead of the usual two. Let's denote the current point in time as $t$, the beginning of the forward contract as $s$, and the maturity of the underlying deposit as $T$. With this basic notation, we may now proceed to construct a forward contract.

There are two basic steps. First, imagine that you borrowed \$1 from time $t$ to $T$ at the zero-coupon rate, $z(t, T)$. This would provide you with $\delta(t, T)$ in funds now and an obligation to pay \$1 at time $T$. Second, imagine further that you take these

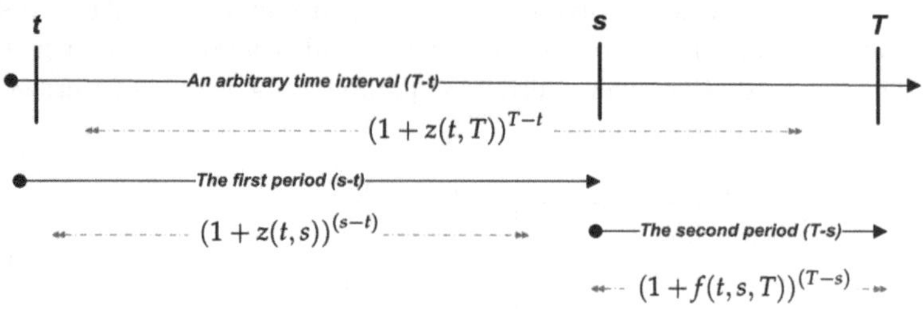


**Fig. 5.7** Rates in the future. It is also possible to determine today the rate for a transaction occurring from one point in the future to another. These are termed *implied* forward rates

proceeds and invest them from time $t$ to $s$ at rate, $z(t,s)$. You will thus receive the following proceeds at time $s$,

$$\underbrace{\delta(t,T)}_{\substack{\text{Loan}\\ \text{proceeds}}}\underbrace{(1+z(t,s))^{s-t}}_{\substack{\text{Rate of return}\\ \text{from } t \text{ to } s}} \tag{5.9}$$

It may not feel like it, but we have just created a forward contract. Through the simultaneous borrowing and lending of \$1 at time $t$, we have

- a receipt of funds (i.e., Eq. (5.9)) at time $s$; and
- an obligation to pay \$1 at time, $T$.

This is precisely what we desired; to borrow money at some future time, $s$, and pay it back at another future point in time, $T$.

The forward rate is the rate of return (or rather the cost) of this strategy or, simply put, the ratio of the final payment to the initial obligation. Recalling that $\delta(t,T) = \frac{1}{(1+z(t,T))^{T-t}}$, we can derive an expression for the forward rate as,

$$\begin{aligned}(1+f(t,s,T))^{T-s} &= \underbrace{\frac{1}{\delta(t,T)(1+z(t,s))^{s-t}}}_{\text{Equation (5.9)}}, \\ &= \frac{1}{\frac{(1+z(t,s))^{s-t}}{(1+z(t,T))^{T-t}}}, \\ f(t,s,T) &= \left(\frac{(1+z(t,T))^{(T-t)}}{(1+z(t,s))^{(s-t)}}\right)^{\frac{1}{(T-s)}} - 1, \\ &= \left(\frac{\delta(s,t)}{\delta(t,T)}\right)^{\frac{1}{(T-s)}} - 1.\end{aligned} \tag{5.10}$$

The forward rate is merely a time-adjusted ratio of pure-discount bond prices or spot rates. While hardly surprising, since we have constructed it using zero-coupon bonds, it underscores the point that forward rates are merely a transformation of pure-discount bond prices or yields. The underlying shaded section provides a second equivalent approach for deriving the implied forward rate using the concept of a break-even yield.

A second and equivalent way to deriving the forward interest rate follows from the exposition in Fig. 5.7. It is predicated on the idea that the forward rate is a break-even rate.

There are two steps. First, ones invest \$1 from time $t$ to $s$ at the known rate, $z(t,s)$. The proceeds of this investment—which are known to be $(1+z(t,s))^{s-t}$—are subsequently reinvested at the prevailing interest rate from $s$ to $T$. This rate is not known at time $t$, but only at time $s$. The forward interest rate is defined as the rate from time $s$ to $T$ that makes one indifferent between the following two strategies:

1. investing from $t$ to $s$, then reinvesting at prevailing rates from $s$ to $T$; or
2. investing directly from $t$ to $T$.

This so-called break-even rate is determined by solving the underlying equation for $f(t,s,T)$,

$$(1+z(t,s))^{(s-t)}\,(1+f(t,s,T))^{(T-s)} = (1+z(t,T))^{(T-t)}, \tag{5.11}$$

$$\boxed{f(t,s,T)} = \left(\frac{(1+z(t,T))^{(T-t)}}{(1+z(t,s))^{(s-t)}}\right)^{\frac{1}{(T-s)}} - 1,$$

$$= \left(\frac{\delta(s,t)}{\delta(t,T)}\right)^{\frac{1}{(T-s)}} - 1,$$

which is consistent with Eq. (5.10). Whether we think about the forward rate as a portfolio of zero-coupon bonds or a break-even rate, the result is identical. In both cases, the forward interest rate is merely a simple transformation of zero-coupon bonds or prices.

### 5.2.5 Bringing It All Together

The key point of this section is that if you know one of the *four* key interest-rate elements, you can derive the others. Figure 5.8 graphically underscores this idea. The corollary of this fact is that, from a yield-curve fitting perspective, you may seek to fit discount factors, zero-coupon bond prices, par yields, or forward rates. The choice lies with the analyst. By fitting any one of them, you are essentially fitting them all. Not surprisingly, therefore, the yield-curve fitting literature contains a broad range of models that attempt to fit the relationship between bond yields and their tenors from each of these four possible starting points.

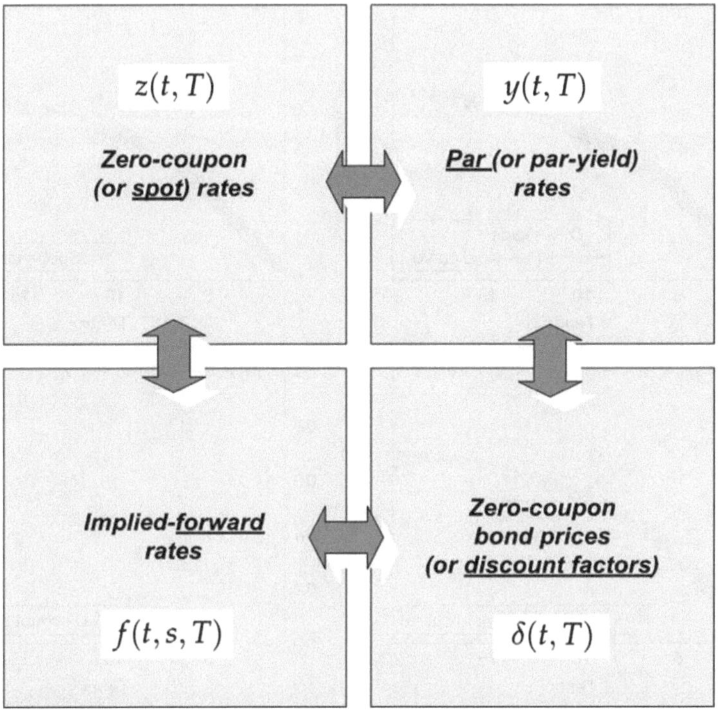


**Fig. 5.8** Four key interest-rate elements. There are four key interest-rate elements: pure-discount prices, zero-coupon rates, implied forward rates, and par-bond yields. With knowledge of any of one element, the other three may easily be derived

What do these yield-curve elements look like in real life? Let's return to the US Treasury yield environment on 31 May 2009. Figure 5.9 illustrates each of the four building blocks for the original set of US Treasury bonds yields originally displayed in Fig. 5.1 on page 114. The top left-hand quadrant displays the actual bond yields along with the par-yield curve. The top right-hand adds the zero-coupon rates, while the bottom left-hand quadrant also includes the forward rates. The final graph, in the bottom right-hand quadrant, displays the pure-discount bond prices associated with this collection of bond yields.

We have so far, of course, said nothing about how one actually succeeds in extracting these elements from the set of bond yields. Our discussion has been descriptive and preparatory. Obtaining these elements is precisely the focus of our subsequent discussion.

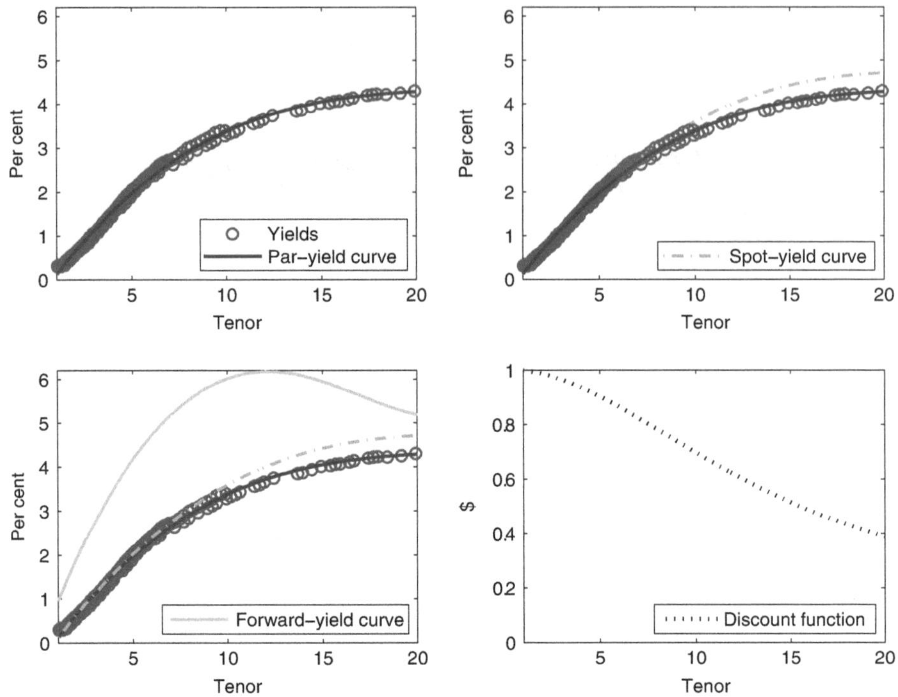


**Fig. 5.9** Par, spot, forward rates and discount factors. These four graphics display each of the four interest-rate elements for the US Treasury bond yields displayed in Fig. 5.1

## 5.3 Curve-Fitting

It is hard to get excited about curve fitting. In addition to being a rather dry technical subject, it feels rather more like engineering than finance. Yield curves, however, lie at the heart of everything one does in fixed-income analytics. Building sensible yield curves, therefore, is absolutely critical.

There are many alternative ways to fit a yield curve ranging from dead simple to extraordinarily complex approaches. It is natural to ask what approach should be employed. While we cannot answer the question for you, we can highlight a few alternative approaches in the context of a few concrete examples and provide a bit of guidance. The choice of fitting technique should be tailored to one's final application. Different applications require varying levels of precision, parsimony, or curve smoothness. Different model requirements, in turn, are associated with different levels of complexity. In this way, the final application of the yield curve model is paramount in determining the requisite level of model complexity.

Fitting US Treasury yield curves is generally quite straightforward. The reasons are simple: the market is deep and liquid and the US Treasury has an extremely

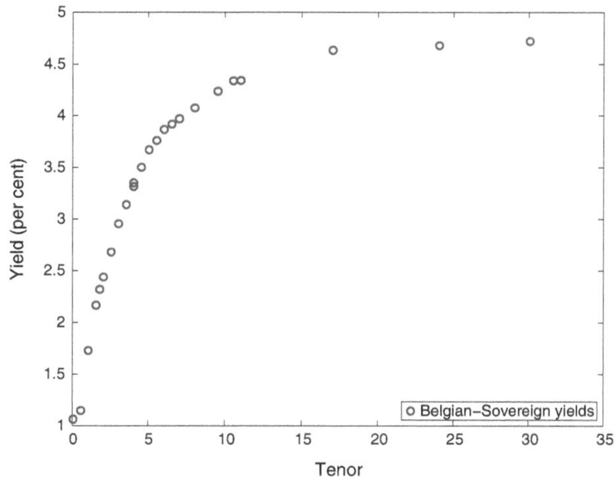


**Fig. 5.10** Belgian government bonds. The underlying figure outlines a collection of Belgian government bonds on 3 March 2011

regular quarterly issuance pattern. The consequence is a strong and consistent relationship between US Treasury yields and their tenors. The choice of yield-curve model is typically unimportant as most approaches generate very similar results.[8] It thus seems a bit unfair to use US Treasury yields as an example as it does not permit us to demonstrate many of the potential pitfalls that can arise.

Let's instead examine a collection of Belgian government bonds on 3 March 2011. The Belgian case is more challenging because of the relatively smaller number of bond issues and the larger space between them. Figure 5.10 outlines a collection of Belgian bond prices as of 3 March 2011. Our task is to find a sensible approach to fitting a *static* yield-curve model to these observed yields.

We will examine *two* basic ways to address this problem: classic and non-classic approaches.[9] The remainder of this chapter is allocated to considering each of these cases.

### 5.3.1 The Classic Approach

The classic approach involves specifying a mathematical form for the yield curve that generates *theoretical* prices that are consistent with *observed* prices.[10] This requires performing a few manipulations to the general bond-price equation. The bond price function for a bond with tenor, $\tau_k$, is given as,

[8]One occasionally suspects that a child could efficiently fit the UST curve with a pencil.

[9]It may be more realistic to think of them as the hard and easy approaches.

[10]Bond yields are the typical focus in casual conversation among financial practitioners. In the classic curve-fitting literature, however, one typically does works with bond prices rather than yields.

$$V(t, y_{\tau_k}) = \underbrace{\sum_{i=1}^{I} \frac{c}{\left(1 + y_{\tau_k}\right)^{t_i - t}}}_{\text{Coupon payments}} + \underbrace{\frac{N}{\left(1 + y_{\tau_k}\right)^{t_I - t}}}_{\text{Principal}}. \tag{5.12}$$

Each cash-flow is discounted with its own yield.[11]

Equation (5.12) discounts each cash-flow with a single bond yield. This is *not* helpful. We want to incorporate one of our basic yield-curve building blocks into Eq. (5.12). To accomplish this, we re-write the price function as,

$$V(t, y_{\tau_k}) = \underbrace{\sum_{i=1}^{I} \frac{c}{(1 + z(t, t_i))^{t_i - t}}}_{\text{Coupon payments}} + \underbrace{\frac{N}{(1 + z(t, t_I))^{t_I - t}}}_{\text{Principal}}. \tag{5.13}$$

Each cash-flow is now discounted by the appropriate zero-coupon rate. This permits us to make progress by specifying a sufficiently tractable mathematical function for describing the zero-coupon curve as a continuous function of tenor. That is, we need something like,

$$\hat{z}_\theta(t, t_i) = g(\theta, t_i). \tag{5.14}$$

for $t_i \in [0, T]$. $T$ denotes the longest tenor of interest in your analysis, $g(\cdot)$ is our relatively well-behaved mathematical function, and $\theta$ are the parameters of the function, $g(\cdot)$.

With such a function $g$, we may re-write the bond price function as,

$$\hat{V}(t, y_{\tau_k}, \theta) = \sum_{i=1}^{I} \frac{c}{(1 + \hat{z}_\theta(t, t_i))^{t_i - t}} + \frac{N}{(1 + \hat{z}_\theta(t, t_I))^{t_I - t}}, \tag{5.15}$$

The only difference between Eqs. (5.14) and (5.15) is that we have replaced the zero-coupon bond function with a parametrized mathematical function of the zero-coupon curve. We thus no longer speak only about the observed price of the bond, but also about its estimated bond price, $\hat{V}(t, y_{\tau_k})$. Ultimately, we wish to make our estimates consistent with the real-world observations.

To do this successfully, we need to:

1. make a choice of $\hat{z}_\theta(t, t_i)$ for $t_i \in [0, T]$; and
2. depending on this choice, determine its parameters so that estimated prices are as close as possible to the observed prices.

[11]Given the price, $V(t, y_{\tau_k})$, one alternatively solves for this yield.

Both of these tasks are inherently mathematical and have relatively little to do with finance. We will consider *three* alternative choices of $\hat{z}_\theta$ and apply them to our collection of Belgian bonds. In doing this, we will only touch briefly on the details of the associated parameter optimization.[12]

Let's begin with a toy example to see how this might work. A dead-simple pure-discount price function would have the following form,

$$\hat{\delta}_a(t, t_i) = e^{-a\cdot(t_i - t)}. \tag{5.16}$$

We've assumed that the discount function is a negative exponential function of the tenor with a parameter, $a$.[13] Some reflection reveals that this is not a completely crazy choice. If the tenor of the bond is zero—that is $t_i = t$—then the pure-discount bond price is equal to unity. Conversely, as tenor gets very large—or mathematically as $t_i \to \infty$—then the pure-discount bond price tends to zero. Our proposed pure-discount function, therefore, takes a series of smooth values between one and zero, which is precisely what we would expect from such a function.

To transform Eq. (5.16) into the zero-coupon function, we need only apply Eq. (5.3).[14] This provides us with,

$$\begin{aligned} \hat{z}_a(t, t_i) &= -\frac{\ln \hat{\delta}_a(t, t_i)}{t_i - t}, \\ &= -\frac{\ln \overbrace{e^{-a\cdot(t_i - t)}}^{\text{Equation (5.16)}}}{t_i - t}, \\ &= a. \end{aligned} \tag{5.17}$$

The result is a bit surprising and certainly simple: our mathematical choice of discount function is equivalent to assuming a constant value, $a$, for the zero-coupon function. Such a choice is not likely to provide an enormous amount of flexibility in fitting our set of observed bonds yields, but it should help us to understand the basic idea.

At this point, we should stress that computation of the estimated bond prices required working with the individual cash flows of each bond in one's sample. This requires detailed security level data—at the very least, one needs coupon rates, maturity dates, day-count conventions, and payment frequencies—and an algorithm for generating the associated collection of cash flow dates and amounts for each

[12]The optimization details obviously matter, but gaining a conceptual understanding of the approach is much more important.

[13]In other words, our parameter set is extremely simple: $\theta = \{a\}$.

[14]We use continuously compounded rates for mathematical simplicity, although you have approximately the same result using Eq. (5.2).

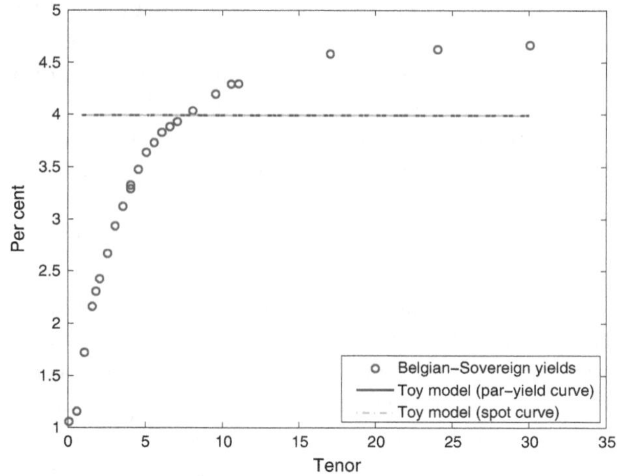


**Fig. 5.11** A toy model. This figure illustrates the optimal fit of our toy model—as described in Eqs. (5.16) and (5.17)—to our Belgian bond data. Beyond describing the basic idea, this model does not appear terribly successful

bond. This is not complicated, but tedious and time consuming and necessitates a flexible computer program.

We now proceed to the second step involving finding a value of $a$ that best fits our set of observed Belgian bond yields. This requires an optimization algorithm. We seek a value of $a$ that minimizes the distance between the estimated and observed prices. Mathematically, this leads to the following optimization problem,

$$\min_{a} \left\| \underbrace{\begin{bmatrix} \hat{V}_1(t, y_{\tau_1}, a) \\ \vdots \\ \hat{V}_m(t, y_{\tau_m}, a) \end{bmatrix}}_{\substack{\text{Estimated} \\ \text{prices}}} - \underbrace{\begin{bmatrix} V_1(t, y_{\tau_1}) \\ \vdots \\ V_m(t, y_{\tau_m}) \end{bmatrix}}_{\substack{\text{True} \\ \text{prices}}} \right\| \tag{5.18}$$

We have a vector of observed prices and a vector of estimated prices that depend on the choice of our parameter, $a$. The solution to this optimization problem will be the single *best* choice of $a$ that provides the best representation of bond prices.[15]

The optimal level of $a$, which we will denote as $a^*$, turns out to be equal to 0.4. Figure 5.11 provides a graphic view of the Belgian yields and the resulting zero-coupon and par-yield curves coming from our toy model.

The result is unsurprisingly poor. The par and spot rate curves are, as expected, a constant horizontal line. While a flat yield-curve model might prove to be reasonable

[15]There are myriad details involved, for non-trivial choices of $g_\theta$, in efficiently and accurately solving this optimization problem. There are many good references for help on this point. See Bolder and Stréliski [10], Bolder and Gusba [7], Cairns [11], Cairns [12], or Bliss [4] to get started.

in some environments, it is naive to expect that a constant zero-coupon curve could work generally as a yield-curve model. While we clearly need a more mathematically flexible and complex choice of $\hat{z}_\theta$, the actual approach applied by the next *two* classical yield-curve fitting models remains conceptually unchanged.

Our next model has been a popular choice, particularly among central bankers. It represents the work of two academics, Nelson and Siegel [25], who proposed the following zero-coupon curve,

$$\hat{z}_\theta(t, t_i) = \alpha_0 + \alpha_1 \left( \frac{1 - e^{-\lambda(t_i - t)}}{\lambda(t_i - t)} \right) + \alpha_2 \left( \frac{1 - e^{-\lambda(t_i - t)}}{\lambda(t_i - t)} - e^{-\lambda(t_i - t)} \right), \tag{5.19}$$

for all $t_i > t$. There are four parameters collected as follows,

$$\theta = \{\alpha_0, \alpha_1, \alpha_2, \lambda\}, \tag{5.20}$$

This may appear somewhat ugly, but it turns out that this was not a random choice. It can be shown that the Nelson–Siegel model is constructed from the first two Laguerre polynomials.[16] These are a sequence of flexible orthogonal polynomial functions employed in statistics, engineering and physics.

Mathematical origins aside, what matters is how well this model succeeds in describing the yield curve. Figure 5.12 illustrates the observed prices of our collection of Belgian bonds (i.e., the squares) along with the estimated prices (i.e., the x's) from an optimized set of parameters for the Nelson–Siegel model. For most tenors, the distance between the observed and estimated prices is relatively modest. For some tenors, particularly at longer tenors, the distance attains as much as EUR 5.

Figure 5.13 provides more financial insight into the Nelson–Siegel model, by illustrating the original Belgian sovereign bond yields along with the par-yield and zero-coupon rate curves. This model, with its more flexible choice of $\hat{z}_\theta$, dramatically outperforms our toy model.

The capacity of the Nelson–Siegel model to closely fit the observed bond yields and prices appears nonetheless limited. This is not a coincidence. The Nelson–Siegel was conceived to provide a parsimonious representation of the zero-coupon, forward, discount function, and par-yield curves. It was a response to other models with significantly more parameters that closely fit prices, but led to uneven, occasionally bumpy, and generally unrealistic looking yield curves.[17] Smoothness and parsimony are thus a general characteristic of the Nelson–Siegel approach.

[16] The interested reader is referred to Hurn et al. [18] or Bolder and Liu [8, Appendix C] for a derivation and appropriate references.

[17] Its stylized view of the yield curve, along with a generally reasonable fit to bond prices, was one of the reasons the Nelson–Siegel model has found favour with central banks. Central banks are generally more interested in understanding general macroeconomic trends than the detailed over- or under-pricing of specific securities.

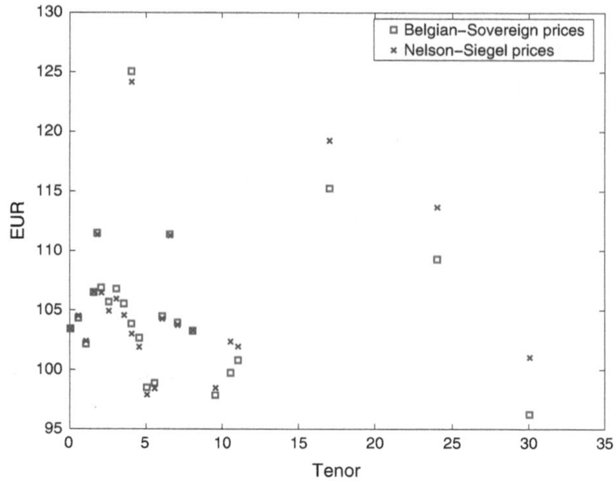


**Fig. 5.12** Fitting bond prices. This figure summarizes actual and fitted prices from the Nelson–Siegel model to our collection of sovereign Belgian bonds. Note that it is rather hard to interpret

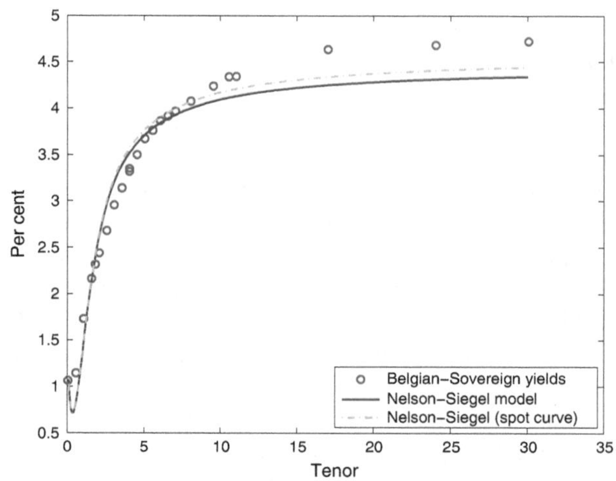


**Fig. 5.13** Nelson–Siegel curves. This figure outlines the par and spot yield curves derived from fitting the Nelson–Siegel model to our collection of Belgian bonds. Such a representation is much easier to understand and interpret

This highlights the key trade-off associated with classical yield-curve fitting models: the tension between a smooth, parsimonious model and a more flexible, better fitting, although less smooth and perhaps less economically convincing approach. If one was performing rich-cheap analysis for the purpose of bond trading, then this model would not be sufficient. Its estimation errors are simply too large. If, conversely, one seeks to extract expectations of future interest rates from market yields, then the Nelson–Siegel model is quite attractive. Let's now turn to examine an alternative model on the other side of this spectrum.

Our next approach operates directly on the discount factor and was suggested by Li et al. [22], Shea [29], and Vasicek and Fong [30]. It postulates that the discount function can be represented as a linear combination of exponential functions; this has led to calling it the exponential-spline model. A very popular, and competing, fitting model, which we will completely ignore, is the cubic-spline model. Cubic splines employ piecewise combinations of cubic functions—hence the name—to fit the yield curve.[18] While it is a solid and useful approach, it is conceptually identical to the two highlighted techniques and its complexity would take us too far afield.

The exponential-spline model does not employ ordinary exponentials, but rather orthogonalized exponentials.[19] This is basically a dramatically more complete version of our toy model. The exponential is a very sensible choice for the discount function, but it requires flexibility to capture the rich variety of yield-curve shapes.

Mathematically, the discount function of the exponential-spline model has the following form,

$$\hat{\delta}_\theta(t, t_i) = \sum_{k=1}^{n} \xi_i \underbrace{e_k(\alpha, t_i - t)}_{\substack{\text{Orthogonalized} \\ \text{exponentials}}}, \tag{5.21}$$

for all $t_i > t$ and where the parameter set is,

$$\theta = \{\xi_1, \ldots, \xi_n, \alpha\}. \tag{5.22}$$

An optimization algorithm is required to find the values of these $n + 1$ parameters that are most consistent with observed bond prices. This is an admittedly somewhat complex approach, but conceptually it is identical to the Nelson–Siegel model. The underlying complexity of the approximating function used to describe the discount factors is a secondary or even tertiary consideration. Where the model falls on the smoothness versus goodness-of-fit spectrum is the important conceptual aspect.

Figure 5.14 demonstrates the observed and fitted prices from the exponential-spline model. There is relatively little distance, at most tenors, between the observed and estimated values. The largest difference looks to be about EUR 1, which represents a significant improvement over the previous model.

Figure 5.15 highlights the estimated par- and zero-coupon yield curves for our Belgian sovereign example. The estimated par yield curve appears rather more consistent with the observed Belgian bond yields across all maturities. It looks particularly good for tenors out to 10 years, where the fit is almost perfect. At the

[18] Numerous good references, for the interested reader, include McCulloch [23], Fisher et al. [16], Anderson and Sleath [2], Shumaker [28], Eilers and Marx [15], Wegman and Wright [32], Ahlberg and Nilsen [1], Nürnberger [26], and Bolder and Gusba [7].

[19] Without getting into the gory details, this essentially amounts to a certain degree of independence between these functions making them more efficient for our application. For more detail on the exponential spline model in general, and orthogonal functions in particular, the reader is referred to Li et al. [22] and Bolder and Gusba [6].

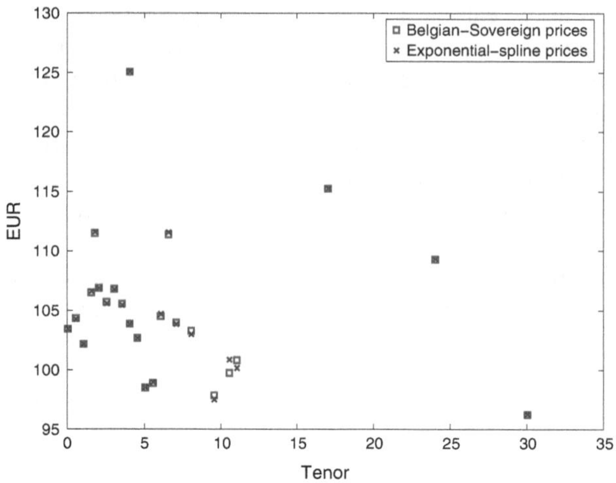


**Fig. 5.14** Fitting bond prices, again. This figure summarizes actual and fitted prices from the exponential-spline model applied to our collection of sovereign Belgian bonds. While challenging to interpret, it can usefully be compared to Fig. 5.12

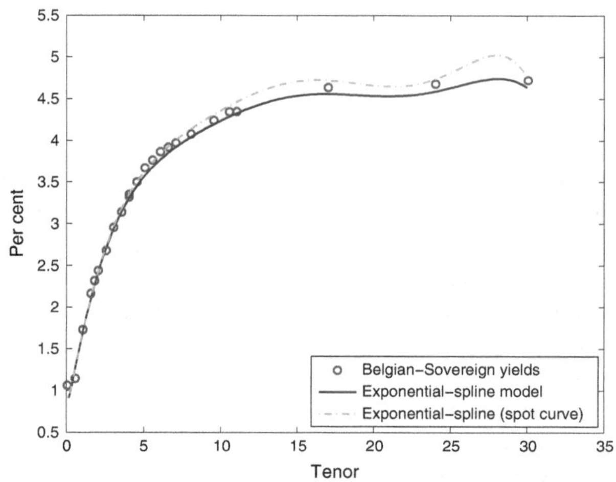


**Fig. 5.15** Exponential-spline curves. This figure outlines the par and spot yield curves derived from fitting the exponential-spline model to our collection of Belgian bonds. Note, by comparison to Fig. 5.13, that the exponential-spline model offers a visibly better, albeit less smooth, fit to the observed Belgian bond yields

same time, however, it demonstrates slightly oscillatory behaviour at the longer bond tenors. This highlights the tension between the goodness of fit and the parsimony of a model.

We could easily continue and consider a dozen or more alternative specifications for the zero-coupon, forward, discount, or par curves. Three examples should nonetheless suffice given that, despite dramatic differences in complexity, there are no conceptual differences among these approaches.

Recalling that we are not trying to become experts at yield-curve fitting, we should ask what have we actually learned from examining a few classical approaches. The main points are summarized as:

- the classical approaches are rather *complicated* as they involve non-linear optimization and manipulation of rather complex mathematical descriptions of the *four* yield-curve building blocks;
- fitting the yield curve requires substantial information about each individual bond including prices, yields, durations, coupons, and cash-flows implying a significant resource investment for its construction; and
- there is a tension between the smoothness, or economic realism, of a yield curve estimate and its capacity to accurately fit observed bond prices.

Although there is certainly debate about the *best* classical approach, it is difficult to argue against the use of a classic approach to fitting the yield curve. It is definitely the *correct* approach for yield-curve fitting. It is, however, not without costs. In what remains of this chapter, we will turn our attention to alternative approaches that attempt to replicate what is accomplished with the classical methods, albeit in a simpler manner.

### 5.3.2 Non-Classical Approaches

Is there an alternative to the complicated classical approach outlined in the previous section? The short answer is yes. The longer answer is, of course, it depends. To this point, we have examined some fairly complex models. For pricing fixed-income securities or derivative contracts, there really is no choice. One requires an accurate and robust method to fit the yield curve.

For portfolio analytics, the situation is less definitive. Bond prices and yields are typically provided through an external source—that is, one's yield-curve model is not being used to price securities. The yield is a reference point for computation of spreads or equivalent treasury yields as well as an input into one's risk computations. One may well argue—and in this context, we do—that a lower degree of accuracy is required and, consequently, that less complicated models may be employed. We thus examine three alternative approaches that could be considered for use in a portfolio-analytic application.

One of the disadvantages of the classical approaches is the necessity of fitting directly to bond prices. This requires determining the timing and magnitude of the cash-flows for a typically sizeable collection of bonds. While not conceptually difficult, such a determination requires substantial information about the bond and reasonably robust computer systems to keep careful track of that information. The simpler approaches discussed in the following pages work directly with bond yield and tenors.[20]

[20]This immediately reduces the data burden thus leading to greater simplicity. Many of these computations can easily be implemented in a simple spreadsheet.

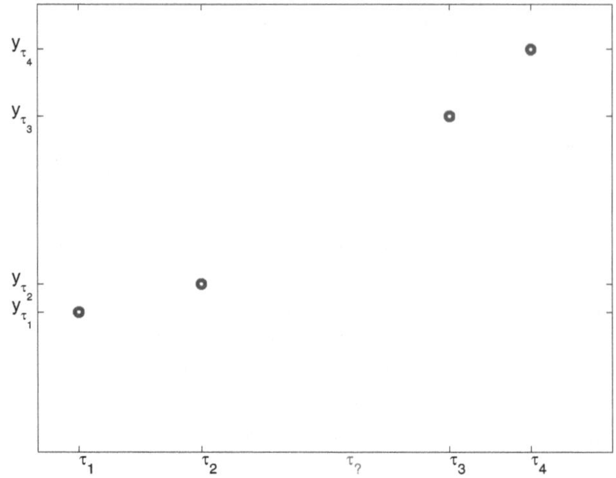


**Fig. 5.16** Zooming in. Imagine that we zoom in on a few tenors and bond yields from our collection of Belgian or US Treasury yields. If we did, it would look something like this

The first approach is the mathematical equivalent of drawing lines through a sequence of points. Despite its conceptual simplicity, it nevertheless has a fancy name: linear interpolation.[21] As always, we need to introduce some notation to permit a rigorous definition of this method. Let's summarize the tenors as,

$$\tau = \begin{bmatrix} \tau_1 \cdots \tau_m \end{bmatrix}, \tag{5.23}$$

and the bond yields as,

$$y = \begin{bmatrix} y_{\tau_1} \cdots y_{\tau_m} \end{bmatrix}. \tag{5.24}$$

Linear interpolation seeks to approximate the relationship between bond yields and tenors. It does not accomplish this, as did the classical approaches, in a global fashion by trying to fit the best curve to all observations. Instead, it works from point to point.

We begin by zooming in on the collection of sovereign Belgian yields and tenors as shown in Fig. 5.16. We want to draw a straight line through each of the points. Thus, it must be that our line will pass through each of the points in Fig. 5.16.

Let us focus on the yields between tenors $\tau_2$ and $\tau_3$. How do we determine these intermediate yield points? We know the yield at the two points $\tau_2$ and $\tau_3$: they are simply the observed values $y_{\tau_2}$ and $y_{\tau_3}$, respectively. We simply need a mathematical definition for an arbitrary value, let's call it $y_{\tau_?}$, in between the two known points.

The plan is merely to create a linear function between $\tau_2$ and $\tau_3$—to draw a line, in plain English. To draw a straight line through the points $y_{\tau_2}$ and $y_{\tau_3}$, we first

[21]See Ralston and Rabinowitz [27] for a formal description of the general notions of interpolation.

require its slope. The slope of a straight line through two given points, of course, is merely

$$b \equiv \frac{\Delta y}{\Delta \tau} = \frac{y_{\tau_3} - y_{\tau_2}}{\tau_3 - \tau_2}, \tag{5.25}$$

or the ratio of the change in the yield associated with a given change in tenor. The intercept is simply the value of the initial yield at $\tau_2$ or,

$$a = y_{\tau_2}. \tag{5.26}$$

This is the mathematical equivalent of drawing a line through the points. $a$ is the intercept and $b$ is the slope of the line between the two nearest points. The intermediate value is now given as follows,

$$y_{\tau_?} = \overbrace{\underbrace{y_{\tau_2}}_{a} + \underbrace{\left(\frac{y_{\tau_3} - y_{\tau_2}}{\tau_3 - \tau_2}\right)}_{b} (\tau_? - \tau_2)}^{f(\tau)}. \tag{5.27}$$

Equation (5.27) now provides us with the desired intermediate yield points—all that is required is the tenor. All of the values between $\tau_2$ and $\tau_3$ now fall proportionally along the straight line running between these two points.

Let's verify that Eq. (5.27) makes logical sense by verifying the end points. If $\tau_? = \tau_2$, then

$$\begin{aligned} y_{\tau_2} &= y_{\tau_2} + \left(\frac{y_{\tau_3} - y_{\tau_2}}{\tau_3 - \tau_2}\right) \underbrace{(\tau_2 - \tau_2)}_{=0} \\ &= y_{\tau_2}, \end{aligned}$$

which is exactly what you would expect. If $\tau_? = \tau_3$, then

$$\begin{aligned} y_{\tau_3} &= y_{\tau_2} + \left(\frac{y_{\tau_3} - y_{\tau_2}}{\cancel{\tau_3 - \tau_2}}\right) (\cancel{\tau_3 - \tau_2}), \\ &= y_{\tau_3}, \end{aligned}$$

which is again as expected. It looks like everything works.

Figure 5.17 illustrates the application of the linear interpolation idea to our zoomed-in example. The solid point along the line between $\tau_2$ and $\tau_3$ corresponding

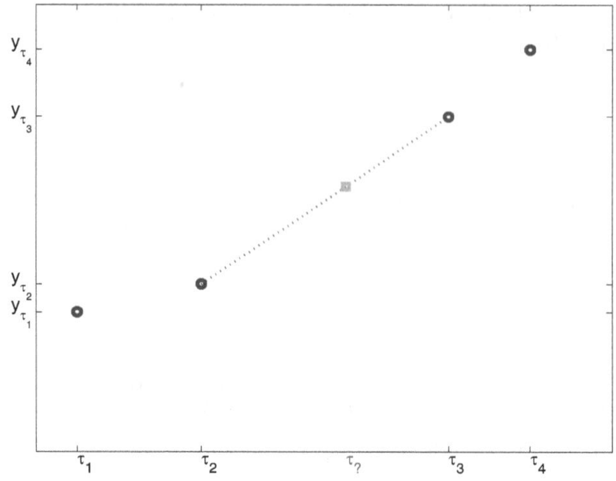


**Fig. 5.17** Linear interpolation in action. This figure applies the notion of linear interpolation to our zoomed-in bond and tenor example described in Fig. 5.16

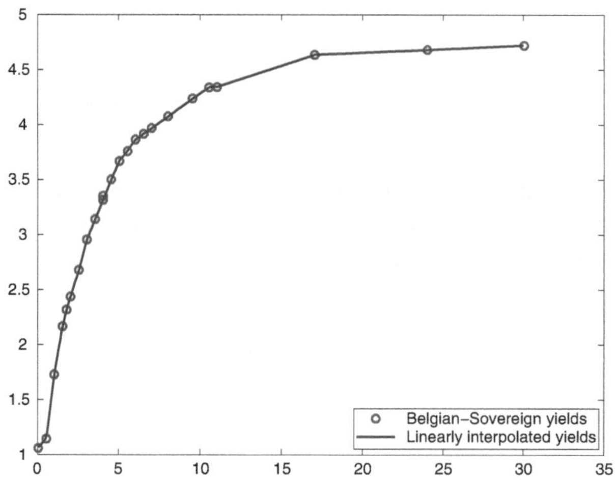


**Fig. 5.18** Linear interpolation in reality. This figure linearly interpolates between each of the bond yields in our collection of Belgian bonds. This seems to produce a very sensible yield-curve estimate

to $\tau_?$ is easily computed using Eq. (5.27). This idea can be applied to each individual pair of adjacent points to create a piecewise linear approximation of the yield curve.

Figure 5.18 applies this fitting algorithm to our Belgian sovereign bond example. The result is very encouraging. Figure 5.18 appears to fit the collection of bond yields and tenors remarkably well.

Have we found a straightforward approach for the more involved classical approaches described in the previous section? The use of linear approximation will often work quite well. It can also fail miserably. The reader may have noticed that the algorithm does not accommodate the possibility of two yields occurring at the same tenor. If two yields do share a common tenor, then one must select one of these points for the interpolation. If the yields are very close, then this will not likely make a large difference. Depending on the liquidity and efficiency of the

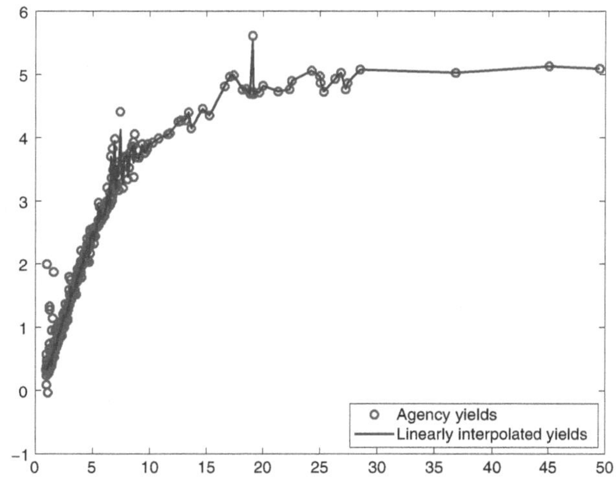


**Fig. 5.19** Fly in the ointment. Here we apply the linear interpolation technique to a collection of US Agency bonds. Clearly, it no longer performs so well. Linear interpolation will clearly have difficulty with noisy data and multiple observations sharing a common tenor

underlying market, differences in yields of bonds at the same date can be important. If the movement from one yield to the next is not smooth, one cannot expect that the resulting linearly interpolated curve will also be smooth.

Figure 5.19 displays the application of linear interpolation to a collection of US Agency bonds. Sadly, the technique does not perform nearly as well for this noisy data set containing multiple yield observations with common tenors. The conclusion, therefore, is that while linear interpolation can be quite effective for well-behaved collections of bond yields, it does not seem appropriate as a general purpose algorithm for fitting yield curves.

The next approach also uses a linear perspective, but applies it globally across all bond yields. This basically means that instead of having a linear function defined for each set of adjacent points, we have a single linear function across the entire spectrum of tenors. Mathematically, this amounts to the following approximation,

$$\underbrace{\begin{bmatrix} y_{\tau_1} \\ \vdots \\ y_{\tau_m} \end{bmatrix}}_{y} = \overbrace{a + b \underbrace{\begin{bmatrix} \tau_1 \\ \vdots \\ \tau_m \end{bmatrix}}_{\tau}}^{f(\tau)}. \tag{5.28}$$

Analogous to the linear interpolation technique, the parameter $a$ is again the intercept, while the parameter $b$ is the slope of the line. Unlike the linear interpolation technique, which operated on pairs of points, these two parameters now apply to *all* yields and tenors.

How do we determine the values of $a$ and $b$? We should try to find $a$ and $b$ such that they provide the best fit to the observed yield-tenor relationship. The distance between observed and estimated yields is termed the error. For any choice of $a$ and $b$, the error in this linear model is given as,

$$\underbrace{\begin{bmatrix} \epsilon_1 \\ \vdots \\ \epsilon_m \end{bmatrix}}_{\epsilon} = \begin{bmatrix} y_{\tau_1} \\ \vdots \\ y_{\tau_m} \end{bmatrix} - \underbrace{\left( a + b \begin{bmatrix} \tau_1 \\ \vdots \\ \tau_m \end{bmatrix} \right)}_{f(\tau)}. \tag{5.29}$$

We could try to minimize the sum of the errors, but generally this does not work very well.[22] A better approach is to minimize the sum of squared errors.

By minimizing the sum of squared approximation errors—or more formally, $\epsilon^T\epsilon$—we find ourselves using the well-known linear-regression model. With a bit of manipulation, we need not perform any numerical optimization. The solution is merely a simple function of the elements found in Eq. (5.29). Re-write Eq. (5.29) as,

$$\underbrace{\begin{bmatrix} \epsilon_1 \\ \vdots \\ \epsilon_m \end{bmatrix}}_{\epsilon} = \underbrace{\begin{bmatrix} y_{\tau_1} \\ \vdots \\ y_{\tau_m} \end{bmatrix}}_{y} - \underbrace{\begin{bmatrix} 1 & \tau_1 \\ \vdots & \vdots \\ 1 & \tau_m \end{bmatrix}}_{X} \underbrace{\begin{bmatrix} a \\ b \end{bmatrix}}_{\beta}, \tag{5.30}$$

$$\epsilon = y - X\beta.$$

The optimal values of the parameters $a$ and $b$, summarized in the vector $\hat{\beta}$, are given as,

$$\hat{\beta} = \left(X^T X\right)^{-1} X^T y. \tag{5.31}$$

This solution can easily be programmed into a spreadsheet. Most spreadsheet software also has built-in functionality allowing one to solve the least-squares, or linear regression, problem without any matrix manipulation.

The solution summarized in Eq. (5.31) comes directly from the first-order conditions associated with the least-squares minimization problem. The curious, or perhaps less trusting reader, can verify that the dimensions of the matrix

(continued)

[22]The reason is that some errors are positive and others are negative and one can have a small sum of errors overall, but sizeable individual errors. This can be resolved by minimizing the sum of the *absolute value* of the errors. This is also challenging. The derivative of the absolute-value function is unfortunately discontinuous at zero and, as such, makes the use of standard optimization methods a bit tricky.

manipulation are consistent as follows,

$$\underbrace{\hat{\beta}}_{(2\times1)} = \underbrace{\underbrace{\left(\underbrace{\underbrace{X^T}_{(2\times m)}\underbrace{X}_{(m\times2)}}_{(2\times2)}\right)^{-1}\underbrace{X^T}_{(2\times m)}}_{(2\times m)}\underbrace{y}_{(m\times1)}}_{(2\times1)}. \tag{5.32}$$

Equation (5.31) did not, unfortunately, fall out of the sky. It is the solution to a quadratic optimization problem. Classically, this involves solving the so-called normal equations. There is, however, a more direct approach involving rather less rigour, which gets us to the same place. Consider our regression equation as the following simple re-arrangement of Eq. (5.30)

$$X\beta = y. \tag{5.33}$$

This is a linear system and, generally, we would be tempted to invert $X$ to solve for $\beta$. The problem is that $X$ is not square and, as such, not invertible. What about $X^TX$? This is a square matrix and, under certain conditions, is invertible. If we multiple both sides by $X^T$, assume that $X^TX$ is invertible, and simplify we have

$$\begin{aligned} X^TX\beta &= X^Ty, \\ \underbrace{(X^TX)^{-1}X^TX}_{I}\beta &= (X^TX)^{-1}X^Ty, \\ \beta &= (X^TX)^{-1}X^Ty, \end{aligned} \tag{5.34}$$

which coincides precisely with Eq. (5.31). This is basically one possible solution to an overdetermined linear system—this simple approach also happens to be equivalent to the ordinary least squares solution. The formal regression framework is valuable and should under no circumstances be ignored, but this simple computation can nonetheless provide some useful intuition. The formality behind the regression framework can be found in many sources—Judge [19], for example, is an excellent starting point.

The next step is to apply our linear regression in Eq. (5.28) to our Belgian sovereign and US Agency datasets and see how well it performs. Figure 5.20

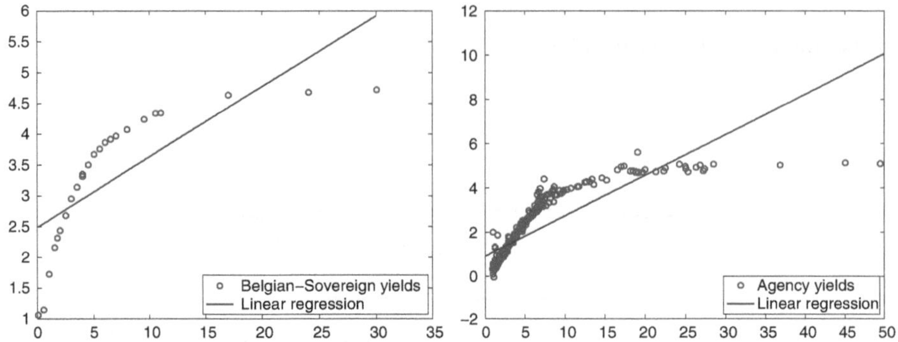


**Fig. 5.20** Linear regression in action. Here we apply the linear regression technique to a collection of Belgian sovereign and US Agency bonds. The results are not terribly satisfactory

illustrates the results of fitting a linear regression to our yield-tenor data. The results do not seem terribly encouraging.

What has gone wrong? We are attempting to fit a straight line to a non-linear function. Although we have found the best possible straight line, the overall fit remains poor. Have we been wasting our time with the linear regression methodology? Not at all. The linear regression approach is very flexible. Let's add a non-linear term to our linear regression as follows,

$$\underbrace{\begin{bmatrix} y_{\tau_1} \\ \vdots \\ y_{\tau_m} \end{bmatrix}}_{y} = \overbrace{a + b \underbrace{\begin{bmatrix} \tau_1 \\ \vdots \\ \tau_m \end{bmatrix}}_{\tau} + c \underbrace{\begin{bmatrix} \tau_1^2 \\ \vdots \\ \tau_m^2 \end{bmatrix}}_{\tau^2} + \underbrace{\begin{bmatrix} \epsilon_1 \\ \vdots \\ \epsilon_m \end{bmatrix}}_{\epsilon}}^{f(\tau)}. \tag{5.35}$$

The third parameter, $c$, is the coefficient for the quadratic term. Re-writing it in the form of Eq. (5.30),

$$\underbrace{\begin{bmatrix} \epsilon_1 \\ \vdots \\ \epsilon_m \end{bmatrix}}_{\epsilon} = \underbrace{\begin{bmatrix} y_{\tau_1} \\ \vdots \\ y_{\tau_m} \end{bmatrix}}_{y} - \underbrace{\begin{bmatrix} 1 & \tau_1 & \tau_1^2 \\ \vdots & & \\ 1 & \tau_m & \tau_m^2 \end{bmatrix}}_{X} \underbrace{\begin{bmatrix} a \\ b \\ c \end{bmatrix}}_{\beta}, \tag{5.36}$$

we observe that the form is identical. We've merely added one more column to our matrix, $X$. The solution in Eq. (5.31) applies equally well. This is a common and straightforward extension to the linear regression framework. It is nonetheless a clever idea, because it preserves the linearity of the model while introducing a non-linear component into the yield-tenor relationship.

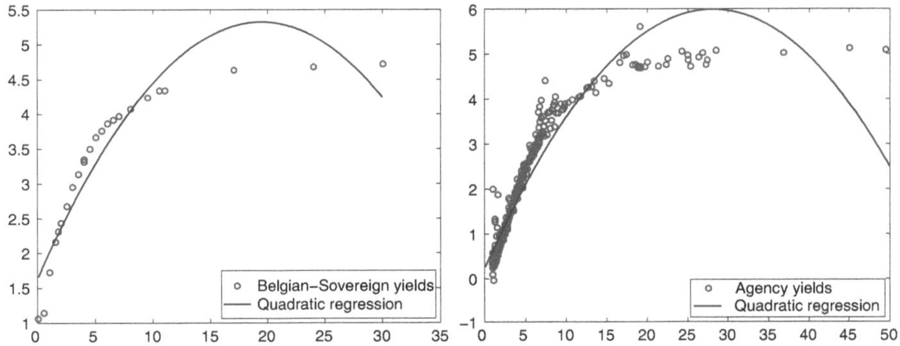


**Fig. 5.21** Quadratic regression in action. Here we apply the quadratic regression technique to a collection of Belgian sovereign and US Agency bonds. The results are still not completely satisfactory, but it looks much better than the linear case in Fig. 5.20

Figure 5.21 outlines the results of this extended linear regression with the quadratic term. The results are not perfect, but they do represent quite a substantial improvement over the strictly linear approach described in Fig. 5.20. The approximation captures some of the upward sloping form of the yield curve, albeit with too much curvature at longer tenors. This is a direct consequence of the parabolic form of the quadratic term introduced into our linear regression.

There is nothing to stop us at a quadratic approximation. We can, of course, go beyond the form of Eq. (5.35). We need only extend the trick that we just introduced. Now that we have the idea, let us merely add a cubic and quartic term to our linear regression. This requires adjusting the linear regression in the following manner:

$$\underbrace{\begin{bmatrix} y_{\tau_1} \\ \vdots \\ y_{\tau_m} \end{bmatrix}}_{y} = \overbrace{a + b\underbrace{\begin{bmatrix} \tau_1 \\ \vdots \\ \tau_m \end{bmatrix}}_{\tau} + c\underbrace{\begin{bmatrix} \tau_1^2 \\ \vdots \\ \tau_m^2 \end{bmatrix}}_{\tau^2} + d\underbrace{\begin{bmatrix} \tau_1^3 \\ \vdots \\ \tau_m^3 \end{bmatrix}}_{\tau^3} + e\underbrace{\begin{bmatrix} \tau_1^4 \\ \vdots \\ \tau_m^4 \end{bmatrix}}_{\tau^4} + \underbrace{\begin{bmatrix} \epsilon_1 \\ \vdots \\ \epsilon_m \end{bmatrix}}_{\epsilon}}^{f(\tau)}. \tag{5.37}$$

We now have two additional parameters, $d$ and $e$, that are the coefficients for the cubic and quartic terms, respectively. The cubic term, given it may take both positive and negative values, provides additional flexibility to our approximation.

The results, summarized in Fig. 5.22 along with the fit from the linear interpolation approach, are quite encouraging. For both the Belgian sovereign and US Agency curves, the quartic regression provides a very respectable fit. For the evenly spaced and well-behaved Belgian sovereign bond yields, the linear interpolation appears to be a slightly better choice, while quartic regression appears to be quite

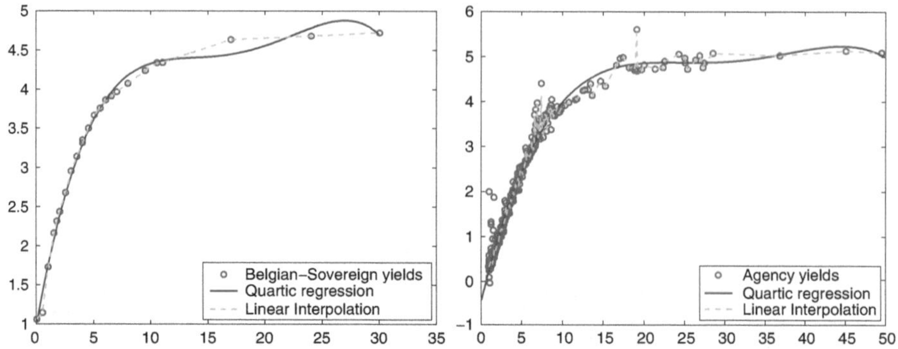


**Fig. 5.22** Quartic regression in action. Here we apply the quartic regression technique to a collection of Belgian sovereign and US Agency bonds. The results look quite promising

a strong alternative for the noisier US Agency data. The strength of the linear regression approach, its global approach, is also its weakness. Fitting a single functional form to a broad range of yields is a difficult task. Linear interpolation achieves substantial flexibility by having a very local focus, but for noisy data this can be a drawback. The tension with these non-conventional methods is again, as with the classical approaches developed previously, basically the trade-off between goodness of fit and smoothness.

Before concluding, we offer a final alternative approximation approach. This technique is slightly more complicated to implement, but is conceptually very similar to linear interpolation and regression. It is, in fact, something of a mixture between these two techniques.

The basic idea is to determine the yield value for a given tenor as a kind of average of the observations around it. More technically, one speaks of the approximated value as being a weighted average of the observations in the neighbourhood of that point. Two important elements need to be determined:

1. What observations belong to the neighbourhood of a given tenor?
2. How much weight is given to each of these observations?

Some notation is necessary. We begin by trying to approximate the yield value for an arbitrary tenor, $\tau$. We define the set $N_h(\tau)$ as a collection of points defined in a neighbourhood of points around $\tau$. The size of the neighbourhood is determined by the parameter, $h$, which is also called the bandwidth. $h$ could be something like 3 months on either side of our current tenor, $[\tau - 0.25, \tau + 0.25]$. Our approximated yield value for the tenor, $\tau$ can be written as,

$$\hat{y}(\tau) = \underbrace{\text{mean}\left(y(\tau_i) \mid \tau_i \in N_h(\tau)\right)}_{f(\tau)}. \tag{5.38}$$

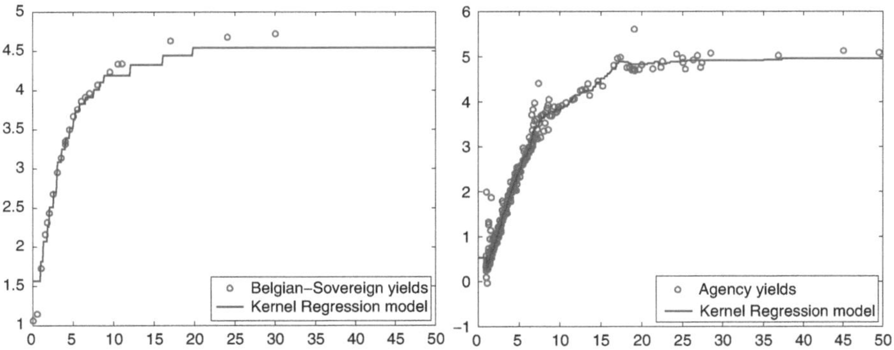


**Fig. 5.23** Kernel regression in action. Here we apply the kernel regression technique to a collection of Belgian sovereign and US Agency bonds. While the results do not look very appealing, the computations are extremely stable and useful for input to a risk model

That is, $\hat{y}(\tau)$, is the simple average of all the observations within the neighbourhood of $\tau$.

This technique is called a *kernel* regression. Other than $h$, it requires no parameters.[23] One can increase the flexibility of this approach by replacing the simple average in Eq. (5.38) with some kind of weighted average.[24] Conceptually, this is not a highly complicated approach and, while a bit more involved than linear interpolation and regression, it can still be implemented in a spreadsheet.

Figure 5.23 provides an example of the application of Eq. (5.38) to our two yield datasets. The results do not look terribly smooth, but the overall fit is quite sensible. Stability is the principal advantage of this approach, which makes it useful for risk applications where one seeks to compute the yield volatilities and correlations of different key tenors across time. This approach also has the strong advantage of permitting estimation of different curves, denominated in different currencies or from different issuers, upon the same *grid* of bond tenors—for example, $\left[\frac{1}{365}, ..., 50\right]$. Some countries (or issuers) have tenors out to 20 years, other to 30 years, and yet other out beyond 50 years. The presented techniques—classical and non-classical alike—all perform quite poorly outside of the range of existing data. This makes it a challenge to computing volatility and correlation values on a single uniform grid. Kernel regressions behave quite robustly beyond the range of observations, making it a valuable technique for exactly this application.

[23]It is thus classified as a non-parametric approach. Most of the effort involved in using kernel regression is finding a reasonable value for $h$. See Nadaraya [24] and Watson [31] for more details.

[24]Often, the weighting is defined as a decreasing function of the distance from the point, $\tau$. This has the logical effect of increasing the importance of nearby yield observations and decreasing the importance of yields that are further away.

## 5.4 Concluding Thoughts

Given the central role of the yield curve in the fixed-income world in general and in portfolio analytics in general, it is important to invest some time towards understanding it. This chapter has focused on describing various approaches to fitting the relationship between yield and tenor at a given point in time. There is no doubt that this topic is somewhat dry and technical, but an understanding of the relative strengths and weaknesses of the various approaches is essential.

After introducing the most fundamental elements of the yield curve, we proceeded to examine two broad families of yield-curve fitting models: classical and non-classical approaches. Classical approaches use non-linear optimization techniques and manipulation of one or more of the basic yield-curve building blocks to find the best possible fit to observed bond prices.

Classical approaches require substantial information about the individual bonds, fairly complicated mathematics, and typically require a reasonably involved computer program for their implementation. Choice of a specific approach is also a challenge given that there is often a tension between the economic realism, or smoothness, of an approximated curve and its capacity to accurately fit observed bond prices.

The non-classical approaches are, by comparison, rather *straightforward*. We examined *three* alternative techniques for fitting the yield-tenor relationship: linear interpolation, linear regression, and kernel regression. Linear interpolation is very appealing, but is probably not sufficient for general application. A naive linear regression is not very useful, but by adding higher-order terms, one quickly arrives at a reasonable approximation. Kernel regressions by contrast are not terribly smooth, but are incredibly stable and quite useful for risk applications.

Although classical approaches are the technically *correct* way to fit the yield curve, non-classical approach warrant a second look. Everything depends on the degree of accuracy required in one's applications. For most portfolio analytic applications, given that one only requires a reasonable approximation of the yield curve, one can safely use an alternative approach and avoid the complexity (and headache) of classical approaches.

## References


1. J.H. Ahlberg, E.N. Nilson, *The Theory of Splines and Their Applications* (Academic Press, New York, 1967)
2. N. Anderson, J. Sleath, New estimates of the uk real and nominal yield curves. Bank of England Working Paper, 2001
3. N. Anderson, F. Breedon, M. Deacon, A. Derry, G. Murphy, *Estimating and Interpreting the Yield Curve* (Wiley, England, 1996)
4. R.R. Bliss, Testing term structure estimation methods. Federal Reserve Bank of Atlanta: Working Paper 96–12, 1996
5. D.J. Bolder, Easing restrictions on the stripping and reconsititution of government of canada bonds. Bank of Canada: Working Paper 1998–8, June 1998

6. D.J. Bolder, Towards a more complete debt strategy simulation framework. Bank of Canada: Working Paper 2002–13, May 2002
7. D.J. Bolder, S. Gusba, Exponentials, polynomials, and fourier series: more yield curve modelling at the bank of canada. Bank of Canada: Working Paper 2002–09, 2002
8. D.J. Bolder, S. Liu, Examining simple joint macroeconomic and term-structure models: a practitioner's perspective. Bank of Canada: Working Paper 2007–49, 2007
9. D.J. Bolder, T. Rubin, Optimization in a simulation setting: use of function approximation in debt-strategy analysis. Bank of Canada: Working Paper 2007–13, 2007
10. D.J. Bolder, D. Stréliski, Yield curve modelling at the bank of canada. Technical Report No. 84 (Bank of Canada, Ottawa, 1999)
11. A.J.G. Cairns, Descriptive bond-yield and forward-rate models for the pricing of british government securities. Department of Actuarial Mathematics and Statistics, Heriot-Watt University: Working Paper, 1997
12. A.J.G. Cairns, A family of term-structure models for long-term risk management and derivative pricing. Math. Finance **14**, 415–444 (2004)
13. P. deBoor, *A Practical Guide to Splines* (Springer, Berlin, 1978)
14. P. Dierckx, *Curve and Surface Fitting with Splines* (Clarendon Press, Oxford, 1993)
15. P.H.C. Eilers, B.D. Marx, Flexible smoothing with b-splines and penalties. Stat. Sci. **11**(89), 102 (1996)
16. M. Fisher, D. Nychka, D. Zervos, Fitting the term structure of interest rates with smoothing splines. U.S. Federal Reserve Board Working Paper, 1994
17. J.H. Friedman, Multivariate adaptive regression splines. Ann. Stat. **19**(1), 1–67 (1991)
18. A.S. Hurn, K.A. Lindsay, V. Pavlov, Smooth estimation of yield curves by laguerre functions. Queensland University of Technology Working Paper, 2005
19. G.G. Judge, W.E. Griffiths, R.C. Hill, H. Lütkepohl, T.C. Lee, *The Theory and Practice of Econometrics*, 2nd edn. (Wiley, New York, 1985)
20. P. Lancaster, K. Salkauskas, *Curve Fitting and Surface Fitting: An Introduction* (Academic, Orlando, 1986)
21. P.A. Lewis, J.G. Stevens, Nonlinear modelling of time series using multivariate adaptive regression splines (mars). J. Am. Stat. Assoc. **86**(416), 864–877 (1991)
22. B. Li, E. DeWetering, G. Lucas, R. Brenner, A. Shapiro, Merrill lynch exponential spline model. Merrill Lynch Working Paper, 2001
23. J.H. McCulloch, Measuring the term structure of interest rates. J. Bus. **44**, 19–31 (1971)
24. E. Nadaraya, On estimating regression. Theory Probab. Appl. **9**(1), 141–142 (1964)
25. C.R. Nelson, A.F. Siegel, Parsimonious modeling of yield curves. J. Bus. **60**, 473–489 (1987)
26. G. Nürnberger, *Approximation by Spline Functions* (Springer, Berlin, 1980)
27. A. Ralston, P. Rabinowitz, *A First Course in Numerical Analysis*, 2nd edn. (Dover Publications, New York, 1978)
28. L.L. Schumaker, *Spline Functions: Basic Theory* (Wiley, West Sussex, 1978)
29. G.S. Shea, Interest rate term structure estimation with exponential splines: a note. J. Finance **40**, 319–325 (1985)
30. O.A. Vasicek, H.G. Fong, Term structure modeling using exponential splines. J. Finance **37**, 339–348 (1981)
31. G.S. Watson, Smooth regression analysis. Sankhya Indian J Stat. Ser. A **26**(4), 359–372 (1964)
32. E.J. Wegman, I.W. Wright, Splines in statistics. J. Am. Stat. Assoc. **78**, 351–365 (1983)

# 6 Modelling Yield Curves

*Essentially, all models are wrong, but some are useful.*

Box and Draper 1987

Yield-curve models fall into two main categories: static and dynamic. Static yield-curve models—addressed in detail in the previous chapter—involve fitting a mathematical function to the yield-tenor relationship. This engineering type exercise—involving relatively little or no economic intuition—requires determination of the parameters of a mathematical function to a collection of bonds at a *single* point in time. Dynamic yield-curve modelling, the focus of this chapter, involves describing how a given yield curve evolves over time.[1] In contrast to the static problem, dynamic yield-curve modelling has an important economic element.

Both static and dynamic yield-curve models are an important part of our portfolio-analytic tool-kit. Fitted yield curves are a critical input into the computation of portfolio exposures, return approximations, portfolio risk computations, performance attributions, and optimizations. Dynamic yield-curve models find application in risk, performance computation and attribution.

Dynamic yield-curve modelling is an intensely mathematical exercise. While this is a widely researched area of finance, it is also, by virtue of its mathematical complexity, a relatively inaccessible area of the literature.[2] Our objective is not to make the reader an expert in the development and estimation of dynamic yield-curve models. Instead, we seek to describe a set of common assumptions shared by all

[1]These two approaches are not independent. To describe the evolution of the yield curve over time, one will need to observe current and historical yield curves. This is turn will involve repeatedly performing the curve-fitting exercise for a large collection of different points in time. Fitted curves are, therefore, an essential input into any dynamic yield-curve model.

[2]Typically one requires graduate-level mathematics and statistics to read, understand, and implement the models found in the finance literature.



D.J. Bolder, *Fixed-Income Portfolio Analytics*,
DOI 10.1007/978-3-319-12667-8_6

yield-curve models. Understanding these assumptions and seeing their application in a series of examples should help to make the reader a better consumer of dynamic yield-curve models.

The principal objective in this chapter is to provide the reader with a set of tools that can be used to critically question and understand dynamic term-structure models proposed by academics, colleagues, subordinates, or vendors. We also offer varying degrees of complexity in our yield-curve model examples. A secondary objective, therefore, is to provide some modelling alternatives for portfolio-analytic purposes where the mathematical burden might not be quite so high.

## 6.1 Why a Dynamic Yield-Curve Model?

A model is a mathematical simplifications of a typically much more complex reality. Building and using a model is a difficult, time consuming and often frustrating undertaking. Simplifying assumptions are required and substantial resources are essential for the construction and implementation of a model. Consumers of models, such as senior management, often complain about model complexity.[3]

Given these challenges, a model should only be employed when it is *absolutely* necessary. In our view, absolutely necessity can be determined by two elements:

1. importance; and
2. complexity.

If something is both important and complex, then it is worthwhile trying to model it. The reasoning is simple. Should something be important, but simple, one can easily avoid constructing a model. Simple intuition should suffice. If something is very complex, but unimportant, then it can safely be ignored. If something is both important and complex then neither sole reliance on intuition nor ignoring it is an option. Constructing a model to better understand and predict it thus is the logical choice.

Convincing you that a dynamic interest-rate model is required, therefore, requires convincing you that interest rates are both important and complex. Let's start with importance. The market value, performance, and risk of any fixed-income portfolio depend, typically in an important manner, on interest rates. One cannot value one's portfolio or compute its exposures without a yield curve. One cannot compute or attribute performance or risk without a yield curve. One can rarely take an active position in a fixed-income without a view on the yield curve. The evolution of the yield curve, therefore, is a key determinant of the risk and return characteristics of

[3]Users also often simultaneously demand for greater model granularity and realism thereby further heightening the complexity. This is part of the natural tension between understandability and usability of any model.

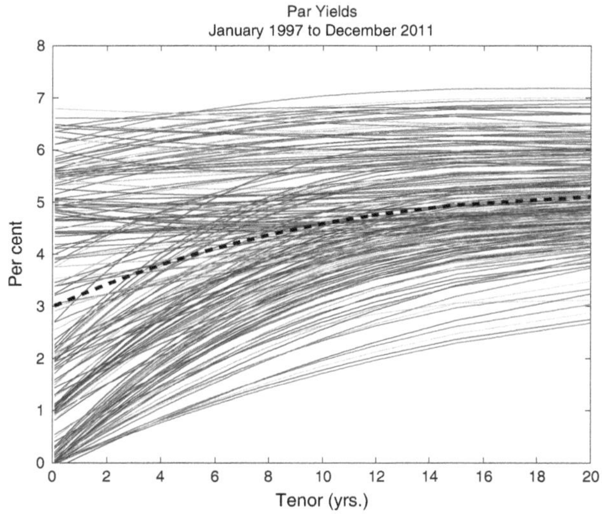


**Fig. 6.1** UST yield curves. This is a collection of US Treasury yield curves ranging from January 1997 to December 2011—this incorporates 180 separate yield-curve observations spanning a period of 15 years. We will use this data throughout the entire chapter

one's fixed-income portfolio. Given its centrality, the yield curve cannot be avoided. In short, the evolution of the yield curve is *important*.

Having established its importance, is the yield curve complicated? This is less obviously accomplished. We could provide a lengthy list of yield-curve characteristics and simply claim that they are complicated. A better approach, however, would be to demonstrate these characteristics. To this end, we will examine a dataset of US Treasury yield curves spanning the period from January 1997 to December 2011.[4] With 180 separate US Treasury yield curves covering a respectable time period of 15 years, we should be able to draw a few conclusions.

Figure 6.1 begins our examination by simply graphing all of our US Treasury yield curves together against their tenor. What can we conclude? Yields may take a wide range of values across all tenors. Short-term rates, for instance, range from basically zero to slightly less than 7 % over this time interval. Long-term rates cover the spectrum from 2.5 to 7 %. The yield curve can also take a variety of different shapes. It may be upward sloping, flat, or downward sloping. Sometimes the curve moves upwards almost linearly, while in other cases it slopes up steeply and then flattens off.[5] The average yield-curve—represented by the dashed black line in Fig. 6.1—is upward sloping with roughly 200 basis-points difference between yield at the longest and shortest tenor. In short, therefore, the yield curve takes a wide range of shapes and levels.

[4]Each of these yield curves was initially fit with a kernel regression and then smoothed out with a quartic regression approach. The consequence are quite smooth, but robust estimates of the US Treasury yield curve at each point in time.

[5]This characteristic is termed *curvature*.

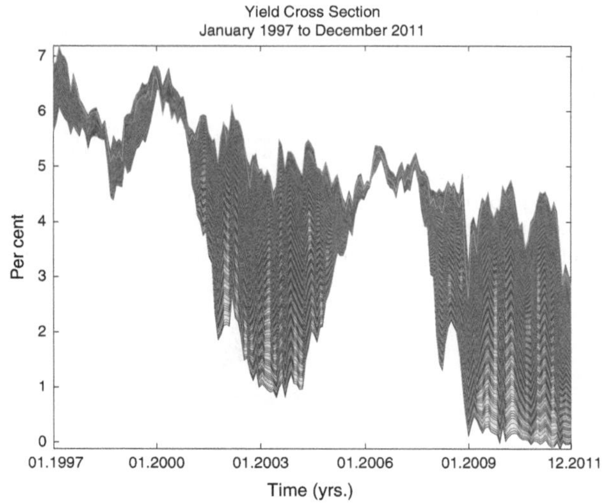


**Fig. 6.2** Another yield-curve perspective. Using the same data as displayed in Fig. 6.1, we examine an alternative perspective: the passage of time

Figure 6.1 tells us little about the relationship of the yield curve from 1 month to the next. We need to add the time dimension. Figure 6.2 accomplishes this by providing the cross section of yields ranging from 1 week to 20 years for each month across our 15-year dataset. At each individual date, the breadth of the yield observations provides insight into the distance between the short- and long-tenor yields. This is typically termed the slope (or the steepness) of the yield curve. There are lengthy periods where the yield curve is quite flat followed by extended periods of steeper yield curves. Sometimes this transition occurs rather smoothly, but in other cases it can occur more quickly. We do not, however, observe yield curves that oscillate from flat to steep every month. On the contrary, if a yield curve is flat during a given month, it is quite probable that the yield curves in the previous and subsequent periods will also be relatively flat. This is termed *persistence*.

With 180 separate curves, it is challenging to see a common pattern. Figure 6.3 attempts to extract such a pattern by computing the average yield, zero-coupon, and implied 1-year forward curves across the entire period. These average curves demonstrate that, on average, the US Treasury yield curve was upward sloping over the 15-year period under investigation. An upward sloping yield curve implies upward sloping zero-coupon and implied forward curves.

An upward sloping implied forward curve is an interesting phenomenon. The implied forward rate is also, in a world where market agents are risk neutral, the expected future zero-coupon curve. Given that the implied forward curve lies above the zero-coupon curve, this implies that on average, when abstracting from risk, yields are expected to increase. This is puzzling since yields have fallen secularly over our time period. This is not a new observation. Academics have spent much time trying to understand why yield curves typically exhibit an upward sloping shape. The general answer relates to risk—market participants are, of course, not

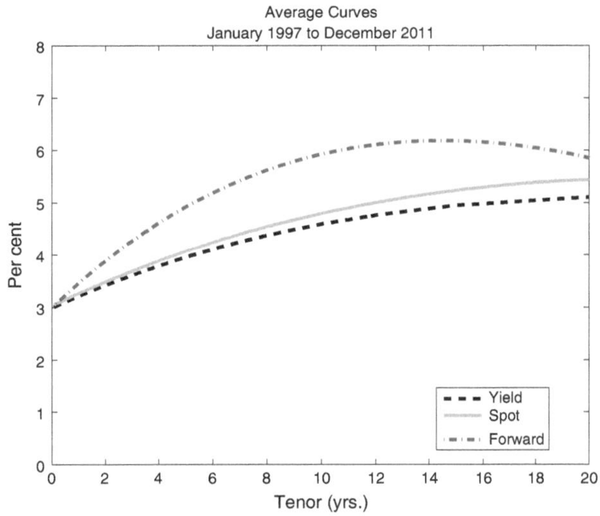


**Fig. 6.3** Average yield curves. Again using our 15-year UST yield-curve dataset, we compute and illustrate the average par-yield, zero-coupon, and forward interest rate curves

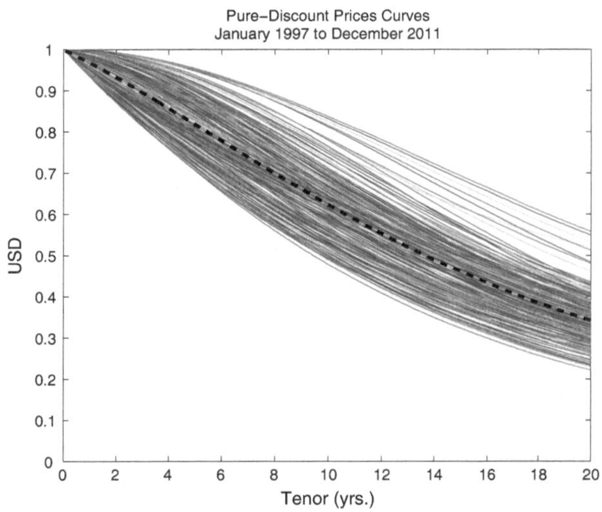


**Fig. 6.4** Pure-discount bond prices. One may also compute the implicit pure-discount prices, or discount factors, associated with each of our 180 US Treasury yield curves

neutral towards risk, but rather need to be compensated for taking it. We return to this point shortly.

Figure 6.4 displays the discount factors, associated with each of our US Treasury curves. Interpretation of discount factors is not terribly obvious. The bread of possible discount factors is rather impressive: the 10-year discount factor appears to range from 0.5 to 0.85. This implies that over the last 15 years, the present value of USD 1,000,000 paid out in 20 years has ranged roughly from USD 500,000 to USD 850,000. That is quite a significant difference.

The pure-discount price function is also always a decreasing function of tenor. This means that the pure-discount price at time $t$ is less than the associated price

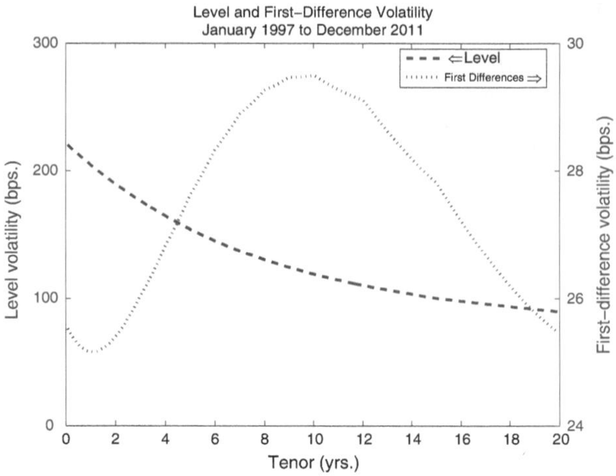


**Fig. 6.5** Monthly yield volatility. This figure displays the average yield curve volatility over our 180-month dataset from two perspectives: by yield-curve level and by monthly yield curve changes. Each perspective implies rather different volatility outcomes

at time $t + 1$. This holds for all $t$ and is a critical characteristic of the yield curve. If the discount factor were to remain flat over a given period, this would imply zero interest rates. This is not impossible, although it occurs rarely. If, however, the discount factor increases over time this would imply negative forward rates and, consequently, arbitrage. Strictly decreasing discount factors are one element of an arbitrage-free yield-curve system.[6]

It is natural to consider the standard deviation, or volatility, of the yield curve. There are two different ways that one can think about yield-curve volatility. One may directly compute the volatility of the yield levels or compute the volatility of yield-curve changes. Both are summarized in Fig. 6.5. Direct volatility of the yield curve is quite substantial with short-term values exceeding 200 basis points as compared with long-term values of approximately 100 basis points. A clear characteristic is that short-term yields are typically substantially more volatile than their longer-term counterparts.[7]

The dashed line in Fig. 6.5 highlights the volatility of monthly yield-curve changes (i.e., first differences) from our 180 US Treasury curves. The corresponding structure of yield-change volatility is rather different. Overall, short-, mid-, and long-term yield volatility are reasonably flat ranging from roughly 25–30 basis points. The volatility of yield changes nevertheless demonstrates somewhat less volatility at the short- and long-end of the curve compared to the centre of the curve.

[6]While arbitrage opportunities do occasionally exist, they typically disappear rapidly. When building a model of yield-curve dynamics, therefore, it is generally a good idea to construct it so that arbitrage is avoided.

[7]These are both notions of unconditional volatility. Loosely speaking, this means volatility estimated over a long time period, without reference to the given time period. Volatility may also be measured conditionality, which involves explicitly taking into account current market conditions. When we consider the measurement of risk, we will have much more to say on this idea.

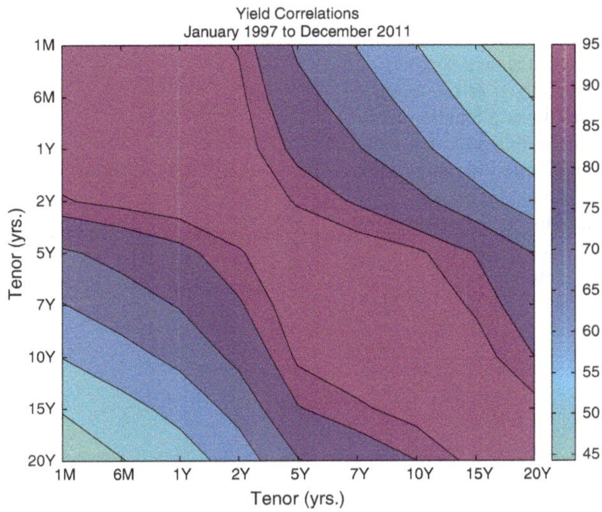


**Fig. 6.6** Yield correlations. This figure displays a heat map describing the correlation of yield changes between nine selected points along the US Treasury curve

It appears to reach a maximum around the 10-year yield. While both perspectives are interesting, the volatility of yield movements is probably the most interesting for the construction of a dynamic yield-curve model.

It is interesting, and enlightening, to see how yield movements are correlated. Figure 6.6 attempts to graphically illustrate the correlation of yield changes for *nine* selected points along the US Treasury curve ranging in tenor from 1-month to 20 years. The technique used is called a heat-map, which uses colour to describe the level of correlation between two separate US Treasury yields. Pink values denote highly correlated values, whereas various shades of blue describe lower levels of correlation. To interpret the figures, one must read the figure as a matrix. Select a tenor on the vertical axis, say 7 years, and compare it to another tenor on the horizontal axis such as 6 months. For these two choices of tenor, the colour is the lightest shade of blue corresponding to about 0.5 on the colour scale found on the right-hand side of Fig. 6.6. The correlation of each yield level with itself will naturally be 1.0, which explains why there is a diagonal swath of bright pink running through the figure from the upper left to lower right corner.

While it takes a bit of getting used to, Fig. 6.6 provides useful insight into the correlation between yield movements. The closer the tenor between two yields, the higher the level of correlation. 1- and 6-month or 5- and 7-year yields exhibit high levels of correlation. 1-month and 10-year yield movements, however, have relatively lower levels of correlation. Overall, the level of correlation across all yield changes remains large and positive. The smallest correlation coefficient occurs between the 1-month and 20-year tenor, but is still significantly positive at about 0.5. Adjacent tenors routinely demonstrate yield-change correlations exceeding 0.85 or 0.9. We conclude that the yield-curve is *not* an independent collection of yields, but rather an organized and correlated system.

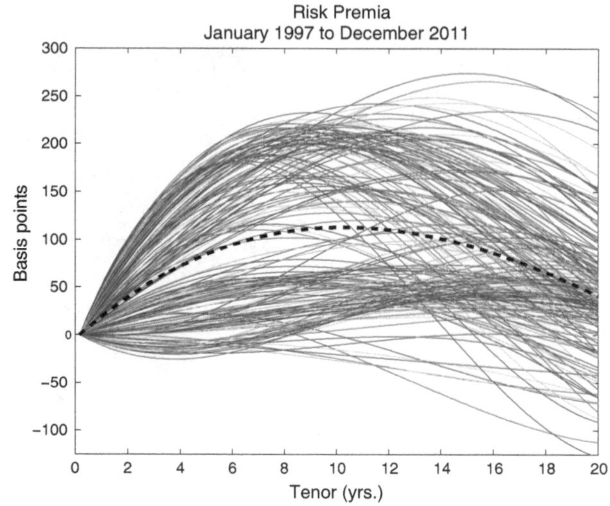


**Fig. 6.7** Risk premia. This figure outlines the difference between annual forward rates and realized zero-coupon rates. Over a lengthy period of time, this provides some insight into the risk premia demanded by market participants for holding US Treasury securities at different tenors

Let's return to the mystery associated with a positively sloped average US Treasury curve. We can use the fact the implied forward rates are the risk-neutral expectation of future rates to perform a simple experiment. We compute the 1-year implied forward curve for each of the 180 months in our dataset. We then move 1 year forward and compare this *expected* curve to the actual realized curve. If one performed this exercise for a small number of periods, then it would be unreasonable to assume that the forward and realized zero-coupon curves should coincide. Repeating this exercise for each month in our dataset, we can reasonably hope to extract something interesting.

Figure 6.7 provides the results of this exercise. It displays the difference between *expected* 1-year forward yields and the actual observed yields. If the forward curve was a good predictor of the future zero-coupon curve, we would expect to see the average difference to be roughly null. Instead, we note that the difference is overwhelmingly positive. The *expected* forward value is very often significantly larger than the realized value. There are some negative differences, but they appear to represent only about 10 % of the cases.

Why is this the case? The answer relates to the fact that bondholders take risk. A number of risks are undertaken, but the most important probably relates to unexpected inflation.[8] Bondholders are aware of this and thus demand a slightly

[8]Imagine that you purchase a nominal bond and expect inflation to remain about 2 % over its life. If inflation unexpectedly jumps to 5 %, the value of your bond is eroded. The longer the tenor of the bond, the greater the erosion of value. Inflation risk is not the only risk for which bond holders must be compensated. Liquidity and market-specific factors are also important.

higher rate of return on their bonds for longer tenors. This is called a term premia. The presence of this term premia implies, on average, an upward-sloping yield curve. The consequence is also higher implied forward rates, which are not necessarily consistent with expected future bond yields.

The dashed black line in Fig. 6.7 represent a rather noisy estimate of the average risk premia demanded by bondholders for tenors ranging from 1 month to 20 years. It suggests that there is a modest risk premia demanded for short-tenor securities, but that it increases to approximately 100 basis points at the 10-year sector before falling smoothly as tenors increase. Given its imprecision, it is better not to take these estimated risk-premia figures too seriously. The conclusion nonetheless remains clear. Bondholders demand extra compensation for the risks associated with holding US Treasury securities and this risk premia is neither constant with regard to time nor tenor.

Bringing these points together, the yield curve:

- may take a wide range of shapes and levels;
- demonstrates considerable *persistence* over time[9];
- has an average upward-sloping shape consistent with market participants demanding compensation for the risk of holding bonds;
- is subject to arbitrage constraints; and
- displays correlation, volatility, and risk premia that vary both by tenor and over time.

We have thus established that the yield curve is an intricate, high-dimensional, time-varying system. It is, simply put, complicated. Since it is both important *and* complicated, then you really do not have much of a choice—you need to have a model.[10]

## 6.2 Building a Model

Despite the technical complexity of dynamic yield-curve models, the situation is not all bad. Modelling is essentially about simplification and organization. Most of the popular dynamic models of the yield curve, despite the broad literature in this area, have essentially the same conceptual form. Abstracting from the technical details, there are, in fact, really a small number of key assumptions involved in

[9]Whether interest rates revert to a long-term mean is a matter of some debate. Practically and economically, it appears logically obvious, but it is empirically hard to prove. See Ball and Torous [5].

[10]For a more technical review of the stylized facts about yields, see Leippold and Wu [60], Fisher [35], or Bolder et al. [10].

the modelling process. Every dynamic yield-curve model must make *three* basic assumptions. They include:

- $\mathscr{A}_1$: What are the set of state variables that describe interest rates?
- $\mathscr{A}_2$: How do these state variables move through time? What are their dynamics?
- $\mathscr{A}_3$: How does one map/transform the state variables into the term structure of interest rates?

Answering, in some reasonable way, each of these three questions will lead to a yield-curve model. This applies to a wide range of well-known models. A conceptual framework is essential, because the list of extant models is simply dizzying. It includes Heath et al. [41, 42], Ho and Lee [43], Hull and White [47–49], Richard [73], [50], Brennan and Schwartz [13], Longstaff and Schwarz [63], Backus et al. [3], [64], Dai et al. [24], Langetieg [58], Chen [17], Schaefer and Schwartz [76], Brace et al. [12], Christensen et al. [19], Diebold et al. [27], Flesaker and Hughston [36], Hördahl et al. [45], Leippold and Wu [59], Cochrane and Piazzesi (2006, Decomposing the yield curve, unpublished), and Rogers [74, 75]. We desperately need some way to understand and evaluate their reasonableness.

This bold claim aside, this list of assumptions probably does not immediately clarify things very much. A significant amount of explanation and description is still required. By the end of this chapter, however, this three-assumption framework should become a useful tool for understanding and assessing dynamic yield-curve models.

### 6.2.1 $\mathscr{A}_1$

When reading the three key yield-curve assumptions, an undefined term almost certainly jumped out at the reader: the *state variable*. Although we are not quite yet ready to define it, let's begin moving in this direction. Our starting point is to create a list of the various factors that have an influence on the yield curve. A reasonable list would include:

- macroeconomic variables such as output, inflation, and unemployment;
- the current and expected stance of monetary policy;
- the relative liquidity of different sectors of the curve;
- the relative creditworthiness of the underlying debt-issuer;
- the general volatility of bonds and related equity or commodity markets;
- the political situation in the country;
- market-specific conditions such as proximity to bond auctions, general conditions in the repo market, and collateral regulations;
- any legislation requiring that certain important investors, such as pension funds or insurance companies, hold securities with particular tenors;
- the level or volatility of the exchange rate.

While certainly incomplete, we note that this list of possible factors influencing the yield curve is actually depressingly long. It includes a wide range of macroeconomic, financial, and behavioural variables. One approach to the construction of a dynamic yield-curve model would involve carefully and laboriously describing and understanding the interactions between the yield curve and this list of key driving factors. Constructing a model with such a large number of factors would, however, quite clearly make life *very* difficult. Indeed, it may not even be possible.

Is there a better way? The answer is yes. It involves reducing the number of factors. This brings us to our first assumption: the choice of state variables. The concept of a *state variable* comes originally from the physical sciences. State variables describe the *state* of some dynamic system.[11] They allow one to infer the overall state of the system from a, typically lower-dimensional, set of variables.

An excellent example is the weather. An exhaustive list of weather conditions would be quite long. It would include temperature, relative humidity, wind speed and direction, precipitation, cloud cover, and barometric pressure. A meteorologist would certainly make it even longer. To simplify things, we *routinely* use state variables to describe weather conditions. Temperature is a state variable. We often summarize the weather with only the temperature. Stating that it is 20 °C is very different from claiming, usually with some apprehension, that it is −20 °C. This single state variable immediately helps you understand the general state of the weather. If we add an additional state variable such as cloud cover or humidity, we further improve our understanding of the situation. State variables allow one to infer the overall state of the system from a typically lower-dimensional set of variables. The point of a state variable is thus to succinctly describe a complicated system.

There are two types of state variables: observable or unobservable. Observable state variables can be directly seen or measured. Weather examples include temperature, wind speed, or humidity. Unobservable state variables must be computed or inferred from the dynamic system. Measures such as wind-chill factor or the humidity index would be considered unobservable state variables. The wind-chill factor, for example, stems from the idea that during the winter months it *feels* colder when it is windy. By combining wind speed and temperature—according to a pre-determined formula—one can construct an unobservable state variable.[12]

The reader is probably thinking that there is an inordinate amount of discussion about the weather for a text on portfolio analytics. This may be a fair comment, but the intent is to demonstrate that the exotic-sounding notion of a state variable is, in fact, something that one uses on a daily basis.

Returning to the problem at hand, we seek state variables to help us infer the current position of the yield curve. These may either be readily observable factors like output, inflation, or the exchange rate. They may also be unobservable implying

[11] Temperature, pressure, and entropy are common state variables used in a physical setting.

[12] Similarly, the humidity index is a measure that addresses the fact that during the summer it *feels* hotter when the relative humidity is high. Combining temperature and relative humidity, according to a pre-determined formula, leads to another unobservable weather-related state variable.

that they must be inferred from yield-curve data. Unobservable state variables are sufficiently important that we will dedicate an entire section to their development later in this chapter.

The first key assumption required in the construction of any yield-curve model is the choice of state variables. Every yield curve model requires at least one state variable. As we will see in the following examples, the logic behind the choice of state variable may not be immediately obvious. In all cases, however, the state variables serve to simplify the description of our complicated dynamic system: the yield curve.

### 6.2.2 $\mathscr{A}_2$

Knowledge of the current values of the state variables provides one with an understanding of the current state of the dynamic system. The operative word, however, is *dynamic*. To understand the dynamics of the system, one has to have an understanding of how the state variables move over time.

At the risk of testing the reader's patience, let's return to our weather example. Temperature, cloud cover, and relative humidity can provide you with a good understanding of the weather at a given point in time. To predict the weather tomorrow, next week, or next month one needs to understand how these three weather-related state variables are likely to change over the prediction period.

Prediction is a challenging task. The further one moves into the future, the more difficult it becomes to predict the precise value of one's state variables.[13] Although predicting precise values is difficult, one can typically do a better job of predicting the *distribution* of possible state-variable outcomes. Stating, for example, that the maximum temperature next Wednesday will be 21.5 °C is less likely to be successful than suggesting that it will fall in a range of 20–24 °C with the most probable value being 22 °C.[14]

This simple discussion has brought us quite naturally to the realm of statistics. Describing the dynamics of the state variables of a given system, therefore, amounts to constructing a statistical time-series model. To build such a statistical model, let's introduce some mathematical notation. Let us define $X_t \in \mathbb{R}^n$ as an $n$-dimensional vector of state variables with the following form,

$$X_t = \begin{pmatrix} X_{1,t} \\ X_{2,t} \\ \vdots \\ X_{n,t} \end{pmatrix}. \tag{6.1}$$

[13]The reason is simple: more can happen over a longer time horizon than over a shorter period.

[14]The first prediction is termed a point estimate, whereas the second prediction includes a confidence interval.

In Eq. (6.1) each of the state variables is indexed with a $t$. In many statistical settings, one is faced with a collection of random variables, say $Y_1, Y_2, \ldots, Y_n$, that are not indexed by time. Often this is because there is no relationship stemming from the order of the variables—they may be drawn from the same distribution, but their order does not matter.[15] In a time-series setting each random outcome is linked by time. Technically, we say that a time series or random process is a sequence of random variables. In simpler language, the state variable observed today is related to both yesterday's and tomorrow's values.

The temperature today is often closely related to the value from yesterday. The same applies to yield curves. Movements in the yield curve typically occur gradually over time.[16] The state variables of our yield-curve models should also exhibit this behaviour. To capture this statistically, we require two elements:

1. a slowing moving non-random trend; and
2. a component whose size and direction is uncertain (or random).

Splitting state-variable movements into random and non-random movements is an extremely useful and powerful idea.[17]

The non-random element is typically described relative to the long-term average values for the state variables. If, at any current time, the value of the state variables deviates from these long-term values, then it will be slowly pulled back to these values.

One might reasonably ask: once we get these levels, how do we ever get away from them again? This is where the random part comes in.[18] Rates may be, for example, very low and every one expects them to gradually rise over time. The non-random trend, therefore, would be working hard to force increases in the general levels of yields. The random component has no knowledge of this tendency; it is merely random. For any period, therefore, the random element could either lead to a faster increase or a further decrease of yields.

It is useful to actually write down how these two elements work together. Let $\{X_t, t \geq 0\}$ represent a sequence of vector-valued state variables. We can write

[15]This is not always the case, but it is rarely true with time series variables.

[16]In some periods, the overall level of interest rates is low, the yield curve is very steep, or short- and long-term rates are very close in value. These situations typically persist for significant periods of time and typically change only gradually.

[17]This idea is certainly *not* unique to yield-curve modelling, it is a foundational idea in the study of stochastic processes.

[18]This random part has many names: random noise, shocks, innovations, or diffusion. Each name tries to capture the notion of *unpredictability* in the movement of our state variables.

down how these state variables change, as a system, from one period to the next as,

$$\underbrace{X_t - X_{t-1}}_{\substack{\text{Change in}\\ \text{state variable}\\ \Delta X_t}} = \overset{\text{Non-random}}{\text{contribution}} + \overset{\text{Random}}{\text{contribution}}, \tag{6.2}$$

$$= \text{Drift} + \text{Diffusion}.$$

The important point from Eq. (6.2) is that the non-random and random—or, more technically, the drift and diffusion components—work together additively to describe the change in the state-variable vector over the interval, $[t-1, t]$.

One concrete example is,

$$X_t = \underbrace{C + FX_{t-1}}_{\text{Drift}} + \underbrace{\Sigma \nu_t}_{\text{Diffusion}}, \tag{6.3}$$

where $\nu_t \sim \mathcal{N}(0, I_n)$. Examining Eq. (6.3), observe that the current value of our state variables is a linear function of the previous value plus a normally distributed shock. This is a linear regression; more specifically it is a vector-autoregressive (VAR) model. While the specification of the dynamics for one's yield-curve state variables may be quite involved, Eq. (6.3) demonstrates that they need not be extremely complicated. Indeed, for the reader willing to brave the following optional discussion, we will demonstrate that Eq. (6.3) is actually consistent with a rather more involved approach.

The classical approach for the description of the state variables in a dynamic yield-curve model is a mathematical object termed a stochastic differential equation. As the name suggests, it is a differential equation with a stochastic component. Using our previous notation, we can write the infinitesimal dynamics of $\{X_t, t \geq 0\}$ as follows,

$$dX_t = \underbrace{\Phi(\theta - X_t)dt}_{\text{Drift}} + \underbrace{\Sigma dW_t}_{\text{Diffusion}}, \tag{6.4}$$

where $\theta \in \mathbb{R}^{n\times 1}$ denotes the long-term mean values of the state variables, $\Phi \in \mathbb{R}^{n\times n}$ is a matrix describing the speed that the system returns to these long-term means, and $\Sigma \in \mathbb{R}^{n\times n}$ is a matrix summarizing the volatility and correlation between the state variables.

The final element, $dW_t$, is the complicated part. It represents the infinitesimal changes in a so-called $n$-dimensional independent collection of Wiener processes, $\{W_t, t \geq 0\}$. To be technically correct, one needs to define a

(continued)

sample space and a filtration upon which our Wiener process resides. Given the already extremely loose nature of this discussion, we will dismiss with these (admittedly important) formalities. The curious reader is referred to the classic reference, Karatzas and Shreve [56].

While a Wiener process is a complicated animal, it does have some remarkable properties. One of which will prove very useful; an increment of a Wiener process, $W_t - W_s$ where $t > s$ is normally distributed as follows,

$$W_t - W_s \sim \mathcal{N}\left(\vec{0}, (t-s)I\right), \tag{6.5}$$

where $I \in \mathbb{R}^{n \times n}$ is an identity matrix. The point is that a Wiener increment has zero expected value and a volatility that is proportional to the square root of the passage of time.

Using this fact, we proceed to discretize Eq. (6.4). This means that we will, using a number of approximations, move from a continuous to a discrete representation of our state variable dynamics. This is termed an Euler discretization. When doing so, we arrive at

$$\begin{aligned} dX_t &= \underbrace{\Phi(\theta - X_t)dt}_{\text{Drift}} + \underbrace{\Sigma dW_t}_{\text{Diffusion}}, \qquad (6.6)\\ X_t - X_{t-1} &\approx \Phi(\theta - X_{t-1})\Delta t + \Sigma(W_t - W_{t-1}),\\ X_t &\approx \Phi\theta\Delta t + (I - \Phi)X_{t-1}\Delta t + \Sigma(W_t - W_{t-1}). \end{aligned}$$

This does not appear to be much progress. Let's thus do *three* things:

- normalize our time interval such that $\Delta t = 1$;
- simplify our notation by setting $C = \Phi\theta$ and $F = I - \Phi$; and
- make use of the property of the Wiener increment described in Eq. (6.5).

Applying these three points, we can simplify Eq. (6.6) as follows,

$$X_t \approx C + FX_{t-1} + \Sigma\nu_t, \tag{6.7}$$

where $\nu_t \sim \mathcal{N}\left(\vec{0}, I_n\right)$. This is precisely the form of the VAR representation in Eq. (6.3). We see that a very complex object—a specific stochastic differential equation called the Ornstein–Uhlenbeck process in Eq. (6.4)—can with a few assumptions be reasonably approximated with a rather more familiar and manageable object. The curious reader looking for more information on the topic of stochastic differential equations is referred to Oksendal [70].

### 6.2.3 $\mathscr{A}_3$

Having defined our state variables and described how they move through time, all that is missing is an explicit link between these state variables and the yield curve at each instant in time. We need to map, or transform, our state variables into a yield curve.

There is fortunately nothing terribly mysterious about this step. In all cases, the yield curve is simply a mathematical function of the *current* state variables. More specifically,

$$Y(t, T) = g(T, X_t), \tag{6.8}$$

for $T$ is the maturity (or tenor) of the zero-coupon bond. Armed with the function $g$ and a collection of tenors, say $T \in (0, 30)$, one can easily construct a yield curve for time $t$. The trick, of course, is finding a sensible choice for $g$.

The choice of $g$ may be either complicated or straightforward; it depends on the dynamic yield-curve model under discussion. Broadly speaking, however, yield-curve mappings fall into one of *two* categories:

- no-arbitrage; and
- empirically based.

The difference is significant. If you opt for the no-arbitrage approach, it implies that you are constructing a mapping between your state variables and the yield curve that precludes the existence of arbitrage opportunities. It might seem indeed rather odd to do otherwise. Choosing an no-arbitrage mapping does, however, reduce one's flexibility. The mapping must be derived from one's choice of state variables and their associated dynamics.[19] Although we may all agree that no-arbitrage mappings are desirable, this choice is not without practical challenges.

An empirically based mapping is a rather more flexible link between state variables and the yield curve. It is not selected to explicitly avoid arbitrage, but instead to fit the observed (i.e., empirical) behaviour of the yield curve. This does not immediately imply less complex mathematics, but one has rather more control over the final outcome.

There is also a reasonable amount of academic evidence suggesting that empirically-motivated mappings may actually outperform—or, at least, represent a challenge to—their no-arbitrage counterparts in out-of-sample forecasting of yield-curve movements.[20] Various hypotheses have been proposed for this out-

[19] One embarks on a sequence of mathematical computations and, generally after some heavy lifting, arrives with a potentially very complex mapping. It is also possible to write down choices of state variables and their dynamics for which no arbitrage-free mapping exists.

[20] See, for example, Christensen et al. [19], Diebold and Rudebusch [29], Bolder [8] and Bolder and Liu [9].

performance and, as usual, there is some conflicting evidence in the literature.[21] In the discussion that follows, we will examine both alternatives in some detail.

### 6.2.4 Bringing it All Together

The state variables determine (almost) everything in a dynamic term-structure model. First, we summarize the current, and complex, state of our yield curve system with a relatively modest number of state variables. Second, describing how the yield curve moves, boils down to describing the state-variable movements or dynamics. Finally, the current state of the yield curve is determined by a mapping, or transformation, of the current state variables; this mapping may either be no-arbitrage or empirically motivated.

Once we have selected sensible state variables, understand how they move through time, and have decided how to use them to make a yield curve, then we have succeeded in constructing a dynamic yield-curve model. There is thus no real conceptual mystery to a dynamic yield-curve model. The mathematics can be very complicated, but these three assumptions must always be present.

As a consumer of these models, you should always try to understand the inherent choices in any model. One can easily turn around these assumptions into three questions for anyone attempting to convince you that they have a sensible yield-curve model. One can always legitimately ask:

1. What are your choice of state variables?
2. How do you describe the movement of your state variables over time?
3. What specific type of mapping have you employed between your state variables and the yield curve?

The owner of the model should be able to easily and confidently answer these questions and defend his or her choices. If the owner cannot answer these questions or you find the response unconvincing, you will be well counselled to find another approach. Understanding these three elements is a useful tool to improving your consumption of yield-curve models.

It sounds relatively easy, doesn't it? Conceptually, it is relatively straightforward. In practice, however, there are a number of challenges. In the following discussion, we will examine *three* alternative examples to make these ideas more concrete. Before we do that, however, we will take a brief statistical digression.

---

[21] We will not resolve this question in this document. The point, however, is that there are reasonable arguments for each choice of mapping.

## 6.3 A Statistical Digression

Like the wind-chill factor and the humidity index in a weather-related setting, the most common state variables used to describe the yield curve are not directly observable and must be extracted from the data. These unobservable yield-curve state variables are so important that they merit an entire section for their discussion.

These state variables began as a remarkable result from the academic paper, Litterman and Scheinkman [62]. This paper demonstrated—using an interesting statistical technique—that *three* statistically determined factors actually describe a very large proportion of the variability in yield-curve movements. These statistical factors are naturally an excellent choice for one's state variables. These factors show up repeatedly, in different forms, in both practice and academic literature.[22]

The statistical technique employed by Litterman and Scheinkman [62] is termed principal component analysis (PCA).[23] Mathematically, the idea behind the approach is to find a subset of orthogonal (i.e., uncorrelated) factors that explain the maximum amount of variance in a system. PCA can transform a high-dimensional system with strong correlation and reduce it into a representative, uncorrelated, and generally smaller system.[24] As we learned in our introductory overview, the yield curve:

- is a system that varies by both tenor and time;
- has significant yield correlation among different tenors; and
- the state of the yield curve appears to persist for extended periods.

These characteristics make the yield curve uniquely suited for the application of PCA. To avoid treating this technique as a *black box*, we work through its basic derivation in the following shaded section. Readers uninterested in these details can simply skip forward to the application of PCA to our yield dataset.

Let's roll up our sleeves and examine the PCA technique. We will begin with a dataset and a simple idea. Imagine that $X \in \mathbb{R}^{d \times N}$ denotes a collection of data for a $d$-dimensional vector-valued random variable. In our case, $X$ will be 15 years of monthly yield-curve movements at tenors ranging from 1-month to

(continued)

[22]This chapter will be no exception.

[23]PCA was by no means developed for yield-curve applications, but is a powerful and more than 100-year-old technique used by statisticians and engineers for understanding data and reducing dimensionality.

[24]See Jolliffe [54] for much more detail on this technique.

20 years. We then define a real-valued vector, $a \in \mathbb{R}^{d \times 1}$ and modify $X$ as follows,

$$a^T X \in \mathbb{R}^{1 \times N}. \tag{6.9}$$

This is a linear projection, or transformation, of $X$ into a *single* dimension. We have projected our $d$-dimensional system into one dimension; this is termed a *projection*. For more on the notion of projections, see Huber [46], Chen [16] or Hall [38].

With one simple matrix multiplication, we have reduced a system with $d$ dimension down into single dimension. This is precisely what we are looking for in a state variable. It is nevertheless hard to believe that $a$, an arbitrary projection of $X$, will be interesting. How do we determine the *most* interesting projection? While there are a number of possibilities—we could seek the most noisy, non-linear, or non-Gaussian projection—PCA defines interesting as those projections that describe the most variability within the system. Variable and noisy projections are, from the perspective of the PCA technique, of the greatest interest.

Mathematically, the variance of the projection described in Eq. (6.9) is simply defined as,

$$\begin{aligned} \text{var}\left(a^T X\right) &= a^T \text{var}(X) a, \\ &= a^T \Omega a, \end{aligned} \tag{6.10}$$

where we define $\Omega$ as the variance-covariance matrix of our interest-rate dataset. What we are really interested in is finding the vector, $a$, that explains the *most* variance of $X$.

This is basically an optimization problem. We must find a choice of $a$ that maximizes Eq. (6.10) or more formally,

$$\begin{aligned} &\max_a a^T \Omega a \\ &\text{subject to } a^T a = 1. \end{aligned} \tag{6.11}$$

We constrain $a$ to have unit length—hence the condition, $a^T a = 1$—to make the result easier to interpret. With a bit of simple calculus, we find that the first-order conditions of the constrained problem in Eq. (6.11) are given as,

$$(\Omega - \lambda I) a = 0, \tag{6.12}$$

(continued)

where $\lambda$ arises from the constant in the method of Lagrange multipliers. This is a familiar problem in matrix algebra. The vector, $a$, that solves this equation is the eigenvector associated with the largest eigenvalue of $\Omega$. This eigenvector helps us to determine the first principal component or rather the projection of $X$ that explains the largest amount of variance of our yield-curve system.

Using the same logic, one may proceed to compute not only the projection explaining the most variance, but also the projection describing the second most variance and so on. The answer, of course, will be the eigenvector associated with the second largest eigenvalue and then the third and so on. Eigenvectors of covariance matrices like $X$ have a remarkably useful characteristic; they are orthogonal. This means that the variance described by the second principal component is not explained by the first principal component and vice versa. There is no overlap among the principal components, which is a desirable feature in a state variable–it implies that you are not double-counting any aspect of the dynamic system.

Let's compute the first five principal components associated with our US Treasury data. Table 6.1 illustrates the eigenvalue and the proportion of yield curve variance explained by each of the first three principal components.

What can we conclude from this analysis? The first principal component explains almost 90 % or, quite simply, the lion's share of the variance in yield-curve movements. The first three factors also describe virtually all of the variance of yield-curve movements—anything beyond the first three factors can be safely ignored. This is a remarkable result. Our statistical technique has reduced the dimension of the yield curve to *three* state variables.

**Table 6.1** Factor explanation

| PCA factor | Eigenvalue | Variance explained (%) |
|---|---|---|
| First | 24.16 | 88.66 |
| Second | 2.35 | 8.61 |
| Third | 0.58 | 2.13 |
| Fourth | 0.16 | 0.60 |
| Fifth | 0.00 | 0.00 |
| Total | n/a | 100 |

This table outlines the eigenvalues—multiplied by $10^4$ so as to be readable—and proportion of the variance of yield-curve movements explained by the first three principal components of $X$. The first factor explains a large amount of the variance in yield-curve movements.

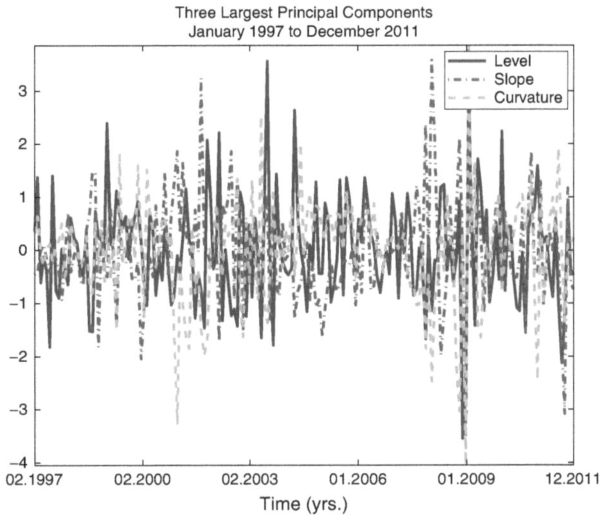


**Fig. 6.8** Principal components. Applying principal components analysis to our US Treasury yield-curve dataset, we arrive at the following three principal components

Table 6.1 is precisely the Litterman and Scheinkman [62] result, but there is one drawback with the use of principal components analysis.[25] Although we have extracted three orthogonal state variables explaining virtually all of the variance of yield-curve movements, we do *not* know what these factors are! They are just simply statistical factors. Figure 6.8 illustrates these statistical factors over our data period.

The principal components are indeed statistically created objects, but they are likely to be correlated with real-world elements. To shed some light on these principal components, we can examine the factor loadings. The factor loadings determine how each principal component impact each individual tenor in the yield curve. Figure 6.9 illustrates the factor loadings for each principal component.[26]

The factor loading of the first factor is essentially a straight line. This implies that the change in the yield curve for all tenors is basically the same. When the yields, for all tenors, change along the yield curve in the same way, we term this a parallel shift. It represents a wholesale movement in the level of the yield curve. For this reason, the first principal component is often termed the *level* factor. Interestingly, as evidenced by Table 6.1, about 90 % of the variance of yield-curve movements can be explained by parallel movements in the level of bond yields.

[25]This result is also very robust: it has been repeatedly reproduced using different markets and time periods.

[26]For those who have read the technical description of PCA, these are the eigenvectors associated with the three largest eigenvalues of $\Omega$.

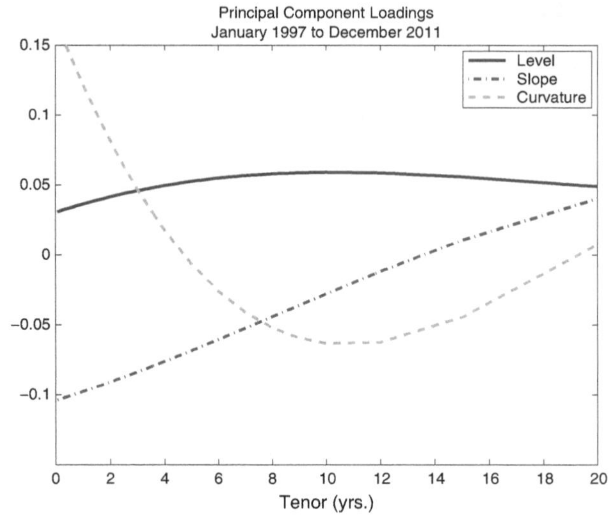


**Fig. 6.9** Factor loadings. Here we plot how the individual principal components affect the observed yield curves. These so-called factor loadings provide substantial insight into the relationship between the principal components and the yield curve

The second factor impact yield movements of short-term tenors in a different way than their long-term counterparts. If the short-end of the yield curve moves more (or less) than the long end of the curve, then the *steepness* of the yield curve will change. Accordingly, the second principal component is typically called the *slope*. Slope movements account for almost 10 % of the variance in yield-curve movements.

The final factor, describing about 2 % of the variance of yield-curve changes, is a bit more complicated. The factor loadings appear to be similar at short and long yield tenors, while it dips down for intermediate tenors. If long and short yields remain (approximately) fixed, while the middle part of the curve increases, then the yield becomes more curvy. The slope has not changed nor has the curve moved in a parallel fashion. Instead, the relative curvature or bend of the curve has increased. The third yield-curve factor is thus termed *curvature*.[27]

This is the origin of the so-called level, slope, and curvature factors that one often hears about during discussions of the yield curve. PCA analysis has thus assisted us in identifying three unobservable state variables that describe virtually all of the variation in yield-curve movements. This also makes them interesting candidates as yield-curve-model state variables. The underlying shaded section takes this a step further and demonstrates how we might approximate this statistical factors with linear combinations of observed bond yields.

[27]It is also occasionally called *butterfly*. The reason is that looking at the factor loading—and with a bit of imagination—one can see two wings at short and long tenors and body in the intermediate part of the curve.

To avoid the complex computations involved with PCA, we can use linear combinations of actual observed yields to proxy the level, the slope, and curvature of the curve. A natural proxy for the level of the yield curve is the 20-year yield (i.e, $Y_{20Y}$). The slope can be proxied with the difference between a long-term and a short-term yield; one typically uses the differences between the 20-year and 3-month yield (i.e., $Y_{20Y} - Y_{3M}$). Curvature is more complicated. We could measure curvature as the difference between:

- an intermediate rate and short-term rate; or
- an intermediate rate and a long-term rate.

Even better, we could use both. Imagine that we computed the sum of the these two elements: again using the 3-month and 20-year tenors for the short and long yields and the 5-year rate for the intermediate yield. This generates,

$$\begin{aligned} \text{Curvature} &\approx (Y_{5Y} - Y_{3M}) + (Y_{5Y} - Y_{20Y}), \\ &\approx 2 \cdot Y_{5Y} - (Y_{3M} + Y_{20Y}). \end{aligned} \tag{6.13}$$

This quantity will be large when there is substantial curvature in the yield curve and small when the curvature is modest; in the limit when the intermediate point sits exactly in between the long- and short-term rates, it will take a value of zero when the curve is flat (i.e., no curvature at all).

How do these proxies compare to the PCA factors from our dataset? The first principal component and the 20-year yield exhibit an almost perfect positive correlation of 0.96. This strongly supports the notion that our first statistical factor describe the general level of yields. The second PCA factor also exhibits very strong positive correlation of 0.83 with the slope of the yield curve underscoring its common name, the slope factor. Finally, the curvature also demonstrates a strong negative correlation of −0.66 with the linear proxy defined in Eq. (6.13). This is hardly problematic as only a negative sign separates these two state variables.

Armed with these definitions, one can readily employ these three proxies for the three principal components. Instead of performing complicated statistical analysis, therefore, one merely needs to compute linear combinations of yields.

## 6.4 Model Examples

Practice makes perfect. Examining some practical examples and extracting the three main assumptions will underscore the ideas presented in the first part of this chapter. We consider *three* alternative examples of a dynamic yield-curve model:

1. a (completely invented) toy model;
2. a very complicated model; and
3. a popular, and not so complicated, model.

Each choice has a purpose. The toy model introduces some key ideas with a minimum of mathematical overhead. The second, and complicated, choice introduces (as gently as possible) the mathematical difficulty possible in this area of finance. The final, less difficult, model is intended to provide the reader with an overview of a common and thoroughly implementable model.

### 6.4.1 A Toy Example

As the moniker *toy model* suggests, this is *not* a real model. It does nevertheless highlight some interesting ideas. The first step, of course, is to select our state variables. To make our life easy, we opt to employ the three PCA factors: level, slope, and curvature.[28] Let us define these three state variables as,

$$\begin{aligned} \text{Level:} &\quad \{l_t, t = 1, \ldots, N\}, \\ \text{Slope:} &\quad \{s_t, t = 1, \ldots, N\}, \\ \text{Curvature:} &\quad \{c_t, t = 1, \ldots, N\}, \end{aligned} \tag{6.14}$$

where we have $N$ yield-curve observations.

We may now proceed to determine the next two key model assumptions: $\mathscr{A}_2$ and $\mathscr{A}_3$. We will be relatively unimaginative and straightforward in our definition of the state-variable dynamics. We specify the state-variable dynamics as an independent system of auto-regressive ordinary least-squares (OLS) equations,

$$\begin{aligned} \text{Level } (l_t) \Rightarrow l_t &= \alpha_1 + \gamma_1 l_{t-1} + \varepsilon_{1,t}, \\ \text{Slope } (s_t) \Rightarrow s_t &= \alpha_2 + \gamma_2 s_{t-1} + \varepsilon_{2,t}, \\ \text{Curvature } (c_t) \Rightarrow c_t &= \alpha_3 + \gamma_3 c_{t-1} + \varepsilon_{3,t}. \end{aligned} \tag{6.15}$$

[28] Should we wish to avoid the statistical computations of the PCA factors, we could easily have proxied them with our straightforward linear combinations of observed yields.

These amount to three separate independent regression equations.[29] One could maybe do a little bit better by constructing a vector auto-regression, adding multiple lags, or a moving average term. This is nevertheless a toy model, so it's probably better to keep things as simple as possible.

With our choice of state variables and their associated dynamics, we need to map a given set of state variables into a yield curve. We will not use an no-arbitrage mapping.[30] Instead, we will describe the change in the bond yield as a linear combination of our three state variables. Reflecting this mathematically, we have

$$\Delta Y(t,T) = \underbrace{\beta_1 l_t + \beta_2 s_t + \beta_3 c_t}_{\substack{\text{Linear combination of} \\ \text{our state variables}}}, \tag{6.16}$$

where $T$ denotes the tenor of the yield. This does not look much like a mapping. With a bit of manipulation, we can expand it to

$$\begin{aligned} \overbrace{Y(t+1,T) - Y(t,T)}^{\Delta y(t,T)} &= \beta_1 l_t + \beta_2 s_t + \beta_3 c_t, \\ Y(t+1,T) &= \underbrace{Y(t,T) + \beta_1 l_t + \beta_2 s_t + \beta_3 c_t}_{f(t,T,l_t,s_t,c_t)}. \end{aligned} \tag{6.17}$$

The basic idea is that a bond yield is equal to its value in the previous period plus a linear combination of our three state variables. This mapping has three parameters, $\{\beta_i, i = 1,..,3\}$, that describe the importance of each of the state variables for each point along the yield curve. It not the most convincing mapping that you will ever see, but please remember, this is only a toy model.

The principal advantage of our toy model is its ease of comprehension—no complicated mathematics are involved and, should one wish, even the principal components analysis can be avoided. It is also straightforward to estimate—the model parameters, for our US Treasury dataset, are summarized in Table 6.2. With only *nine* parameters, one would be hard pressed to find a simpler dynamic yield-curve model.

Simplicity is clearly a virtue, but is this model any good? There are a few ways to measure the effectiveness of a model. We will examine the in-sample fit of the model. One could easily write another chapter about various techniques for

---

[29]Using independent regression equations is entirely reasonable given the uncorrelated (i.e., orthogonal) nature of the state variables. If one opted for the linear combination of yields, then a vector auto-regression might be preferable.

[30]The reason will become obvious once you see the complexity of the next example.

**Table 6.2** Toy model parameters

| Yield-curve factor | Mapping $\beta$ | Dynamics $\alpha$ | $\gamma$ |
|---|---|---|---|
| Level | 0.03 | 0.00 | 0.07 |
| Slope | −0.05 | 0.00 | 0.05 |
| Curvature | 0.05 | 0.00 | 0.02 |

This table summarizes the estimated dynamic and mapping parameters for our toy yield-curve model.

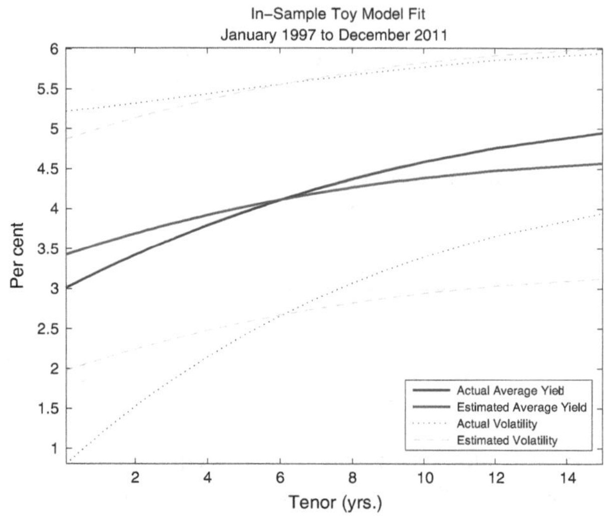


**Fig. 6.10** Average Toy-model fit. This figure outlines how well our toy yield-curve model fits the average US Treasury data and volatility in our dataset. Note that it neither captures the average shape of the US Treasury yield curve nor the dispersion around this average

judging the effectiveness of a yield-curve model.[31] Our objective in this chapter is nonetheless to help one better understand the working of yield-curve models. As such, we take a relatively limited view of a model's fit to the historical data used to estimate it.

In-sample fit is, quite simply, a measure of how well your yield-curve mapping describes the observed historical data used to estimate it. Figure 6.10 describes how successfully our yield-curve model fits on average, after optimization, the observed yields and their volatility.[32] It plots the observed average yield curve over our data sample relative to the estimated yield-curve over the same sample period. It also show the volatility of the actual and estimated curves. While our estimated curves

[31]Better measures examine the capacity of the model to fit the observed historical data out of sample. This essentially amounts to predicting future interest-rate outcomes. As the thinking goes, if a model can predict future outcomes relatively well, then it is probably a good model. See Duffee [32] for more on this idea.

[32]How is this done? One take the state variables for each period and applies the mapping Eq. (6.17) to it. This provides an estimated yield curve, which is immediately compared to the actual yield curve.

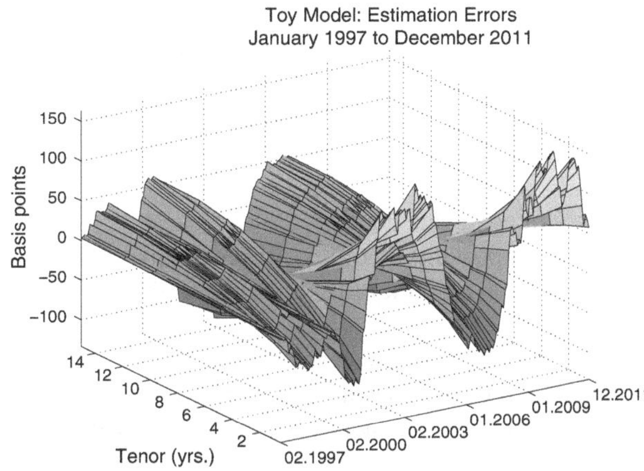


**Fig. 6.11** Global Toy-model fit. This figure displays, in basis points, the difference between observed US Treasury yields and the in-sample values suggests by our toy model. Note that, in some periods, the differences exceed 100 basis points

does a reasonable job of describing the general levels of interest rates, it seems to have difficulty capturing the slope of the yield curve and appears to misrepresent overall yield volatility.

Figure 6.11 takes out our examination of the in-sample model fit one step further. It shows the estimation error along the dimensions of yield, tenor and time. In some periods, our toy model misrepresents the actual yield curve by more than 100 basis points in both directions. Sadly, our toy model is not quite ready for mass consumption and application. Publication of our toy model in a leading finance journal, however, was never our objective. The goal was to highlight the three key dynamic yield-curve assumptions with a minimum of complexity. In the remaining sections, we turn to examine real dynamic yield-curve models and try to keep these lessons in mind. It is also interesting that with an extremely naive set of model assumptions, we have not tabled a completely ridiculous dynamic yield-curve model.[33]

### 6.4.2 A Complex Example

We move directly from an easily accessible toy model to a decidedly more complex version of the dynamic yield-curve model. The classic dynamic term-structure model is the so-called *affine* model. Historically, this model originated from option-pricing theory. Vasicek [78] basically took the Black–Scholes framework and cleverly twisted it to fit interest rates and, in doing so, created the class of so-called the *affine* yield-curve models.[34]

[33]This is not to say that it is going to win any awards.

[34]Vasicek [78] is a landmark finance paper that thoroughly warrants reading. The Black–Scholes idea was to create a replicating portfolio for an option. Vasicek [78] noticed that a zero-coupon

In a few short pages, we hope to delve into the world of affine yield-curve models. This is not nearly enough space to do this important topic justice, but it will hopefully suffice to stress our *three* key assumptions and demonstrate that these complex models also fall into our conceptual framework. To be successful, we will need three main ingredients:

1. a new interest-rate concept;
2. a (very important) theorem from mathematical finance; and
3. a non-trivial amount of trust in faith in the author .

The final ingredient is necessary because we will be skipping the majority of the mathematical derivations and all of the rigour.[35] This is not a chapter on graduate-level mathematics. As such, I hope that you will trust that the following results are true and correct.

The first step is to define our state variable. For an affine yield-curve model, this requires the definition of a new interest-rate concept. We wish to make use of the calculus—this includes both the regular, deterministic, variety and the stochastic calculus. Both types of calculus operate on infinitesimally small time intervals. We need a fixed-income contract that spans an infinitely short horizon. In the real world, interest-rates have a finite time horizon—1 year, 1-month or 1 week. Pushing this to the very limit brings us to the overnight loan, which is the shortest possible interest-rate contract.

Although this is progress, overnight contracts are still *not* short enough. We require something infinitely short like,

$$r(t) = \lim_{T \to t} z(t, T). \tag{6.18}$$

This strange object, which of course does not really exist, is the rate of interest (or yield) demanded for a loan lasting an infinitesimally short period of time. It is called the instantaneous short rate and is merely a theoretical construct.[36] Equation (6.18) opens up an entire world of mathematical techniques that would otherwise be unusable. More importantly, the instantaneous short rate ($r(t)$) is the state variable for the affine yield-curve model.[37]

---

bond is merely a kind of contingent claim (i.e., option) on an interest rate and adjusted the overall framework accordingly.

[35]The curious, and masochistic reader, is referred to James and Webber [52], Brigo and Mercurio [14] or Bolder [7] and the excellent references they contain.

[36]Like other non-existent theoretical concepts such the notion of perfect competition in microeconomics, it is quite useful.

[37]If the instantaneous short rate actually existed, then an investment of $K$ units of currency in a bank account, $B$, over $[t, T]$ would return,

$$B(t, T) = Ke^{\int_t^T r(u)du}. \tag{6.19}$$

Moving on to the second ingredient, we introduce a result that underpins modern asset-pricing theory. It requires a few definitions and substantial explanation. We start with a contingent claim, which is a financial contract whose value depends on some other financial variable.[38] Imagine now that you have a contingent claim, $\phi(\cdot)$, whose value is determined at some future time, $T$. There is a financial theorem stating that the value of this claim at time, $t$, can be written as,

$$\phi(t) = \mathbb{E}^{\mathbb{Q}}\left(e^{-\int_t^T r(u)du}\Phi(T)\bigg|\mathscr{F}_t\right) \tag{6.20}$$

This is often termed the fundamental theorem of asset pricing.[39] While this expression is a bit intimidating, we must notice that it includes our state variable, $r(t)$. Equation (6.20) holds that the current value of the contingent claim (i.e., $\phi(t)$) is merely equal to its discounted expected future pay-off. That's all. The future expected pay-out is the expectation of $\Phi(T)$ and the discount factor is the integral of the instantaneous short rate (i.e., $e^{-\int_t^T r(u)du}$) over the contract's remaining life, $[t, T]$.

This result becomes really interesting when we apply it to a pure-discount bond. A bit of refection reveals that a pure-discount bond is actually a contingent claim on an interest rate—as interest rates change, so does the value of the pure-discount bond. Applying our theorem, the current value of $P(t, T)$ can be written as its expected future pay-off (i.e., $P(T, T)$ or \$1) discounted back to the present. Mathematically, we have

$$\begin{aligned} P(t,T) &= \mathbb{E}^{\mathbb{Q}}\left(e^{-\int_t^T r(u)du}\underbrace{P(T,T)}_{=1}\Bigg|\mathscr{F}_t\right), \qquad (6.21)\\ &= \mathbb{E}^{\mathbb{Q}}\left(e^{-\int_t^T r(u)du}\bigg|\mathscr{F}_t\right). \end{aligned}$$

Since there is no uncertainty about the final pay-off of a pure-discount bond—it always returns one unit of currency \$1 (or $P(T, T)$)—the value of the pure-discount bond depends on the expected path of the discounted factor. We now have a

---

The cumulative return on one's bank account is merely the integral of the instantaneous short rate over the period. Conversely, if you wish to determine the discount rate over a given period, one need only use the negative of the integral in Eq. (6.19): $e^{-\int_t^T r(u)du}$. This may seem a bit crazy, but it avoids the unwieldy geometric sums typically employed for computing fixed-income returns.

[38] An option is a good example; the value of a commodity option on corn, for example, depends importantly on the underlying value of corn.

[39] This short expression warrants significant explanation, but it is quite technical. $\mathbb{Q}$, for example, is termed the equivalent martingale measure induced by using the money-market account as the numeraire asset. The still interested reader is referred to a broad mathematical finance literature. Some good starting points include Panjer et al. [71], Karatzas and Shreve [57], Duffie [31], Neftci [67], Musiela and Rutkowski [66] and Bjork [6].

mathematical expression that links our state variable, $r(t)$, to any arbitrary pure-discount bond. We have a link between the instantaneous short rate and the whole yield curve. This is the foundation upon which the affine yield-curve model is constructed.

Having established that the yield curve depends on the expected evolution of the instantaneous short rate, we can alternatively state that we have our mapping between the yield curve and the state variable. The first and third fundamental yield-curve assumptions have been established.

In early models, such as Vasicek [78], the instantaneous short rate was a simple, single-dimensional variable. This did the job, but restricted the capacity of the model to capture the richness of actual yield-curve movements.[40] In modern affine yield-curve models, the short rate remains the state variable, but it is assumed to be a linear combination of three factors,

$$r(t) = \sum_{i=1}^{3} \delta_i X_{i,t}. \tag{6.22}$$

There are not observable state variables. They must be extracted from observed yield-curve data, but the extraction and estimation of the mapping occurs in a single step.

The final assumption relates to the dynamics of our state variable, $r(t)$. These are typically described by the following stochastic differential equation,

$$dX_t = \alpha(t, X_t)dt + \sigma(t, X_t)dW_t. \tag{6.23}$$

There are naturally a number of restrictions on the form of the coefficients, $\alpha$ and $\sigma$.

We have arrived at the point of the discussion, where faith and trust are required. A wide range of loose ends need to be tied up and computations are necessary. One needs to solve Eq. (6.23) for $X_t$. One then uses the definition of the state variable and the mapping in Eq. (6.21) to actually solve the expectation for the form of the pure-discount bond-price function. With some additional assumptions, several pages of horrible computations,[41] and lots of patience, you will arrive at the following yield-curve mapping,

$$Y(t,T) = A(t,T) + B(t,T)^T X(t). \tag{6.24}$$

$A(t,T)$ and $B(t,T)$ are complicated, but non-random, functions. This is our mapping, which ensures the absence of arbitrage. The state variable is a linear combination of factors summing to the instantaneous short rate, these state variables

[40] Recall that we need *three* factors to fully explain the variance of yield-curve movements.

[41] See Bolder [8] for the gory details.

follow a stochastic differential equation, and the yield curve can be written as an, admittedly complicated, linear function of the state variables.[42]

Estimating an affine model is also, sadly, a complicated affair. There are two principal challenges with estimating these models. First, a multivariate affine yield-curve model has a relatively large number of parameters and many of these parameters enter into the model in a non-linear manner.[43] Finding the optimal parameters requires the use of non-linear optimization techniques. While there exist many possible approaches for non-linear optimization, relatively few of them can guarantee that one finds the global minimum.[44] The second issue is that the estimation of the model must simultaneously

- extract the unobserved state variables; and
- estimate the model parameters.

This is a tall order and does not make for straightforward implementation. It can be done, but it is not simple and could easily consume one, or more chapters, on its own. Indeed, the list of references for estimating classical no-arbitrage yield-curve models is about as depressingly long as the collection of models. The following papers are but a short sample: Abken [1], Babbs and Nowman [2], Backus et al. [4], Chan et al. [15], Chen and Scott [18], de Jong [25], Geyer and Pichler [37], Honoré [44], Jeffrey et al. [53], Linton et al. [61], Lund [65], Nowman [69], Duan and Simonato [30], and Pearson and Sun [72].

There is a point to this intimidating list of references on statistical estimation of these models. Returning to our ultimate perspective as a consumer of dynamic yield-curve models, anyone depending on an in-house or external affine model would be well advised to question and gain confidence about the estimation technique employed. If your internal analysts or external vendors cannot provide you with clear and convincing answers, then you would be well advised to reconsider your modelling arrangements. It is also reasonable to ask if there exist simpler alternative models.

Sparing the reader the details, we have proceeded to estimate a three-factor dynamic affine term-structure model to our US Treasury dataset.[45] Figure 6.12 highlights the three state variables extracted by the estimation algorithm. The three affine state variables are typically difficult to interpret. Very often they are related to the level, slope, and curvature factors stemming from principal components

[42]To complete the story, *affine* is an old-fashioned mathematical term used to describe a linear function. Given the simple linear form of the mapping in Eq. (6.24), this is the genesis of the name *affine* model.

[43]There is fortunately a burgeoning literature seeking to ease the estimation of these models through the use of linear regressions. See Diez [26] for more details.

[44]Typically, the ability to find a global minimum for a non-linear optimization problem depends on the characteristics of one's objective function. Rarely does the objective function used in the estimation of affine yield-curve models exhibit these characteristics.

[45]Again, the interested reader is referred to Bolder [8] for more detail.

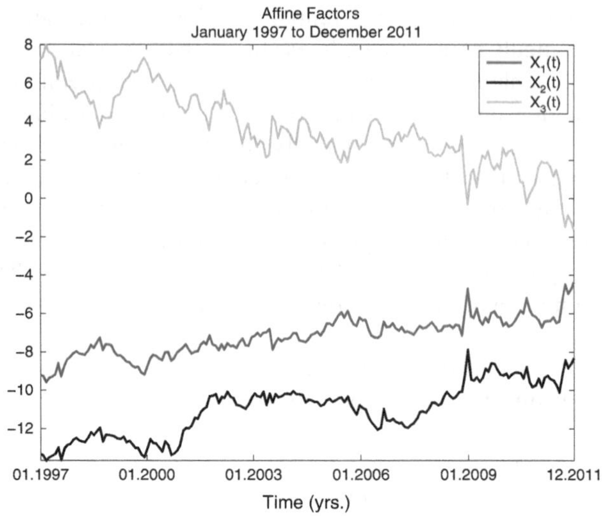


**Fig. 6.12** Affine factors. The estimation algorithm for the affine model extracts the implied factors associated with the yield curve at each point in time. This figures illustrates the evolution of these unobserved affine *state variables* over the length of our dataset

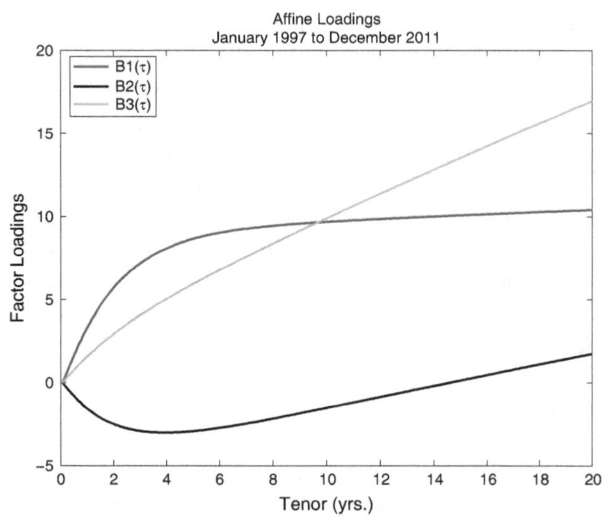


**Fig. 6.13** Affine factor loadings. This figures outlines the loadings for each of the three affine-model factors. Observe that, while not identical, they do have a certain resemblance with the principal component loadings

analysis. The reason is that these factors, explaining virtually all of the variance of yield-curve movements, are a fundamental element of yield-curve dynamics. Almost all dynamic yield-curve models incorporate these factors in one way or another and affine models are no exception.

The extent of this link to the principal components can be seen more clearly in Fig. 6.13, which illustrates the factor loadings for the affine state variables or factors. Recall that the factor loadings explain how the individual factors impact the various tenors along the yield curve. The affine factors need not be so cleanly defined and ordered by relative importance as with the principal components. This is because the optimization algorithm is not constructed to do so.

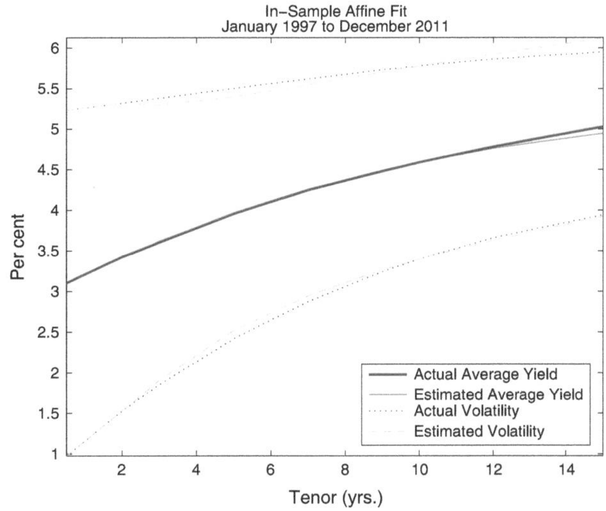


**Fig. 6.14** Average affine fit. This figure outlines how well our three-factor affine yield-curve model fits the average US Treasury data and volatility in our dataset. Note that it quite closely captures the average in-sample behaviour of the US Treasury yield curve

What can we glean from Fig. 6.13? The third factor loading looks suspiciously like the slope factor in Fig. 6.9. It effects short- and long-tenor bond yields in a different ways leading to changes in the relative steepness of the yield curve—in short, this appears to be a slope factor. The first and second factors are less obvious. If we add these two factor loadings together, however, we arrive at something close to a straight line that impacts all yields in, more or less, the same manner. This would be a level factor. It appears, therefore that the first and second factors work together to form a single level factor.[46] Given that the optimization algorithm is under no obligation to create orthogonal, or uncorrelated, state variables, this is not so surprising. Given that the curvature factor has a limited capacity to explain yield-curve variance, it is not surprising that it doesn't play such a prominent role.

To complete our examination of the affine yield-curve model, we examine the in-sample fit to our yield-curve data. Figure 6.14 outlines, in precisely the same way as with our toy model, the average fit of the estimated curve to the observed yield curves over our 15-year data sample.

Relative to our toy model, the affine model provides a precise fit the in-sample yield-curve data. While it perhaps demonstrates a bit of difficulty in fitting the long end of the curve, the average effect looks to be only a few basis points.

Figure 6.15 examines the fit across each individual yield curve over the sample period. The largest deviations from the observed curve look to fall into a range of plus or minus 20-odd basis points.[47] The model seems to have a reasonable amount

[46]This is further supported by the fact that the parameters on these two factors—$\delta_1$ and $\delta_2$ from Eq. (6.22)—are similar in magnitude, but have the opposite sign.

[47]At some tenors, the estimated curve exactly fits the observed curve across the entire data sample. The estimation algorithm permits the model to exactly fit one tenor for each state variable. With three state variables, this implies that three yield tenors are fitted perfectly.

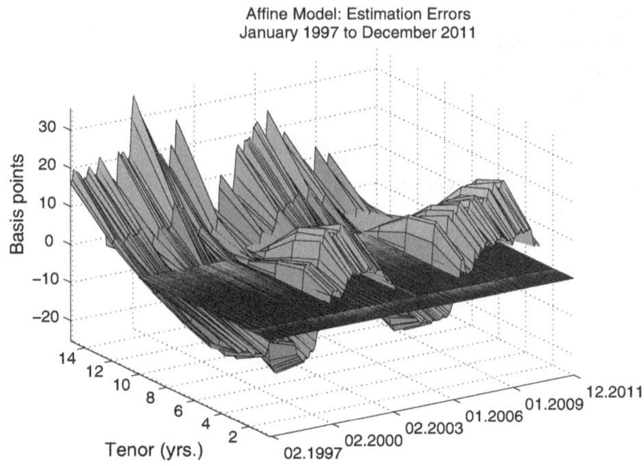


**Fig. 6.15** Global affine fit. This figure displays, in basis points, the difference between observed US Treasury yields and the in-sample values suggested by the three-factor affine model. The differences rarely exceed about 20 basis points

of difficulty fitting the short-end of the curve. This effect is hidden in Fig. 6.14 because, on average, the over- and under-estimates of the observed curve appear to cancel out.

While the class of affine yield-curve models is complex, they are no exception to our three-assumption framework.[48] The instantaneous short rate is the state variable and represented as the sum of three unobserved factors. These factors have dynamics described by a vector-valued stochastic differential equation. The mapping, derived from first principles to avoid arbitrage, is a linear function of the three state variables. The underlying mathematics of the model and the econometrics required to estimate it are good reasons to be intimidated by affine yield curve models.[49] Conceptually, however, given our three-assumption framework, there is no reason to be concerned about this class of dynamic yield-curve model.

### 6.4.3 A Simpler Example

The first example was a bit too simple, whereas the second was probably too complicated. Can we make our life simpler, but maintain a reasonable description of yield-curve dynamics? The short answer is yes. We offer an alternative that involves substantially less overhead and can be used in risk and performance computations. This general approach has found substantial favour in recent years from individuals seeking to simplify their lives.

[48]There are many different flavours of affine term-structure models. Dai and Singleton [22, 23] provide, in their seminal papers, a definitive specification of the various types of affine model. Duffie [33], and Duffie et al. [34], and Cox et al. [20, 21] are also excellent references.

[49]We've been almost criminally brief and imprecise in the development, but hopefully there is enough here to form an idea about the approach.

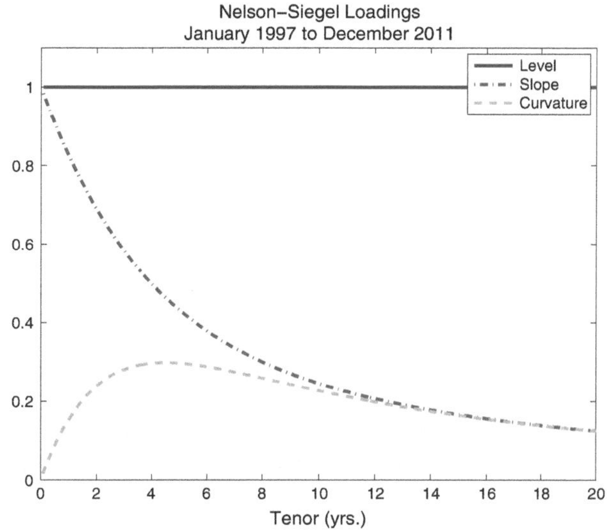


**Fig. 6.16** Nelson–Siegel factor loadings. This figures outlines the first three Laguerre polynomials; these are, in fact, the Nelson–Siegel factor loadings. While not identical, they demonstrate a strong resemblance with the principal component loadings

We begin with a bit of history. Nelson and Siegel [68] wrote a paper where they proposed the following yield-curve function,

$$y(t,T) = a + b\left(\frac{1-e^{-\lambda(T-t)}}{\lambda(T-t)}\right) + c\left(\frac{1-e^{-\lambda(T-t)}}{\lambda(T-t)} - e^{-\lambda(T-t)}\right), \quad (6.25)$$

where $a, b, c,$ and $\lambda$ are parameters. At first glance, it does not seem terribly exciting. Technically, it is a collection of exponential functions. In fact, the first *three* Laguerre polynomials. This was not a haphazard choice. Laguerre polynomials, part of a family of orthogonal polynomials, are often used in engineering applications for function approximation.[50]

The Nelson–Siegel model was originally intended to help fit the yield curve at a given point in time. Indeed, we used this model in the previous chapter to fit Belgian sovereign and US Agency yield curves. For the first 15-odd years of this model's life, it was used to fit yield curves at a given point in time.[51]

It would have likely stayed this way if Diebold and Li [28] had not made a useful and clever observation in an interesting, influential, and path-breaking academic paper. They recognized the link between the Laguerre polynomials and the first three principal components. Figure 6.16 illustrates the individual components of the Nelson–Siegel model in Eq. (6.25) when plotted against tenor—these are, in fact, a type of factor loading.

[50]See Hurn et al. [51] for much more detail.

[51]It was used extensively by central bankers and even extended by Svensson [77], because it provided a sensible, parsimonious description of the yield curve.

Examination of Fig. 6.16 reveals a relationship between yield tenor and factor loadings that should by now look quite familiar. The factor loadings can be classified as having a level, slope, and curvature effect in a manner almost analogous to the PCA factor loadings. We see again why any discussion of yield-curve dynamics would be incomplete without the PCA factors.

Having made this important observation, Diebold and Li [28] proceeded to slightly modify Eq. (6.25) as follows,

$$y(t,T) = a_t + b_t\left(\frac{1-e^{-\lambda(T-t)}}{\lambda(T-t)}\right) + c_t\left(\frac{1-e^{-\lambda(T-t)}}{\lambda(T-t)} - e^{-\lambda(T-t)}\right), \tag{6.26}$$

where $\{a_t, b_t, c_t, t = 1, \ldots, N\}$ are now linked to time.[52] This slight modification transformed a static curve-fitting technique into a dynamic yield-curve model. The parameters of the model, once static and now dynamic, are the state variables of the Diebold and Li [28] model. That is, the state variables are the time-varying parameters,

$$\{a_t, b_t, c_t, t = 1, \ldots, N\} \tag{6.27}$$

To obtain these state variables, one need only fit the Nelson–Siegel model to one's yield-curve for every date over one's sample period. We have done precisely this with our US Treasury dataset. Figure 6.17 illustrates the estimated coefficients, $\{a_t, b_t, c_t, t = 1, \ldots, N\}$, for each date in our sample period. These are the state variables for the so-called Nelson–Siegel model.[53]

Reviewing Eq. (6.26), we see not only the choice of state variables, but also the mapping between them and the yield curve. Equation (6.26) provides both the first and third assumptions in our conceptual framework.

The only missing element, for a complete description of this model, are the yield-curve dynamics. Here one has a bit of flexibility. Generally speaking, a very simple approach is employed where the state-variable dynamics are described by a vector auto-regression as,

$$\begin{bmatrix} a_t \\ b_t \\ c_t \end{bmatrix} = \begin{bmatrix} c_0 \\ c_1 \\ c_2 \end{bmatrix} + \begin{bmatrix} F_{1,1} & \cdots & F_{1,3} \\ \vdots & \ddots & \vdots \\ F_{3,1} & \cdots & F_{3,3} \end{bmatrix} \begin{bmatrix} a_{t-1} \\ b_{t-1} \\ c_{t-1} \end{bmatrix} + \begin{bmatrix} \varepsilon_{a,t} \\ \varepsilon_{b,t} \\ \varepsilon_{c,t} \end{bmatrix}, \tag{6.28}$$

$$X_t = C + FX_{t-1} + \varepsilon_t.$$

[52] $\lambda$, given its non-linear form, is typically fixed and forgotten about.

[53] By all rights, of course, this should be called the Diebold–Li model for all of their hard work and cleverness. Nevertheless, life is not always fair, and this model is predominately termed the Nelson–Siegel model.

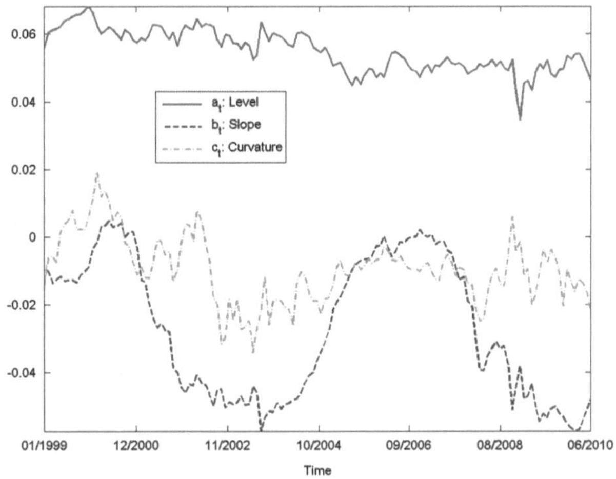


**Fig. 6.17** Nelson–Siegel factors. If one estimates the Nelson–Siegel model for each month of our dataset and collects the results, one will have the set of state variables for the Nelson–Siegel model. This figures illustrates the results of performing just such an exercise

A vector auto-regression is a relatively simple and flexible statistical model that essentially assumes that the conditional distribution of the state variables is Gaussian,

$$X_t \mid X_{t-1} \sim \mathcal{N}\left(C + FX_{t-1}, \text{var}(\varepsilon_t)\right), \tag{6.29}$$

Simply put, the current state variables are a linear function of *previous* state variables plus some normally distributed noise.[54] One could also use a collection of independent regression equations, as with our toy model, but this would ignore the correlations between our state variables. Although the three factors resemble the first three principal components, they were not constructed to be orthogonal. As such, capturing their relative correlation with a vector auto-regression is likely to provide a better description of their dynamics.

To round up our discussion of the Nelson–Siegel model, we need to examine its in-sample fit. Figure 6.18, therefore, demonstrates how well our yield-curve model fits on average, after optimization, the observed yields. The overall average performance seems essentially similar to the fit exhibited by the three-factor affine model examined in the previous section of this chapter. On average the model provides a sensible in-sample fit. This should hardly be a surprise given that it was originally intended to do precisely that: fit the yield curve at a given point in time.

Figure 6.19 provides a final look at the estimation errors over time. Again, the overall fit, ranging from plus or minus about 25 basis points, looks quite acceptable.

The Nelson–Siegel dynamic yield curve model, proposed by Diebold and Li, offers a relatively straightforward and robust alternative to the theoretically more complete, but more complex, affine yield-curve models. It also fits easily into our

[54]These are flexible models used extensively in practice. Judge [55], Harvey [40] and Hamilton [39] are excellent references on this topic (among other things).

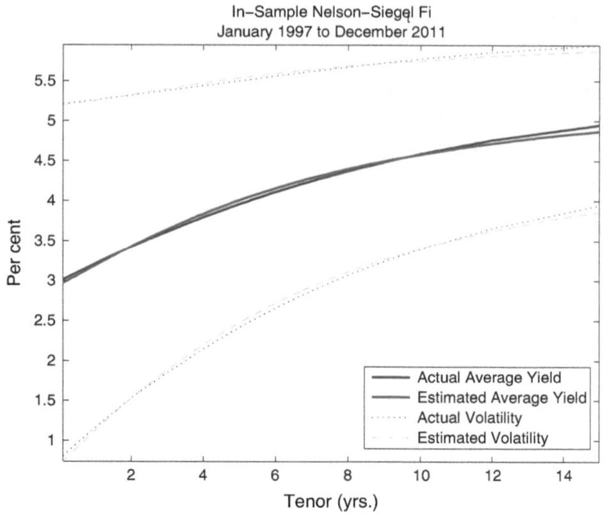


**Fig. 6.18** Average Nelson–Siegel fit. This figure outlines how well our Nelson–Siegel yield-curve model fits the average US Treasury data and volatility in our dataset. As with the affine model, it closely mirrors the average in-sample behaviour of the UST yield curve

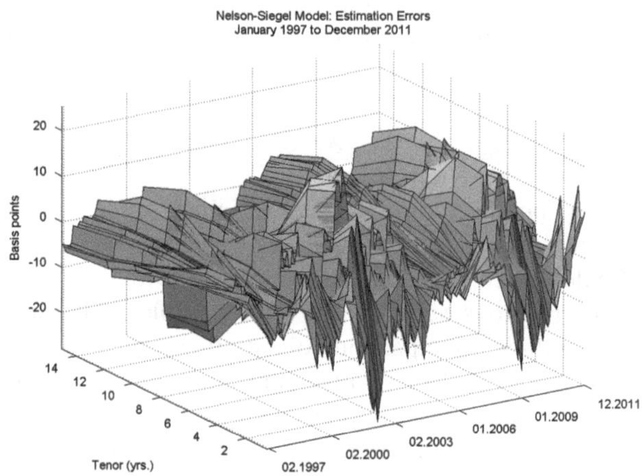


**Fig. 6.19** Global Nelson–Siegel fit. This figure displays, in basis points, the difference between observed US Treasury yields and the in-sample values suggests by the Nelson–Siegel model. Note that, once again, the differences rarely exceed about 20 basis points. One may conclude that the in-sample fit is quite similar between the three-factor affine and Nelson–Siegel models

conceptual three-assumption framework. The state variables and the associated mapping are both specified by a slight modification to the original Nelson–Siegel yield equation. The state-variable dynamics are typically described using a vector auto-regression. At the cost of foregoing a no-arbitrage mapping between the yield curve and the state variables, one obtains a reasonable alternative.[55]

[55]It is also possible to adjust the Diebold–Li model to preclude arbitrage opportunities. See Diebold and Rudebusch [29] for more details.

## 6.5 Concluding Thoughts

The widespread use of these models in performance and risk computations and the fundamental role of the yield curve in all fixed-income portfolios makes this an important area of study. Although this is a technical subject, we hope to have demonstrated—with ample examples, graphics and a conceptual framework—that it is not entirely inaccessible. It will still remain the domain of quantitative analysts, but it is too important to ignore. This chapter has hopefully helped you to become a better *consumer* of these models.

There are three key assumptions involved in the construction of virtually every yield-curve model. Each of these three assumptions is critically important and include:

1. a collection of observable and/or unobservable state variables;
2. a description of the dynamics of these state variables; and
3. a mapping between these state variables and the yield curve, which may be theoretically motivated and constructed so as to avoid arbitrage opportunities or constructed solely based upon empirical considerations.

These three assumptions lead logically to the construction of a dynamic yield-curve model. Constructing a *sensible* and *useful* model, however, is rather more difficult as we saw through our experience with a toy model.

Table 6.3 closes the chapter with a summary of the differences and similarities between the three examples presented as a chapter. We hope this will serve as a useful summary of our conceptual framework and a reminder to employ these ideas—without forgetting the requisite *three* key assumptions—whenever assessing a dynamic yield-curve model.

**Table 6.3** Model comparison

| | Toy model | Affine model | Nelson–Siegel |
|---|---|---|---|
| $\mathscr{A}_1$ | The first three principal components: $l_t, s_t$, and $c_t$ | Three unobserved factors that sum to the short rate: $r(t) = \sum_{i=1}^{3} X_{i,t}$ | The three parameters from the Nelson–Siegel model: : $a_t, b_t$, and $c_t$ |
| $\mathscr{A}_2$ | Use of three independent linear regressions | A multidimensional stochastic differential equation | A vector auto-regression |
| $\mathscr{A}_3$ | A bit silly: $y(t,T) = \gamma \ln(T-t) + \beta_1 l_t + \beta_2 s_t + \beta_3 c_t$ | Complicated, but no-arbitrage: $y(t,T) = A(t,T) + B(t,T)^T X_t$ | From model, but permits arbitrage: $y(t,T) = a_t + b_t \left(\frac{1-e^{-\lambda(T-t)}}{\lambda(T-t)}\right) + c_t \left(\frac{1-e^{-\lambda(T-t)}}{\lambda(T-t)} - e^{-\lambda(T-t)}\right)$ |

This table outlines, along our three main assumptions, the main differences and similarities between the three example models considered in this chapter.

## References


1. P.A. Abken, Innovations in the modeling of the term structure of interest rates. Fed. Reserv. Bank Atlanta Econ. Rev. **10**, 2–27 (1990)
2. S.H. Babbs, K.B. Nowman, Kalman filtering of generalized vasicek term structure models. J. Financ. Quant. Anal. **34**, 115–130 (1999)
3. D. Backus, S. Foresi, C. Telmer, *Discrete-Time Models of Bond Pricing*. Graduate School of Industrial Administration, Carnegie Mellon University (1998)
4. D. Backus, C. Telmer, L. Wu, *Design and Estimation of Affine Yield Models*. Graduate School of Industrial Administration, Carnegie Mellon University (1999)
5. C.A. Ball, W.N. Torous, Unit roots and the estimation of interest rate dynamics. J. Empir. Financ. **3**, 215–238 (1996)
6. T. Björk, *Arbitrage Theory in Continuous Time*, 1st edn. (Oxford University Press, Oxford, 1998)
7. D.J. Bolder, Affine term-structure models: theory and implementation. Bank of Canada: Working Paper 2001-15 (2001)
8. D.J. Bolder, Modelling term-structure dynamics for portfolio analysis: a practitioner's perspective. Bank of Canada: Working Paper 2006-48 (2006)
9. D.J. Bolder, S. Liu, Examining simple joint macroeconomic and term-structure models: a practitioner's perspective. Bank of Canada: Working Paper 2007-49 (2007)
10. D.J. Bolder, G. Johnson, A. Metzler, An empirical analysis of the canadian term structure of zero-coupon interest rates. Bank of Canada: Working Paper 2004-48 (2004)
11. G.E.P. Box, N.R. Draper, *Empirical Model-Building and Response Surfaces* (Wiley, New York, 1987)
12. A. Brace, D. Gatarek, M. Musiela, The market model of interest rate dynamics. Math. Financ. **7**, 127–155 (1997)
13. M.J. Brennan, E.S. Schwartz, A continuous time approach to the pricing of bonds. J. Bank. Financ. **3**, 133–155 (1979)
14. D. Brigo, F. Mercurio, *Interest-Rate Models: Theory and Practice*, 1st edn. ( Springer, Berlin, 2001)
15. K.C. Chan, G.A. Karolyi, F.A. Longstaff, A.B. Sanders, An empirical comparison of alternative models of the short-term interest rate. J. Financ. **XLVII**, 1209–1227 (1992)
16. H. Chen, Estimation of a projection-pursuit type regression model. Ann. Stat. **19**(1), 142–157 (1991)
17. R.-R. Chen, A two-factor preference-free model for interest rate sensitive claims equilibrium model of the term structure of interest rates. J. Futur. Mark. **15**, 345–372 (1995)
18. R.-R. Chen, L. Scott, Maximum likelihood estimation for a multifactor equilibrium model of the term structure of interest rates. J. Fixed Income **3**, 14–31 (1993)
19. J.H.E. Christensen, F.X. Diebold, G.D. Rudebusch, The affine arbitrage-free class of Nelson–Siegel term structure models. J. Econ. **164**, 4–20 (2011)
20. J.C. Cox, J.E. Ingersoll, S.A. Ross, An intertemporal general equilibrium model of asset prices. Econometrica **53**, 363–384 (1985)
21. J.C. Cox, J.E. Ingersoll, S.A. Ross, A theory of term structure of interest rates. Econometrica **53**, 385–407 (1985)
22. Q. Dai, K.J. Singleton, Specification analysis of affine term structure models. J. Financ. **55**, 1943–1978 (2000)
23. Q. Dai, K.J. Singleton, Expectation puzzles, time-varying risk premia, and dynamics models of the term structure. J. Financ. Econ. **63**, 415–441 (2002)
24. Q. Dai, A. Le, K.J. Singleton, Discrete-time dynamic term structure models with generalized market prices of risk. University of Stanford Working Paper (2006)
25. F. de Jong, Time-series and cross-section information in affine term structure models. Department of Financial Management, University of Amsterdam (1998)

26. A.D. de los Rios, A new linear estimator for gaussian dynamic term structure models. Bank of Canada Working Paper No. 13–10 (2013)
27. F.X. Diebold, L. Ji, C. Li, A three-factor yield curve model: non-affine structure, systemic risk sources, and generalized duration. University of Pennsylvania Working Paper (2004)
28. F.X. Diebold, C. Li, Forecasting the term structure of government bond yields. University of Pennsylvania Working Paper (2003)
29. F.X. Diebold, G.D. Rudebusch, *Yield Curve Modeling and Forecasting: The Dynamic Nelson–Siegel Approach* (Princeton University Press, Princeton, 2013)
30. J.-C. Duan, J.-G. Simonato, Estimating and testing exponential-affine term structure models by kalman filter. Technical report, Centre universitaire de recherche et analyze des organizations (CIRANO) (1995)
31. D. Duffie, *Dynamic Asset Pricing Theory*, 2nd edn. (Princeton University Press, Princeton, 1996)
32. G.R. Duffee, Term premia and interest rate forecasts in affine models. J. Financ. **57**, 405–443 (2002)
33. D. Duffie, R. Kan, A yield-factor model of interest rates. Math. Financ. **6**, 379–406 (1996)
34. D. Duffie, D. Filipovic, W. Schachermayer, Affine processes and applications in finance. Ann. Appl. Probab. **13**, 984–1053 (2003)
35. M. Fisher, Forces that shape the yield curve: parts 1 and 2. U.S. Federal Reserve Board Working Paper (2001)
36. B. Flesaker, L. Hughston, Positive interest. Risk **9**, 46–49 (1996)
37. A.L.J. Geyer, S. Pichler, A state-space approach to estimate and test multifactor cox-ingersoll-ross models of the term structure of interest rates. Department of Operations Research, University of Economics Vienna (1998)
38. P. Hall, On projection pursuit. Ann. Stat. **19**(1), 142–157 (1989)
39. J.D. Hamilton, *Time Series Analysis*, Chapter 22 (Princeton University Press, Princeton, 1994)
40. A.C. Harvey, *Forecasting, Structural Time Series Models and the Kalman Filter*, 1st edn. (Cambridge University Press, Cambridge, 1990)
41. D. Heath, R. Jarrow, A. Morton, Bond pricing and the term structure of interest rates: a discrete time approximation. J. Financ. Quant. Anal. **25**, 419–440 (1990)
42. D. Heath, R. Jarrow, A. Morton, Bond pricing and the term structure of interest rates: a new methodology for contingent claims valuation. Econometrica **60**, 77–105 (1992)
43. T.S.Y. Ho, S.-B. Lee, Term structure movements and pricing contingent claims. J. Financ. **XLI**, 1011–1029 (1986)
44. P. Honoré, Maximum likelihood estimation of non-linear continuous term structure models. Technical report, The Aarhus School of Business (1995)
45. P. Hördahl, O. Tristani, D. Vestin, A joint econometric model of macroeconomic and term structure dynamics. European Central Bank Working Paper No. 405 (2004)
46. P.J. Huber, Projection pursuit. Ann. Stat. **13**(2), 435–475 (1985)
47. J. Hull, A. White, Pricing interest-rate derivative securities. Rev. Financ. Stud. **3**, 573–592 (1990)
48. J. Hull, A. White, One-factor interest rate models and valuation of interest rate derivative securities. J. Financ. Quant. Anal. **28**, 235–254 (1993)
49. J. Hull, A. White, Numerical procedures for implementing term structure models I: single-factor models. J. Deriv. **2**, 7–16 (1994)
50. J. Hull, A. White, Numerical procedures for implementing term structure models II: two-factor models. J. Deriv. **2**, 37–48 (1994)
51. A.S. Hurn, K.A. Lindsay, V. Pavlov, Smooth estimation of yield curves by laguerre functions. Queensland University of Technology Working Paper (2005)
52. J. James, N. Webber, *Interest-Rate Modelling* (Wiley, Chichester, 2000)
53. A. Jeffrey, O. Linton, T. Nguyen, Flexible term structure estimation: which method is preferred. Yale International Centre for Finance: Discussion Paper No. ICF-00-25 (2000)
54. I.T. Jollliffe, *Principal Component Analysis* (Springer, New York, 2002)

55. G.G. Judge, W.E. Griffiths, R.C. Hill, H. Lütkepohl, T.-C. Lee, *The Theory and Practice of Econometrics*, 2nd edn. (Wiley, New York, 1985)
56. I. Karatzas, S.E. Shreve, *Brownian Motion and Stochastic Calculus*, 2nd edn. (Springer, Berlin, 1991)
57. I. Karatzas, S.E. Shreve, *Methods of Mathematical Finance* (Springer, New York, 1998)
58. T.C. Langetieg, A multivariate model of the term structure. J. Financ. **XXXV**, 71–97 (1980)
59. M. Leippold, L. Wu, Quadratic term-structure models. Swiss Institute of Banking and Finance Working Paper (2000)
60. M. Leippold, L. Wu, Empirical performance of quadratic term structure models. Swiss Institute of Banking and Finance Working Paper (2001)
61. O. Linton, E. Mammen, J. Nielsen, C. Tanggaard, Estimating yield curves by kernel smoothing methods. Yale International Centre for Finance: Discussion Paper (1999)
62. R. Litterman, J. Scheinkman, Common factors affecting bond returns. J. Fixed Income **1**, 54–61 (1991)
63. F.A. Longstaff, E.S. Schwartz, Interest rate volatility and the term structure: a two-factor general equilibrium model. J. Financ. **XLVII**, 1259–1282 (1992)
64. F.A. Longstaff, E.S. Schwartz, A two-factor interest rate model and contingent claims evaluation. J. Fixed Income **2**, 16–23 (1992)
65. J. Lund, Econometric analysis of continuous-time arbitrage-free models of the term structure of interest rates. Department of Finance, The Aarhus School of Business (1997)
66. M. Musiela, M. Rutkowski, *Martingale Methods in Financial Modelling*, 1st edn. (Springer, Berlin, 1998)
67. S.N. Neftci, *An Introduction to the Mathematics of Financial Derivatives* (Academic Press, San Diego, 1996)
68. C.R. Nelson, A.F. Siegel, Parsimonious modeling of yield curves. J. Bus. **60**, 473–489 (1987)
69. K.B. Nowman, Gaussian estimation of single-factor continuous time models of the term structure of interest rates. J. Financ. **LII**, 1695–1706 (1997)
70. B.K. Oksendal, *Stochastic Differential Equations*, 4th edn. (Springer, Berlin, 1995)
71. H.H. Panjer, P.P. Boyle, S.H. Cox, D. Dufresne, H.U. Gerber, H.H. Mueller, H.W. Pedersen, S.R. Pliska, M. Sherris, E.S. Shiu, K.S. Tan, *Financial Economics* (The Actuarial Foundation, Schaumberg, 1998)
72. N.D. Pearson, T.-S. Sun, Exploiting the conditional density in estimating the term structure: an application to the cox-ingersoll-ross model. J. Financ. **XLIX**, 1279–1304 (1994)
73. S.F. Richard, An arbitrage model of the term structure of interest rates. J. Financ. Econ. **6**, 33–57 (1978)
74. L.C.G. Rogers, Which model for term-structure of interest rates should one use? IMA Vol. Math. Appl. **67**, 93–115 (1995)
75. L.C.G. Rogers, The potential approach to the term structure of interest rates and foreign exchange rates. Math. Financ. **7**, 157–176 (1997)
76. S.M. Schaefer, E.S. Schwartz, A two-factor model of the term structure: an approximate analytical solution. J. Financ. Quant. Anal. **4**, 413–424 (1984)
77. L.E.O. Svensson, Estimating and interpreting forward interest rates: Sweden 1992–1994. International Monetary Fund: Working Paper No. 114 (1994)
78. O. Vasicek, An equilibrium characterization of the term structure. J. Financ. Econ. **5**, 177–188 (1977)

# Part III
# Performance

Computing the return of a fixed-income security or portfolio provides limited information on its performance over the return horizon. This information is essentially restricted to the size and magnitude of the return. While useful, it is not completely sufficient. We also seek to understand *where* this performance came from. The following chapters use the additive risk-factor-based breakdown of return to perform a range of different performance attributions that each seek to answer this question.

# 7 Basic Performance Attribution

*Measure what is measurable, and make measurable what is not so.*

Galileo Galilei

Performance and risk attribution are the heart of this book. After several chapters of preparation, we are finally in a position to begin with the performance dimension. Performance attribution centres around *one* principal question: where did the return on my portfolio come from? This is an easy question to ask, but it turns out that it is surprisingly difficult to answer. It is difficult because it essentially involves breaking down, or decomposing, the return of a portfolio over a given period into different buckets. Each of these buckets must describe an alternative dimension of the return.

The challenge is that there is *no* unique set of buckets for the breakdown of return.[1] We can, generally speaking, agree on the return over the period. How exactly to decomposse it is another question. Two reasonable people can reasonably disagree on how to decompose the return and thus generate two different performance attributions.

Performance attribution is conceptually similar to assessing the outcome of an ice hockey match. At the end of the game, everyone observes the objective final score. While not everyone may be pleased with this final tally, it is a clear and objective measure, which is analogous to the performance of a portfolio over a given period of time. If one were asked to describe how each team played and why one team was successful, however, the discussion would become rather more subjective. Different people have alternative perspectives and, as a consequence, would likely describe (or attribute) the success (or performance) of each team in a different way. Use of

[1]The additive risk-factor-based decomposition developed in a previous chapter provides the starting point for such a decomposition. As we will see in the following discussion, there are numerous alternative directions that can be taken within this framework.



D.J. Bolder, *Fixed-Income Portfolio Analytics*,
DOI 10.1007/978-3-319-12667-8_7

objective measures to make this judgement is unlikely to completely improve the situation. One could focus on a number of different dimensions such as the time of puck possession, the number of times the puck was given away, the number of shots on net, the relative face-off wins and losses, or the number and severity of penalties. The choice of measures and their relative importance will still have a somewhat subjective flavour. Nevertheless, examination of a wide range of measures in an organized manner should provide, to a wide range of hockey fans, a good idea of what happened.

Our point is not to write a chapter on ice hockey, as interesting as that may be, but rather to highlight the fact there is no single, correct approach to the attribution of a portfolio's performance. An element of subjectivity will always remain; this should not imply that it is neither useful nor necessary. We will describe one principal approach in this chapter and offer a few possible variations, drawing from the lessons learned in previous chapters, but the reader should be aware that there are numerous alternatives.

It is natural to ask why it is even necessary to attribute the performance of one's portfolio. The answer is that performance analysis is an essential element of the investment process. Figure 7.1 provides a simplified schematic of a generic investment process. We view it as four stages that operate in a circular fashion. First, one analyzes the economic environment. This analysis permits one to formulate multiple views, which are then implemented as positions. These positions will either generate a gain or create a loss. At the end of one's performance or investment period, one can determine if there was a gain or an loss on the overall portfolio. What this overall performance does not tell you, however, is how the individual views and positions of the portfolio manager performed.

An attribution of the portfolio's performance is extremely useful in assessing the success or failure of a portfolio manager's views and, by extension, the efficiency of one's investment process. One could, for example, hope to generate consistently positive returns based on views of future yield-curve movements. An attribution would indicate if these yield-curve positions have actually contributed

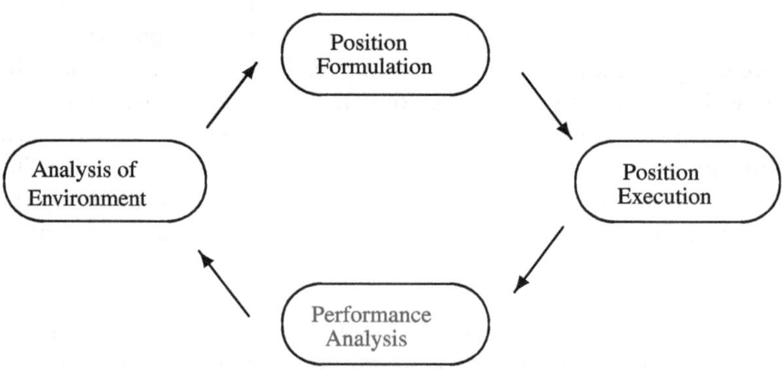


**Fig. 7.1** An attribution schematic. This figure schematically describes the role of performance attribution in the investment process

in a consistently positive manner to the portfolio returns and help distinguish this element of the active returns from other risk factors, such as credit or currency. In short, a performance attribution is essentially a *report card* for the portfolio manager and the investment process. Failure to regularly assess the success (or failure) of one's active investment positions makes is basically impossible to judge or improve one's active-management process.

Although performance attribution can help judge the efficiency of one's investment process, this usefulness comes at a cost. A proper performance attribution requires significant amounts of information about one's portfolio and benchmark. This is because proper performance attribution must be performed at the *instrument* level. In other words, one must perform a full attribution for each instrument in both the portfolio and the benchmark and aggregate this attribution across all instruments. It is not enough to focus on weighted average quantities for one's portfolio and benchmark.

An instrument-level approach is intensive in terms of both data and computational resources. It might thus be tempting to attempt to compute the performance of the portfolio at the aggregate level using weighted average duration movements and weighted average yields—such an approach would certainly lighten the computational burden. It is nonetheless rather dangerous to do so.

To convince you of these dangers, we start with a US-Treasury bond example, where the use of average yield changes and average modified duration provides an inaccurate picture of changes in the return of a fixed-income portfolio. Table 7.1 outlines the basic security details for the three US Treasury bonds in our example portfolio. The second table, Table 7.2, describes the relevant market data for the

**Table 7.1** A simple example #1

| ISIN | Maturity | Coupon (%) | Duration | Notional |
|---|---|---|---|---|
| US912828PW43 | 31 Dec 2012 | 0.625 | 1.976 | \$100,000,000 |
| US912828PM60 | 31 Dec 2015 | 2.125 | 4.713 | \$125,000,000 |
| US912828PC88 | 15 Nov 2020 | 2.625 | 8.536 | \$75,000,000 |

Here is the static data for a sample UST portfolio.

**Table 7.2** A simple example #2

| | Yield | | Dirty price | | Market value | | Weight | |
|---|---|---|---|---|---|---|---|---|
| ISIN | $y_{t_1}$ | $y_{t_2}$ | $P_{t_1}$ | $P_{t_2}$ | $V_{t_1}$ | $V_{t_2}$ | $\omega_{t_1}$ (%) | $\omega_{t_2}$ (%) |
| US912828PW43 | 0.637 | 0.645 | 99.981 | 99.967 | 99,981,417 | 99,967,378 | 33.8 | 33.8 |
| US912828PM60 | 2.064 | 2.051 | 100.305 | 100.372 | 125,381,625 | 125,465,488 | 42.4 | 42.4 |
| US912828PC88 | 3.367 | 3.354 | 94.165 | 94.278 | 70,623,639 | 70,708,516 | 23.9 | 23.9 |
| Sum/average | 1.893 | 1.887 | 98.642 | 98.687 | 295,986,681 | 296,141,382 | 100.0 | 100.0 |

Here are the yield, price, and market values for two different days. The final column indicates the average or sum of this dynamic data.

portfolio on two adjacent days, denoted $t_1$ and $t_2$. Using the weighted average modified duration of 4.7, we may approximate the expected return due to 1-day yield changes at the aggregate level as

$$\begin{aligned}
r_{\text{Yield, Aggregate Level}} &\approx \text{Portfolio Duration}_{t_1}\left(\text{Portfolio Yield}_{t_2} - \text{Portfolio Yield}_{t_1}\right) \qquad (7.1)\\
&\approx \left(\sum_{i=1}^{3} \omega_{t_{1,i}} D_{M,t_{1,i}}\right)\left(\underbrace{\sum_{i=1}^{3} \omega_{t_{1,i}} y_{t_{2,i}}}_{\bar{y}_{t_2}} - \underbrace{\sum_{i=1}^{3} \omega_{t_{1,i}} y_{t_{1,i}}}_{\bar{y}_{t_2}}\right),\\
&\approx \left(\underbrace{\sum_{i=1}^{3} \omega_{t_{1,i}} D_{M,t_{1,i}}}_{\bar{D}_M}\right)\left(\underbrace{\sum_{i=1}^{3} \omega_{t_{1,i}} \left(y_{t_{2,i}} - y_{t_{1,i}}\right)}_{\Delta\bar{y}}\right),\\
&\approx \bar{D}_M \cdot \Delta\bar{y},\\
&\approx -4.701 \cdot (1.893\,\% - 1.887\,\%),\\
&\approx -4.701 \cdot (-0.0059\,\%),\\
&\approx 2.78 \text{ basis points.}
\end{aligned}$$

The estimated 1-day curve-related return, using aggregate portfolio measures, is about 2.8 basis points.

Alternatively, we can perform the equivalent 1-day curve return computation using individual instrument-level data. The result is,

$$\begin{aligned}
r_{\text{Yield, Instrument Level}} &\approx \sum_{i=1}^{n} \begin{matrix}\text{Security}\\ \text{Weight}_{t_{1,i}}\end{matrix} \begin{matrix}\text{Security}\\ \text{Duration}_{t_{1,i}}\end{matrix} \left(\begin{matrix}\text{Security}\\ \text{Yield}_{t_{2,i}}\end{matrix} - \begin{matrix}\text{Security}\\ \text{Yield}_{t_{1,i}}\end{matrix}\right) \qquad (7.2)\\
&\approx \sum_{i=1}^{3} \omega_{t_{1,i}} D_{M,t_{1,i}} \left(\underbrace{y_{t_{2,i}} - y_{t_{1,i}}}_{\Delta\bar{y}_i}\right),\\
&\approx 0.338 \cdot 1.976 \cdot (0.645 - 0.637) + \cdots\\
&\quad 0.424 \cdot 4.713 \cdot (2.051 - 2.064) + \cdots\\
&\quad 0.239 \cdot 8.536 \cdot (3.354 - 3.367),\\
&\approx 4.71 \text{ basis points.}
\end{aligned}$$

There is almost two basis points of difference between the aggregate- and instrument-level computations.

Which one is correct? This is relatively easy to verify as the actual return, including accrued interest through use of the dirty-price derived market values, is simply

$$\begin{aligned} r_{\text{Market Value}} &= \frac{\$296,141,382 - \$295,986,681}{\$295,986,681}, \\ &= 5.23 \text{ basis points.} \end{aligned} \tag{7.3}$$

The return computed using the individual-instrument-level details is thus much closer to the true yield return over the period than the use of the aggregate-level approach.[2] The reason stems from a simple mathematical property of sums and products: namely, the sum of the products is not equal to the product of the sums. More specifically,

$$\begin{aligned} \text{Equation (7.1)} &\neq \text{Equation (7.2)}, \\ r_{\text{Yield, Aggregate Level}} &\neq r_{\text{Yield, Instrument Level}}, \end{aligned} \tag{7.4}$$

$$\underbrace{\left(\sum_{i=1}^{3} \omega_{t_{1,i}} D_{M,t_{1,i}}\right)\left(\sum_{i=1}^{3} \omega_{t_{1,i}} \left(y_{t_{2,i}} - y_{t_{1,i}}\right)\right)}_{\substack{\text{Product} \\ \text{of sums}}} \neq \underbrace{\sum_{i=1}^{3} \omega_{t_{1,i}} D_{M,t_{1,i}} \left(y_{t_{2,i}} - y_{t_{1,i}}\right)}_{\text{Sum of products}}.$$

In brief, a proper performance attribution must be performed at the individual-security level. See the underlying shaded box for an alternative description of why use of portfolio aggregates typically fails in performance approximation.

Let's repeat the previous computation without all of the bond details. Imagine that you have two sequences of numbers, $a$ and $b$, defined as,

$$\begin{aligned} a &= \{1, 2, 3\}, \\ b &= \{4, 5, 6\}. \end{aligned} \tag{7.5}$$

Now if we compute the product of the sums of $b$ and $a$, we have

$$\left(\sum_{i=1}^{3} b_i\right) \cdot \left(\sum_{i=1}^{3} a_i\right) = \underbrace{(4+5+6)}_{15} \cdot \underbrace{(1+2+3)}_{6}, \tag{7.6}$$

(continued)

[2]It is even closer when the overall return is adjusted for the approximately 0.5 basis points of 1-day carry return.

$$= 90. \tag{7.7}$$

If, however, we compute the sum of the products of $b$ and $a$, we arrive at an entirely different result,

$$\sum_{i=1}^{3} b_i \cdot a_i = \underbrace{(4 \cdot 1)}_{4} + \underbrace{(5 \cdot 2)}_{10} + \underbrace{(6 \cdot 3)}_{18}, \tag{7.8}$$
$$= 32.$$

The obvious conclusion is that,

$$\left(\sum_{i=1}^{3} b_i\right) \cdot \left(\sum_{i=1}^{3} a_i\right) \neq \sum_{i=1}^{3} b_i \cdot a_i, \tag{7.9}$$

or the sum of products is *not* equal to the product of the sums. This fact is often misunderstood and overlooked leading to computational errors. It is nonetheless a clear mathematical fact that applies equally well to this simple example and to more complex situations involves hundreds of security returns.

We have established that performance attribution centres around *one* principal question: where did the return on a given portfolio come from? Moreover, we've seen that answering this question provides a helpful tool for judging the efficiency of your investment process. In the remainder of this chapter we will introduce the principal ideas by computing a performance attribution for:

- a single instrument; and
- a portfolio with an explicit benchmark.

Along the way, we will make extensive use of examples to render these notions as concrete, and useful, as possible for the reader.

## 7.1 A Single Security

If one performs some background research on the topic of performance attribution, one will find a wide variety of both academic and practitioner literature. The large majority of these papers unfortunately focus on attribution analysis for equity portfolios. This is hardly surprising since the roots of performance attribution are to be found in the equity literature and rely on algebraic manipulations to isolate different aspects of a portfolio's return. The seminal paper in this area, Brinson et al. [3], provided a framework for breaking return into the following *two* categories:

**Asset Allocation** Return associated with over- or under-weighting different *asset-category* weights relative to the strategic benchmark.

**Security Selection** Return associated with over- or under-weighting different *individual-security* weights relative to the strategic benchmark.

Asset allocation simply represents a top-down tactical choice—over- or under-weighting equities, for example, in a multi-asset class portfolio. Security selection, conversely is a bottom-up tactical choice. Does one, for example, when seeking US Agency exposure, purchase Fannie Mae *or* Freddie Mac securities? Both are different ways to implement a general position: US Agencies.

Asset allocation and security selection are interesting and useful concepts that, to a certain extent, may be applied to fixed-income portfolios.[3] Many aspects of fixed-income returns, however, are quite different and are not readily accommodated by this framework. The principal driver of high-quality fixed-income returns is the movement of the underlying sovereign yield curve. An asset-allocation and security-selection breakdown does not capture this dimension and, as such, is not well-suited to fixed-income performance attribution.

What then are we to do? If the security return is our starting point, we should start with the return for a given security over a given time interval, $[t, T]$. More precisely, if the security value is $V_t$ at the beginning of the period and $V_T$ at the end, then

$$r(t, T) = \frac{V(T) - V(t)}{V(t)}. \tag{7.10}$$

For those readers who have read the previous chapters, this should look quite familiar. Our additive risk-factor-based return decomposition will be very helpful in this regard. There is, unfortunately, some additional complexity relating to injections and withdrawals (i.e., external cash-flows) that must first be considered. Dealing with external cash-flows is essential, because failure to do so can lead to sizeable problems with one's performance figures. In the subsequent section, we will take some time to consider how one might deal with this practical issue.

### 7.1.1 Dealing with Cash-Flows

We've indicated that a portfolio's return over a given interval is an unequivocal, objective figure. Sadly, although often true, this is *not* always the case. External cash-flows create a number of challenges for the computation of performance.[4]

[3] In a subsequent chapter, we will introduce a version of this idea and apply it to long and short duration positions relative to the benchmark.

[4] Our treatment of external cash flows is, more or less, illustrative. For a more exhaustive discussion, see Bacon [1].

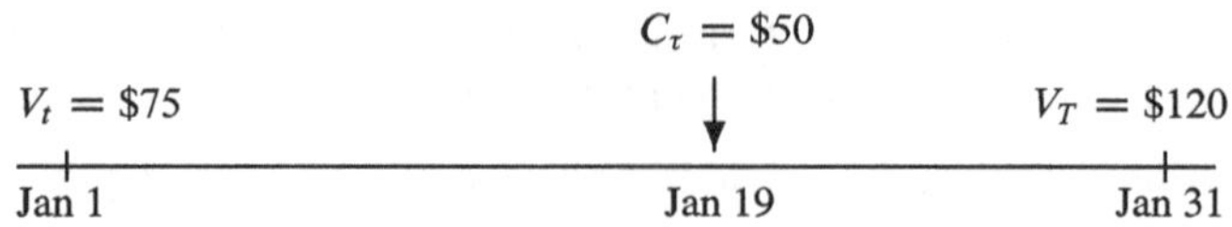


**Fig. 7.2** An injection example. This figure outlines a simple example of a portfolio with an injection over the performance period

The basic formulae quite simply no longer provide reasonable results. An example of such a situation is summarized in Fig. 7.2.

Imagine that we start with \$75, inject \$50 slightly past mid-month, and end the period with a portfolio value of \$120. Naively computing return in the usual way, we obtain

$$\begin{aligned} r(t,T) &= \frac{V_T - V_t}{V_t}, \\ &= \frac{120 - 75}{75} = 60\,\%. \end{aligned} \tag{7.11}$$

This is *not* a reasonable result.[5]

Without some adjustment, the injection of \$50 on January 19 is incorrectly treated as strong positive performance. Any approach to resolve this issue will need to break down the contribution (or weight) of the return before and after the injection. There are *two* main techniques for determining the relative contribution of the pre- and post-cash-flow return: value- and time weighting.[6] One could easily write a chapter discussing variations, disadvantages and advantages of these two methods. Instead of allocating an entire chapter to the topic, however, we will consider a few simple examples of value- and time-weighting return computations. This should nevertheless provide the reader with a good feel for the main issues in handling external cash flows.

The basic idea of value-weighting is that the weight of the pre- and post-cash flow returns is determined by the amount (or value) of the funds under investment. It has two steps.[7] First, the cash-flow (i.e., injection) is used to adjust the investment profit-and-loss on the portfolio. Thus, the value-added over the period is reduced by the injection as follows,

$$\begin{aligned} V_T - V_t - C_\tau &= 120 - 75 - 50, \\ &= -5. \end{aligned} \tag{7.12}$$

[5]Without an adjustment, one could fabricate any desired portfolio return by merely arranging for the necessary injection or withdrawal of funds into the portfolio.

[6]It is important to note that within each approach, there are a number of different variations and sub-methods.

[7]The value-weighted approach that we will consider is called the modified-Dietz method.

Second, the starting investment is adjusted by the external cash-flow as follows,

$$V_t + \left(\frac{T-\tau}{T}\right) C_\tau = 75 + \left(\frac{31-19}{31}\right) 50, \qquad (7.13)$$
$$= 94.4.$$

We can see, therefore, that the starting investment is increased by a factor of $\frac{12}{31}$, which reflects the fraction of the month these funds were available to the portfolio manager. The modified-Dietz return is merely the ratio of Eqs. (7.12) and (7.13),

$$r_{\text{MD}} = \frac{V_T - V_t - C_\tau}{V_t + \left(\frac{T-\tau}{T}\right) C_\tau}, \qquad (7.14)$$
$$= \frac{-5}{94.4},$$
$$= -5.30\,\%,$$

which appears to be a more reasonable result.

The idea behind the time-weighting approach is that each period is given equal weight regardless of how much is invested. The time-weighted approach also requires additional information regarding the value of the portfolio on the injection date. This information is provided in Fig. 7.3.

The value of the portfolio on the injection date, excluding the \$50 injection, was \$74. This permits us to compute the return for each sub-period. The return in the first period is,

$$\frac{V_\tau - C_\tau - V_t}{V_t} = \frac{74-75}{75}, \qquad (7.15)$$
$$= -1.33\,\%.$$

The return for the second period is,

$$\frac{V_T - V_\tau}{V_\tau} = \frac{120-124}{124}, \qquad (7.16)$$
$$= -3.23\,\%.$$

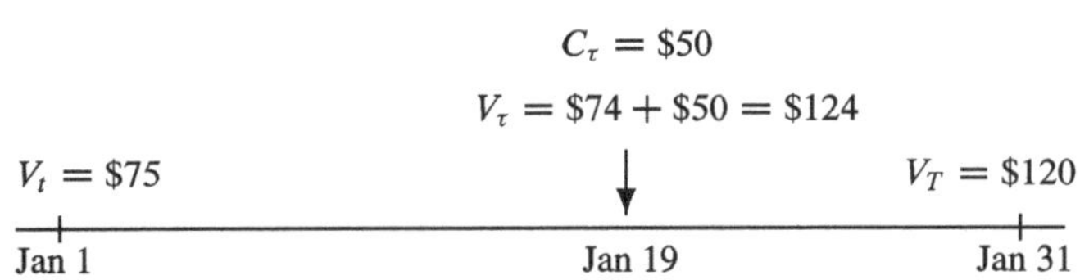


**Fig. 7.3** A time weighted schematic. This figure outlines the injection example with the additional information required for the computation of a time-weighted return

Linking them together geometrically, we arrive at the return for the entire period,

$$\left(1+\underbrace{\frac{V_\tau - C_\tau - V_t}{V_t}}_{\substack{\text{Equation}\\ (7.15)}}\right)\left(1+\underbrace{\frac{V_T - V_\tau}{V_\tau}}_{\substack{\text{Equation}\\ (7.16)}}\right) - 1 = (\mathbf{1\text{-}1.33\,\%})\cdot(\mathbf{1\text{-}3.23\,\%}) - 1,$$
$$= -4.52\,\%, \tag{7.17}$$

which is also a more reasonable figure.

Having now examined two ways to compute the performance of a portfolio with an injection occurring in the middle of the performance period, it is clear that a reasonable adjustment for external cash-flows is possible. The two approaches, however, yield different results. Although, the value and time-weighted return computations both generate return values in the neighbourhood of $-5\,\%$, they differ by roughly 80 basis points.[8]

Even worse, the differences between the two methods can sometimes be dramatic. Figure 7.4 describes a situation with a starting portfolio value of \$10, an injection of \$100 slightly past mid-month, and a final market value of \$105. For the time-weighted method, the market value of the portfolio on the injection date was, not including the injection, \$11.

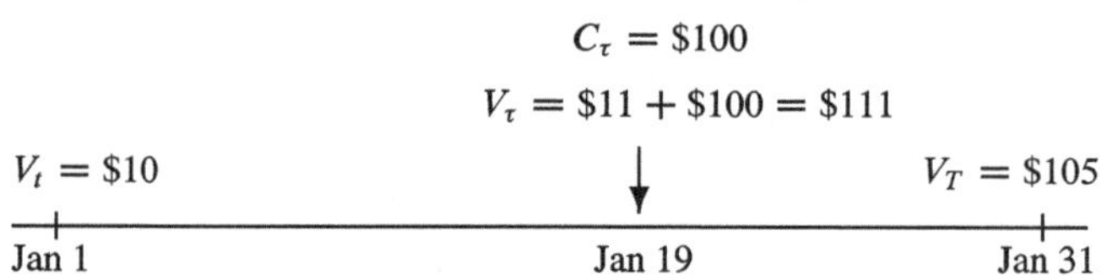


**Fig. 7.4** A pathological example. This figure outlines a simple, but rather pathological, example where the difference between the two approaches is substantial

[8]Moreover, more information is required for the time-weighted approach. This may seem like a trivial point, but some external managers provide only monthly valuations. In such a situation, computation of time-weighting returns is *not* an option.

**Table 7.3** Pathological results

| Approach | Return (%) |
|---|---|
| Value-weighted | −7.71 |
| Time-weighted | 4.05 |

This table summarizes the results from the pathological example described in Fig. 7.4.

Table 7.3 summarizes the results of applying the previously described methods to the rather extreme example in Fig. 7.4.

As advertised, the results are dramatically different. The value-weighted approach generates a return of −7.71 %, while the time-weighted approach suggests that the return was 4.05 %. This difference of almost 12 percentage points between the two methods is disconcerting.

What is going on? Here we see, in an extreme way, the differences between the two weighting methods. The portfolio returned 10 % during the first period—from 1 to 19 January—on the original investment of \$10. It returned, however, −5.41 % in the second period—from 19 to 31 January—on an original investment of \$111. When weighting the return by *value*, the second period receives more weight simply because more funds (or value) are invested during this period. Consequently, the value-weighted approach is −5.41 % due to the relatively large and negative return in the second period.

When weighting by *time* the two periods receive an equal billing. Consequently, the positive return in the first period offsets and surpasses the negative return in the second to generate an overall positive return. This is a fairly extreme, and maybe even slightly unfair, example intended to highlight the differences between the two methods. The fact remains that differences can and do arise.[9]

Which method is preferable? While there is no *correct* approach, most modern performance systems typically compute daily time-weighted returns. One reason is that time weighting is deemed to provide fairer returns for portfolio managers who do not control external cash-flows. Moreover, when time-weighted is performed on a daily basis, the pathological situations described in Fig. 7.4 are not quite as severe. This point is discussed in more detail in the next chapter. Time-weighting, it should be stressed, is not without its own costs. It requires a robust daily valuation of all portfolio and benchmark positions. While most modern portfolio-management systems provide this functionality, it is nonetheless no trivial task.

[9] As a general remark, the differences between the two approaches will tend to be larger when the injection or withdrawal is large relative to the size of the underlying portfolio.

### 7.1.2 Revisiting Our Risk-Factor Decomposition

Recalling the central result from our additive risk-factor-based return approximation framework, we have that our fixed-income return is well approximated by the sum of the following factors,

$$r \approx \overbrace{\underbrace{y\Delta t}_{\substack{\text{Carry}\\ \text{return}}} + \mathbb{I}_{\text{ILB}} \underbrace{\pi(t,t+1)}_{\substack{\text{Inflation}\\ \text{carry}}} - \underbrace{\sum_{i=1}^{\nu} \kappa_i \Delta y_{\text{TRE},i}}_{\substack{\text{Treasury curve}\\ \text{return}}} - \underbrace{D_{\text{S}} \Delta s_{\text{OAS}}}_{\substack{\text{Credit}\\ \text{return}}} + \underbrace{\frac{1}{2} C(\Delta y)^2}_{\substack{\text{Convexity}\\ \text{return}}}}^{\text{Local currency return}}$$

$$+ \underbrace{\sum_{i=1}^{\alpha} \mathbb{I}_{FX_i} \left( \frac{E_{i,1} - E_{i,0}}{E_{i,0}} \right)}_{\text{FX return}}, \tag{7.18}$$

where the FX exposure is,

$$\mathbb{I}_{FX_i} = \begin{cases} 0 : \text{Not exposed to currency } i \\ 1 : \text{Exposed to currency } i \end{cases}, \tag{7.19}$$

and the exposure to inflation is,

$$\mathbb{I}_{ILB} = \begin{cases} 0 : \text{A nominal bond} \\ 1 : \text{An ILB} \end{cases}. \tag{7.20}$$

Even if you've read the previous chapters, Eq. (7.18) can be difficult to digest. The logic behind Eq. (7.18) can be conceptualized in a very simple manner. We can think of the return associated with an arbitrary risk factor, let's call it the $i$th factor, as

$$\text{Return to } i\text{th Factor} = \text{Exposure}_i \cdot \text{Change in } i\text{th Factor}. \tag{7.21}$$

The return to the $i$th risk factor, in our base approximation, is defined as the product of the security's exposure to that factor and the change in the factor over the performance period. The total return is correspondingly merely the sum of the return associated with each of the $n$ risk factors or,

$$\text{Total Return} = \sum_{i=1}^{n} \underbrace{\text{Return to } i\text{th Factor}}_{\text{Equation (7.21)}}. \tag{7.22}$$

**Table 7.4** Basic risk factors

| Return | Factor | Exposure |
|---|---|---|
| Carry | Time ($t$) | Yield to maturity ($y$) |
| Curve | Sovereign yield ($y_T$) | Modified duration ($D_M$) |
| Credit | OA spread ($s_{\text{OAS}}$) | Spread duration ($D_S$) |
| Convexity | Yield $(\Delta y)^2$ | Convexity ($C$) |
| FX | FX return ($r_{\text{FX}}$) | Indicator variable ($\mathbb{I}_{FX_i}$) |

This table outlines the risk factors and exposures computed in a basic performance attribution.

This is merely a heuristic way to represent Eq. (7.1) that should help conceptualize Eq. (7.18). Table 7.4 provides a summary of the basic risk factors and how we approximate their exposures. The true approximation, which is the basis of our risk and performance computations, is found in Eq. (7.18) and described in great length in the previous chapters.

Its complexity aside, Eq. (7.18) is a remarkably useful approximation. It can incorporate a wide range of fixed-income securities including nominal bonds, notes, and bills, inflation-linked bonds, interest-rate swaps, foreign-exchange swaps and forwards, and bond and rate futures. The additive decomposition of the fixed-income return in Eq. (7.18) is the foundation of our performance attribution.

Each of these return categories can be further decomposed without making dramatic changes to the approximation in Eq. (7.18). In particular, there are a variety of ways to further decompose treasury-curve and spread returns. Our objective in this chapter is to bring this decomposition, and some of the variations upon it, to life through its application to two different practical examples.

For those less comfortable with the mathematical representation in Eq. (7.18), it is worth reviewing each of these return components in detail.

**Carry (or Yield) Return** This is the *time* return of a fixed-income security. If one merely buys and holds a fixed-income security, one will be compensated through coupon payments. Since the coupon payments are relative to the price one paid for the bond, it is more accurate to describe the carry return using the bond's yield to maturity. Higher yielding bonds will generate a higher time return relative to lower yielding equivalents. Inflation-linked bonds have an additional carry component which stems from the inflation compensation provided by these instruments. Conveniently the inflation compensation over a given time interval, is the same for all inflation-linked bonds linked to the same inflation index. In the US TIPS market, for example, all bonds are linked to a specific measure of US inflation. This is not always the case. The French Treasury, for example, has issued two distinct types of inflation-linked bonds: a group of these bonds are linked

(continued)

to European inflation whereas another group are compensated for French inflation. When dealing with French inflation-linked bonds, therefore, extra caution must be exercised to ensure that the appropriate inflation rate is used for the computation of inflation carry return.

**Treasury Curve Return** This is the return associated with changes in the underlying risk-free, or Treasury, curve. The movement of the Treasury curve in turn reflect expectations of market participants about current and future macroeconomic variables such as inflation, output, and monetary policy stance.

**Credit Return** The return associated with changes in the spread of the fixed-income security relative to the underlying Treasury yield curve. This spread is generally expected to summarize movements in the credit quality of the specific instrument relative to the *risk-free* issuer in the economy. It is, however, more accurate to state that the spread is a combination of relative credit views and liquidity. For our purposes, given the complexity of such a task, we assign the entire spread movement to credit. It is nonetheless important to understand that some component of the credit spread is generally related to liquidity differences between the specific fixed-income security and the underlying Treasury market.

**Convexity Return** This is merely a correction term to adjust for the linear approximation inherent in the use of modified duration for the treasury-curve and credit returns. Typically, the contribution of the convexity term is small. In situations where there are large movements in the underlying Treasury curve or in credit spreads, the convexity term can become quite large.

**Foreign-Exchange Return** This is the return associated with movements in the security's exchange rate. Every portfolio must have a base currency. If an instrument in the portfolio is denominated in this base currency, then it gives rise to no foreign-exchange return. Only securities denominated in currencies other than the base currency contribute to the foreign-exchange return. As this rate moves, it can have important implications for a foreign-denominated security's value, all else equal, in the base currency.

## 7.2 Attribution of a Single Fixed-Income Security

The most effective way to understand how performance attribution works is to actually do it.[10] This takes us from abstract discussion directly to the implementation of Eq. (7.18). We will begin with the attribution of the performance of a single

[10]There are naturally other possible directions one may take to attributing return—here we show you a general approach with a number of alternatives. Alternative approaches are legion. A few

**Table 7.5** An example bond

| Characteristic | Value |
|---|---|
| Issuer | Fannie Mae |
| Position | $1,000,000 |
| Maturity date | Oct 15, 2014 |
| Tenor | 5.79 years |
| Issue date | Sep 15, 2004 |
| Coupon | 4.625 % |
| ISIN | US31359MWJ88 |

This table provides the basic information for an investment in a US Agency bond.

fixed-income security over a 30-day period. Performance attribution of a single instrument may seem like overkill, but it will help us get to the key ideas without undue clutter and complexity. Table 7.5 describes the basic characteristics of the single US Agency security that we have selected for our first example.

Let us imagine that we purchased this bond on 31 December 2008, held it for 30 days, and then sold it on 30 January 2009. This is a 30-day investment in a high-quality, approximately 6-year plain vanilla bond. Note, however, that it will *not* have the same yield as a US Treasury because it has a lower credit quality—it is not the US Government itself, but rather a US Agency with a government guarantee.

At the end of your investment period, you would like to answer two related questions:

- Did this investment make or lose money?
- What are the reasons behind any gains and losses?

The first question, with the correct information, is quickly answered whereas the second question requires a bit more effort. Table 7.6 provides us, however, with most of the information required to answer both of our questions.[11]

The total return, given that there were thankfully no injections or withdrawals over the period, is simply

$$
\begin{aligned}
r &= \frac{V_{30\text{ Jan }09} - V_{31\text{ Dec }08}}{V_{31\text{ Dec }08}}, \qquad (7.23)\\
&= \frac{110.48 - 111.72}{111.72},\\
&= -110.99\text{ bps}.
\end{aligned}
$$

possibilities may be found in Campisi and Spaulding [4], Menchero and Hu [14], Hansen and Andersen [10], Knight and Satchell [12] or, when faced with limited data, Colin [6].

[11] See Colin [5] for another, complementary perspective on fixed-income performance attribution.

**Table 7.6** Agency bond details

| Characteristic | Dec 31, 2008 | Jan 30, 2009 | Change |
|---|---|---|---|
| Yield (%) | 2.61 | 2.88 | 0.27 |
| USD/CAD exchange rate | 1.2188 | 1.2296 | 0.0108 |
| **Dirty price** | **\$111.72** | **\$110.48** | **−\$1.24** |
| OA spread (bps) | 111.18 | 95.96 | −15.23 |
| Equivalent treasury yield (%) | 1.50 | 1.92 | −0.42 |
| Tenor (years) | 5.79 | 5.71 | −0.08 |
| Modified duration | 5.07 | 4.98 | −0.09 |
| Spread duration | 5.07 | 4.98 | −0.09 |
| Convexity | 0.29 | 0.30 | 0.01 |
| Number of days | 30 | | |

This table summarizes the necessary market information—at the beginning and end of our performance period—for the example bond introduced in Table 7.3.

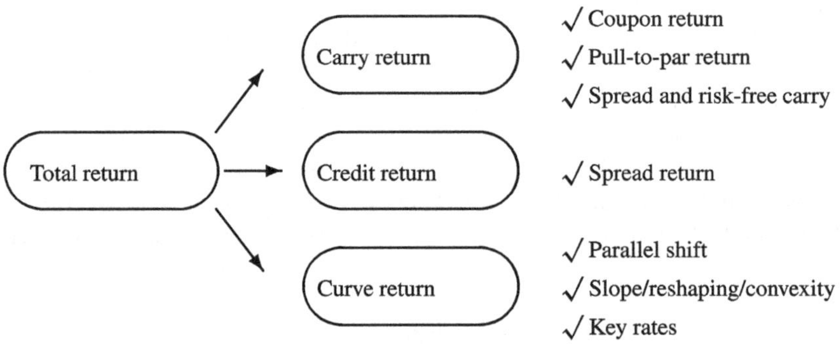


**Fig. 7.5** An attribution schematic. The underlying figure schematically describes the different elements in a basis performance attribution

We have already answered our first question. This investment has *not* been very successful. We have lost approximately 111 basis points.

This brings us to the second question: what are the reasons for this loss? Glancing quickly at Table 7.6 does not provide an immediate response. To answer this question, therefore, we make use of the additive decomposition in Eq. (7.18). Figure 7.5 outlines schematically the main return elements that we will consider for this simple US Agency bond example. While the carry-, credit-, and curve-return categories have already been discussed thoroughly, these base return categories can be further decomposed in different ways. We will consider each of these return categories and their respective decompositions in turn.

### 7.2.1 Carry Return

Carry is the positive return associated with the passage of time. Using the Eq. (7.18), we compute the carry return for our US Agency bond as,

$$\begin{aligned}\text{Carry return} &= y\Delta t, \\ &= 2.61\,\% \cdot \left(\frac{30}{365}\right), \\ &= 21.45 \text{ bps.}\end{aligned} \tag{7.24}$$

Despite our $-111$ basis-point loss on our position, the carry contributes a positive return of about 21 basis points.[12] This is because, as computed in Eq. (7.24), carry return can only be negative if the yield is negative.[13]

We use the yield at the beginning of the period to compute the carry over the entire 1-month period. When computing daily carry, this is a fairly harmless assumption. When computing the carry over an entire month, however, it is somewhat less innocuous. This is because the yield at the end of the period is about 27 basis points higher at 2.88 %. One possible alternative would be to use the average yield over the period. This would amount to,

$$\begin{aligned}\text{Carry return} &= \frac{y_1 + y_2}{2}\Delta t, \\ &= \frac{2.61\,\% + 2.88\,\%}{2} \cdot \left(\frac{30}{365}\right), \\ &= 22.56 \text{ bps.}\end{aligned} \tag{7.26}$$

(continued)

[12]If we use the precise exact carry term from the Taylor expansion derived in the previous chapters, we arrive at

$$\begin{aligned}\text{Carry return} &= \ln(1 + y)\Delta t, \\ &= \ln(1 + 2.61\,\%) \cdot \left(\frac{30}{365}\right), \\ &= 21.18 \text{ bps.}\end{aligned} \tag{7.25}$$

We see, therefore, that the difference associated with our approximation is only a fraction of a basis point.

[13]This is assuming, as we do, that time increments are always positive!

Thus, using an average yield generates about one basis-point of additional carry.

As we will see in the subsequent chapter, it is common to compute *daily* performance attributions and aggregate them over time as desired. In this case, therefore, whether one uses the beginning yield or the average yield over the day makes only a marginal difference in the results. For performance attributions performed over longer periods, however, it can make an important difference. Note, however, that performance attributions performed on a monthly horizon are also typically complicated by transactions occurring during the month. This will also be discussed in much greater detail in the next chapter. For the purposes of the examples in this introductory discussion, however, we will use the values at the beginning of the period for our computations.

One can extend the notion of carry return. If we take the carry return and simultaneously add and subtract the instrument's coupon, which is denoted as $\tilde{C}$, we have

$$\begin{aligned} \text{Carry return} &= y\Delta t, \qquad (7.27)\\ &= \left(y + \underbrace{\tilde{C} - \tilde{C}}_{=0}\right)\Delta t,\\ &= \underbrace{\tilde{C}\Delta t}_{\substack{\text{Coupon}\\ \text{return}}} + \underbrace{(y - \tilde{C})\Delta t}_{\substack{\text{Pull-to-par}\\ \text{return}}}. \end{aligned}$$

The consequence of this small algebraic trick is to break out the carry return into two components: the return stemming from the coupon and the return associated with the *pull-to-par* effect. The coupon aspect is fairly self explanatory—the larger the coupon, the larger the contribution to the coupon return. The pull-to-par effect arises because, at maturity, all bonds repay their face value. The premium bonds will experience, all else equal, a sequence of capital losses as the premium is eroded. Discount bonds, conversely, will tend to experience capital gains, again all else equal. This aspect of return is a function of the size of the discount or premium, the current shape of the yield curve, and the distance to the bond's maturity.

Figure 7.6 uses Eq. (7.27) to compute the coupon and pull-to-par returns, respectively. We observe that the coupon return is positive, as it must be, and stands at approximately 38 basis points. The pull-to-par return is almost −17 basis points. This seems logically reasonable as our bond is currently trading at a premium and we should expect, when looking at this aspect in an isolated manner, that the

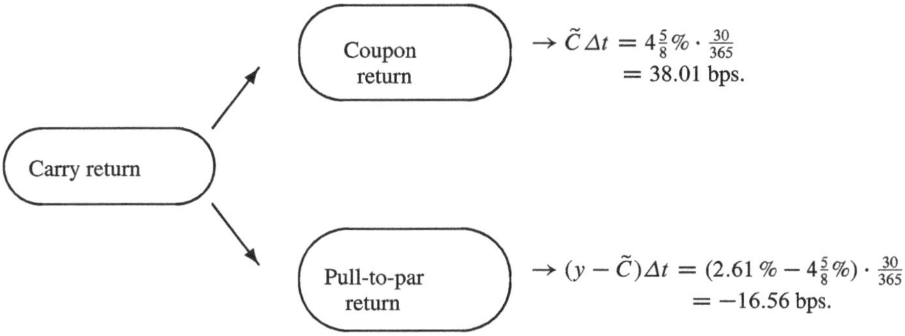


**Fig. 7.6** One possible carry return decomposition. This figure describes one way to further decompose the carry return for the US Agency bond described in Tables 7.5 and 7.6

bond-holder experiences some capital losses as this bond moves towards maturity. We should be clear, however, that merely because a bond is trading at a premium, it does not mean that for any given period it will experience a capital loss. The pull-to-par effect, for any given short period of time, is likely to be relatively small and overwhelmed by other elements of the return such as movements in credit spreads, the underlying treasury yield curve, or foreign-exchange rates.

Another way that one can think about the carry return is also useful in understanding the credit return. Examining Table 7.6, we note that on December 31, 2008 the US Treasury bond with the same tenor as our Fannie Mae bond had a yield of approximately 1.50 %. This quantity is termed the *equivalent Treasury* (ET) yield. The difference between our bond's actual yield, of 2.61 %, and the ET yield is called the *OA spread*.[14] We will also call it the credit spread, or simply the spread. In other words, it is the constant spread over the ET yield that is consistent with the price of our Fannie Mae bond. This spread summarizes the credit uncertainty—and to a certain extent the relative liquidity—of the bond. In our example, the OA spread was 111 basis points on 31 December 2008. The yield can be re-written as,

$$\begin{aligned}\text{Bond Yield} &= \text{ET Yield} + \text{OA Spread}, \\ y &= y_{\text{TRE}} + s_{\text{OAS}}.\end{aligned} \tag{7.28}$$

[14]OA spread stands for *option-adjusted spread*. There need not be any optionality in the underlying bond, although the term "option-adjusted" is used because the concept originated from instruments with embedded optionality such as mortgage-backed securities and callable bonds—in the presence of optionality, a model is required to compute the OA spread.

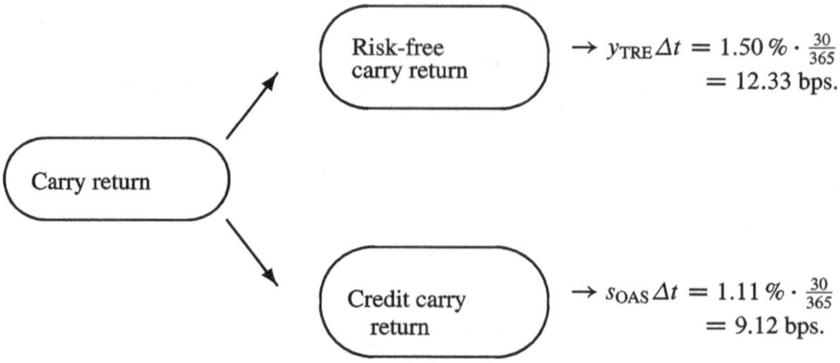


**Fig. 7.7** Another possible carry return decomposition. This figure describes an alternative approach to decompose the carry return for the US Agency bond described in Tables 7.5 and 7.6

If we return to the carry return and replace the yield with the result of Eq. (7.28), we have

$$\begin{aligned} \text{Carry return} &= y\Delta t, \\ &= \underbrace{(y_{\text{TRE}} + s_{\text{OAS}})}_{\substack{\text{Equation}\\ (7.28)}} \Delta t, \\ &= \underbrace{y_{\text{TRE}}\Delta t}_{\substack{\text{Risk-free}\\ \text{carry}\\ \text{return}}} + \underbrace{s_{\text{OAS}}\Delta t}_{\substack{\text{Credit}\\ \text{carry}\\ \text{return}}}. \end{aligned} \tag{7.29}$$

The logic behind this decomposition is quite simple. When one purchases a bond that has greater credit risk relative to the risk-free Treasury, one assumes an additional element of risk; one has risk exposure to movements in the credit spread. Additional risk requires additional compensation. The spread earned over the risk-free Treasury is, in this respect, an important part of the compensation for additional risk. The larger the spread over the treasury curve, the greater the risk of a negative credit event and, thus, the greater the compensation. By decomposing the carry return into the risk-free and credit components, one sees explicitly the additional compensation being earned for the increased credit risk.

Figure 7.7 applies this idea to our US Agency bond. For our US Agency bond, more than 40 % of the carry return can be attributed to the credit spread. Such a decomposition is particularly interesting for larger portfolios with a combination of credits, since it provides some insight into the average compensation received for the increased credit risk taken over risk-free Treasury securities.[15]

[15]Of course, there may be some additional compensation for favourable movements in the credit spread. This *compensation* may also be unfavourable. The credit-carry compensation, in contrast, is always positive.

### 7.2.2 Credit-Spread Return

The second category of return is credit, or spread, return. This is the return associated with movements in the credit spread. Returning to Eq. (7.18), we have that

$$\begin{aligned} \text{Credit return} &= -\text{Spread Duration} \cdot \text{Change in OA Spread}, \\ &= -D_S \cdot \Delta s_{\text{OAS}}, \\ &= -5.07 \cdot -0.15\,\%, \\ &= 76.05 \text{ bps.} \end{aligned} \tag{7.30}$$

Over the period, the credit tightened by 15 basis points from 111 to 96 basis points. This implies that there has been, over the period, some improvement in the credit quality of our US Agency bonds relative to its US Treasury equivalent. Although the overall return on our US Agency bond was approximately −111 basis points, both carry and credit have contributed positively to the return over the performance period.

For a single fixed-income instrument, there is little more to say about the credit return. It can, however, be further broken down into the sources of the credit return by allocating the credit return to different credit classes or ratings. If one had a portfolio of government-agency, supranational, and corporate bonds one could compute the credit return for each of these sub-categories. In the upcoming portfolio example described in the second half of this chapter, we will provide such a decomposition in the context of a European sovereign portfolio.

### 7.2.3 Treasury-Curve Return

The third category of return that we consider is treasury-curve, or simply curve, return. This is the return associated with movements in the underlying risk-free Treasury yield curve. In some cases, this curve in unequivocal. In the United States, for example, it would be hard to argue against the use of the US Treasury curve. In Europe, however, it is a bit more complex. One person might reasonably decide to use the German Treasury, or Bund, curve while another person might argue for the use of the French Treasury curve. The underlying principle, however, is that the treasury curve is associated with the lowest risk issuer in a given economy.[16]

It is useful to underscore that this so-called curve describes the common *curve* element associated with all fixed-income securities in a given currency. Imagine you are investing in the Eurozone and concentrating on, say, Portuguese government

[16]The notion of a risk-free issuer has come increasingly under fire and, thus, one could imagine situations where one might even be tempted to use the swap or Overnight Interest-Rate Swap (OIS) curve as the reference curve in a given economy.

bonds. You will certainly be interested in looking at the Portuguese sovereign yield curve. Conversely, an investor in Dutch government bonds would naturally be interested in the Dutch sovereign curve. Both curves, by themselves, are difficult to compare, because they contain idiosyncratic elements related to the specific creditworthiness of their respective sovereign. The Bund curve, conversely, represents a relatively *clean* view of the European interest rate environment. Breaking down the return contribution from this base Treasury perspective and the incremental element related to credit risk permits easier comparison in portfolio with multiple issuers with diverse credit profiles.

Once again, if we return to the description in Eq. (7.18) we define

$$\begin{aligned}\text{Curve return} &= -\text{Modified Duration} \cdot \text{Change in ET Yield}, \\ &= -D_M \cdot \Delta y_{\text{TRE}}, \\ &= -5.07 \cdot 0.42\,\%, \\ &= -214.22 \text{ bps.}\end{aligned} \tag{7.31}$$

The culprit for the loss on our US Agency bond position has been identified. An increase in the underlying equivalent treasury yield contributed to a sizeable −214 basis-point loss on our investment.

We'd now like to further break down the treasury curve return into the various sectors of the yield curve. We know, for example, that we lost about 214 basis points due to the movement of the sovereign yield curve. The question is what part of the yield curve? Admittedly, this is a bit of an artificial question for a single approximately 6-year bond. A more detailed decomposition is not required to answer this question. The simple answer is we lost due to movement in the yield curve in the 5- to 6-year sector. In a larger portfolio with multiple securities, however, the answer may not be quite as obvious. It is easier, given the smaller number of moving parts, to see how the alternative approaches work in a simple single-security setting.

All methods for decomposing the yield return attempt to do essentially the same thing. Given two treasury yield curves at two different points in time, one seeks a useful way to describe these movements over the period and sensibly allocate the associated curve return. Figure 7.8 provides just such a description of the movement in the underlying US Treasury curve from 31 December 2008 to 30 January 2009. We will consider *three* possible approaches for further breaking down curve return in this chapter, but be aware that we do not claim to exhaustively review all the possible alternatives.[17]

[17]There is no clear *correct* way to accomplish this task. Instead, there exists a number of reasonable alternatives.

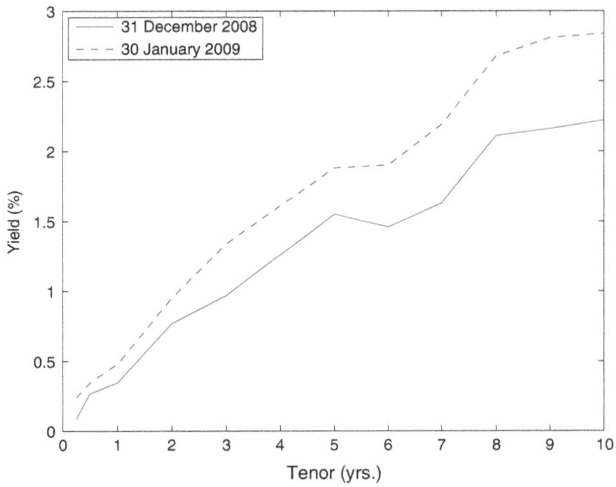


**Fig. 7.8** Underlying treasury curves. This figure describes the movement of the underlying US Treasury curve over the performance period. It is repeated from Fig. 2.6 in Chap. 2, where we discussed some pitfalls in the use of key-rate durations

The *three* alternatives include:

**Using a Model** The idea is to use a term-structure model, as described in the previous chapter to break down the yield curve into different risk factors. The risk factors will be the choice of state variables in the underlying model. In the example we consider, these state variables, or risk factors, will be level, slope, and curvature.

**Using Key-Rate Durations** One can merely follow Eq. (7.18) and use the key-rate durations to understand how the movement of different key sectors of the yield curve influence the overall curve return. As we will see in the discussion, a bit of caution is required in performing these calculations.

**Using an Ad Hoc Definition** There are a number of approaches, some of which are popular in commercial software packages, that use what we would consider to be *ad hoc* definitions of movements in the yield curve. We will consider one possible such definition.

To repeat, this is not an exhaustive list, but rather a description of some possible, and hopefully sensible, ways to perform a curve decomposition.

The first, and most complicated, approach makes use of a dynamic yield-curve model to further decompose the curve return. As a practical example, we will use the so-called Nelson-Siegel model described in detail in Diebold and Li [7], Diebold and Rudebusch [8], and Bolder [2]. The use of other models will follow a similar pattern, although each individual model is likely to have its own intricacies.[18]

[18]The following discussion is a bit involved, but recall that one of our basic principles was to avoid the use of black boxes.

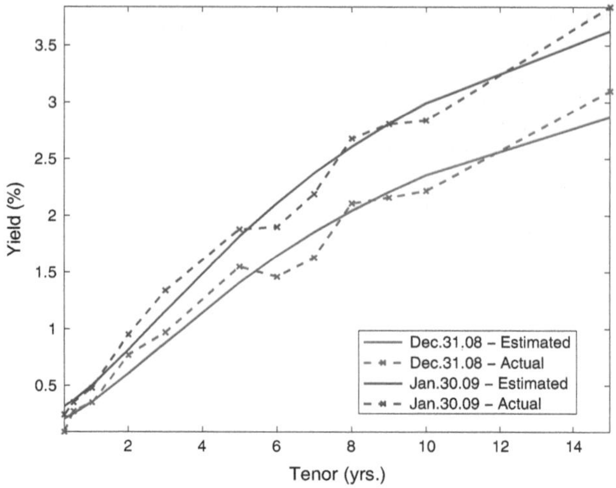


**Fig. 7.9** Nelson-Siegel curves. This figure describes the fit of the Nelson-Siegel curves to the true underlying US Treasury curves. The smooth mathematical functions used in the construction of the Nelson-Siegel model are not sufficiently flexible to perfectly capture the uneven form of the true US Treasury curve. As such, we will have to deal with the corresponding estimation error

Recall that the Nelson-Siegel model describes, at time $t$, a yield with tenor $T$ as,

$$y(t,T) = x_{0,t} + x_{1,t}\left(\frac{1-e^{-\lambda(T-t)}}{\lambda(T-t)}\right) + x_{2,t}\left(\frac{1-e^{-\lambda(T-t)}}{\lambda(T-t)} - e^{-\lambda(T-t)}\right), \tag{7.32}$$

where $\lambda$ is a fixed parameter and $\{x_{0,t}, x_{1,t}, x_{2,t}\}$ are the set of time-varying parameters. These time-varying parameters are, in the Nelson-Siegel setting, the risk factors and are defined as level, slope, and curvature, respectively. We can ease the notation somewhat by making the following replacements,

$$\begin{aligned} y(t,T) &= x_{0,t}\underbrace{1}_{\text{Level}} + x_{1,t}\underbrace{\left(\frac{1-e^{-\lambda\tau}}{\lambda(T-t)}\right)}_{\text{Slope}} + x_{2,t}\underbrace{\left(\frac{1-e^{-\lambda(T-t)}}{\lambda(T-t)} - e^{-\lambda(T-t)}\right)}_{\text{Curvature}}, \qquad (7.33)\\ &= x_{0,t}f_0(t,T) + x_{1,t}f_1(t,T) + x_{2,t}f_2(t,T). \end{aligned}$$

These functions $\{f_0(t,T), f_1(t,T), f_2(t,T)\}$ are termed the factor loadings and can be interpreted as the sensitivity, or exposure, of the yield to a movement in one of the risk factors. This is a very helpful representation. It describes any arbitrary yield, and thus the entire yield curve, as a linear combination of three risk factors. Figure 7.9 demonstrates the fit of the Nelson-Siegel model to the true underlying US Treasury curves from our example as shown in Fig. 7.8.[19]

[19]We can see from Eq. (7.33) that the Nelson-Siegel model is constructed with smooth mathematical functions. As a consequence, it will never be sufficiently flexible to capture the uneven form

Having examined how the Nelson-Siegel model describes the US Treasury curve at the beginning and end of our performance period, we now turn to see how the model describes the movement in the yield of our US Agency bond. In general, to describe the change in the yield of a security with tenor, $T$, over the interval, $[t_1, t_2]$, we make use of Eq. (7.33) as follows,

$$\underbrace{y(t_2,T) - y(t_1,T)}_{\Delta y} = \underbrace{\left(x_{0,t_2} f_0(t_2,T) + x_{1,t_2} f_1(t_2,T) + x_{2,t_2} f_2(t_2,T)\right)}_{y(t_2,T)} - \ldots \\ \underbrace{\left(x_{0,t_1} f_0(t_1,T) + x_{1,t_1} f_1(t_1,T) + x_{2,t_1} f_2(t_1,T)\right)}_{y(t_1,T)}. \tag{7.34}$$

At first glance, this expression cannot be readily simplified. If we assume, however, that the level, slope, and curvature terms—that is, $f_0(t,T)$, $f_1(t,T)$, and $f_2(t,T)$—remain fixed over the time period, then we can make some progress.[20] We will use, in a manner similar to our other calculations, the value of the level, slope, and curvature functions at the beginning of the period as,

$$\begin{aligned}\Delta y &\approx \underbrace{(x_{0,t_2} f_0(t_1,T) + x_{1,t_2} f_1(t_1,T) x_{2,t_2} f_2(t_1,T))}_{\approx y(t_2,T)} \\ &\quad - \underbrace{(x_{0,t_1} f_0(t_1,T) + x_{1,t_1} f_1(t_1,T) + x_{2,t_1} f_2(t_1,T))}_{y(t_1,T)}, \\ &\approx f_0(t_1,T)\underbrace{(x_{0,t_2} - x_{0,t_1})}_{\Delta x_0} + f_1(t_1,T)\underbrace{(x_{1,t_2} - x_{1,t_1})}_{\Delta x_1} + f_2(t_1,T)\underbrace{(x_{2,t_2} - x_{2,t_1})}_{\Delta x_2}, \\ &\approx \sum_{i=1}^{3} \overbrace{f_i(t_1,T)\underbrace{(x_{i,t_2} - x_{i,t_1})}_{\Delta x_i}}^{\Delta y_i}, \\ &\approx \sum_{i=1}^{3} \text{Exposure to Yield Factor}_i \cdot \text{Change in Yield Factor}_i.\end{aligned} \tag{7.35}$$

Using the Nelson-Siegel model, we have broken down the change in yield into a familiar form. It is now the sum of the product of the exposure to the various

---

of the true US Treasury curve—this is typical and it is to be expected that all models will lead to some level of estimation error.

[20]This is not such a strong assumption for relatively small time intervals as these are smooth functions. Indeed, for $f_0(t,T)$, there is no assumption involved as it is a constant function: $f_0(t,T) = 1$ for all value of $t$ and $T$. Moreover, since the first factor accounts for roughly 90 % of yield-curve movements in the Diebold and Li [7] dynamic Nelson-Siegel setting, we probably shouldn't be too worried.

**Table 7.7** Model-based yield movement decomposition

| Factor | Factor loading $(f_i)$ | Risk factors $x_{t_1}$ | $x_{t_2}$ | $\Delta x_i$ | Yield change $(f_i \Delta x_i)$ |
|---|---|---|---|---|---|
| Level | 1.0 | 0.040 | 0.051 | 0.010 | 104.0 |
| Slope | 0.4 | −0.039 | −0.048 | −0.009 | −35.6 |
| Curvature | 0.3 | 0.040 | 0.051 | −0.007 | −21.2 |
| Total | n/a | n/a | n/a | n/a | *46.2* |

This table outlines the various calculations in the application of our Nelson-Siegel model for the decomposition of the change in the 6-year US Treasury bond yield from 31 December 2008 to 30 January 2009.

yield-curve factors and the respective changes in these factors over the performance period. Incidentally, this is also precisely the form of our additive risk-factor decomposition.

Table 7.7 uses Eq. (7.35) to compute the yield change for our US Agency bond over the month and allocate the yield changes into the contribution from movements in the level, slope, and curvature of the US Treasury curve, respectively. What is immediately evident is that the predicted change in yield is 46 basis points, whereas the actual yield movement was, in fact, 42 basis points. One has, at this point, a choice. If one uses the raw predicted yield from the model, there will be an error in the performance figures on the order of about 20 basis points.[21] Conversely, one may attempt to adjust, or normalize, the three yield changes in such a manner as to ensure that the yield movement sums to 42 basis points.

We would recommend the normalization to avoid additional confusion in interpreting the error term and understanding that fact that there will always be estimation error involved in fitting models to the yield curve—Fig. 7.9 demonstrates this quite clearly. A simple approach to normalizing the yield would be the following,

$$\Delta \tilde{y}_i = \frac{f_i(t_1, T)\Delta x_i}{\sum_{k=1}^{3} f_k(t_1, T)\Delta x_k} \Delta y, \tag{7.36}$$

where $\Delta \tilde{y}_i$ is the yield-change for the $i$th Nelson-Siegel risk factor and $\Delta y$ is the true yield change over the period for the security. Table 7.8 summarizes the results of applying this normalization of the factor yield change contributions to our simple example.

[21] With a bond duration of 5 and a yield error of four basis points, the discrepancy will be about 20 basis points.

**Table 7.8** Model-based curve return decomposition

| Factor | Yield change ($f_i \Delta x_i$) | Normalized yield change | Curve return ($-D_M f_i \Delta x_i$) |
|---|---|---|---|
| Level | 104.0 | 95.1 | −482.2 |
| Slope | −35.6 | −33.5 | 169.7 |
| Curvature | −21.2 | −19.4 | 98.3 |
| Total | *46.2* | *42.2* | −214.2 |

This table outlines the various calculations in the application of our Nelson-Siegel model based decomposition of the Treasury curve return for our simple US Agency bond example.

Now, to see how we use this model-based information to decompose the curve return, we need to go back to the definition of the curve return,

$$
\begin{aligned}
\text{Curve return} &= -D_M \cdot \Delta y_{\text{TRE}}, \qquad (7.37)\\
&= -D_M \underbrace{\left(f_0(t_1,T)\Delta x_0 + f_1(t_1,T)\Delta x_1 + f_2(t_1,T)\Delta x_2\right)}_{\text{Equation (7.35)}},\\
&= -D_M \underbrace{\left(\Delta y_0 + \Delta y_1 + \Delta y_2\right)}_{\text{Equation (7.35)}},\\
&\approx -D_M \underbrace{\left(\Delta \tilde{y}_0 + \Delta \tilde{y}_1 + \Delta \tilde{y}_2\right)}_{\text{Equation (7.36)}},\\
&\approx \sum_{i=1}^{3} -D_M \Delta \tilde{y}_i .
\end{aligned}
$$

The model-based composition is thus conceptually quite straightforward. One merely breaks down the yield change into a linear combination of the yield movements associated with each of the individual model risk factors. The curve return associated with each of these yield movements is approximated in the usual manner through the use of the modified duration.

Table 7.8 computes the contribution of each curve factor based on the final development in Eq. (7.37). The Nelson-Siegel model suggests that, of the −214 basis points attributable to curve return, there is actually a loss of −482 basis points associated with movements in the level of the yield curve. This is offset by a 169 and 98 basis-point *gains* stemming from the slope and curvature factors.

If one does not regularly use the Nelson-Siegel model and it is not part of one's overall investment process, these results are quite difficult to interpret. A guiding principle for selecting one's treasury curve decomposition is to use an approach that is consistent with how your institution conceptualizes yield-curve movements. If a model-based approach is used then, by all means, a model-based decomposition is very useful. If not, then choice of specific model might, in some cases, prove confusing.

The second approach for decomposing the curve return is the use, as is described in Eq. (7.18), of the key-rate durations. Conceptually, this is very straightforward, but again, it is not as practically obvious to implement as one might imagine. The question is which yield changes should one use: the yield changes in the individual rates or the yield change of the security? Let's examine each in turn, in the context of our simple example, to understand the implications of this choice. If we decompose the return as follows,

$$\begin{aligned}\text{Curve return} &= -\text{Modified Duration} \cdot \text{Change in ET Yield}, \qquad (7.38)\\ &\approx -\sum_{i=1}^{n} \text{Key-rate duration}_i \cdot \text{Change in } i\text{th Key-Rate Yield},\\ &\approx -\sum_{i=1}^{n} \kappa_i \cdot \Delta y_{\text{TRE},i},\end{aligned}$$

then one is using *both* the key-rate durations and the key-rate yield movements. Note that we have added an approximation sign. This is because, despite the fact that the sum of the key-rate durations is generally very close to the modified duration, there is no guarantee that the first and second lines of Eq. (7.40) will be equal. This is due to the vagaries of the movement of the yield curve. In other words, the individual key-rate yields can move in such a way that the average movement is *not* equal to the change in the specific bond yield.[22]

Figure 7.10 reproduces our two US Treasury curves along with the changes in the individual key-rate durations as well as the change in our US Agency bond. We first verify that the yield of our bond—with a 5.79-year tenor—moved by 42 basis points. For this example, we have identified the key-rates as the 6-month, 2-year, 5-year, and 7-year nodes on the US Treasury yield curve. Observe that these yields, as well as the curve in general, both moved upwards and steepened. This is evidenced by an eight basis-point movement in the 6-month rate and a 56 basis-point increase in the 7-year sector.

Table 7.9 summarizes the results of applying Eq. (7.38) to our specific example. If we use the individual key-rate movements in conjunction with the key-rate durations, we arrive at a total curve return of −182 basis points. This is a substantial 32 basis points away from the estimated curve return of −214 basis points. There are situations where this error will be small, but others where it may be even larger.[23] Consequently, we recommend—in a manner similar to the model error in

[22]Again, the solution to this problem descends into a high level of detail, but we promised you no *black boxes*.

[23]Imagine a situation where the curve is essentially flat across all sectors except for a slight increase of 10 basis points in the 10-year sector—such a yield curve movement is unlikely, but entirely possible. The contribution of all other key-rate durations will be essentially zero as the yield change is zero. Only the 10-year sector will contribute to the curve return leading to an under- or overestimate of the overall curve return.

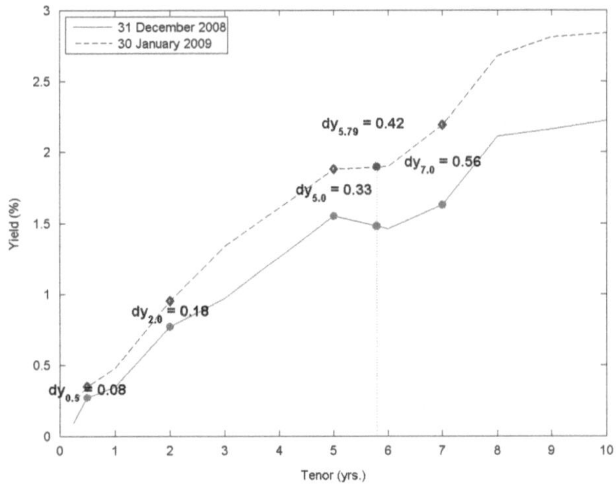


**Fig. 7.10** Curve return with key-rate durations. This figure describes the key-rate durations and the key-rate movements over the 1-month performance period for our US Agency example. The weighted average key-rate yield movements need not be equal to the yield movement of the actual bond

**Table 7.9** A word on key-rate durations

| Tenor | $\kappa_i$ | $\Delta y_i^{\text{TRE}}$ (%) | $\Delta y^{\text{TRE}}$ | $-\kappa_i \cdot \Delta y_i^{\text{TRE}}$ (bps) | $-\kappa_i \cdot \Delta y^{\text{TRE}}$ |
|---|---|---|---|---|---|
| 6M | 0.02 | 0.08 | 0.42 | (0.20) | **(0.90)** |
| 2Y | 0.14 | 0.18 | 0.42 | (2.60) | **(6.00)** |
| 5Y | 4.16 | 0.33 | 0.42 | (137.20) | **(175.60)** |
| 7Y | 0.75 | 0.56 | 0.42 | (42.00) | **(31.77)** |
| Total | 5.07 | | | (182.00) | **(214.22)** |

This table demonstrates that, if we use key-rate movements as described in Fig. 7.8, we will not necessarily reconcile to the overall curve change. This is because the individual key-rate movements need not correspond to the actual yield change of the underlying bond.

the previous example—the replacement of the individual key-rate yield movements with the yield movement of the individual security as follows,

$$
\begin{aligned}
\text{Curve return} &= -\text{Modified Duration} \cdot \text{Change in ET Yield}, \qquad (7.39)\\
&= -\sum_{i=1}^{n} \text{Key-rate duration}_i \cdot \text{Change in ET Yield},\\
&= -\sum_{i=1}^{n} \kappa_i \cdot \Delta y_{\text{TRE}}.
\end{aligned}
$$

The only possible discrepancy arises from differences in the modified duration and the sum of the key-rate durations—which is typically very small. Table 7.9 also summarizes the results for the curve decomposition described in Eq. (7.39). The results are extremely similar, although the total curve return is equal to the −214 basis points arising from the simple modified-duration calculation.

The third category of curve-return decomposition involves an *ad hoc* description of yield-curve movements. This is, in fact, a collection of approaches rather than a single, unique method for decomposing curve return.[24] We will demonstrate one possible approach that has attained a certain amount of popularity, but the reader should be aware that other techniques are also entirely possible. We consider an *ad hoc* approach that appears to be motivated by the idea that a large part of the variability of yield-curve movements can be explained by parallel movements of the yield curve.[25] With this idea in mind, it seeks to find a way to isolate the parallel movement of the curve from the other yield-curve movements. While there is no single unique way to accomplish this task, it can be reasonably approximated.

The approximation begins through the selection of a pivot point. The idea is that the parallel movement of the curve is defined as the distance between the pivot point at the beginning and ending date. Imagine, as is often the case in practice, that the 10-year yield is selected as the pivot point. On 30 January 2009 the 10-year US Treasury yield was 2.84 % while it was 2.22 % on 31 December 2008. The change in the 10-year yield over the period was 62 basis points. This is defined as the parallel movement in the yield curve over the period. One then uses the starting curve, on 31 December 2008, and computes an intermediate curve by adding 62 basis points to every point along the curve.

Figure 7.11 illustrates this computation. The lower curve represents the original curve on 31 December 2008, while the intermediate curve is described in a lighter tone. The difference between the intermediate curve and the final, true, yield curve on 30 January 2009 is thus assumed to describe the other non-parallel-movement

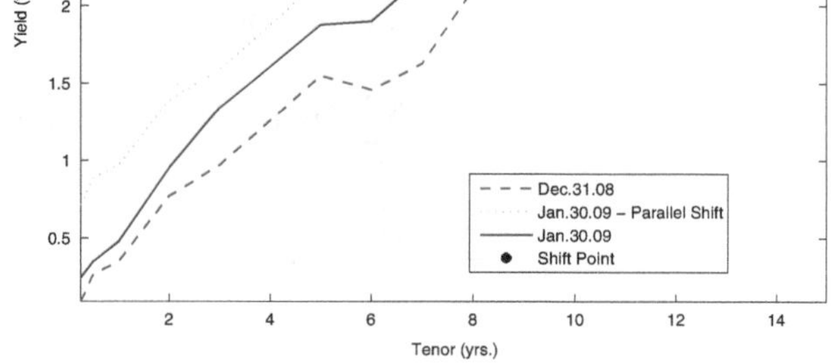


**Fig. 7.11** Ad hoc curve decomposition. This figure summarizes one possible *ad hoc* approach for describing the movement of the yield curve from one period to another. In this approach, one selects a pivot point, here it is 10 years, and breaks down the movement into two steps: a parallel movement along the pivot point and a second movement from the intermediate curve to the final curve

[24]See, for example, Zambruno [16], Gillet and Hommolie [9], Hansen and Sogaard-Andersen [10], or Murira and Sierra [15].

[25]This notion, originating from the work of Litterman and Scheinkman [13], is discussed in detail in Bolder [2].

yield curve effects—typically, this is assumed to represent relative changes in slope and curvature of the yield curve.

Let's make this somewhat more precise. We define the change in the treasury yield of US Agency security over the period as,

$$\Delta y_{\text{TRE}} = y_{\text{30 Jan}}^{\text{TRE}} - y_{\text{31 Dec}}^{\text{TRE}}. \tag{7.40}$$

If we further define the interpolated yield for our security from the intermediate curve—computed using the relative change in the 10-year yields—as $y_{\text{INT}}^{\text{TRE}}$ we can manipulate Eq. (7.40) algebraically as follows,

$$\begin{aligned}
\Delta y_{\text{TRE}} &= \underbrace{y_{\text{30 Jan}}^{\text{TRE}} - y_{\text{31 Dec}}^{\text{TRE}}}_{\text{Equation (7.40)}} + \left( \underbrace{y_{\text{INT}}^{\text{TRE}} - y_{\text{INT}}^{\text{TRE}}}_{=0} \right), \qquad (7.41)\\
&= \underbrace{\left(y_{\text{INT}}^{\text{TRE}} - y_{\text{31 Dec}}^{\text{TRE}}\right)}_{\substack{\text{Yield change from}\\ \text{parallel shift}}} + \underbrace{\left(y_{\text{30 Jan}}^{\text{TRE}} - y_{\text{INT}}^{\text{TRE}}\right)}_{\substack{\text{Yield change from}\\ \text{other effects}}},\\
&= \Delta y_{\text{Parallel}}^{\text{TRE}} + \Delta y_{\text{Other}}^{\text{TRE}}.
\end{aligned}$$

Armed with this convenient decomposition, the corresponding curve return is merely

$$\begin{aligned}
\text{Curve return} &= -D_M \cdot \Delta y_{\text{TRE}}, \qquad (7.42)\\
&= -D_M \underbrace{\left(\Delta y_{\text{Parallel}}^{\text{TRE}} + \Delta y_{\text{Other}}^{\text{TRE}}\right)}_{\text{Equation (7.41)}},\\
&= -D_M \underbrace{\Delta y_{\text{Parallel}}^{\text{TRE}}}_{\substack{\text{Curve return}\\ \text{from parallel}\\ \text{movements}}} - D_M \underbrace{\Delta y_{\text{Other}}^{\text{TRE}}}_{\substack{\text{Curve return}\\ \text{from other}\\ \text{movements}}}.
\end{aligned}$$

Table 7.10 uses Eqs. (7.40) to (7.42) to compute this *ad hoc* curve decomposition for our simple US Agency security. The result, not entirely dissimilar to the result from the model-based decomposition, is a loss of −314 basis points associated with a parallel shift over the period. This is offset by a gain of 110 basis points arising from the other yield-curve movements to bring the total curve return to the familiar −214 basis points.

To summarize, there are a variety of possible approaches for the further decomposition of treasury curve return. We have examined *three* possible alternatives: model-based, key-rate durations, and a possible *ad hoc* approach. Figure 7.12 outlines graphically the results of these three methods. None of these methods is superior to the others, but it bears repeating that the selected method should be consistent with one's investment process. If one's portfolio managers think about

**Table 7.10** Ad hoc curve return

| Movement type | $\Delta y_{TRE}$ Formula | Calculation (%) | Curve return |
|---|---|---|---|
| Parallel shift | $y^{\text{TRE}}_{\text{INT}} - y^{\text{TRE}}_{31\,\text{Dec}}$ | $2.12 - 1.50 = 0.62$ | $\mathbf{-5.07 \cdot 0.62\,\% = -314.34\ bps}$ |
| Other | $y^{\text{TRE}}_{30\,\text{Jan}} - y^{\text{TRE}}_{\text{INT}}$ | $1.92 - 2.12 = -0.20$ | $\mathbf{-5.07 \cdot (-0.20\,\%) = 100.40\ bps}$ |
| Total | $y^{\text{TRE}}_{30\,\text{Jan}} - y^{\text{TRE}}_{31\,\text{Dec}}$ | $1.92 - 1.50 = -0.42$ | $\mathbf{-5.07 \cdot 0.42\,\% = -214.22\ bps}$ |

This table implements the curve-return decomposition described in Eqs. (7.40) to (7.42) for our simple US Agency security.

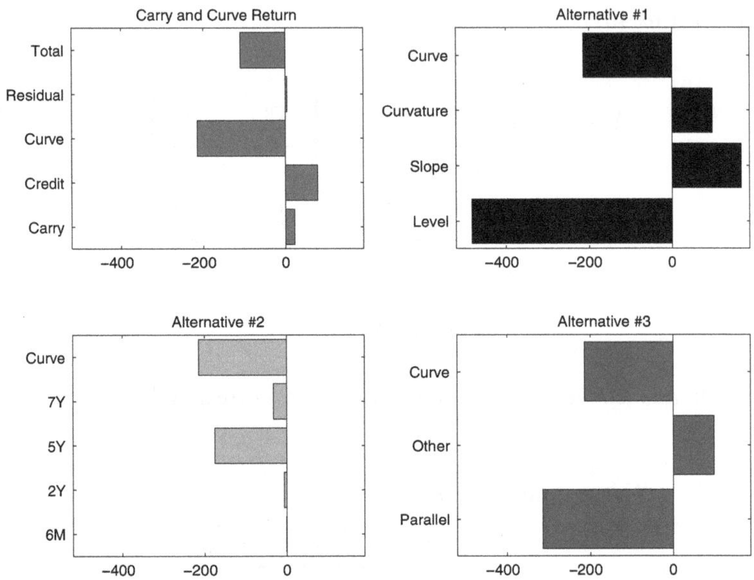


**Fig. 7.12** Curve return at a glance. This figure graphically describes the three alternative approaches to the decomposition of curve return described in the text

yield-curve movements in a specific manner, then it seems very sensible that, to the extent possible, one's curve decompositions should be structured in the same way. This will foster a more efficient assessment of one's investment positions—it should also lead to a broader acceptance and use of one's performance analysis.

### 7.2.4 Convexity Return

We have established that the curve return for our US Agency bond over the performance period was approximately −214 basis points. The true figure is slightly larger. The reason is that the computation in Eq. (7.31) is based on a linear approximation. For a large yield movement of 42 basis points, it will lack somewhat in terms of accuracy. This is precisely the role of the convexity return or correction. Returning to Eq. (7.18), we have that

$$\begin{aligned}\text{Convexity return} &= \frac{1}{2}\text{Convexity} \cdot (\text{Change in Security Yield})^2, \quad (7.43)\\ &= \frac{C}{2}(\Delta y)^2,\\ &= \frac{0.29}{2} \cdot \underbrace{(2.61\,\% - 2.88\,\%)}_{-0.27\,\%}{}^2,\\ &= 1.05 \text{ bps.}\end{aligned}$$

As a consequence, there is an additional one basis-point correction to the linear approximation made through the use of modified and spread duration.[26] The addition of the convexity correction does not imply that the approximation is now perfect, but rather that it is somewhat closer than suggested by Treasury curve and credit returns.

### 7.2.5 Foreign-Exchange Return

There has been, as yet, no mention of the foreign-exchange return. This is because we have assumed that the base currency of the investor is, in fact, US dollars (USD). In this situation, foreign-exchange rates are irrelevant as the investor has no exposure to the currency market. If, however, the investor's base currency is not USD, but instead Canadian dollars (CAD), then the situation changes.[27] Table 7.6 on page 210 has cleverly predicted exactly such a situation and has provided us with the actual USD/CAD exchange rates at the beginning and end of our performance period. This permits us to compute the CAD return of this investment as,

$$\begin{aligned}r_{\text{CAD}} &= \frac{S\left(\frac{\text{CAD}}{\text{USD}}\right)_{30\text{ Jan }09} V_{30\text{ Jan }09} - S\left(\frac{\text{CAD}}{\text{USD}}\right)_{31\text{ Dec }08} V_{31\text{ Dec }08}}{S\left(\frac{\text{CAD}}{\text{USD}}\right)_{31\text{ Dec }08} V_{31\text{ Dec }08}}, \quad (7.44)\\ &= \frac{1.2296 \cdot 110.48 - 1.2188 \cdot 111.72}{1.2188 \cdot 111.72},\\ &= -23.36 \text{ bps,}\end{aligned}$$

[26] Note that although one can easily use the modified and spread duration to separate the curve and credit return, such a decomposition is a bit trickier with convexity. This is due to the fact that the yield movement enters non-linearly in the convexity computation.

[27] A good source for the general discussion on attributing foreign-exchange rate movements can be found in Karnosky and Singer [11].

where $S\left(\frac{\text{CAD}}{\text{USD}}\right)_{\text{dd mmm yy}}$ denotes the spot USD-CAD exchange rate for a given date. Interestingly, although we have still lost money on this bond, the loss is only −23 basis points instead of the original −111 basis points. The reason is simple: as a Canadian investor, there has been a gain on the investment in a USD asset relative to one's base currency of CAD.[28] The following Taylor's series approximation permits the additive treatment of the foreign-exchange return,

$$\begin{aligned} r_{\text{CAD}} &\approx \underbrace{\frac{V_{30\text{ Jan }09} - V_{31\text{ Dec }09}}{V_{31\text{ Dec }09}}}_{\text{Equation (7.23)}} + \underbrace{\frac{S\left(\frac{\text{CAD}}{\text{USD}}\right)_{30\text{ Jan }09} - S\left(\frac{\text{CAD}}{\text{USD}}\right)_{31\text{ Dec }08}}{S\left(\frac{\text{CAD}}{\text{USD}}\right)_{31\text{ Dec }08}}}_{\text{CAD/USD foreign-exchange return}}, \\ &\approx \underbrace{\frac{110.48 - 111.72}{111.72}}_{-110.99\text{bps.}} + \underbrace{\frac{1.2296 - 1.2188}{1.2188}}_{88.61\text{bps.}}, \\ &\approx -22.38 \text{ bps.} \end{aligned} \tag{7.45}$$

We observe that a foreign-exchange gain of approximately 89 basis points offset the −111 basis-point loss on our US Agency security generating a total return approximation of −22 basis points. As Eq. (7.46) is an approximation, it deviates from the true return by almost a basis point.[29]

### 7.2.6 Pulling It All Together

Having patiently worked through all of the minute details of a performance attribution, we are now in a position to see the full picture. Given that we have seen a number of different ways to perform the attribution, we need to take a few

[28] The multiplicative form of Eq. (7.44) is not terribly practical for performance attribution purposes. It simply does not permit one to isolate the return contribution from foreign-exchange movements from the other risk factors.

[29] Recalling the precise, but perhaps less intuitive, expression from Chap. 3, we have,

$$\begin{aligned} r_{\text{CAD}} &\approx \underbrace{\frac{110.48 - 111.72}{111.72}}_{-110.99\text{bps.}} + \underbrace{\frac{1.2296 - 1.2188}{1.2188}}_{88.61\text{bps.}} \\ &\quad + \underbrace{\left(\frac{110.48 - 111.72}{111.72}\right) + \left(\frac{1.2296 - 1.2188}{1.2188}\right)}_{0.98\text{bps}}, \\ &\approx -23.36 \text{ bps,} \end{aligned} \tag{7.46}$$

which coincides exactly with Eq. (7.44). The additional interaction term is typically ignored because it is typically small and generally hard to interpret.

**Table 7.11** A possible performance attribution

| Return factor | Return contribution | |
|---|---|---|
| Risk-free carry | 12.3 | |
| Spread carry | 9.1 | |
| *Total carry return* | | *21.5* |
| *Credit return* | | *76.1* |
| Level | −482.2 | |
| Slope | 169.7 | |
| Curvature | 98.3 | |
| *Total curve return* | | *−214.2* |
| *Convexity return* | | *1.1* |
| *Foreign-exchange return* | | *88.7* |
| *Unexplained return* | | *3.4* |
| *Total return* | | *−23.4* |

This table attributes the performance of our US Agency bond, from the perspective of a Canadian investor, to the time, credit, curve, convexity, and foreign-exchange risk factors. It also breaks down the treasury curve return through the use of the Nelson-Siegel term structure model and allocates the carry return into the contribution from risk-free and spread yield.

choices to display the full results. For the purposes of exhibition, therefore, let's look at our investment from the perspective of a Canadian investor who uses the Nelson-Siegel model and prefers to break down the carry return into risk-free and spread contributions. Table 7.11 summarizes the overall results.

There should be no surprises in Table 7.11. We lost 23 basis points on our investment despite positive contributions from carry, credit-spread movements, and the appreciation of the USD relative to the CAD. The loss can be attributed almost exclusively to the substantial parallel shift upwards of the underlying US Treasury yield curve over this period. When all of the individual return contributions are brought together, there is a residual of approximately $3\frac{1}{2}$ basis points that is not explained. This is perfectly natural and, indeed, to be expected given that we have used approximations at a number of points along the way. Exact precision in not the goal, but rather a more extensive sense of what contributed to the return of one's security (or, more generally, portfolio) over the period. We hope the reader will agree that Table 7.11 provides a rather better sense of what happened to our US Agency bond investment over the period than the raw-return figure in Eq. (7.23).

## 7.3 Attribution of a Fixed-Income Portfolio

Having seen how one might attribute the performance of a single fixed-income security to a collection of risk factors, we turn to look at a more realistic portfolio example. There are a number of ways that we can construct a portfolio example,

**Table 7.12** High-level portfolio/benchmark comparison

| Characteristic | Benchmark | Portfolio |
|---|---|---|
| # of securities | 259 | 12 |
| Tenor (years) | 8.41 | 8.49 |
| Coupon (%) | 4.49 | 4.45 |
| Yield (%) | 3.28 | 3.12 |
| ET yield (%) | 2.58 | 2.56 |
| Spread (bps) | 70.60 | 56.18 |
| Modified duration | 6.17 | 6.23 |
| Convexity | 72.28 | 75.87 |
| Spread duration | 6.17 | 6.23 |

This describes, along a number of key dimensions, the differences between our sample portfolio and its underlying benchmark as of 30 January 2009.

but we will begin with the definition of the portfolio's benchmark. This seems like a reasonable place to start and immediately provides us with the general risk characteristics of the portfolio. The benchmark for our portfolio will be the JP Morgan EMU benchmark. Table 7.12 provides the basic details of this benchmark for February 2009. In brief, the benchmark is comprised of 259 securities issued from 14 different European sovereigns across the entire European yield curve. The average tenor of these securities is almost 8.5 years with a corresponding modified duration of slightly more than 6 years. The ET yield of 2.58 % is defined as the average German Bund yield associated with an identical tenor for each of the 259 bonds in the benchmark. The credit spread of 71 basis points represents the average spread of the benchmark bonds over the German Bund yield. This makes for an interesting choice of benchmark, because of the wide variety of different credits and the coverage of the entire European sovereign yield curve.

Now that we have a benchmark, the next step is to create a portfolio. While there are, of course, many ways to invest the funds, we elected for a portfolio generated using an optimization approach.[30] Table 7.12 provides a high-level comparison of the portfolio and the benchmark. Observe that the portfolio has only 12 securities with a slightly longer duration than the benchmark. With only 12 securities, it is not possible to invest in all of the 14 sovereigns in the benchmark. As we will see later in more detail, the portfolio is only invested in five of the 14 available sovereigns.

How do we make the transition from a single security to a portfolio of securities. Conceptually, it is very easy. We continue to make use of the general expression in Eq. (7.18) on page 206 for each individual security. Each term, however, needs to be

[30]The optimization approach is based on a linear-programming framework described in Appendix B.

appropriately weighted by its market value. We define the market weight of the $i$th bond as,

$$\omega_i = \frac{v_i}{\sum_{k=1}^{n} v_k}, \tag{7.47}$$

where $v_i$ is the market value of the $i$th bond in a portfolio of $n$ securities.

The weighting convention in Eq. (7.47) is quite sensible until one adds derivative positions into one's portfolio. If one introduces derivative positions—a good example would be rate or bond futures—computation of the market proportional weight is complicated by the fact that many derivative contracts typically have market values very close to zero. The consequence is a zero weight assigned to futures positions. As such Eq. (7.47) would lead to ignoring futures profit-and-loss. In this case, it is common to use the bond-equivalent exposure of the futures contract, defined as,

$$\text{Bond Equivalent Exposure} = (\text{\# of Contracts}) \cdot (\text{Contract Size}) \cdot (\text{Futures Price}). \tag{7.48}$$

If one places the futures exposure in both the numerator and denominator of the market-weight computation in Eq. (7.47), however, one ignores the leverage effect associated with the use of futures. The leverage effect occurs because, with bond or rate futures, one can increase or decrease one's portfolio duration (or alter key-rate durations) with no up-front cash investment. A reasonable approach for the computation of weights in a portfolio with bond futures, therefore, is to set;

$$\omega_i = \frac{\text{Exposure}_i}{\sum_{k=1}^{n} v_k}, \tag{7.49}$$

where the exposure of a plain-vanilla instrument, such as a bond, is equal to its market value. With such an approach, it is no longer the case that the portfolio weights sum to unity.

**Table 7.13** Portfolio active return

| Element | Portfolio ($r_p$) | Benchmark ($r_b$) | $r_p - r_b$ |
|---|---|---|---|
| Approximated return | 40.1 | 84.5 | −44.4 |
| Residual | 1.0 | 0.0 | 1.0 |
| True return | 41.1 | 84.5 | −43.4 |

Here is the portfolio, benchmark, and active return over the 1-month performance period.

Applying the weights in Eqs. (7.47) or (7.49) to the individual securities in one's portfolio, we arrive a reasonable approximation of our portfolio return, $r_p$, as follows,

$$\underbrace{\sum_{i=1}^{n}\omega_i r_i}_{r_p} = \underbrace{\sum_{i=1}^{n}\omega_i y_i \Delta t}_{\text{Carry return}} + \underbrace{\sum_{i=1}^{n}\omega_i \mathbb{I}_{\text{ILB}_i}\pi(t,t+1)}_{\text{Inflation carry return}} - \underbrace{\sum_{i=1}^{n}\omega_i \sum_{j=1}^{\nu}\kappa_j \Delta y_{\text{TRE},i}}_{\text{Treasury curve return}}$$
$$-\underbrace{\sum_{i=1}^{n}\omega_i D_{i,S}\Delta s_{\text{OAS},i}}_{\text{Credit return}} + \underbrace{\frac{1}{2}\sum_{i=1}^{n}\omega_i C_i(\Delta y)^2}_{\text{Convexity return}} \tag{7.50}$$
$$+\underbrace{\sum_{i=1}^{n}\omega_i\left(\sum_{k=1}^{\alpha}\mathbb{I}_{FX_{k,i}}\left(\frac{E_{k,t+1}-E_{k,t}}{E_{i,t}}\right)\right)}_{\text{FX return}}.$$

Equation (7.50) provides us with all of the information required to compute the return of the portfolio and its associated benchmark over the month of February 2009. Table 7.13 summarizes the approximated return from Eq. (7.50), the true return computed using the actual price movement in the securities, and the residual.

We draw two principal conclusions from Table 7.13. First, our approximation appears to be reasonably accurate at a portfolio level. The difference between the approximated and true returns for the portfolio and benchmark are both less than a single basis point.[31] Certainly, they are not exact, but perfect precision is not to be expected. The second point is that, despite a positive portfolio return of about 41 basis points, our investment choices have underperformed the benchmark by 43 basis points. Our task for the remainder of this chapter will be to use the

[31]The error on the benchmark approximation is almost zero. This is quite often, although of course not always, the case due to the large number of securities in the benchmark. With so many securities of different types, the various approximation errors often act to cancel one another out. This cancellation effect is smaller for a typical portfolio, as it is generally constructed with a relatively smaller number of securities.

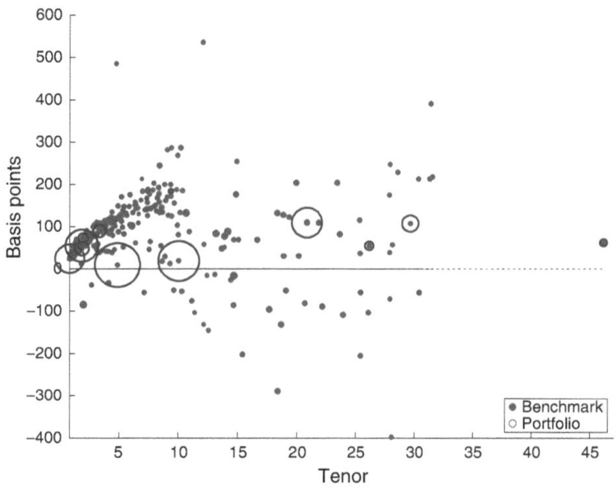


**Fig. 7.13** Total return by security. This figure describes the total return for each of the securities in the portfolio and the benchmark over the performance period. Note that the size of the observation denotes the relative size of the position in the portfolio or benchmark, respectively

performance-attribution methodology outlined in the previous section to determine the source of this under-performance.

From this point forward, there will be no further mathematical equations, calculations, or derivations. With a benchmark consisting of more than 250 securities, mathematical calculations are likely to prove extremely repetitive and tedious. Instead, we will make extensive use of figures and tables to successively examine each of the various return components. One can imagine that the following analysis is representative, albeit in perhaps a bit greater detail, of the type of reporting that one might provide to a senior management group for a given portfolio.

Figure 7.13 provides a scatter-plot of the total return incorporating each of the individual securities in the benchmark and the portfolio. We will use this figure repeatedly for each of the return types. The solid and open circles denote the benchmark and portfolio observations, respectively. The size of the circle, however, indicates the relative size of the position. Not surprisingly, therefore, we observe that the benchmark is comprised of a large number of small, almost identically sized positions. The portfolio, conversely, has a small number of positions of varying sizes.

Examination of Fig. 7.13 reveals that, in contrast to the benchmark, all of the 12 portfolio positions have generated a positive return over the performance period. Moreover, there is a substantially greater dispersion in the return of the individual benchmark positions with extreme returns in a range of more than plus or minus 400 basis points. There are a few large portfolio positions with relatively small returns relative to what appears to be the average benchmark return at an equivalent tenor.

The first return element of our performance attribution is the carry return. Table 7.14 provides a summary of the carry return for the portfolio as compared to the total return and the residual.

The portfolio has a slightly lower carry relative to the benchmark, although it has only contributed approximately one basis point to the under-performance of

**Table 7.14** Portfolio active return

| Return type | Portfolio ($r_p$) | Benchmark ($r_b$) | $r_p - r_b$ |
|---|---|---|---|
| Risk-free carry | 19.6 | 19.8 | −0.2 |
| Spread carry | 4.3 | 5.4 | −1.1 |
| *Total carry* | *23.9* | *25.2* | *−1.3* |
| *Other* | *16.2* | *59.4* | *−43.2* |
| *Residual* | *1.0* | *0.0* | *1.0* |
| *True total* | *41.1* | *84.5* | *−43.4* |

Here is the carry return for the portfolio, benchmark, and the difference over the period. All figures are in basis points.

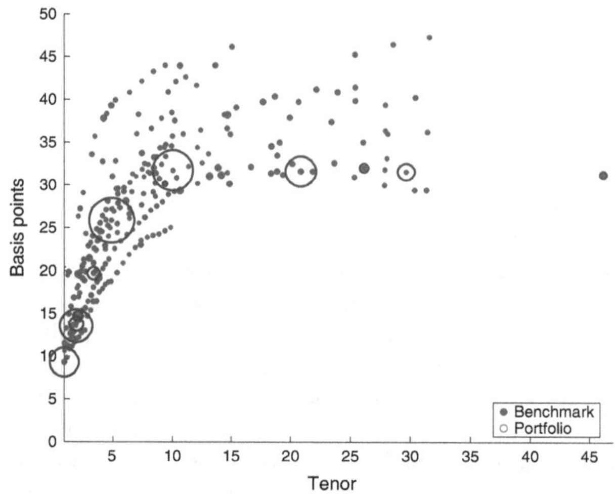


**Fig. 7.14** Carry by security. This figure describes the carry return for each of the securities in the portfolio and the benchmark over the performance period

the portfolio. On an overall basis, we would expect that the benchmark's carry return would be slightly greater than that of the portfolio simply by virtue of the benchmark's higher yield in Table 7.12. Although since the yield difference is only about 16 basis points, this is not a strong effect.

Table 7.14 further decomposes the carry return into the contribution from the risk-free yield—as measured by the equivalent German Bund yield—and the credit spread contribution. The majority of the one basis-point portfolio under-performance appears to stem from a slightly lower spread-carry return. This would suggest that, on average, the portfolio has a somewhat lower exposure to spreads or, at the very least, that the average credit quality of the portfolio is slightly better than the benchmark.

Figure 7.14 summarizes the carry return for each security over the performance period. Here we see quite clearly how the carry return contributes positively to the return of all securities over the period. It is also fairly clear that the

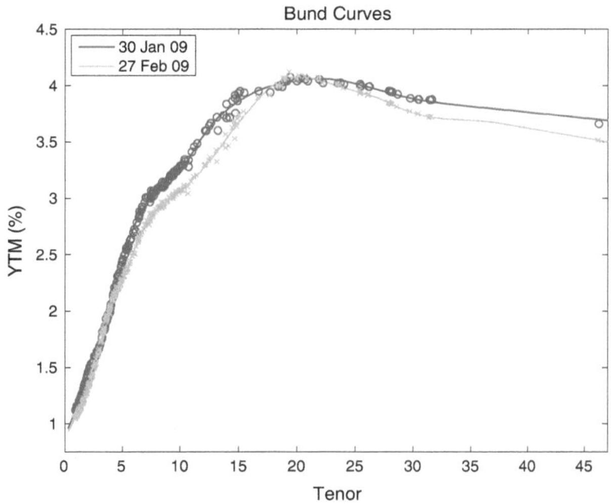


**Fig. 7.15** Movement in the German Bund yield curve. This figure illustrates the movement in the German Bund yield curve over the 1-month performance period

under-performance of our portfolio was not generated by differences in carry return between the portfolio and benchmark.

We now move to Treasury-curve return. The first step is to examine, in Fig. 7.15, what happened to the German Bund yield curve over the performance period. The German Bund yield curve appears relatively unchanged at the short end, but seems to have steepened somewhat over the 5- to 20-year sector. Rates came down, or remained essentially flat, at virtually all points along the curve with the sole exception of the 20-year sector, where yields have slightly increased. This would appear to be a reasonably positive environment for the treasury curve return.

The next step is to examine how the portfolio manager is positioned within this yield-curve environment. Figure 7.16 introduces an extremely useful graphic used to identify, at a glance, a portfolio manager's yield curve positioning relative to the benchmark. It provides the benchmark and portfolio key-rate durations along with the differences between the two for each of the seven key-rate tenors. The portfolio is overall slightly long duration with a modest short position in the 2- and 5-year sectors offset by a long position in the 10- and 20-year sectors. Such a position benefits when the yield curve flattens.

This portfolio is thus positioned for a flattening of the Bund curve.[32] This does not bode particularly well given that Fig. 7.15 suggests that the curve has not flattened and, if anything, may have slightly steepened.

[32]A curve flattening typically involves increases in the yield curve at the short end offset by decreases in the long end. The short duration position, at the short end, will outperform the benchmark in the face of yield increases associated with a curve flattening. Simultaneously, the long duration position at the long end of the yield curve will outperform the benchmark as rates fall.

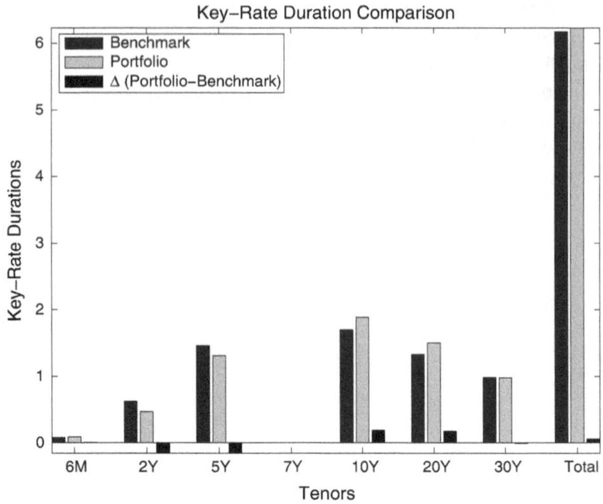


**Fig. 7.16** Key-rate comparison. This bar chart compares the key-rate durations of the portfolio to the benchmark. Observe that the portfolio has a slight curve-flattening position

**Table 7.15** Curve return

| Return type | Portfolio ($r_p$) | | Benchmark ($r_b$) | | $r_p - r_b$ | |
|---|---|---|---|---|---|---|
| Risk-free carry | 19.6 | | 19.8 | | −0.2 | |
| Spread carry | 4.3 | | 5.4 | | −1.1 | |
| *Total carry* | | *23.9* | | *25.2* | | *−1.3* |
| 6M | 0.8 | | 0.8 | | 0.0 | |
| 2Y | 6.8 | | 5.6 | | −1.2 | |
| 5Y | 21.8 | | 22.2 | | −0.4 | |
| 10Y | 38.3 | | 33.9 | | 4.3 | |
| 20Y | −2.2 | | 6.2 | | −8.2 | |
| 30Y | 9.6 | | 9.7 | | −0.1 | |
| *Total curve* | | *75.2* | | *78.3* | | *−3.2* |
| *Other* | | *−59.7* | | *−19.7* | | *−40.0* |
| *Convexity* | | *0.7* | | *0.8* | | *−0.1* |
| *Residual* | | *1.0* | | *0.0* | | *1.0* |
| *True total* | | *41.1* | | *84.5* | | *−43.4* |

This table summarizes the portfolio, benchmark, and active return over the performance period. It also breaks out the return by their key rates.

With this understanding of the movement in the German Bund curve and the positioning of the portfolio relative to the benchmark, we can now turn to examine the treasury curve return. Table 7.15 provides a decomposition of the curve return by key-rate duration. We immediately see that the portfolio generated 75 basis points of curve return relative to 78 basis points for the benchmark. This amounts to a modest under-performance of three basis points.

We draw two broad conclusions from Table 7.15. First, as expected, the generally positive movements in the German Bund curve over the period generated a fairly

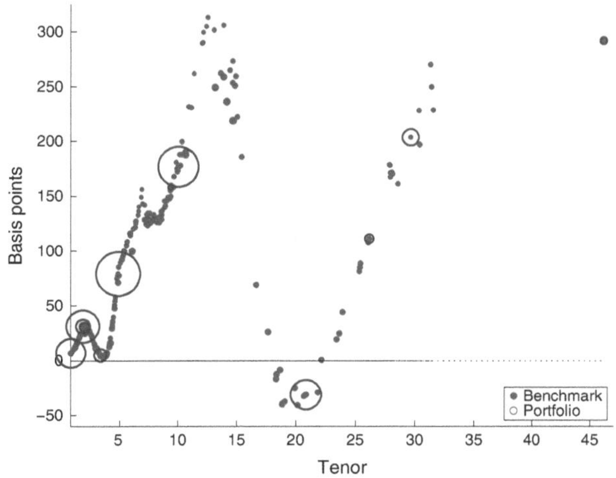


**Fig. 7.17** Curve by security. This figure describes the curve return for each of the securities in the portfolio and the benchmark over the performance period

sizeable curve return for both the portfolio and the benchmark. Second, the curve flattening position taken by the portfolio manager was not exactly successful, but with an under-performance of only three basis points, it also did not particularly harm the portfolio.

Examining the key-rate duration decomposition of the curve return, we see that most of the absolute return was generated by downward movements in 5-, 10-, and 30-year German Bund yields. On a relative basis, the portfolio outperformed slightly in the 10-year sector, by virtue of its long duration position in this part of the curve, but lost fairly significantly in the 20-year sector. Figure 7.17 graphically supports these conclusions. We see clearly, for example, that virtually all instruments, with the exception of those immediately around the 20-year German Bund yield, experienced positive curve return over the performance period. This is interesting and illuminating, but we have not yet found the reason behind the more than 40 basis-point under-performance of the portfolio relative to the benchmark.

The final return element that we will consider in the context of our portfolio example—the spread return—will solve this problem. It should now be clear that—with a lack of foreign-exchange exposure and only marginal under-performance in carry and curve return—movements in credit spreads will be the principal reason behind our portfolio's under-performance. Let us begin with an examination of the movements in the credit spreads. Figure 7.18 summarizes the movement in the spread for each of the individual securities in both the portfolio and benchmark.

Over the period, the majority of benchmark bonds, particularly from the 5-year sector and beyond, appear to have experienced a spread widening. This would suggest an overall negative contribution from credit return—spread widening at the long-end is particularly problematic because of the larger spread duration of these

**Fig. 7.18** OA spread movements by security. This figure describes the movements in the OA spread for each of the securities in the portfolio and the benchmark over the performance period

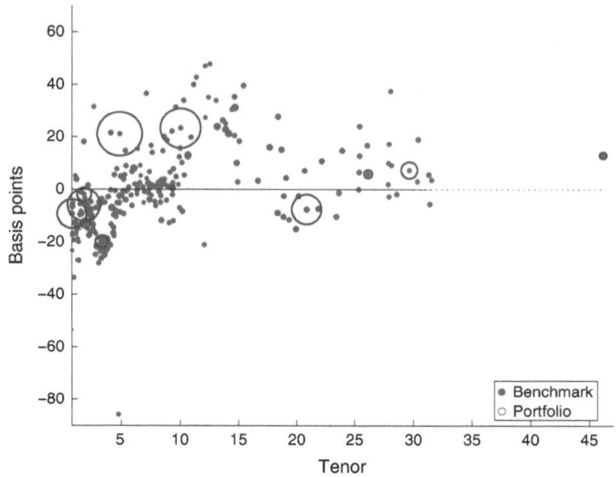


**Table 7.16** Sovereign portfolio positions

| Issuer | Portfolio (%) | Benchmark (%) | $\Delta$ (%) |
|---|---|---|---|
| Austria | 49.6 | 3.8 | *45.8* |
| France | 20.1 | 20.1 | −0.1 |
| Netherlands | 5.2 | 5.2 | 0.0 |
| Finland | 1.2 | 1.2 | 0.1 |
| Germany | 24.0 | 23.6 | 0.4 |
| Other countries | 0.00 | 46.2 | *−46.2* |
| *Total* | *100.0* | *100.0* | *0.0* |

This table summarizes the proportion of the portfolio, relative to the benchmark, held in the various sovereign government markets.

positions.[33] The portfolio has a number of positions with spread tightening, but two quite sizeable positions with substantial spread *widening* in the 5- and 10-year sectors, respectively.

Let us now turn to examine the relative credit positioning of the portfolio relative to the benchmark. Table 7.16 outlines the proportion, in market value terms, of the portfolio and the benchmark allocated to a number of key sovereign issuers in the benchmark.

The portfolio is only invested in five of the 14 sovereign issuers found in the benchmark. It is heavily overweighted to Austria with almost half of the portfolio allocation invested in Austrian government bonds. It is, however, very closely matched to four other important issuers: France, Germany, Finland, and the Netherlands. This is admittedly a fairly extreme example that a prudent portfolio

[33]This is the reason that it is useful to understand the overall spread duration, or spread sensitivity, in one's portfolio. Spread exposure to short-duration bonds is *not* equivalent to similar exposure to long-duration bonds.

**Table 7.17** Adding spread return

| Return type | Portfolio ($r_p$) | | Benchmark ($r_b$) | | $r_p - r_b$ | |
|---|---|---|---|---|---|---|
| Risk-free carry | 19.6 | | 19.8 | | −0.2 | |
| Spread carry | 4.3 | | 5.4 | | −1.1 | |
| *Total carry* | | *23.9* | | *25.2* | | *−1.3* |
| 6M | 0.8 | | 0.8 | | 0.0 | |
| 2Y | 6.8 | | 5.6 | | −1.2 | |
| 5Y | 21.8 | | 22.2 | | −0.4 | |
| 10Y | 38.3 | | 33.9 | | 4.3 | |
| 20Y | −2.2 | | 6.2 | | −8.2 | |
| 30Y | 9.6 | | 9.7 | | −0.1 | |
| *Total curve* | | *75.2* | | *78.3* | | *−3.2* |
| Austria | −68.5 | | −6.1 | | −62.4 | |
| Belgium | 0.0 | | −1.3 | | 1.3 | |
| Germany | 13.0 | | 7.8 | | 5.2 | |
| Spain | 0.0 | | 4.8 | | −4.8 | |
| Finland | 0.3 | | 0.3 | | 0.0 | |
| France | −5.9 | | −4.3 | | −1.7 | |
| Greece | 0.0 | | −1.4 | | 1.4 | |
| Ireland | 0.0 | | −2.3 | | 2.3 | |
| Italy | 0.0 | | −14.6 | | 14.6 | |
| Netherlands | 1.4 | | −0.7 | | 2.2 | |
| Portugal | 0.0 | | −2.0 | | 2.0 | |
| Other | 0.0 | | 0.0 | | 0.0 | |
| *Total spread* | | *−59.7* | | *−19.7* | | *−40.0* |
| *Convexity* | | *0.7* | | *0.8* | | *−0.1* |
| *Residual* | | *1.0* | | *0.0* | | *1.0* |
| *True total* | | *41.1* | | *84.5* | | *−43.4* |

This table now shows the complete performance attribution for our sample portfolio.

manager would likely seek to avoid.[34] Nevertheless, it makes for an interesting example.

Table 7.17 provides us with the full performance attribution for our portfolio, relative to its benchmark, over the month of February 2009. Note that, in addition to a full analysis of credit-spread return, we have also quietly added the convexity return. Given relative modest movements in yields over the period, the convexity return for both the portfolio and the benchmark are less than one basis point.

More than 90 %, or 40 basis points, of the portfolio's 43 basis-point underperformance can be explained by adverse movements in credit spreads. In particular, the overweight position in Austrian government bonds proved disastrous by

[34]The ex-ante tracking error of this portfolio would, quite likely, be quite sizeable given the large spread exposure not only to Austria, but also to the other un-invested sovereigns in the benchmark.

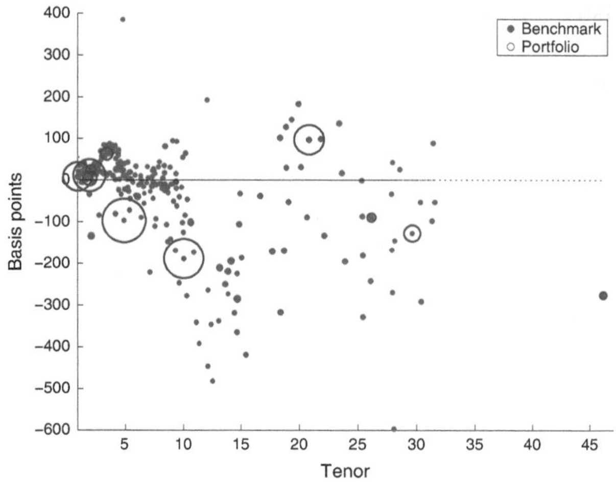


**Fig. 7.19** Spread return by security. This figure describes the movements in the spread return for each of the securities in the portfolio and the benchmark over the performance period

contributing almost 69 basis points of under-performance during the period. While the widening in Austrian credit spreads relative to the German Bund market was not enormous—Fig. 7.18 suggests it was about 20 basis points—the extreme allocation to these Austrian bonds implied a large negative credit return. The Austrian position was slightly offset by favourable movements in German Bunds and the failure to invest in Italian government bonds. Italian government bonds were the largest single negative contributor to the benchmark under-performance at almost −15 basis points. A zero weight in Italian bonds, therefore, was positive for the portfolio.

Figure 7.19 concludes this section with an illustration of the credit return for each of the positions in the portfolio and benchmark. The two large portfolio positions, in Austrian bonds, clearly contributed negatively to return. We also observe a few small negative positions and a couple of modest positive credit returns.

After this lengthy analysis, we have substantially more knowledge about the reasons behind the 43 basis-point under-performance of our portfolio than we did at the beginning. We learned that carry and treasury curve returns were both positive in absolute terms and marginally small in relative terms. The carry return suggests that the portfolio had a slightly higher credit quality than the benchmark. We also learned that the portfolio manager had positioned the portfolio for a curve flattening and although it was not a successful position, it ultimately did little or no damage to the portfolio relative return. In the end, we found that adverse credit movements were the principal culprit of the portfolio's under-performance. More specifically, an extreme position in Austrian government bonds generated almost 70 basis of under-performance relative to the benchmark. Armed with this information, senior management gains a much better understanding of the underlying reasons for a portfolio's under- or over-performance and is well positioned to manage the portfolio managers.

## 7.4 Closing Thoughts

This chapter brought to life, through two practical examples, the additive risk-factor decomposition derived through the application of the Taylor series expansion. We saw how, with the use of generally available data and some fairly simple computations, one can gain significant insight into the performance of one's portfolio over a given period of time. As always, there is more to the story. In the subsequent chapter, we will expand on the ideas in this chapter and address some of the most important additional complexities involved in performance attribution that were not immediately obvious in our *two* practical one-period examples.

## References


1. C.R. Bacon, *Practical Portfolio Performance Measurement and Attribution* (Wiley, New York, 2008)
2. D.J. Bolder, Modelling term-structure dynamics for portfolio analysis: a practitioner's perspective. Bank of Canada: Working Paper 2006–48, Ottawa, 2006
3. G.P. Brinson, L.R. Hood, G.L. Beehower, Determinants of portfolio performance. Financ. Anal. J. **42**(4), 39–44 (1986)
4. S. Campisi, D. Spaulding, *Readings in Fixed Income Performance Attribution* (TSG Publishing, Somerset, 2007)
5. A. Colin, *Fixed Income Attribution* (Wiley, West Sussex, 2005)
6. A. Colin, Fixed income attribution with minimum raw material. J. Perform. Meas. **11**, 50–57 (Winter 2007)
7. F.X. Diebold, C. Li, Forecasting the term structure of government bond yields. University of Pennsylvania Working Paper, 2003
8. F.X. Diebold, G.D. Rudebusch, *Yield Curve Modeling and Forecasting: The Dynamic Nelson-Siegel Approach* (Princeton University Press, Princeton, 2013)
9. P. Gillet, B. Hommolie, Fixed-income attribution: a combined methodology. J. Perform. Meas., 64–77 (Summer 2006)
10. L.B. Hansen, P. Søgaard Andersen, Fixed-income performance attribution: a flexible approach. J. Perform. Meas., 6–21 (Spring 2006)
11. D.S. Karnosky, B.D. Singer, *Global Asset Management and Performance Atrribution* (The Research Foundation of the Institute of Chartered Financial Analysis, Charlottesville, 1994)
12. J. Knight, S. Satchell, *Performance Measurement in Finance: Firms, Funds, and Managers* (Butterworth-Heinemann, Woburn, 2002)
13. R. Litterman, J. Scheinkman, Common factors affecting bond returns. J. Fixed Income **1**, 54–61 (1991)
14. J. Menchero, J. Hu, Errors in transaction-based performance attribution. J. Perform. Meas., 57–69 (Fall 2003)
15. B. Murira, H. Sierra, Fixed-income attribution: a unified framework—part I. J. Perform. Meas. **11**(1), 23–35 (Fall 2006)
16. G.M. Zambruno, Some refinements on fixed-income performance attribution. Eur. J. Oper. Res. **185**, 1574–1577 (2006)

# Advanced Performance Attribution 8

*Things should be made as simple as possible, but not any simpler.*

Albert Einstein

The previous chapter introduced the basic ideas behind fixed-income performance attribution, but also left part of the overall story untold. Our previous development was predicated on the simple assumption that all positions in one's portfolio remain fixed over the performance horizon. This is termed the *buy-and-hold* assumption.

The buy-and-hold assumption may, at times, be very close to reality. A portfolio manager that passively manages a fixed-income portfolio against an external benchmark will typically not engage in many trades over a monthly reporting horizon. Over a 2-month horizon, however, even for a passive portfolio manager, it is less likely to be realistic. The vast majority of fixed-income benchmarks is revised on a monthly basis and even a passive manager will need to adjust his or her portfolio positions to maintain benchmark replication.

The buy-and hold assumption may also, at times, be seriously violated. An active portfolio manager who takes duration and credit positions relative to his benchmark, may change his positions on a weekly, or even daily, basis. A buy-and-hold approach to performance attribution will not take the time-varying nature of such a portfolio into consideration.

Given the relatively high likelihood of this assumption being violated, it is important that it be relaxed. While our basic approximation framework will fortunately be preserved, there are a number of thorny issues that arise when one attempts to relax this assumption. These issues are sufficiently important that one would not be in a position to implement a reasonable performance attribution system in one's organization without explicitly addressing them. As a consequence, this chapter's principal focus centres around a discussion of the challenges—and some potential solutions—associated with a relaxing the buy-and-hold assumption.



D.J. Bolder, *Fixed-Income Portfolio Analytics*,
DOI 10.1007/978-3-319-12667-8_8

As we have seen, our performance attribution framework is founded on an approximation, which inevitably gives rise to residual terms. On average, this unexplained, or residual, return should be small. Ensuring that this is indeed the case, however, is best not left to chance. A secondary focus of this chapter, therefore, is to consider a few techniques for determining the long-term efficiency, or accuracy, of one's performance attributions through systematic examination of these residual terms.

## 8.1 Truth in Advertising

Performance attribution is *not* an exact science. Perfect allocation of a portfolio's return to a meaningful collection of risk factors is simply *not* possible. We could, in the development of our additive decomposition, have used a 6th order Taylor Series expansion with all possible cross terms and achieved an extremely high degree of accuracy. The individual terms, however, would have been virtually impossible to interpret. What makes performance attribution challenging, therefore, is finding the right trade-off between accuracy and interpretability. Some subjectivity is inevitable in this process. Whether it relates to the return approximation of a given security to a set of risk factors or the choice of how one presents the results, choices must be made. Under the buy-and-hold assumption, performance attribution requires a number of:

- approximations;
- assumptions; and
- a few decisions.

In this respect, relaxation of the buy-an-hold assumption is conceptually similar—it also requires a new set of approximations, assumptions, and decisions.

Let us return to the fundamental result from the previous chapters. Our additive decomposition of the return of a fixed-income portfolio is the result of a Taylor series expansion of the security price function. This approximation operates on the individual security level. At the portfolio level, the individual security-return decompositions must be aggregated. A weighting scheme, therefore, is necessary.

Consider the time interval, $[t, t+1]$ and define the market weight of the $i$th position as,[1]

$$\omega_i(t) = V_i(t) \bigg/ \sum_{k=1}^{n} V_k(t)\,, \tag{8.1}$$

[1]Recalling that the market weights are slightly different when dealing with bond or rate futures. In this case, the numerator is the bond-equivalent exposure whereas the denominator remains the market value of the portfolio.

where $V_i(t)$ is the marked-to-market value of the $i$th bond at time $t$. Armed with these weights, the portfolio return, $r_p$, is well approximated by,

$$\underbrace{\sum_{i=1}^{n}\omega_i(t)r_i}_{r_p} = \underbrace{\sum_{i=1}^{n}\omega_i(t)y_i\Delta t}_{\text{Carry}} + \underbrace{\sum_{i=1}^{n}\omega_i(t)\mathbb{I}_{\text{ILB}_i}\pi(t,t+1)}_{\substack{\text{Inflation carry}\\ \text{return}}}$$
$$\underbrace{-\sum_{i=1}^{n}\omega_i(t)D_{i,\text{M}}\Delta y_{\text{TRE},i}}_{\text{Curve}} \underbrace{-\sum_{i=1}^{n}\omega_i(t)D_{i,\text{S}}\Delta s_{i,\text{OAS}}}_{\text{Credit}}$$
$$\underbrace{+\frac{1}{2}\sum_{i=1}^{n}\omega_i(t)C_i(\Delta y)^2}_{\text{Convexity}} + \underbrace{\sum_{i=1}^{n}\omega_i(t)\left(\sum_{k=1}^{\alpha}\mathbb{I}_{FX_{k,i}}\left(\frac{E_{k,t+1}-E_{k,t}}{E_{i,t}}\right)\right)}_{\text{FX}}. \tag{8.2}$$

The return of the portfolio over a given time interval is merely the weighted average of the returns of the individual positions in the portfolio.

This should be completely familiar. Let us now pose a simple question. What happens if one or more of the positions change—through purchases, sales, or simply maturity—over the performance interval, $[t, t+1]$? The simple answer is: it makes things significantly more complicated.

The complexities arise on, at least, *two* levels. First, it is no longer clear exactly how the returns over a given period should be weighted. If the portfolio's position in a given security at time $t$ is different from the value at time, $t+1$, does one employ the $t$ or $t+1$ value?[2] It's also entirely possible that the position did not even exist at the start of the performance period. Resolving the weighting of the individual returns is a key challenge associated with handling changing positions.

The second challenge is slightly more subtle, but equally difficult to handle. If a position changes over the reporting period, it is reasonable to ask what exposures and factor changes should be used for the return approximations. Should one use the traditional factor changes—computed using starting and ending values—even if there were significant differences between these values and the market conditions at the time of any transactions? What if a position exists in the portfolio at the beginning of the reporting period, but was sold at some point over the reporting horizon? Should one continue to approximate the return with the factor changes over the entire period? *Probably not* is the answer to each of these questions.

[2]We have used the term *position* and not market value. This is because, even in a buy-and-hold setting, the market value of a given position is very likely to be different at the beginning and ending of the performance period. Thus, questions arise regarding the appropriate weights even in the buy-and-hold setting.

To summarize, handling the changing positions arising from relaxation of the buy-and-hold assumptions implies that it is *not* always obvious:

- which position weight (i.e., $\omega_i(t)$) should be used;
- what factor changes (i.e., treasury-yield, OA-spreads, and exchange-rate movements) to use; and
- which factor sensitivities (i.e., yields, modified, spread, and key-rate durations) should be employed.

These three points would seem to suggest that the relatively straightforward expression in Eq. (8.2) can no longer be generally applied once we decide to relax the buy-and-hold assumption.

The situation is fortunately not quite so dire as it might appear. There is a relatively simple solution that attenuates these challenges, although it does not entirely resolve them.

## 8.2 Daily Attribution

A reasonable solution to addressing the aforementioned challenges associated with relaxation of the buy-and-hold assumption is to perform *daily* performance attributions. Why is this helpful? It reduces complexity and provides clarity about the set of possible outcomes. From one day to the next, only *three* things can happen to a portfolio instrument. A portfolio position may be

- unchanged;
- reduced; or
- increased.

These three outcomes, of course, also describe the set of possible movements in a portfolio over periods longer than one day. Nevertheless, for longer periods, the time dimension becomes more complicated. A position in a given instrument may be increased over a weekly time interval, but one also needs to know exactly *when* it was increased. Over a single business day, we know exactly on what date the transaction occurred, although we may not know the exact time of day.[3] Over lengthy periods, a position in an instrument may possibly be added after the beginning of the performance interval and sold before the end. The analyst computing the performance attribution would have no knowledge of the position and ignore it completely. Even with knowledge of a purchase and sale occurring during

[3]Daily attributions will not be able to reasonably handle intra-day trading. That is, a purchase and resale of a position within a single business day. As this situation neither occurs within our institution nor, to our knowledge, within other official institutions, we feel confident in ignoring this possibility. If this is an important activity, then one would require a more nuanced approach.

the middle of the performance period, it is not obvious how its impact could be incorporated into our framework. A single business-day horizon simplifies matters by providing a clear description of the possible outcomes.

The three possible position outcomes over a single business day are generally related to actions of the portfolio manager. In the first case where the position is unchanged, it is related to the inaction of the portfolio manager. The second case involves the reduction of a current position. This essentially involves the sale of some, or indeed, all of a position.[4] The third, and final, case involves an increase in the security's position. In other words, it is a purchase of more of an existing position or the creation of a new position in the portfolio.

We now need to be very precise about our terminology. An *instrument* represents a specific security—be it a bond, future contract, cash account, or OTC derivative contract. A *position*, or instrument position, denotes the amount that one holds in a given instrument or security. To this point, we have always implicitly assumed that an instrument and a position are inseparable. We now need to think about them separately. This is because our performance-attribution computations occur at the instrument level, but we are trying, though the relaxation of the buy-and-hold assumption, to handle changing positions in these instruments.

Some effort is required to systematically incorporate this logic into our performance-attribution framework. The first step in this process is to use some basic set logic to transform our *three* qualitative outcomes into a more quantitative set of events. To begin this process, let us define the set of instruments in our portfolio at a given point in time. We define,

$$I_t \triangleq \{\text{Portfolio instruments at time } t\}. \tag{8.3}$$

Thus $I_t$ is the collection of instrument positions, in a given portfolio, at a given point in time, $t$—let's call it the instrument set. This object is particularly useful when two instrument sets are combined across two different points in time. If we set our business-day interval as $[t, t+1]$, then the following set

$$I_t \cup I_{t+1}, \tag{8.4}$$

is the superset, or union, of all instrument in one's portfolio on dates, $t$ and $t+1$. This is the collection of all instruments, in our portfolio, at the beginning and ending of a single business-day performance period.

To keep this from being overly abstract, imagine that this set is a collection of unique identifiers for each instrument—such as ISIN, Sedol, or CUSIP numbers. We have access to information on the size of the associated position, but it is not directly

[4]Technically, the reduction in one's position could also result from a maturity of the underlying security. For a garden-variety fixed-income security, one could easily treat this outcome as the sale of the security at par value. Maturity of derivatives contracts are not worth much consideration, because most portfolio managers will close out these positions prior to maturity.

part of the set.[5] We naturally have access to more information for each security to run the actual performance attributions, but knowledge of the instruments and access to their positions will suffice for their classification.

The superset $I_t \cup I_{t+1}$ includes all three possible cases: unchanged, reduced, or increased instrument positions. Our task is now to find a way to isolate the three possible outcomes and determine a way to handle them. We begin by identifying the common set of instruments in the portfolio over the whole period. If we take the intersection of the two instrument sets on each of our two dates as follows,

$$H = I_t \cap I_{t+1}, \tag{8.5}$$

then we have the subset of instruments held in the portfolio at both time $t$ and $t + 1$. We denote this set as $H$ for (unchanged) holdings.

It is tempting to conclude that $H$ is a set of unchanged positions. An instrument appearing in the sets $I_t$ and $I_{t+1}$ does not, however, immediately imply that its position has not changed. A transaction could have occurred on any of the instruments in $H$. What it does imply, however, is that while the position could have changed, it has neither been decreased to zero nor is it an entirely new position. We may thus classify *three* sub-cases within the set $H$: situations where the instrument position remained unchanged, where it has increased, and where it has decreased.

The first, and easiest case, is represented as the set $H_0$, where

$$H_0 = \left\{ \underbrace{I_t \cap I_{t+1}}_{H} : N_{I_t} = N_{I_{t+1}} \right\}. \tag{8.6}$$

Putting this into words, Eq. (8.6) represents the collection of all those instruments observed in the portfolio at time $t$ and $t + 1$ where the notional value, or rather position, is unchanged. These are basically the buy-and-hold instrument positions over $[t, t + 1]$. These are typically the largest set of positions and the easiest to handle, since we may directly employ the buy-and-hold approach from the previous chapter.

The second sub-case of $H$ is defined as,

$$H_1 = \left\{ \underbrace{I_t \cap I_{t+1}}_{H} : N_{I_t} > N_{I_{t+1}} \right\}. \tag{8.7}$$

The portfolio manager has, therefore, a position in these instruments on both business days, but they have decreased. Consequently, some part of the time-$t$ holdings, or instrument positions, have been sold.

[5]The position size would need to be the notional amount for most instruments, but the number of contracts for certain derivative contracts such as futures.

The third and final sub-case associated with set $H$ is,

$$H_2 = \left\{ \underbrace{I_t \cap I_{t+1}}_{H} : N_{I_t} < N_{I_{t+1}} \right\}. \tag{8.8}$$

Here the portfolios' instrument positions have been increased over the time interval, $[t, t+1]$, implying an addition to, or purchase of, additional securities.

The set $H$, can thus be redefined as the union of these three sub-cases,

$$H = \bigcup_{i=0}^{2} H_i. \tag{8.9}$$

If we return to our original *three* cases, the set $H_0$ denotes unchanged positions, whereas $H_1$ and $H_2$ represent decreased and increased positions, respectively. One can, of course, consider the sets $H_1$ and $H_2$ as a combination of a buy-and-hold position and a transaction. We will exploit this fact later in our discussion.

We may now proceed to examine those situations where the instruments are observed in the portfolio on the first day, but not the second day. These are *liquidations* of an instrument position and are represented as,

$$S = I_t \setminus I_{t+1}. \tag{8.10}$$

In set notation, this implies the instruments found in set $I_t$, but not in set $I_{t+1}$.

The final possibility involves instruments *not* observed in the portfolio on the first day, but are present on the second day. These represent *acquisitions* of a new instrument positions and are described as,

$$B = I_{t+1} \setminus I_t, \tag{8.11}$$

where again, using set notation, these are the instruments found in set $I_{t+1}$, but not in set $I_t$.[6] The sets $S$ and $B$ represent, therefore, the liquidation of a security from one to the day or the acquisition of a new security, respectively.

From one day to the next, there are only three possible cases. Using some relatively simple set logic, we have broken these three situations into three mathematical sets covering the superset of all instruments held in the portfolio at the beginning

[6]Moreover, it is a feature of set arithmetic that the intersection of these two sets,

$$\begin{aligned} S \cap B &= (I_t \setminus I_{t+1}) \cap (I_{t+1} \setminus I_t), \\ &= \emptyset, \end{aligned}$$

is empty. Or, in other words, there is no overlap between these two sets: an instrument is found in only one of these two sets (or neither), but not both.

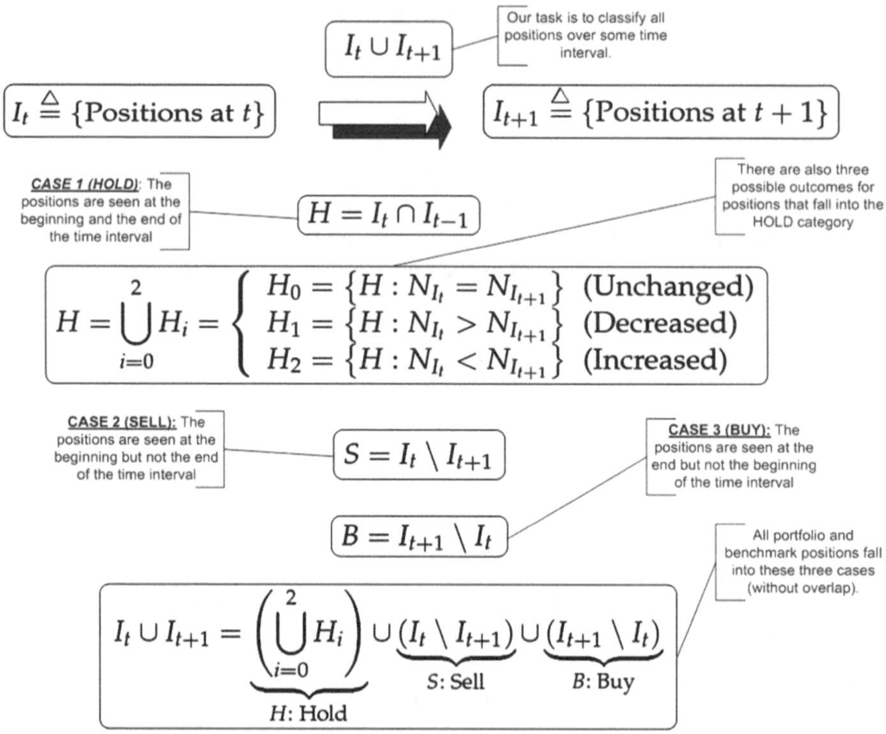


**Fig. 8.1** Set logic to the rescue. This schematic outlines, using set logic, the various possible outcomes for a portfolio position over the course of a one-day period

and ending of the performance period. The superset, $I_t \cup I_{t+1}$, can be thus written as the union of our three sets:

$$I_t \cup I_{t+1} = \underbrace{\left(\bigcup_{i=0}^{2} H_i\right)}_{\text{Holdings}} \cup \underbrace{(I_t \setminus I_{t+1})}_{\text{Sales}} \cup \underbrace{(I_{t+1} \setminus I_t)}_{\text{Purchases}}. \tag{8.12}$$

In this manner, no instruments are ignored and all are correctly classified and eagerly waiting for their performance to be attributed.

Figure 8.1 provides a graphical schematic of our three possible cases. This detailed mathematical breakdown may seem like overkill to some readers, but this exercise was intended to ensure that we have a logically consistent view of the problem. While alternative approaches might be employed, it is essential to ensure that we have not overlooked any possible outcome. Other readers may disagree with the form of the disaggregation of the three separate cases. One might, quite reasonably, for example, argue that part of the changed holdings (i.e., sets $H_1$ and $H_2$) should be organized with the sales and purchases (i.e., $S$ and $B$). Our response would be that there are a number of possible ways to organize the three possible cases and that one is advised to organize them in one's preferred manner. We have chosen to perform the breakdown in a particular manner that seemed both sensible and helpful in organizing our computer programs.[7]

## 8.3 A Simple Practical Example

Having appropriately classified our instrument positions from one business day to the next, the next order of business is to determine how to handle the return and attribution computations for each category. To facilitate this task, let us consider a practical example of these abstract ideas that also underscores the importance of incorporating transactions into one's performance attributions. While relatively small, this example will cover most of the challenges associated with incorporating transactions. Despite its size, it remains reasonably complicated. This complexity underscores the importance of having a well organized logical view of what can, and will, happen over the performance period.

We begin with a collection of instruments in our portfolio at time $t$, which we have already denoted as $I_t$, summarized in Table 8.1. We see that the portfolio is comprised of *three* Canadian government bonds and a CAD cash account. The portfolio has an average duration of slightly more than 5 years and a market value of CAD 172.18. The actual time interval is the single business day running from 12–13 April 2011—we will, however, use the usual notation of $t$ and $t + 1$.

The natural starting point is to compute the one-day return on this portfolio using our approximation approach. In this simple example, we can safely ignore foreign-exchange, inflation-accrual, convexity, and credit-spread effects. Given the information found in Table 8.1, we can easily approximate the return for each

**Table 8.1** Instruments at time $t$: $I_t$

| ISIN | $N$ | Tenor | $D_M$ | $Y_t$ (%) | $Y_{t+1}$ (%) | Price | $V_t$ |
|---|---|---|---|---|---|---|---|
| CA135087(*YR94*) | 100.0 | 8.1 | 6.90 | 3.30 | **3.35** | \$103.19 | \$103.19 |
| CA135087(*ZD99*) | 35.0 | 1.9 | 1.84 | 1.85 | **1.82** | \$99.81 | \$34.93 |
| CA135087(*ZL16*) | 25.0 | 5.1 | 4.80 | 2.79 | **2.75** | \$96.24 | \$24.06 |
| Cash | 10.0 | 0.00 | 0.00 | 0.00 | **0.0** | \$100.00 | \$10.00 |
| Total | 170.0 | 6.0 | 5.18 | 2.74 | **2.70** | \$101.28 | \$172.18 |

This table summarizes some key information about a small portfolio on the first business day of our performance period. The portfolio is composed of three CAD government bonds and a small amount of CAD cash. All values are as at time $t$ with the exception of the bold values, which represent the yield of these securities at time $t + 1$. After this table, we will dispense with the repetitive CA135087 prefix to the CAD sovereign security ISIN.

[7]This is not to imply, however, that other approaches are not possible nor that they may not even be superior to this presented methodology.

instrument. The resulting computation is summarized in Table 8.2 and results in an approximated return for our portfolio of 24 basis points, which is consistent with the observed open-to-close yield tightening of around five basis points across the curve and the portfolio duration of roughly 5 years.

Had the portfolio remained unchanged over our performance period, this would be a sensible approximation. The portfolio, however, did change over the interval, $[t, t+1]$. There were *three* transactions each summarized in Table 8.3—including one sale and two purchases. The entire position in the 8-year YR94 bond was liquidated and replaced with a roughly equivalent holding in a new 9-year YZ11 bond. These two transactions look to be a single operation, which is a typical roll from a shorter maturity into a longer maturity bond. The holding in the 2-year YD99 bond was simultaneously increased by CAD 10 to a new position of CAD 45. Table 8.3 summarizes the transaction prices (and yields), computes the proceeds, and determines the impact on the portfolio's cash balance. The net cash effect of these trades was to reduce the cash balance by CAD 7.91. In summary, there is quite a bit of action over this one-day interval. We need, therefore, to use our instrument-set logic to untangle the various moving parts.

Combining the original holdings in Table 8.1 with the transactions in Table 8.3, we can describe the portfolio composition at time $t+1$. These holdings, along with the market data from the close of time $t+1$, are illustrated in Table 8.4.

**Table 8.2** Buy-and-hold returns

| | Return (bps.) | | | | |
|---|---|---|---|---|---|
| ISIN | Yield | Carry | Total | Weight (%) | Contribution |
| *YR94* | 32.3 | 0.9 | 33.2 | 59.9 | 19.9 |
| *ZD99* | 6.0 | 0.5 | 6.5 | 20.3 | 1.3 |
| *ZL16* | 22.1 | 0.8 | 22.9 | 14.0 | 3.2 |
| Cash | 0.0 | 0.0 | 0.0 | 5.8 | 0.0 |
| Total | – | – | – | 100 | *24.4* |

This table summarizes the approximated return applying the buy-and-hold approach to the time $t$ holdings summarized in Table 8.1. We use the basic approximation $r_i \approx y_i \Delta t - D_M \Delta y$.

**Table 8.3** Transactions

| ISIN | Amount | Yield (%) | Price | Proceeds |
|---|---|---|---|---|
| *YR94* | 100.00 | 3.30 | $103.15 | $103.15 |
| *ZD99* | (10.00) | 1.85 | $99.82 | (9.98) |
| *YZ11* | (100.00) | 3.36 | $101.08 | ($101.08) |
| Total (Cash impact) | (10.00) | – | – | ($7.91) |

This table describes the three transactions, and their associated cash impact, occurring during our single business-day performance period, $[t, t+1]$. Negative proceeds indicate a purchase, whereas positive proceeds represent a sale of a security.

**Table 8.4** Instruments at time $t+1$: $I_{t+1}$

| ISIN | $N$ | Tenor | $D_M$ | $Y_t$ (%) | $Y_{t+1}$ (%) | Price | $V_t$ |
|---|---|---|---|---|---|---|---|
| *ZD99* | 45.00 | 1.9 | 1.84 | *1.85* | 1.82 | \$99.87 | \$44.94 |
| *ZL16* | 25.00 | 5.1 | 4.80 | *2.79* | 2.75 | \$96.45 | \$24.11 |
| *YZ11* | 100.00 | 9.1 | 7.68 | *3.41* | 3.37 | \$101.04 | \$101.04 |
| Cash | 2.14 | 0.0 | 0.00 | *0.00* | 0.00 | \$100.00 | \$2.18 |
| Total | 172.14 | 6.6 | 5.66 | *2.87* | 2.84 | \$100.05 | \$172.27 |

This table summarizes some key information about our portfolio on the *second* business day of our performance period. All values are as at time $t+1$ with the exception of the italicized values, which represent the yield of these securities at time $t$.

Taking the market values from our portfolio at time $t$ and $t+1$, we may easily compute the true daily return. The result is

$$\begin{aligned}\text{True Return} &= \frac{V(t+1) - V(t)}{V(t)}, \\ &= \frac{172.27 - 172.18}{172.18}, \\ &= 5.1 \text{ basis points.}\end{aligned} \tag{8.13}$$

This is *not* consistent with the 24 basis points computed in Table 8.2 using the buy-and-hold assumption and, quite clearly, the transactions are the culprit.

There are two aspects to the problem. First, the transaction prices are not entirely consistent with the closing values of those securities. Second, not all securities are being considered in the computation—the purchased bonds are not considered, for example, because they are simply not included in the time $t$ data. These two issues ultimately bring us to questions of what risk factor values to use in our computations and how to weight the return on each individual security.

Before addressing these important questions and attempting to compute a more reasonable return approximation, let us first break down the various instruments into the three categories outlined in the previous section. Table 8.5 illustrates each of our previously defined sets along with their definition and the final result. The four instruments in our portfolio over the performance period are easily allocated to our three cases: holdings, liquidations, and new acquisitions.[8] In this example, we have one liquidation, one acquisition, and an increased position in an existing holding.

[8]This may seem like substantial effort for a small example. With real portfolios and benchmarks comprised of hundreds of securities, however, it is essential that one have a consistent and robust technique for identifying which category an instrument falls into and how to process it.

**Table 8.5** Categorizing the instruments

| Set | Definition | Description | Result |
|---|---|---|---|
| $I_t$ | $I_t$ | Instruments at $t$ | $\{YR94, YD99, YL16\}$ |
| $I_{t+1}$ | $I_{t+1}$ | Instruments at $t+1$ | $\{YD99, YL16, YZ11\}$ |
| $I_t \cup I_{t+1}$ | $I_t \cup I_{t+1}$ | All instruments | $\{YR94, YD99, YL16, YZ11\}$ |
| $H$ | $I_t \cap I_{t+1}$ | Holdings | $\{YD99, YL16\}$ |
| $H_0$ | $\left\{ \underbrace{I_t \cap I_{t+1}}_{H} : N_{I_t} = N_{I_{t+1}} \right\}$ | Unchanged holdings | $\{YL16\}$ |
| $H_1$ | $\left\{ \underbrace{I_t \cap I_{t+1}}_{H} : N_{I_t} > N_{I_{t+1}} \right\}$ | Decreased holdings | $\emptyset$ |
| $H_2$ | $\left\{ \underbrace{I_t \cap I_{t+1}}_{H} : N_{I_t} < N_{I_{t+1}} \right\}$ | Increased holdings | $\{YD99\}$ |
| $S$ | $I_t \setminus I_{t+1}$ | Liquidations | $\{YR94\}$ |
| $B$ | $I_{t+1} \setminus I_t$ | Acquisitions | $\{YZ11\}$ |

This table applies the set-theoretic ideas from the previous section to our small example. We ignore cash in all of our computations as it does not contribute to return.

**Table 8.6** Buy-and-hold and transaction returns

| | Return (bps.) | |
|---|---|---|
| ISIN | Buy-hold | Transaction |
| YR94 | 33.2 | (2.6) |
| ZD99 | 6.5 | 1.9 |
| ZL16 | 22.9 | 0.0 |
| YZ11 | 33.4 | 5.4 |
| Cash | 0.0 | 0.0 |

This table illustrates the buy-and-hold and transaction returns for each of the instruments in the superset, $I_t \cup I_{t+1}$.

The actual return over the period is the combination of a few transactions and some unchanged holdings. Successfully relaxing the buy-and-hold assumption will involve combining the transaction and buy-and-hold returns with the appropriate weights to determine a sensible return approximation. To this end, Tables 8.6 and 8.7 provide a summary of the returns for the three transactions. While the closing yields were about three to five basis points lower than the previous day's close, the transactions occurred at significantly different levels. This is an important point— there is no guarantee that end-of-day and transaction prices will coincide. The majority of the yield reduction must have occurred at some point in the day after the

**Table 8.7** Transaction returns

| | Return (bps.) | | |
|---|---|---|---|
| ISIN | Yield | Carry | Total |
| *YR94* | (3.5) | 0.9 | (2.6) |
| *ZD99* | 1.4 | 0.5 | 1.9 |
| *YZ11* | (−4.5) | 0.9 | 5.4 |

This table summarizes, using transaction data, the approximated return decomposition for each the three transactions in Table 8.3.

transactions were performed. It is precisely these price differences and the resulting changes in the cash balance that create the return differences.

Table 8.6 also summarizes the buy-and-hold and transaction returns for each of the securities present in the portfolio at both the beginning and ending of our single-business-day performance period.[9]

What remains is to determine the appropriate weight to use for each of these returns. There are, at least, three alternatives ways to weight these returns: using the position values at the *start* of the period, using the position values at the *end* of the period, or something in between these two extremes. While we will use the mid-points in this example, each of the alternatives are described in the underlying box.

Table 8.6 illustrates these three alternative weighting schemes. We have defined the *in-between* market value weighting scheme as occurring at the mid-point. Mathematically, therefore, the mid-weight of the $i$th position, $\omega_{i,\text{Mid}}$, in a portfolio of $n$ total positions is given as,

$$\omega_{i,\text{Mid}} = \frac{V_i(t) + V_i(t+1)}{\sum_{k=1}^{n} \left(V_k(t) + V_k(t+1)\right)}. \tag{8.14}$$

The idea is relatively simple. We do not know whether to use the starting or ending market values for return weighting, so we use both. There is nothing particularly scientific about this approach, but it is reasonably appealing when we observe, from Table 8.6, that using starting or ending values can lead to a zero weight on some positions. More specifically, if we use the market value of the starting portfolio to compute our return weights, then we will,

(continued)

[9]This is certainly not the only way to handle transactions. There is a small, but interesting, literature on the role of transactions in performance attribution. The interested reader is directed to Spaulding [9], Laker [5], Bonafede and McCarthy [3], and Menchero and Hu [8].

by definition, ignore the new purchases. By the same logic, use of the ending weights leads use to ignore liquidations. The mid-point weights, despite their seemingly arbitrary definition, permit us to include both holdings, purchases, and liquidations. We recommend, therefore, the use of the mid-point weights.

Table 8.8 indicates how much weight we could to allocate to each of the securities. The remaining task is to determine how much weight to place on the buy-and-hold return and transaction returns, respectively. The reason is simple: if there has been a transaction related to a particular position over the one-day performance horizon, the return is some linear combination of the buy-and-hold and transaction returns.

How, therefore, should we allocate the weights to the buy-and-hold and transaction returns? The first two cases are relatively straightforward:

1. If the holding is unchanged, then 100 % of the weight is applied to the buy-and-hold return.
2. If it is an acquisition or a liquidation, then 100 % is allocated to the transaction return.

The increased and decreased positions remain to be addressed. An increased or decreased position can always be separated into two parts: an unchanged holding and a transaction. The unchanged position determines the weight on the buy-and-hold return, whereas the transaction element determines the weight on the transaction return. How might this be done? Security ZD99 has a $45 position at $t+1$ with a market value of $44.94. $35 of this position is unchanged, whereas $10 relates to a new acquisition. Thus, $\frac{\$35}{\$45}$ or 78 % of this position relates to the buy-and-hold return, with the remaining 22$ allocated to the transaction return. There are more complex ways to make this separation—see the underlying shaded box—although this very simple approach is both robust and easy to understand.

**Table 8.8** Possible weighting schemes

| | $V$ | | Weights | | |
|---|---|---|---|---|---|
| ISIN | $t$ | $t+1$ | Start (%) | Mid (%) | End (%) |
| YR94 | 103.19 | 0.00 | 59.9 | **30.0** | *0.00* |
| ZD99 | 34.93 | 44.94 | 20.3 | **23.2** | 26.1 |
| ZL16 | 24.06 | 24.11 | 14.0 | **14.0** | 14.0 |
| YZ11 | 0.00 | 101.04 | *0.0* | **29.3** | 58.7 |
| Cash | 10.00 | 2.18 | 5.8 | **3.5** | 1.3 |
| Total | 172.18 | 172.27 | 100.0 | **100.0** | 100.0 |

This table outlines three possible ways to compute the weights to apply to the returns on individual securities.

A more general framework for the allocation between these two returns can be constructed. The security weight, mid-point or otherwise, tells us how much importance that we want to allocate to a given position's return. We are, therefore, basically trying to separate the weight into two components: one related to the unchanged, or buy-and-hold, component and the other to the transaction component. Any approach, therefore, should

1. ensure that we do *not* allocate more than the mid-point weight to any individual position;
2. be sufficiently flexible to handle *both* sales and purchases;
3. allocate zero weight to transaction returns when no transaction occurs; and
4. allocate zero weight to buy-and-hold returns when there is a liquidation or the acquisition of a new instrument.

We offer one alternative that meets these four criteria. In this approach, the weight on the buy-and hold (B&H) return is described as,

$$\omega_{i,\text{B\&H}} = \underbrace{\left( \frac{\max\left(V_i(t), V_i(t+1)\right) - \left|\begin{matrix}\text{Transaction}\\ \text{proceeds}\end{matrix}\right|}{\max\left(V_i(t), V_i(t+1)\right)} \right)}_{\substack{\text{Weight on the unchanged}\\ \text{position component}}} \omega_i . \tag{8.15}$$

whereas the return on the transaction (TRN) return is,

$$\omega_{i,\text{TRN}} = \underbrace{\left( \frac{|\text{Transaction proceeds}|}{\max\left(V_i(t), V_i(t+1)\right)} \right)}_{\substack{\text{Weight on the}\\ \text{transaction component}}} \omega_i . \tag{8.16}$$

Does this approach meet the previously outlined criteria? Inspection of Eqs. (8.15) and (8.16) reveals that their sum is equal to $\omega_i$. This satisfies the first criterion—using this method will not lead to exceeding the mid-point weight. The second criterion is handled by taking the absolute value of the transaction proceeds. The logic of this approach is that the maximum of the position values at the start and end of the period is taken. This denominator value is basically the position before a sale or after a purchase. Now the weight on the transaction return is merely the absolute value of the transaction proceeds over our denominator, while the weight on the buy-and-hold return is simply one minus this value.

(continued)

**Table 8.9** Weighting buy-hold and transaction returns

| | Return (bps.) | | Weights | | Contribution | |
|---|---|---|---|---|---|---|
| Id | B&H | TRN | B&H (%) | TRN (%) | B&H | TRN |
| YR94 | 33.2 | (2.6) | *0.0* | 30.0 | 0.0 | (0.8) |
| ZD99 | 6.5 | 1.9 | 18.0 | 5.2 | 1.2 | 0.1 |
| ZL16 | 22.9 | 0.0 | 14.0 | *0.0* | 3.2 | 0.0 |
| YZ11 | 33.4 | 5.4 | *0.0* | 29.3 | 0.0 | 1.6 |
| Cash | 0.0 | 0.0 | 3.5 | 0.0 | 0.00 | 0.0 |
| Subtotal | – | – | 35.6 | 64.4 | 4.4 | 0.9 |
| Total | – | – | *100* | | *5.3* | |

This table describes one possible approach for the allocation of mid-point weights across buy-and-hold (B&H) and transaction (TRN) returns.

The third criterion is easily satisfied. If the transaction proceeds are zero, then the weight on the transaction return, as described by Eq. (8.16), is obviously zero. The fourth criterion is only approximately satisfied. If we have a new acquisition or a liquidation, then the maximum market value at the start and end of the performance period, $\max(V_i(t), V_i(t+1))$, will be quite close to the transaction proceeds implying that the weight on the buy-and-hold return, $\omega_{i,\text{Buy/Hold}}$, will be approximately zero.

This idea is applied to our small example in Table 8.9, which summarizes our approach for allocating the mid-point weights between buy-and-hold and transaction returns.[10] Table 8.9 collects the buy-and-hold and transaction returns illustrated from the previous tables and used the basic logic described in the previous paragraphs to compute the appropriate weights.

The final two columns of Table 8.9 apply these buy-and-hold and transaction weights to their respective returns and subsequently arrive at the total return contribution of each position. The result is a total return approximation of 5.3 basis points, which compares favourably to the true return of 5.1 basis points computed from the change in the market-to-market value of our portfolio over the last business day. We may now proceed to use the normal performance attribution approach, outlined in the previous chapter, separately on the buy-and-hold and transaction returns. For the buy-and-hold returns, this is straightforward. For the transactions returns, however, it is a bit more complicated. This is discussed in the following section.

[10]Whether one uses the simple ratio of notional amounts or the more complex expressions in Eqs. (8.15) and (8.16), the results are virtually identical.

### 8.3.1 The Very Fine Print

Conceptually, this is a relatively easy and appealing approach to incorporating transaction returns into one's daily attributions. There is nonetheless a slight practical problem. Although we observe the transaction price, and consequently, store key transaction information in our internal systems, we generally do *not* have access to the equivalent treasury yield, option-adjusted spread, or foreign-exchange rate at the point of transaction. The simple reason is that we generally only have curve information for a small number of points of time during the day.

Indeed, most institutions have only a single market-data capture at the close of business each day. Transactions, conversely, may occur at any point of the course of the working day. We find ourselves, therefore, in a difficult position. To resolve this problem, there are two alternatives:

- Ignore the transaction information and use opening and closing prices; or
- Use transaction prices, but make some assumptions regarding the additional market data required for one's attributions.

Neither approach is perfect, be we argue for the use of the latter approach. The first approach offers the advantage of always using defensible inputs and having a clear understanding of all market data. It has the important drawback of ignoring the information and impact of transaction prices. In the previous section, we showed that this can be quite important. The importance of transaction prices pushes us towards the latter approach.

Use of the latter approach requires us to make a few assumptions. We start by assuming a set of exposures. Since it is only a one-day time period, use of either begin or end of day exposures should not typically make an enormous difference. If it is an acquisition, then exposures will probably only be available for the end of the period (i.e., at the close). For all other situations, the beginning (i.e., open) exposures may be employed.

The second issue involves obtaining the requisite factor changes for our attribution. Let's consider each in turn:

$\Delta y$ The overall yield change is merely the difference between the transaction yield and, depending on the type of transaction, the open or closing yield value.

$\Delta s_{\text{OAS}}$ and $\Delta y_{\text{TRE}}$ The credit spread and equivalent treasury yield are more complicated. We assume that the credit spread is unchanged from its previous closing value—this basically amounts to the assumption that the spread remained unchanged up until the time of the transaction. Applying this spread to the transaction yield, one can easily compute an equivalent treasury yield. In situations where one purchases a new bond (i.e., an acquisition), one may not have access to bond specific information from the previous close. In this case, one could employ the credit spread from the current day's close. Again, one is assuming that the spread remains fixed from the transaction time to that day's close. Neither of these assumptions are particularly defensible, but it is hard to find a superior solution given the general lack of intra-day market data.

$\Delta E$ For normal sales or purchases of a bond, the foreign-exchange movement can be approximated with the open-to-close change. If you are a USD investor and perform a transaction in a EUR-denominated bond, then the foreign-exchange risk can be attributed normally and separately within the buy-and-hold and transaction returns. If it is an explicit foreign-exchange transaction, however, then you will have an intra-day transaction rate to use for your computations.

Under the preceding set of assumptions, we have identified the weights, exposures, and factor changes necessary to compute the performance attribution associated with the buy-and-hold and transaction returns. Combining them together across all securities in the portfolio consequently provides one with the final daily performance attribution.

In contrast to one's portfolio, ignoring transaction data is perfectly acceptable when dealing with benchmarks. On the vast majority of days during a given month, most fixed-income benchmarks will not change since the individual constituents of most benchmarks are updated once a month.[11] A number of transactions occur in the benchmark once a month, but the good news is that these transactions occur at the closing prices prevailing on the rebalance date. Consequently, the transaction price is equal to the closing price. This implies that, for the benchmark, the buy-and-hold and transaction returns are identical easing the benchmark return computations.[12]

## 8.4 A Complicated Practical Example

We have examined a simple, although reasonably involved, daily performance-attribution example. This provided some insight into the main challenges and their possible resolution. In this section, we move beyond this simple setting and examine how this approach applies to multiple portfolios over a lengthy time horizon.

### 8.4.1 An Experiment

We have examined a simple example with four individual bonds over a single business day. Over the course of working through this example, we found that using mid-point weights, buy-and-hold returns, and transaction returns, we could sensibly relax the buy-and-hold assumption and handle a changing portfolio composition.

[11]Changes generally involve the introduction of some new bonds, the extraction of some existing bonds, and a general updating of the relative weights of a large number of existing constituents.

[12]One still needs to determine the implied transactions associated with the benchmark rebalancing. This is accomplished using the framework outlined in earlier sections. Once this is done, one need only compute the buy-and-hold returns and apply the mid-point weight to determine the benchmark return for the rebalancing date.

While hopefully compelling, it represents only a single case. It would be more useful, not to mention far more convincing, to apply this approach across a sizeable number of actual portfolios and a lengthy range of different business dates.

This is precisely our task in this section. We will perform an experiment. For 12 separate portfolios, we will use the previously described methodology to compute daily performance attributions over 1,600 calendar days running from 31 December 2009 up to, and including, 2 May 2014.[13] Our experiment should include sufficient attributions to gain a representative picture of the reasonableness of our approach.

How do we plan on gauging the reasonableness of this approach across so many observations? This is not an easy question to answer. On one level, the ultimate measure of a performance attribution methodology is its usefulness to the individuals—portfolio managers, senior management, and clients—using this information to understand their respective portfolios. Usefulness is rather difficult, if not impossible, to objectively measure. Instead we will focus on a few perhaps less helpful, but easier to describe, measures of reasonableness. We will consider *three* alternative approaches including:

1. an examination of the formal statistical relationship between actual and approximated returns;
2. an *ad hoc* computation intended to estimate the percentage of actual return explained by the performance attribution; and
3. various summary statistics related to the approximation error.

In short, the focus is principally on the residual, or unexplained, term in our attributions. After a thorough examination of our data from these perspectives, we should have a better idea of the robustness and accuracy of our performance-attribution framework.

### 8.4.2 Regression Analysis

The data collected for our 12 portfolios over a 4 + year time period lends itself readily to statistical analysis. In particular, we have a collection of true daily returns,

$$\{r_k, k = 1, \ldots, n\}, \tag{8.17}$$

where in our case $n \approx 1,100$ as well as the approximated returns denoted as,

$$\{\tilde{r}_k, k = 1, \ldots, n\}. \tag{8.18}$$

[13]This time interval includes 1,113 business days, which amounts to approximately 13,400 daily performance attributions. Given that each portfolio, and its associated benchmark, has on average several hundred fixed-income instruments, this involves several million individual security attributions.

Ordinary least squares, or linear regression, provides us with a formal framework to understand the relationship between these two collections of observations. Specifically, it posits the following relationship between actual and approximated returns,

$$r_k = \alpha + \beta \tilde{r}_k + \varepsilon_k, \tag{8.19}$$

for $k = 1, \ldots, n$.

There are, at least, *three* interesting questions that may be posed:

1. Is the intercept term, $\alpha$, statistically different from zero? If so, it would suggest a systematic bias in the approximation of the actual returns.
2. Is the $\beta$ coefficient statistically and economically different from one? With more than 13,000 observations, we should be able to make a fairly precise estimate of the coefficients and answer the statistical question. An important related question: is it economically different from unity?
3. How strong is the relationship between the actual and approximated returns? This could be determined by the $R^2$ of the regression and the root-mean squared residual term.

The approximations can be examined from three perspectives: the portfolio returns, the benchmark returns, and the active returns.[14] Before starting with the formal analysis, we can make a few predictions about the relative accuracy of these three types of approximations. Given the virtual absence of transactions, one would expect that the benchmark estimations to be the most accurate. The active return approximations, given that they are the difference between two possibly noisy approximations, should probably be expected to be the least accurate. This leaves us to predict that the accuracy of the portfolio-return attributions lies somewhere in between.

Figure 8.2 uses a scatter-plot to outline the relationship between the approximated and true portfolio, benchmark, and active returns for our dataset. Each graphic in Fig. 8.2 includes a 45° line; should all of the observations fall on this line, the relationship between the actual and approximated returns would naturally be perfect. With some exceptions, we see that, for each of the return types, the majority of the observations do indeed fall along this line. The active returns do however appear to be the noisiest, although outliers are also evident among the portfolio- and benchmark-return approximations.

[14]We could have collected all of the portfolio, benchmark and active returns together and examined them as a single group. It is nonetheless more interesting to perform the analysis for these three different return types.

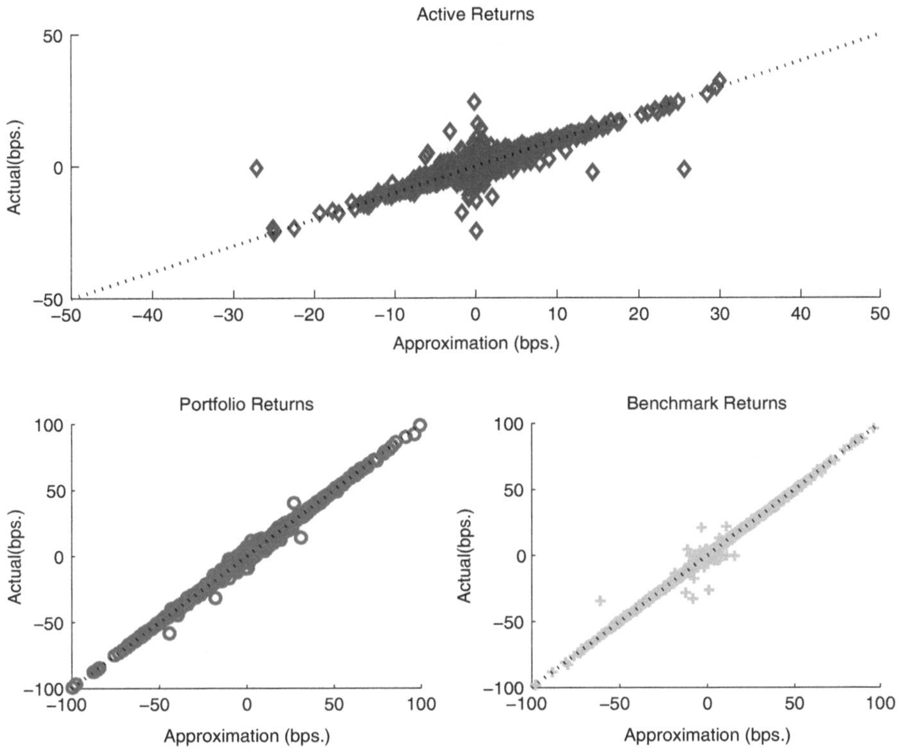


**Fig. 8.2** Actual vs. approximated returns. This underlying graphic outlines the relationship between the approximated and true portfolio, benchmark, and active returns

**Table 8.10** Regression results

| | Return type | | |
|---|---|---|---|
| Summary statistic | Portfolio | Benchmark | Active |
| Correlation: $\rho(r, \tilde{r})$ | 0.997 | 0.998 | 0.906 |
| Regression fit: $R^2$ (%) | 99.3 | 99.6 | 82.1 |
| Regression coefficient: $\beta$ | 0.998 | 1.006 | 0.901 |

This table summarizes the results of a simple linear regression of actual returns, organized by return type, against their approximations.

Table 8.10 provides some key elements of a simple linear regression of actual returns, organized by return type, against their approximations. While each regression was initially run with an intercept, it did not prove statistically significant for any of the data types and had no influence on the final $R^2$ values; as a consequence, it was suppressed. This answers our first question by suggesting that there is no systematic bias in the approximation of the actual return—our performance-attribution approach using our additive risk-factor decomposition and transaction methodology does not appear to systematically over- or under-estimate the return.

Our second question related to the statistical significance of the $\beta$ coefficient. Statistically, we may reject the null hypothesis that $\beta = 1$ for both portfolio and active returns. Only for benchmark returns can this hypothesis not be rejected. Economically, however, the difference between our $\beta$ estimates and unity appears quite small—in all cases, the estimate is virtually identical to one.[15] We may conclude that there is a statistically significant approximation error, but that it is practically quite small.

The final question has also been answered visually by Fig. 8.2 and supported by the numerical analysis in Table 8.10. The correlations and $R^2$ figures are extremely high, with values approaching unity. Our intuition regarding the data types has also been verified. Benchmark return approximations exhibit the strongest link to actual returns, while the active-return estimates demonstrate the weakest relationship with their true returns. The portfolio-return approximations lie somewhere in between.

### 8.4.3 An Invented Measure

How much of the return is explained by the performance attribution? For a given attribution, this natural question is often relatively easy to answer. On a systematic basis, however, it is somewhat more challenging. The reason is that one would typically take the ratio of the approximated return to the actual return. When the actual daily return approaches zero—since it is the denominator—this ratio tends to the rather unhelpful value of infinity. A second problem is that sometimes the approximation overestimates the actual return, while on other occasions it underestimates it. These over- and understatements may actually offset one another, thereby overstating the accuracy of the approximation.[16]

To avoid these problems, we constructed our own measure using a few building blocks beginning with the average absolute actual return,

$$r_{\text{abs}} = \frac{1}{N} \sum_{k=1}^{N} |r_k|, \tag{8.20}$$

and the average absolute approximate return,

$$\tilde{r}_{\text{abs}} = \frac{1}{N} \sum_{k=1}^{N} |\tilde{r}_k|, \tag{8.21}$$

[15]The large number of observations gives us the necessary power to statistically reject the null hypothesis of $\beta = 1$ with a confidence level of 99 %.

[16]Not to mention that stating that one's approximation explains 103 % of the actual return is difficult to interpret.

**Table 8.11** Explained return

| | Return type | | |
|---|---|---|---|
| Summary statistic | Portfolio | Benchmark | Active |
| Percentage explained | 99.1 % | 99.5 % | 91.7 % |

This table summarizes the results of the measure, described in Eq. (8.23), for our three return types across all of the portfolio in our sample.

The absolute distance between these two values is a possible approximation of the approximation error,

$$|\tilde{r}_{\text{abs}} - r_{\text{abs}}| \,, \tag{8.22}$$

that is independent of the sign of the underlying returns.

Using these inputs, we proceed to define the percentage of return explained by the approximation as,

$$\% \text{ Return Explained} = 1 - \frac{|\tilde{r}_{\text{abs}} - r_{\text{abs}}|}{r_{\text{abs}}}. \tag{8.23}$$

This essentially compares the average absolute approximation error to the average absolute return. While somewhat unattractive, it resolves the aforementioned problems providing a robust measurement that can be used to answer our initial question. It nevertheless needs to be applied over a reasonably lengthy time period to permit sensible conclusions.

Table 8.11 summarizes the results of the measure in Eq. (8.23) for our three return types across all of the portfolio in our sample. The results appear to be consistent with the regression analysis in the previous section. The daily approximations appear to explain the majority of the actual returns, although the lowest level of explanation is observed for the active returns.

### 8.4.4 Approximation Errors

We now examine the unexplained component of the return, which is defined as the distance between the true return of the portfolio and the sum of the individual factor returns from our additive decomposition. This is essentially the approximation error associated with our daily performance attribution. We would expect, and hope, that if we are reasonably successful in handling transactions, that this approximation error would be relatively small. Optimally, it would be zero, but given the number of approximations involved in our process, we would expect that it would typically fall within a few basis points.

Table 8.12 highlights a number of summary statistics for the daily approximation error stemming from our 13,400 separate performance attributions. The average *absolute* approximation error for the portfolio, benchmark, and active return ranges

**Table 8.12** Experiment at a glance

| | Return type (bps.) | | |
|---|---|---|---|
| Summary statistic | Portfolio | Benchmark | Active |
| Mean | 0.3 | 0.2 | 0.3 |
| Median | 0.1 | 0.1 | 0.1 |
| Volatility | 0.9 | 0.7 | 1.1 |
| **Inter-quartile range** | **0.3** | **0.2** | **0.2** |
| **99th percentile** | **1.5** | **1.4** | **1.8** |
| 1st percentile | −1.4 | −1.2 | −1.7 |

This table summarizes the results of our simple daily attribution experiment.

somewhere between 0.2 and 0.3 basis points.[17] The median daily *absolute* approximation arising from our approximation is even smaller, at roughly 0.1 basis points, indicating that a small number of relatively large errors are impacting the mean. Although not quite zero, these results nevertheless suggest that our approximation approach provides, on average, a reasonable approximation of the true daily return.

The second set of summary statistics looks at the dispersion of the absolute value of the approximation errors. The volatility, or standard deviation, looks to be about a single basis point whereas the inter-quartile range varies between about 0.2 and 0.3 basis points.[18] These results further suggest that the approximation errors are relatively tightly grouped around the mean and median values, respectively.

The final two rows of Table 8.12 attempt to understand the tails of the error distribution. We examine the 1st and 99th percentile of the daily attribution errors and find that approximately 98 % of the approximation errors fall between about ±2 basis points. Again, this is fairly encouraging.

Figure 8.3 describes the evolution of the approximation errors over our 1,100+ consecutive business days of daily performance attribution for 12 separate portfolios. The vast bulk of the approximation errors visually fall between roughly ± 2 basis points. On occasion, as evidenced by the outliers in Fig. 8.3, the deviations are

[17]The mean absolute approximation error (MAE) is defined as follows,

$$\begin{aligned} \text{MAE} &= \frac{1}{N}\sum_{k=1}^{n} |r_k - \tilde{r}_k|, \\ &= \frac{1}{N}\sum_{k=1}^{n} |\varepsilon_k|, \end{aligned} \tag{8.24}$$

where there are $N$ daily performance attributions and $\varepsilon_k$ is the approximation error of the $k$th attribution. We examine absolute values of the approximation errors so that the positive and negative errors do *not* cancel each other out.

[18]Recall that the inter-quartile range is the difference between the 75th and 25th percentiles.

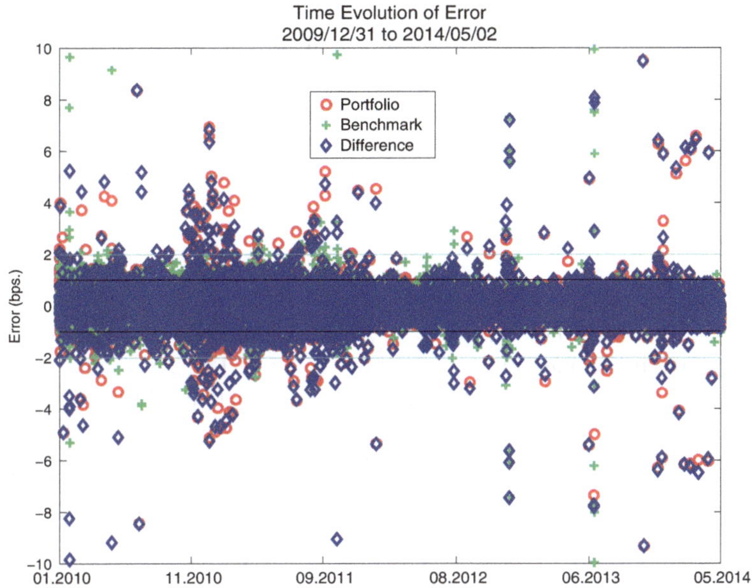


**Fig. 8.3** Time evolution of approximation errors. This figure describes the evolution of the approximation errors over our 1,113 consecutive business days of daily performance attributions for 12 separate portfolios

more significant. This highlights the fact that, on some occasions, particularly in the face of large market movements, our approximations will have difficulty.[19]

Our experiment indicates that our attribution framework—involving our additive decomposition coupled with daily performance attributions and incorporation of transactions—appears to be generally applicable beyond our simple four-instrument example. In the subsequent section, we will consider how combines these *daily* attributions over time.

## 8.5 Some Frustrating Mathematical Facts

Many of the complexities of relaxing the buy-and-hold assumption can be mitigated by computing daily performance attributions. Despite their benefits, daily performance attributions are sadly not without their own challenges. The principal remaining challenge is: how do we transform a sequence of daily attributions into

[19]A good example occurred on 10 May 2010 where a number of relatively large approximation errors were present in the benchmark approximation. These were not data errors, but instead represented a enormous tightening of Greek sovereign spreads in the benchmark of some European sovereign debt benchmarks. By virtue of their sheer size, they were not fully captured by our linear duration approximation and convexity correction.

weekly, monthly, or annual attributions? Can we, for example, merely sum the daily attributions across time? Stated differently, can we add the individual daily risk-factor returns for each business date in a given month to arrive at the monthly collection of risk-factor returns? The unfortunate answer is no.

The reason relates to some unpleasant facts from basic arithmetic. Let's consider a simple daily attribution where the active return of one portfolio is broken down into two components as follows,

$$p_t - b_t = \mathscr{A}_t + \mathscr{B}_t, \tag{8.25}$$

for each business day where $\{p_t, t = 1, \ldots, n\}$ are the portfolio returns, $\{b_t, t = 1, \ldots, n\}$ are the benchmark returns, and $\{\mathscr{A}_t, \mathscr{B}_t : t = 1, \ldots, n\}$ are the two return-attribution categories.[20] To compute the cumulative active return over the period $\{t = 1, .., n\}$, we need to take the difference between the geometric sum of the portfolio and benchmark returns as follows,

$$\begin{aligned} \text{Active Return} &= \prod_{t=1}^{n}(1 + p_t) - \prod_{t=1}^{n}(1 + b_t), \\ &\neq \prod_{t=1}^{n}\underbrace{(1 + p_t) - (1 + b_t)}_{p_t - b_t}. \end{aligned} \tag{8.26}$$

We cannot, however, take the geometric sum of the active returns for each period, because the product of the differences is *not* equal to the difference of the products. This means that we may not combine the active portfolio returns across time—although this is a generally well-known fact, it is nonetheless makes our life more difficult.

Let's repeat the previous computation in a generic manner without all of the portfolio and benchmark details. Imagine that you have two sequences of numbers, $a$ and $b$, defined as,

$$\begin{aligned} a &= \{1, 2, 3\}, \\ b &= \{4, 5, 6\}. \end{aligned} \tag{8.27}$$

(continued)

[20] Of course, in reality there would be a few more, but we only use two to keep the notation under control. All of the ideas, however, generalize seamlessly to an arbitrary number of return categories.

Now if we compute the product of the differences between $b$ and $a$, we have

$$\prod_{i=1}^{3} b_i - a_i = \underbrace{(4-1)}_{3} \cdot \underbrace{(5-2)}_{3} \cdot \underbrace{(6-3)}_{3}, \tag{8.28}$$

$$= 27. \tag{8.29}$$

If, however, we compute the differences of the products of $b$ and $a$, we arrive at an entirely different result,

$$\prod_{i=1}^{3} b_i - \prod_{i=1}^{3} a_i = \underbrace{(4 \cdot 5 \cdot 6)}_{120} - \underbrace{(1 \cdot 2 \cdot 3)}_{6}, \tag{8.30}$$

$$= 114.$$

The consequence is that,

$$\prod_{i=1}^{3} b_i - a_i \neq \prod_{i=1}^{3} b_i - \prod_{i=1}^{3} a_i. \tag{8.31}$$

or the product of the differences is *not* equal to the difference of the products. This annoying fact is often misunderstood and overlooked leading to computational errors.

Let's take the correct definition of active return and attempt to relate it to the daily performance attributions. The following, while tempting, will also not work

$$\underbrace{\prod_{t=1}^{n}(1+p_t)}_{1+p} - \underbrace{\prod_{t=1}^{n}(1+b_t)}_{1+b} \neq \sum_{t=1}^{n} \mathscr{A}_t + \mathscr{B}_t \tag{8.32}$$

$$(1+p) - (1+b) \neq \sum_{t=1}^{n} \mathscr{A}_t + \mathscr{B}_t,$$

$$p - b \neq \sum_{t=1}^{n} \mathscr{A}_t + \mathscr{B}_t.$$

Quite simply, the difference between the geometric sums of the portfolio and benchmark returns are *not* equal to the arithmetic sum of the individual elements of

**Table 8.13** A frustrating mathematical fact

| Return | $p_t$ | $b_t$ | $p_t - b_t$ | $\mathscr{A}_t$ | $\mathscr{B}_t$ | $\mathscr{A}_t + \mathscr{B}_t$ |
|---|---|---|---|---|---|---|
| $t_1$ (%) | 3.5 | −0.5 | **4** | 1 | 3 | **4** |
| $t_2$ (%) | −2 | −3 | **1** | 2 | −1 | **1** |
| $t_3$ (%) | 1.5 | 6.5 | **−5** | −2 | −3 | **−5** |
| Geo. sum (%) | 2.95 | 2.79 | ***0.16*** | 0.96 | −1.09 | ***−0.13*** |

This table provides the details of a concrete example demonstrating that the product of the differences is *not* equal to the difference of the products.

the performance attribution. This is not terribly surprising as clearly, in Eq. (8.32), we are mixing arithmetic and geometric sums.

We could, therefore, try to compute the geometric sum of the returns associated with each of the factor returns. Unfortunately, this is also a dead-end given that,

$$p - b \neq \underbrace{\left(\left(\prod_{t=1}^{n}(1+\mathscr{A}_t)\right) - 1\right)}_{\mathscr{A}} + \underbrace{\left(\left(\prod_{t=1}^{n}(1+\mathscr{B}_t)\right) - 1\right)}_{\mathscr{B}}. \tag{8.33}$$

Put simply, therefore, the difference between the geometric sum of portfolio and benchmark returns is *not* equal to the arithmetic sum of the geometric sums of the individual factor returns. Mathematically linking active returns and performance attributions over time with standard techniques is, apparently, *not* that straightforward.

If we compute daily attributions, we must be able to combine these attributions over time. We want to easily examine weekly, monthly, month-to-date, year-to-date, or semi-annual performance attributions without worrying about the actual time interval under examination. This happy situation, however, is *not* possible without the ability to sensibly link our attributions across the time dimension.

This depressing situation is also beginning to get somewhat complicated. Let's frame the problem a bit more clearly with a concrete example where we have two factor returns, $\mathscr{A}$ and $\mathscr{B}$, and three periods (i.e., $n = 3$). Table 8.13 summarizes the portfolio, benchmark, and factor returns stemming from our daily performance attributions.[21] The geometric sum of the portfolio returns over these three periods is 2.95 % for the portfolio and 2.79 % for the benchmark leading to an active return of 16 basis points.

If we take the geometric sum of the returns to the two return factors we arrive at 0.96 and −1.09 %, respectively. The sum of these two return factors suggests that the active return is −13 basis points. This is clearly inconsistent with the true return of 16 basis points, despite the fact that our performance attributions are consistent at each individual point in time. This looks to be a significant drawback associated

[21]There is, for simplicity, a zero residual in our performance attributions. This is not required. It can just be considered an additional column, with hopefully very small values, in Table 8.13.

with the use of daily performance attributions. What can we do? The solution is to use a so-called *smoothing* algorithm to ensure that the active returns and the factor attributions are consistent.

## 8.6 Smoothing Returns

We require an approach that permits us to adjust the factor returns in such a way that, when we sum them across time, they are consistent with the true geometric sum of active returns. This might seem like we are contemplating a fundamental change to our return attribution. This is *not* true. While a smoothing approach does involve a slight change to our return attributions, fortunately the change is typically quite small.

We will consider the most popular smoothing methodology proposed from Cariño [4].[22] It requires a number of algebraic manipulations of our basic return expressions. To begin, let's define the portfolio return in the usual way as,

$$1 + p = \prod_{t=1}^{n}(1 + p_t). \tag{8.34}$$

If we now apply the natural logarithm to both sides of Eq. (8.34) and recall that the logarithm of a product is equal to the sum of the logarithms, we have

$$\begin{aligned} \ln(1 + p) &= \ln\left(\prod_{t=1}^{n}(1 + p_t)\right), \\ &= \sum_{t=1}^{n} \ln(1 + p_t). \end{aligned} \tag{8.35}$$

Thus far, we have merely transformed our returns to a logarithmic scale. Repeating this general transformation on benchmark returns, we can now turn to write our active return as

$$\ln(1 + p) - \ln(1 + b) = \sum_{t=1}^{n} \ln(1 + p_t) - \sum_{t=1}^{n} \ln(1 + b_t). \tag{8.36}$$

Our objective is to find a way to write the active return over the entire period as the sum of the active returns in each period, $\{p_t - b_t, t = 1, \ldots, n\}$. To accomplish

---

[22] There are other approaches, but we use and demonstrate this approach because it is relatively straightforward, widely used, and quite effective. See, for example, Menchero [6], Menchero [7], Bonafede et al. [2] or Bacon [1, Chapter 8].

this, we introduce the simple ratio, $\frac{p_t - b_t}{p_t - b_t} = 1$, into the right-hand-side of Eq. (8.36). This ratio, given its value of unity, changes nothing in the overall equation but permits us to re-arrange a few terms. Let's see how this works,

$$\begin{aligned} \ln(1+p) - \ln(1+b) &= \sum_{t=1}^{n} \ln(1+p_t) - \sum_{t=1}^{n} \ln(1+b_t), \\ &= \sum_{t=1}^{n} \underbrace{\left(\frac{p_t - b_t}{p_t - b_t}\right)}_{=1} (\ln(1+p_t) - \ln(1+b_t)), \\ &= \sum_{t=1}^{n} \underbrace{\left(\frac{\ln(1+p_t) - \ln(1+b_t)}{p_t - b_t}\right)}_{\text{Call this } \kappa_t} (p_t - b_t), \\ &= \sum_{t=1}^{n} \kappa_t (p_t - b_t). \end{aligned} \tag{8.37}$$

The introduction of this simple ratio has permitted us to write down our logarithmically adjusted active return as a kind of weighted sum of the active returns in each period with relatively simple weights, $\{\kappa_t, t = 1, \ldots, n\}$.

The final step is to eliminate the logarithmic adjustment in our active return. This can be accomplished by defining the following constant,

$$\kappa = \frac{\ln(1+p) - \ln(1+b)}{p - b}. \tag{8.38}$$

If we divide both sides of Eq. (8.39) by this constant value, $\kappa$, we can make further progress,

$$\begin{aligned} \left(\frac{1}{\kappa}\right)(\ln(1+p) - \ln(1+b)) &= \left(\frac{1}{\kappa}\right)\sum_{t=1}^{n} \kappa_t (p_t - b_t), \\ \left(\frac{p-b}{\cancel{\ln(1+p) - \ln(1+b)}}\right)\left(\cancel{\ln(1+p) - \ln(1+b)}\right) &= \sum_{t=1}^{n} \frac{\kappa_t}{\kappa}(p_t - b_t), \\ p - b &= \sum_{t=1}^{n} \frac{\kappa_t}{\kappa}(p_t - b_t). \end{aligned} \tag{8.39}$$

This is our final result. Our active return is described as the weighted sum of the active returns in each period with weights, $\left\{\frac{\kappa_t}{\kappa}, t = 1, \ldots, n\right\}$.

This result can be applied to any arbitrary performance attribution. We only require, as is the case for each daily performance attribution, that the active return is equal to the individual factor returns at each point in time, or mathematically,

$$p_t - b_t = \mathscr{A}_t + \mathscr{B}_t, \tag{8.40}$$

for all $t = 1, \ldots, n$. If this holds then,

$$\begin{aligned} p - b &= \sum_{t=1}^{n} \frac{\kappa_t}{\kappa} \underbrace{(\mathscr{A}_t + \mathscr{B}_t)}_{\text{Equation (8.40)}}, \\ &= \underbrace{\sum_{t=1}^{n} \frac{\kappa_t}{\kappa} \mathscr{A}_t}_{\text{Factor 1}} + \underbrace{\sum_{t=1}^{n} \frac{\kappa_t}{\kappa} \mathscr{B}_t}_{\text{Factor 2}}. \end{aligned} \tag{8.41}$$

The clever manipulation in Cariño [4] basically solves our original problem. It provides us with a method to write the active return as the weighted sum of the individual factor returns. While clever, we should be aware that there is no economic rationale behind this smoothing algorithm. It merely slightly adjusts the weights on the individual return factors in a way that permits their addition over time.

Returning to our example described in Table 8.13 on page 270, we can apply the result from Eq. (8.41) in Table 8.14. Once these weighting factors have been applied, we merely compute the arithmetic sum of the portfolio, benchmark, and factor returns. The total values are slightly different than the geometric sums computed in our first attempt in Table 8.13. What is evident, however, is that the active return using the portfolio and benchmark returns is now equal to the values computed from our daily performance attributions.[23] One drawback of this approach, which should be mentioned, is that the weights are time dependent. If one computes the

**Table 8.14** Back to our example

| Return | $\frac{\kappa_t}{\kappa} p_t$ | $\frac{\kappa_t}{\kappa} b_t$ | $\frac{\kappa_t}{\kappa}(p_t - b_t)$ | $\frac{\kappa_t}{\kappa} \mathscr{A}_t$ | $\frac{\kappa_t}{\kappa} \mathscr{B}_t$ | $\frac{\kappa_t}{\kappa}(\mathscr{A}_t + \mathscr{B}_t)$ |
|---|---|---|---|---|---|---|
| $t_1$ (%) | 3.55 | −0.51 | 4.05 | 1.01 | 3.04 | 4.05 |
| $t_2$ (%) | −2.11 | −3.17 | 1.06 | 2.11 | −1.06 | 1.06 |
| $t_3$ (%) | 1.48 | 6.43 | −4.95 | −1.98 | −2.97 | −4.95 |
| Arith. sum (%) | 2.92 | 2.76 | **0.16** | 1.15 | −0.98 | **0.16** |

This table returns to our previous example in Table 8.13 and demonstrates how the Cariño approach solves our problem.

[23] In practice, as in this example, the adjusted returns are typically very close to the true return values—often the difference is only a few basis points. The absolute differences are typically even smaller with daily attributions because, unlike this example, the magnitude of daily returns on a fixed-income portfolio rarely exceed ± 20–30 basis points.

required weights over a three-day period, one should not expect identical weights when extending the computation to a four-day interval. This is clearly sub-optimal and explains the existence of alternative techniques, which attempt to resolve this shortcoming. The price, however, is additional complexity.

The smoothing algorithm presented in this section is an essential aspect of the use of daily performance attributions. It allows us, with the application of a relatively simple weighting algorithm, to preserve the additivity of the factor returns in our daily performance attribution over time. This permits computations of daily performance attributions reducing the impact of transactions and making them easier to handle. Then, as required by the user, the aggregation of these returns over any arbitrary time interval is possible—weekly, monthly, month-to-date, year-to-date, and so on. This separates the choice of performance interval from the computation horizon of the performance attributions. This result is quite powerful. One merely computes performance attributions on every single business day for each of one's portfolios and then, as required by the user, presents the results over any desired performance horizon.

## 8.7 Concluding Thoughts

A naive application of our additive risk-factor decomposition is complicated by the presence of transactions. Transactions imply instability in the relative weights and returns associated with the positions in one's portfolio over the reporting period. Ignoring this instability and assuming unchanged positions—termed the buy-and-hold assumption—can lead to significant errors in one's attributions. This chapter demonstrates how the use of daily attributions permits one to confidently categorize the set of possible outcomes to one's portfolio. Use of this categorization and the information regarding the transactions permits us to relax the buy-and-hold assumption and enhance the accuracy of our performance attributions. Finally, to avoid some of the arithmetic vagaries associated with combining daily returns over the time dimension, one may employ a smoothing algorithm to link the daily attributions over any arbitrary time interval.

## References


1. C.R. Bacon, *Practical Portfolio Performance Measurement and Attribution* (Wiley, New York, 2008)
2. J.K. Bonafede, S.J. Foresti, P. Matheos, A multi-period linking algorithm that has stood the test of time. J. Perform. Meas. **Fall**, 15–26 (2002)
3. J.K. Bonafede, M.C. McCarthy, Transaction-based vs. holdings-based attribution: the devil is in the details. J. Perform. Meas. **Fall**, 42–51 (2003)
4. D.R. Cari no, Combining attribution effects over time. J. Perform. Meas. **Summer**, 5–14 (1999)
5. D. Laker, Perspectives on transaction-based attribution. J. Perform. Meas. **Fall**, 10–23 (2003)
6. J. Menchero, An optimized approach to linking attribution effects over time. Finan. Anal. J. **5**(1), 36–42 (2000)

7. J. Menchero, Multiperiod arithmetic attribution. Finan. Anal. J. **60**(4), 76–91 (2004)
8. J. Menchero, J. Hu, Errors in transaction-based performance attribution. J. Perform. Meas. **Fall**, 57–69 (2003)
9. D. Spaulding, A case for attribution standards. J. Perform. Meas. **Winter**, 13–21 (2003)

# Traditional Performance Attribution 9

*Tradition is a guide and not a jailer.*

W. Somerset Maugham

The treasury yield curve is the principal driver of high-quality fixed-income security returns in any economy.[1] Our additive risk-factor decomposition correspondingly provides an explicit consideration of the performance impact of treasury curve movements. Moreover, three possible extensions to this basic computation have been demonstrated in detail.[2] We have thus already developed a reasonably complete insight into the yield curve's influence on our portfolio returns.

It is nonetheless possible and desirable to extend one's understanding of yield-curve effects on one's performance. Other, more traditional, techniques are available for approximating different dimensions of the yield-curve impact. Two of these traditional yield-curve-based perspectives include:

1. the disaggregation of positions related to views on duration and views on the relative shape of the curve; and
2. the so-called *roll-down* effect.

The first perspective is closely related to the standard yield-curve factors of level, slope and curvature. Long or short duration positions, all else equal, are equivalent to taking views on the parallel movements (i.e., level) of the yield curve. One may also simultaneously undertake positions that benefit from changes in the yield curve's slope and curvature. Our general additive risk-factor-based decomposition couples these two effects, whereas techniques exist that seek to consider them separately.

[1]It is unsurprising, therefore, that in the mathematical finance literature so-called risk-free fixed-income securities are treated as contingent claims (i.e., derivatives) on the yield curve.

[2]These included key-rate-duration, model-based, and ad hoc decompositions of the treasury return.



D.J. Bolder, *Fixed-Income Portfolio Analytics*,
DOI 10.1007/978-3-319-12667-8_9

The second perspective concerns itself with the performance effect arising from the general shape of the curve and the passage of time.

These *two* traditional perspectives have been, and remain, important elements of fixed-income performance attribution. While both of these elements are handled within our general framework, their impact is typically somewhat obscured by other return elements. Given their potential usefulness, this chapter addresses this shortcoming by examining these two effects in substantial detail. Indeed, a discussion of fixed-income performance attribution would be incomplete without a consideration of these effects.

## 9.1 Asset Allocation and Security Selection

One of the key tenets of our approach to portfolio analytics was to perform all analysis, wherever possible, at the instrument level. So far, we have been fairly successful in this respect. At times, however, it is sensible to deviate from one's fundamental principles. The traditional approach to performance attribution, introduced by Brinson et al. [2], does *not* operate at the instrument level.[3] It nonetheless offers some useful insights into one's performance. In the first part of this section, we will examine this traditional approach and then extend it to enhance its usefulness to fixed-income portfolios.

The Brinson et al. [2] approach essentially employs a few algebraic tricks to decompose the active return of a portfolio into two principal elements:

- an asset-allocation component; and
- a security-selection component.

These two components allow one to differentiate between high-level and low-level decisions. The asset-allocation component is basically the high-level aspect—it helps one to see the value of a decision to overweight a given sector, such as utilities or manufacturing in a stock portfolio. It also works very well in multiple asset-class settings. The second aspect, security selection, provides insight into how well specific securities are chosen within one's overall asset allocation.

Despite the fact that these ideas do not generally apply to fixed-income portfolios, it is nonetheless useful to walk through some of the basic logic in significant detail. It is not only interesting in its own right, but we will also rely heavily on the underlying ideas to enhance our understanding of a portfolio's yield-curve positioning relative to its benchmark.

To understand this traditional approach, we will need to introduce some basic notation. Imagine that you have $m$ sectors in both your portfolio and benchmark—these could be sectors within a benchmark or, indeed, separate asset classes

[3] Ankrim [1] offers some extensions incorporating risk.

altogether. The portfolio weights in each of these sectors are given as,

$$\omega_p = \left[\omega_{p_1} \cdots \omega_{p_m}\right], \tag{9.1}$$

and the benchmark sector weights are given as,

$$\omega_b = \left[\omega_{b_1} \cdots \omega_{b_m}\right]. \tag{9.2}$$

We subsequently denote the portfolio return associated with each sector as,

$$r_p = \left[r_{p_1} \cdots r_{p_m}\right], \tag{9.3}$$

and benchmark sector returns as,

$$r_b = \left[r_{b_1} \cdots r_{b_m}\right]. \tag{9.4}$$

These are the necessary ingredients for the computation of the active return, $r_a$, of the portfolio as,

$$r_a = \underbrace{\sum_{k=1}^{m} \omega_{p_k} r_{p_k}}_{\text{Portfolio return}} - \underbrace{\sum_{k=1}^{m} \omega_{b_k} r_{b_k}}_{\text{Benchmark return}}. \tag{9.5}$$

The portfolio and benchmark returns, therefore, are merely the weighted-average returns of each sector.

Now comes Brinson et al.'s [2] algebraic trick. They merely proposed that we add and subtract an additional term, $\sum_{k=1}^{m} \omega_{p_k} r_{b_k}$, from Eq. (9.5) and simplify,

$$\begin{aligned} r_a &= \sum_{k=1}^{m} \omega_{p_k} r_{p_k} - \sum_{k=1}^{m} \omega_{b_k} r_{b_k} + \underbrace{\sum_{k=1}^{m} \omega_{p_k} r_{b_k} - \sum_{k=1}^{m} \omega_{p_k} r_{b_k}}_{=0}, \\ &= \underbrace{\sum_{k=1}^{m} \left(\omega_{p_k} - \omega_{b_k}\right) r_{b_k}}_{\text{Asset allocation}} + \underbrace{\sum_{k=1}^{m} \omega_{p_k} \left(r_{p_k} - r_{b_k}\right)}_{\text{Security selection}}. \end{aligned} \tag{9.6}$$

As a consequence of this small adjustment, the active return has been broken down into *two* separate components: the terms are called asset allocation and security selection, respectively. The asset-allocation component holds the sector returns fixed and looks at the differences in the weights. The security selection component, conversely, fixes the weights and examines the changes in returns. The basic idea is that asset allocation involves taking decisions on sector weights whereas security selection requires decision-making at the sector-return level.

This is not the only way to decompose the active return, but this is unimportant for our purposes.[4] The key point is that, at a sector or asset-class level, one can re-arrange the active-return expression to extract information on the success, or lack thereof, of portfolio-management decisions.

How does this apply to us? One way to use these ideas in the fixed-income world would be to compute the portfolio and benchmark weights and returns on the sector level—governments, agencies, supra-nationals, and corporate bonds—and perform the decomposition in Eq. (9.6). This may be a very useful supplement for a portfolio invested in a wide range of fixed-income securities, but would be rather less helpful for a single currency, sovereign benchmark.

There is a second possible application of these ideas in a fixed-income portfolio. Our general framework provides a clear view of the active return arising from yield-curve positioning. It also helps with the identification of the active return stemming from long- and short-duration positions. Our additive risk-factor-based decomposition is not particularly good at decoupling how much return comes from long (or short) duration positions and how much arises from curve-positioning.

Consider two different types of position. The first is duration neutral to the strategic benchmark, but has a curve-steepening trade implemented in the portfolio. In this case, there is no duration effect and all return comes from curve positioning. In the second position, the portfolio is short duration relative to the strategic benchmark, but equally so across all sectors of the curve. In this situation, all return stems from the duration effect with no curve-position return.

These are two extreme and unequivocal cases. One can easily imagine a third situation, however, where one simultaneously implements a short duration *and* a curve-steepening trade. In this case, it is not so obvious as to how one can distinguish between the duration and curve-positioning return. Our key-rate or model-based decompositions of curve return will generally not separate out these two effects. We nevertheless wish to understand how much each component contributed to the overall yield return. Decoupling the duration and curve-position effects, in this manner, would enhance our understanding of the overall portfolio return.

Let's use these ideas from Brinson et al. [2] to see how we might break down the yield curve into two separate elements:

- a duration component; and
- a curve-position component.

---

[4]Slightly different algebraic tricks can be used to achieve slightly different definitions of asset-allocation and security-selection returns. As an exercise for the reader, try adding and subtracting the term, $\sum_{k=1}^{m} \omega_{b_k} r_{p_k}$, and simplifying. One may also use both, giving rise to a difficult-to-interpret interaction term.

It will probably not be a surprise to the reader that we intend to treat the duration component as an asset allocation decision and classify curve-positioning as security selection.

To do this, we again need to introduce (and recall) a bit of notation. The active curve return on a portfolio is approximated in our framework as,

$$\begin{aligned} r_{\text{Yield}} &= -D_p \Delta y_p - (-D_b \Delta y_b), \\ &= \underbrace{-D_p \Delta y_p}_{\substack{\text{Portfolio}\\ \text{yield}\\ \text{return}}} + \underbrace{D_b \Delta y_b}_{\substack{\text{Benchmark}\\ \text{yield}\\ \text{return}}}, \end{aligned} \tag{9.7}$$

where $D_p$ and $D_b$ are the modified duration of the portfolio and benchmark, respectively. We use the overall yield changes on the portfolio and benchmark, $\Delta y_p$ and $\Delta y_b$, although one could also quite reasonably focus on changes in equivalent-treasury yield or credit spreads.[5]

In the same spirit as Brinson et al. [2], we add and subtract an appropriate term to the right-hand side of Eq. (9.7)—in this case, $D_p \Delta y_b$—and rearrange as follows,

$$\begin{aligned} r_{\text{Yield}} &= -D_p \Delta y_p + D_b \Delta y_b + \underbrace{D_p \Delta y_b - D_p \Delta y_b}_{=0}, \\ &= \underbrace{-\big(D_p - D_b\big)\Delta y_b}_{\text{Duration component}} - \underbrace{D_p\big(\Delta y_p - \Delta y_b\big)}_{\substack{\text{Curve-position}\\ \text{component}}}, \end{aligned} \tag{9.8}$$

The interpretation is analogous to the Brinson et al. [2] paper. The duration effect holds the yield change constant, but varies the duration, thus describing the impact of the duration decision relative to one's benchmark. The second component holds the duration constant, but examines the different yield movements. This describes the yield return associated with curve-positioning. This simple computation can be a helpful complement to other performance attribution details provided using previously described techniques.

Some reflection reveals that we can accomplish much the same outcome by adding and subtracting a slightly different term from Eq. (9.7). If we instead use

[5]There are some difficulties in computing yield returns from aggregate modified-duration and yield figures. Recall that the product of the sums is not equal to the sum of the products. We can solve this problem by computing the true yield return at the security level, using the known portfolio (or benchmark) modified-duration values, and solving for the *correct* yield change to ensure a consistent yield return.

the term $D_b \Delta y_p$ we arrive at the following result,

$$\begin{aligned} r_{\text{Yield}} &= -D_p \Delta y_p + D_b \Delta y_b + \underbrace{D_b \Delta y_p - D_b \Delta y_p}_{=0}, \\ &= \underbrace{-\left(D_p - D_b\right) \Delta y_p}_{\text{Duration component}} - \underbrace{D_b \left(\Delta y_p - \Delta y_b\right)}_{\substack{\text{Curve-position}\\ \text{component}}}, \end{aligned} \tag{9.9}$$

What is the difference between these two approaches? The short answer is: not much. In both cases, the duration component holds the yield movement fixed and examines the impact of the differing durations. In both cases, a portfolio that exactly matches the benchmark duration will have the sensible result of zero duration return. The difference, however, is that the fixed yield component varies between the two approaches. In the first case, we use the benchmark return whereas the second case employs the portfolio return.[6]

Which is correct? It depends on the relative importance of the duration and position decisions. If the portfolio manager takes long/short duration decisions first and then incorporates curve positions, then we want the cleanest possible view of the duration decisions. The cleanest possible view would be to use the relatively neutral benchmark yield movement with the duration differential between the portfolio and benchmark. If the duration position is the principal decision, one should use Eq. (9.8).[7]

Both approaches examine the curve-position return by holding the duration constant and considering the impact of the difference in yield movements between the portfolio and benchmark. Again, if the yield change in the benchmark and portfolio are identical, both approaches will suggest a zero curve-position return. The difference between the two approaches is the value used to represent the *fixed* element: in this case, modified duration. In the first approach, the portfolio duration is used while the second approach uses the benchmark duration. The choice of approach again depends on the importance of the decision. If curve-position decisions are most important, then you will want to use the benchmark duration to provide a neutral perspective on the duration decision. This would argue for the use of the second approach.[8]

Table 9.1 provides a brief example of how the two approaches—summarized in Eqs. (9.8) and (9.9)—can be implemented. Imagine that we have a portfolio and benchmark with a modified-duration of 3 and 3.5 years, respectively. The

[6]One could also use both terms, but this creates an interaction term, which is difficult to interpret.

[7]In the Brinson et al. [2] setting, this would be termed a *top-down* decision. See Colin [3] for more details.

[8]In the Brinson et al. [2] setting, this would be termed a *bottom-up* decision. It is not obvious that these terms directly apply to the fixed-income case, but it is nonetheless useful to draw the comparison.

**Table 9.1** Simple example

| | $D_M$ | $\Delta y$ | $r^{\text{TRE}} \approx -D_M \Delta y$ |
|---|---|---|---|
| Portfolio | 3.0 | 0.25 % | −75.0 |
| Benchmark | 3.5 | 0.35 % | −122.5 |
| **Active** | **−0.5** | **−0.1 %** | **47.5** |

These are the high-level details—duration, yield movement, and curve return—for a short-duration portfolio.

**Table 9.2** Duration and position return

| Approach | Active yield return | | |
|---|---|---|---|
| | Duration | Position | Total |
| Equation (9.8) | *17.5* | *30* | 47.5 |
| Equation (9.9) | *12.5* | *35* | 47.5 |

This table provides a concrete example of the application of a more traditional performance attribution to a fixed-income portfolio.

portfolio is, therefore, short duration relative to the benchmark. Over a given time interval, yields on the portfolio and benchmark widened by 25 and 35 basis points, respectively. This smaller widening of portfolio yields indicates different curve positioning. Both the portfolio and benchmark experience negative curve returns, but, by virtue of its short duration and superior positioning, the portfolio outperforms the benchmark by approximately 47.5 basis points.

We may now employ the adjusted-Brinson approach to decouple these two effects. Our two different approaches, as summarized in Table 9.2, provide slightly different return attributions. The first approach suggests that the short duration position contributed 17.5 basis points to the active yield return, while the second approach approximates the duration contribution at 12.5 basis points. The curve-position return indicated by the first approach, conversely, is 30 basis points relative to an estimate of 35 basis points for the second approach.

While the two approaches generate slightly different results, the basic message is the same. The short-duration decision was a good one, but the bulk of the outperformance arose due to clever curve positioning by the portfolio manager. The ultimate choice of approach will depend on the relative importance of the duration and curve-position decisions taken by your portfolio managers.

Algebraic manipulations at the aggregate level have helped us to distinguish between duration and curve-position effects. One may also perform such a decomposition at the instrument level using our typical performance attribution framework. Not surprisingly, it requires a bit of additional effort. To provide some additional clarity, we will demonstrate how this might be done in the context of a simple example. Table 9.3 provides the details of a simple US Treasury (UST) benchmark comprised of 2-, 5-, and 10-year bonds. It further outlines *two* possible portfolios, both positioned for a flattening of the US Treasury yield curve.

**Table 9.3** Curve-flattener example

| Object | Cash | 2Y | 5Y | 10Y | Total |
|---|---|---|---|---|---|
| **Portfolio Weights (%)** | | | | | |
| Benchmark | 0.0 | 40.0 | 30.0 | 30.0 | 100.0 |
| $P_1$ | *19.8* | 14.6 | 30.0 | 35.5 | 100.0 |
| $P_2$ | 0.0 | 34.5 | 30.0 | 35.5 | 100.0 |
| **Duration Contribution (yrs.)** | | | | | |
| Benchmark | 0.0 | 0.79 | 1.46 | 2.7 | 4.95 |
| $P_1$ | 0.0 | *0.29* | 1.46 | *3.2* | 4.95 |
| $P_2$ | 0.0 | *0.68* | 1.46 | *3.2* | *5.34* |

This table provides the details of a simple US Treasury benchmark and two portfolios implementing a curve-flattener trade: one flat duration to the benchmark and the other long duration.

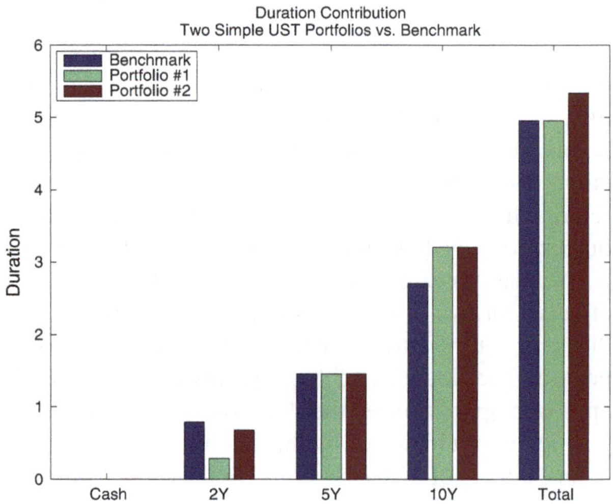


**Fig. 9.1** Curve-flatteners at a glance. This figure displays graphically the overall duration and duration contribution by instrument for the benchmark and the two curve-flattening portfolios in Table 9.3

Although both portfolios have curve-flattening positions, there is a significant difference in their implementation.[9] The first portfolio ($P_1$) is long 10-year bonds and short 2-year bonds, but has a *flat* duration position relative to the benchmark. To accomplish this, it is necessary for $P_1$ to maintain a sizeable cash position of almost 20 %. The second portfolio ($P_2$) is long duration. It has implemented its curve-flattener with a zero cash balance implying a larger 2-year position and a correspondingly higher duration. Figure 9.1 graphically describes the duration contribution of each instrument in these two portfolios relative to the benchmark.

$P_1$ presents *few* conceptual challenges. Whatever happens to the US Treasury yield curve, we fully expect the curve return to be entirely associated with curve

[9]In this example, a curve-flattener is represented as a short 2-year, neutral 5-year, and long 10-year position.

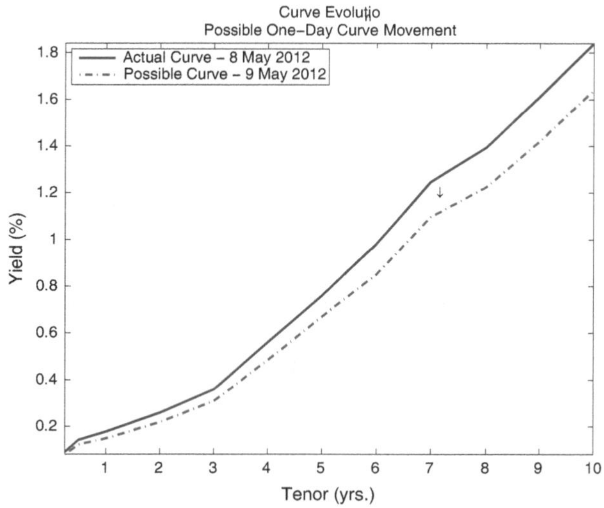


**Fig. 9.2** An actual curve flattening. This figure outlines a possible 1-day movement in the UST yield curve. In this case, we actually observe a flattening of the curve, which should prove advantageous for the two sample portfolios in Table 9.3

**Table 9.4** Typical curve-flattener attribution

| | Benchmark | $P_1$ | $P_2$ | $P_1$ | $P_2$ |
|---|---|---|---|---|---|
| Element | Absolute | | | Active | |
| Carry | 0.2 | 0.3 | 0.3 | **0.0** | **0.0** |
| *Curve* | *70.4* | *78.4* | *79.9* | *8.0* | *9.6* |
| Convexity | 0.6 | 0.7 | 0.7 | **0.1** | **0.1** |
| Residual | 0.2 | 0.3 | 0.3 | **0.0** | **0.1** |
| Total | 71.4 | 79.6 | 81.2 | **8.1** | **9.8** |

This table outlines the typical daily performance attribution for the two curve-flattener portfolios presented in Table 9.3. All figures are in basis points.

positioning. This is because, by maintaining a flat duration position, the portfolio manager has taken a *pure* curve-flattening view. $P_2$ is a bit more complicated. Its curve return will be a combination of the long duration and curve-flattening positions. Not only has the curve comes down across all maturities, it has also flattened somewhat. We've seen how one can algebraically separate these two effects, let's now examine another alternative approach.

Figure 9.2 outlines a possible 1-day movement in the US Treasury yield curve. It illustrates a classic yield-curve flattening that should greatly please both of our fictitious portfolio managers. While perhaps a bit extreme for a 1-day movement, it is nonetheless entirely plausible.

Using this movement in the US Treasury yield curve, we may proceed to perform the usual performance attribution employing our additive decomposition and the approach developed in the two preceding chapters. Table 9.4 summarizes an attribution of the absolute and relative returns for our portfolios and benchmark. The results, not surprisingly, are positive. Each portfolio generated both substantial absolute return and outperformed the benchmark. Also unsurprisingly, given the short time horizon and the nature of the securities, most of the return can be attributed to curve movements.

Table 9.4 indicates that curve return was the dominant contributor to overall return. The next question is: how much of this return can be attributed to duration and curve positioning, respectively? For $P_1$, given the structure of our example, it is obviously due to the curve flattener. For $P_2$ it is not immediately obvious. We seek, therefore, an approach that simultaneously ensures that:

- all of $P_1$'s curve return is attributed to the positioning effect; and
- $P_2$'s curve return is sensibly allocated between the two effects.

Something additional is required given that our general framework does *not* provide this breakdown.

Figure 9.3 provides a possible answer, which should be quite familiar. Using a technique from a previous chapter, it demonstrates how one may break down the curve movement into two elements: a parallel and a non-parallel shift. The basic idea is simple. One selects a pivot point: in this case, it is the 5-year point. One then moves the original curve towards the new curve in a parallel fashion until the two curves become tangent at the pivot point. This leads to an intermediate curve that has, by construction, only moved in a parallel fashion. Consequently, the distance between this intermediate curve and the new curve approximates the non-parallel movement of the yield curve. The distance between the intermediate curve and the new curve thus estimates the non-parallel movement of the curve. This is a simple and—notwithstanding the rather arbitrary nature of the pivot-point selection—surprisingly powerful decomposition of the yield curve.

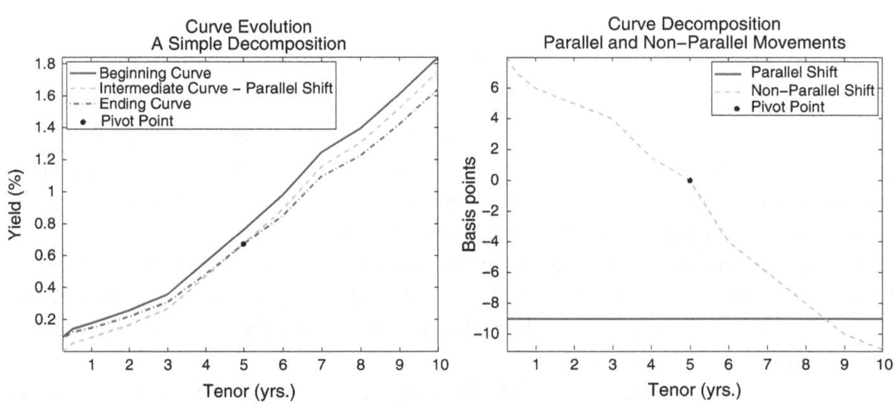


**Fig. 9.3** Yield-curve decomposition. This is a possible decomposition of the yield-curve movement described in Fig. 9.2. The first graphic shows how, using the 5Y-yield as the pivot point, the movement is broken down into parallel and non-parallel movements. The second graphic displays the magnitude of these two distinct yield movements

**Table 9.5** An expanded curve-flattener attribution

| | Benchmark | $P_1$ | $P_2$ | $P_1$ | $P_2$ |
|---|---|---|---|---|---|
| Element | Absolute | | | Active | |
| Carry | 0.2 | 0.3 | 0.3 | **0.0** | **0.0** |
| *Curve (Duration)* | *44.6* | *44.6* | *48.1* | *0.0* | *3.5* |
| *Curve (Position)* | *25.8* | *33.8* | *31.9* | *8.0* | *6.1* |
| Convexity | 0.6 | 0.7 | 0.7 | **0.1** | **0.1** |
| Residual | 0.2 | 0.3 | 0.3 | **0.0** | **0.1** |
| Total | 71.4 | 79.6 | 81.2 | **8.1** | **9.8** |

This table outlines an expanded daily performance attribution for the two curve-flattener portfolios presented in Table 9.3. It now includes a decomposition of the curve return into the duration and curve-position effects.

Before examining the actual mechanics of the calculation, let's turn to Table 9.5 to examine the results using this simple yield-curve decomposition. The eight basis-point out-performance of $P_1$ is attributed, as desired, entirely to its curve-flattening trade. Only about six basis points, however, of $P_2$'s out-performance can be attributed to its curve positioning. The remaining 3.5 basis points of out-performance is allocated to its long duration position. The overall curve return is consistent with our standard framework, it has merely been usefully decomposed into duration and position effects.

How does one actually perform the computations in Table 9.5? It turns out to be surprisingly easy. Let's begin with the usual return decomposition as follows,

$$r \approx y\Delta t - D_M \Delta y_{\text{TRE}} - D_S \Delta s + \frac{C}{2}(\Delta y)^2. \tag{9.10}$$

We merely replace $\Delta y_{\text{TRE}}$ for each individual instrument with the sum of the two parts of the decomposition as,

$$\Delta y_{\text{TRE}} = \Delta y_{\text{TRE,Parallel}} + \Delta y_{\text{TRE,Other}}. \tag{9.11}$$

In other words, the movement in equivalent-treasury yield of each security is merely the sum of the parallel and non-parallel (i.e., other) movements.[10]

This allows us to plug Eq. (9.11) into Eq. (9.10) and expand as,

$$r \approx y\Delta t - D_M \underbrace{\left(\Delta y_{\text{TRE,Parallel}} + \Delta y_{\text{TRE,Other}}\right)}_{\text{Equation (9.11)}} - D_S \Delta s + \frac{C}{2}(\Delta y)^2,$$

[10]The actual decomposition is also trivial. The original and new curves are unchanged, whereas the intermediate curve is merely the original curve plus the change in yield at the pivot point.

$$\approx y\Delta t - \underbrace{D_M \Delta y_{\text{TRE,Parallel}}}_{\substack{\text{Duration}\\ \text{Effect}}} - \underbrace{D_M \Delta y_{\text{TRE,Other}}}_{\substack{\text{Position}\\ \text{Effect}}} - D_S \Delta s + \frac{C}{2}(\Delta y)^2. \tag{9.12}$$

The methodology thus easily extends to incorporate this adjustment. Moreover, if desired, it allows one to use the key-rate duration information to see where precisely along the curve that one is adding (or subtracting) value with respect to the duration and position effects. The result is,

$$r \approx y\Delta t - \boxed{\sum_{k=1}^{\nu} D_k} \Delta y_{\text{TRE,Parallel}} - \boxed{\sum_{k=1}^{\nu} D_k} \Delta y_{\text{TRE,Other}} - D_S \Delta s + \frac{C}{2}(\Delta y)^2. \tag{9.13}$$

The result of this adjustment is an understanding of duration and curve-positioning effects organized across your set of key-rate tenors.

We have now examined two alternative approaches to decoupling the return impact of decisions regarding the duration and curve-positioning of one's portfolio relative to the benchmark. Additional assumptions are required. In the former case, it is necessary to work with aggregate portfolio measures instead of, as is our preference, at the instrument level. In the latter case, we need to perform an ad hoc decomposition of yield-curve movements into parallel and non-parallel elements. In both cases, these additional assumptions permit us to obtain greater insight into the performance of our portfolio. We now turn our attention to another dimension of yield-curve-related return: the roll-down effect.

## 9.2 The Roll-Down Effect

The roll-down effect has long been considered an important contributor to fixed-income portfolio returns. Roll-down returns stem from the typically upward-shaping form of the yield. All else equal, the passage of time in an upward-sloping yield curve environment will generate decreases in yields—and hence positive returns—as one *rolls* down the curve. In our additive risk-factor decomposition this effect is not explicitly considered. Instead, it is embedded in the curve return. We now turn to consider this effect, see how it can be incorporated into our general framework, and try to estimate how important it might be in the overall return of a fixed-income portfolio.

Our additive risk-factor-based return decomposition uses a second-order Taylor series expansion of the bond-price equation to approximate the return of a fixed-income security over the time interval $(t - \Delta t, t)$:

$$r(t - \Delta t, t) \approx y\Delta t - D_M \Delta y + \frac{C}{2}(\Delta y)^2, \tag{9.14}$$

where $y$, $D_M$, and $C$ denote the yield, modified duration, and convexity of the bond, respectively. This approximation attributes the return on a fixed-income security into a carry component, an interest-rate component, and a convexity correction. Each of these elements can, and have been, decomposed further to gain additional insight into the possible sources of return associated with a given security.

To isolate the roll-down effect, we need to look more closely at the interest-rate element: $-D_M \Delta y$. More specifically, we ask what precisely is meant by the term $\Delta y$? It is mechanically defined as,

$$\Delta y = y_t - y_{t-\Delta t}. \tag{9.15}$$

$\Delta y$, in words, represents the change in the yield of the security over the interval $(t - \Delta t, t)$. While this is a sensible definition, it glosses over an important element. It makes *no* mention of the tenor of the security in question. The actual tenor of each security is typically either ignored or assumed to be understood for each security.

A bit of reflection reveals that the tenor of a fixed-income security *changes* with the passage of time. Depending on the horizon of one's analysis, this may be worth considering. If $\Delta t$ is substantial, say 1 month, then the yield change as defined in Eq. (9.15) is perhaps missing something. If you buy a 2-year bond, for example, and hold it for a month, then you have a 1-year and 11-month bond. Equation (9.15) is thus inherently comparing the change in the 2-year yield to the 1-year and 11-month yield.

We wish to explicitly take into account this typically ignored tenor mismatch embedded in $\Delta y$. To accomplish this task, we need to expand our notation. Let's denote the tenor of the fixed-income security at time $t - \Delta t$ as $n$. This implies that, after the passage of $\Delta t$, the security will have a tenor of $n - \Delta t$ years. Applying this notion to Eq. (9.15), we have

$$\Delta y = y_t^{n-\Delta t} - y_{t-\Delta t}^{n}. \tag{9.16}$$

Although certainly a bit difficult to decipher at first, this expanded notation usefully combines the ideas of time and tenor. It holds that if you purchase a bond with $n$-year tenor at time, $t - \Delta t$, is will have a tenor of $n - \Delta t$ at time $t$. Quite simply, as time moves forward, the tenor of all instruments decreases as they move inexorably towards maturity. This is a feature of fixed-income instruments and not an issue in equity markets.

Let's consider a simple example to underscore how Eq. (9.16) might be useful. Imagine that you purchase a bond at time $t$ and hold it for $\Delta t$ periods. Over the interval $(t - \Delta t, t)$, there is *no* change in the yield curve—it remains completely fixed. Have you experienced any interest-rate return over the period? One's first instinct is likely to say: no, of course not, there has been no change in the yield curve over the period. The correct answer, however, is that it depends on the shape of the yield curve. Recall that your bond had a tenor of $n$ years at the beginning of the period and now, after the passage of time, has a tenor of $n - \Delta t$ years. There are *three* possibilities:

$$\underbrace{y_t^{n-\Delta t} - y_{t-\Delta t}^{n}}_{\Delta y} \longmapsto \begin{cases} \Delta y > 0 : \text{Downward-sloping yield curve} \\ \Delta y = 0 : \text{Flat yield curve} \\ \Delta y < 0 : \text{Upward-sloping yield curve} \end{cases} . \tag{9.17}$$

Assuming there has been no change in the yield curve over $(t - \Delta t, t)$, then a yield-change of zero can *only* be associated with a flat yield curve. If the yield curve is upward-sloping—which is the typical case—then the change in the yield curve will be negative and the associated return will be positive.[11] This is termed the *roll-down* effect. Quite simply, all else being equal, an upper-sloping yield curve environment will generate positive interest-rate gains to a security as one *rolls* down the curve through the passage of time.[12]

This logic suggests that there is a tendency, with the passage of time, to enjoy positive interest-rate returns. This tendency, however, is dependent on two assumptions: an upward-sloping yield curve environment and an unchanged yield curve. The former is a typical situation, whereas the latter occurs only very rarely. As long as the yield-curve remains upward sloping, however, this positive-return tendency may even persist in the face of yield-curve volatility.

We need an approach that can (approximately) isolate the roll-down element from the general movement of interest rates. One reasonable alternative is to perform a simple manipulation of Eq. (9.16). We simply add and simultaneously subtract the term $y_{t-\Delta t}^{n-\Delta t}$ and re-arrange as follows,

$$\begin{aligned} \Delta y &= y_t^{n-\Delta t} - y_{t-\Delta t}^{n}, \\ &= y_t^{n-\Delta t} - y_{t-\Delta t}^{n} + \left( \underbrace{y_{t-\Delta t}^{n-\Delta t} - y_{t-\Delta t}^{n-\Delta t}}_{=0} \right), \\ &= \left( \underbrace{y_t^{n-\Delta t} - y_{t-\Delta t}^{n-\Delta t}}_{\substack{\text{Interest-rate}\\ \text{movement}}} \right) - \left( \underbrace{y_{t-\Delta t}^{n} - y_{t-\Delta t}^{n-\Delta t}}_{\substack{\text{Roll-down}\\ \text{effect}}} \right), \\ &= \Delta y^{n-\Delta t} - \Delta\theta, \end{aligned} \tag{9.18}$$

[11]The opposite is, of course, true with an inverted yield curve.

[12]As the yield falls over time, there is a corresponding decrease in the carry return associated with the fixed-income security. We will not explicitly allocate this reduced carry to the roll-down effect, but it is useful to appreciate the interplay between the interest-rate and carry elements.

where $\theta$ denotes the slope of the yield curve around the security's yield. This is a representation of the yield change as *two* separate elements: an interest-rate movement and a roll-down component.

Does this make sense? The first term compares the yield change of an $(n - \Delta t)$-tenor security over the interval $(t - \Delta t, t)$. Comparing a fixed-tenor bond at two points in time seems like a sensible comparison of the *true* yield movement over the period.[13] The second term proceeds to compare the difference between the $n$- and $(n - \Delta t)$-tenor yields at the fixed point in time, $t - \Delta t$. This is basically an approximation of the slope of the yield curve—in the local area around the security's yield—at time $t - \Delta t$. The logic behind this choice is that, should the yield curve remain unchanged over the performance period, this measure of slope will describe the roll-down experienced by the security. To answer the initial question: yes, Eq. (9.18) does appear to make sense.

Figure 9.4 provides a schematic view of the geometry involved in decomposing the roll-down and yield-curve movement effects. It shows very clearly how the traditional computation of the yield change of the period (i.e., $\Delta y$) is simply the *fair*, or fixed-tenor, measure of the change in the yield curve (i.e., $\Delta y^{n-\Delta t}$) less the slope of the curve (i.e., $\Delta\theta$) at the beginning of the period. This is a rather stylized view of this decomposition. The roll-down effect need not be positive and the actual yield-curve movement is almost guaranteed to be more complicated than what is demonstrated in Fig. 9.4.

How do we use these ideas to compute the roll-down effect? The application of this approach is readily deduced from Eq. (9.14). We merely replace the old

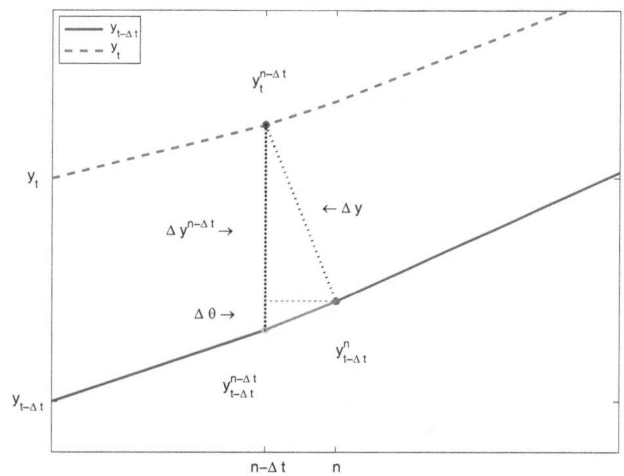


**Fig. 9.4** Isolating the roll-down effect. This figure shows, in a stylized manner, the decomposition of the yield movement into the interest-rate and roll-down components

[13]An implicit choice has been made in this analysis. We chose to compare the $(n - \Delta t)$-tenor security, whereas one could have instead examined the yield change of the $n$-tenor security over the interval $(t - \Delta t, t)$.

definition of $\Delta y$ with the transformed definition in Eq. (9.18) and expand:

$$r(t-\Delta t,t) \approx y\Delta t - D_M \underbrace{\left(\Delta y^{n-\Delta t} - \Delta\theta\right)}_{\text{Equation (9.18)}} + \frac{C}{2}(\Delta y)^2,$$

$$\approx y\Delta t - \underbrace{D_M \Delta y^{n-\Delta t}}_{\substack{\text{Interest-rate}\\ \text{effect}}} + \underbrace{D_M \Delta\theta}_{\substack{\text{Roll-down}\\ \text{effect}}} + \frac{C}{2}(\Delta y)^2. \tag{9.19}$$

The expression is structured such that the roll-down effect has a positive contribution when the slope of the yield curve (i.e., $\theta$) is positive in the neighbourhood of the security's yield. We didn't need to derive Eq. (9.18) this way, but it aids in the interpretation of the results.[14]

One could also, through the replacement of modified durations with key-rate durations, gain an understanding of how the roll-down effect is enjoyed along different points along the yield curve. This would have the form,

$$r(t-\Delta t,t) \approx y\Delta t - \boxed{\sum_{k=1}^{\nu} D_k}\,\Delta y^{n-\Delta t} + \boxed{\sum_{k=1}^{\nu} D_k}\,\Delta\theta + \frac{C}{2}(\Delta y)^2. \tag{9.20}$$

It may be overkill, but it is a possibility and has the potential to provide useful information.

A natural question to ask is whether or not the roll-down effect is important. One can think of at least *three* reasons why it might *not* be an extremely important component of return for a classical portfolio that is rebalanced on a monthly basis.

1. With monthly rebalancing, one is constantly lengthening the tenor of one's holdings thereby giving back the gains associated with the roll-down effect.
2. Since one's benchmark also experiences a certain degree of roll-down and one typically replicates reasonably closely one's benchmark, the net roll-down effect may be relatively modest.
3. Over a monthly holding period with a modest amount of duration and an average yield-curve shape, the mathematical roll-down effect described in Eq. (9.19) is not very large.

In general, it is difficult to verify the first two points. The final point, however, can be tested rather easily in the context of a simple example. Imagine that the yield curve is a straight-line and the slope of the yield curve is defined as the difference

[14]One could also isolate this effect, using the same trick demonstrated in Eq. (9.16), on the equivalent treasury yield or the credit spread.

between the 10-year and overnight yield.[15] For alternative yield-curve slopes, we may compute the actually monthly roll-down. One then merely requires some additional assumptions on the modified duration of one's instrument to approximate the performance impact of these roll-downs.

To begin, we need to apply some simple geometry to determine the magnitude of the roll-down effect. We need to determine the distance between two points along a straight line: $b + mn$ and $b + m(n - \Delta t)$, where $b$ is the intercept and $m$ is the slope. The difference, or the roll-down, is simply $m\Delta t$. The curve slope is roughly,

$$m = \frac{\text{10-Year Yield - Overnight Yield}}{10 - \frac{1}{365}} \approx \frac{\text{Curve Slope}}{10}. \tag{9.21}$$

If we set $\Delta t = \frac{1}{12}$, the roll-down effect (i.e., $m\Delta t$) amounts to,

$$\text{Roll-Down} \approx \frac{\text{Curve Slope}}{120}. \tag{9.22}$$

The roll-down over each month is roughly $\frac{\text{Curve Slope}}{120}$ and is constant across the entire curve.

Employing these assumptions, Table 9.6 computes the roll-down effect for several yield-curve slopes and two different modified durations.[16] For modest yield-curve slopes and short durations, the roll-down effect amounts to a relatively small

**Table 9.6** The magnitude of the roll-down effect

| | | Rolldown effect | |
|---|---|---|---|
| Slope | Roll-down | $D_M = 2$ | $D_M = 5$ |
| 100.0 | −0.8 | 1.7 | 4.2 |
| 200.0 | −1.7 | 3.3 | 8.3 |
| 300.0 | −2.5 | 5.0 | 12.5 |
| 400.0 | −3.3 | 6.7 | 16.7 |

To demonstrate the possible magnitude of the roll-down effect, we assume unchanged linear yield curves over a 1-month period. This allows us to roughly approximate, for two different modified durations, the magnitude of the roll-down effect over a given month. We see clearly that one requires a steep yield curve and sizeable duration for the roll-down effect to obtain a significant roll-down effect.

[15]This is an admittedly extreme example, but it greatly simplifies computations of the yield-curve slope and simultaneously provides a rough idea of the roll-down effect. The actual impact might be slightly larger or smaller depending on the relative curvature of the yield curve.

[16]We have used yield-curve slopes ranging from 100 to 400 basis points. The average daily yield-curve slope over the last 20 years has amounted to roughly 180 basis points, with the steepest curves almost reaching 400 basis points and the largest inversions attaining roughly −100 basis points.

number of basis points. For yield curve with a rather extreme slope of 400 basis points and a modified duration of 5 years, the monthly roll-down effect is almost 17 basis points. Not surprisingly, therefore, we may conclude that one requires a combination of a steep yield curve and a sizeable duration position to realize, on a monthly basis at least, a significant roll-down effect.

This is not to imply that the roll-down effect is always small and unimportant. This analysis merely suggests that the additional effort associated with computing the roll-down effect is probably more worthwhile when the yield-curve is steep and modified durations are reasonably large. When yields are high and the yield-curve is relatively flat, then the roll-down will likely be a relatively small contributor to overall return.

When yields are low and the curve is relatively steep, however, the situation is quite different. Low yields would imply relatively low carry and, as we've seen in our example, yield curve slopes of 200–300 basis points may generate as much as 100–150 basis points of annual roll-down return. Roll-down return may, in fact, be the largest contributor to overall risk. Consequently, although the roll down effect may not always be important, it is probably useful to keep an eye on it. Our general framework, as presented in the previous chapters, does not ignore roll-down return, but does not break it out separately. The previous development rectifies this shortcoming.

## 9.3 Concluding Thoughts

This chapter has considered two traditional perspectives on the impact of yield-curve movements on the return of a fixed-income portfolio: decoupling duration and position views and isolating the roll-down effects. The additional insight provided by these approaches is *not* in conflict with our additive risk-factor decomposition, but instead can be added to our general framework and thereby complement and enhance our overall understanding of curve returns.

## References

1. E.M. Ankrim, Risk-adjusted performance attribution. Financ. Anal. J. **48**(2), 75–82 (1992)
2. G.P. Brinson, L. Randolph Hood, G.L. Beehower, Determinants of portfolio performance. Financ. Anal. J. **42**(4), 39–44 (1986)
3. A. Colin, *Fixed Income Attribution* (Wiley, West Sussex, 2005)

# Part IV
# Risk

Exposures are very useful. They provide us, at a given instant in time, the sensitivity of a fixed-income security or portfolio to a small movement in a given underlying risk factor. Despite their usefulness, they are nonetheless somewhat limited. They neither tell us what a reasonable movement in the risk factor might be nor do they provide any insight into how the collection of risk factors might move together. To answer these questions, one needs to map positions and exposures into risk space. The following chapters examine a set of techniques that seek to combine uncertainty about future risk-factor movements and their co-movements to approximate, from alternative perspectives, the risk associated with a given fixed-income portfolio.

# Introducing Risk

# 10

*Never was anything great achieved without danger.*

Nicolò Machiavelli

Portfolio analytics have three key dimensions: exposure, performance, and risk. Substantial time and effort have been allocated to the treatment of exposure and performance analysis of a fixed-income portfolio. It is now time to consider the third and final dimension: risk.

Risk is the most complex of our three dimensions. It attempts to characterize the *uncertainty* of portfolio returns and allocate, where possible, this uncertainty to the underlying risk factors. Uncertainty is, however, a slippery concept. Characterizing it requires one to formulate a view on the future. This amounts to prediction and it is virtually always a difficult undertaking. Consideration of risk, therefore, not only requires the use of some additional mathematical machinery, it also needs a clear logical structure.

This chapter focuses principally on the necessary logical structure of risk computation. We begin with a formal identification of the two key elements involved in risk computations. Armed with this definition, we then proceed to examine three practical examples. These examples are intended to solidify our understanding of risk analysis and highlight some of the associated challenges. It will also highlight a number of possible risk measures with a particular focus on our workhorse risk measure, the ex-ante tracking error. In this way, we prepare the ground for the next chapter, which deals with the technical aspects of risk measurement and attribution.

## 10.1 Defining Risk

Risk is a commonly used term in everyday speech. While familiarity with the basic idea is helpful, it can also be problematic. The problem is that everyone has their own idea of risk. To discuss the computation of risk in a comprehensive way,



D.J. Bolder, *Fixed-Income Portfolio Analytics*,
DOI 10.1007/978-3-319-12667-8_10

however, we need more clarity about its definition. What, therefore, is risk? The Merriam-Webster dictionary defines it as:

**risk**

1. possibility of loss or injury
2. someone or something that suggests a hazard
3. to expose to hazard or danger

The terms *injury* and *loss* are particularly important. Both words suggest a negative or undesirable outcome. *Possibility*, *hazard*, and *danger* also jump out at the reader in the preceding definition. These words imply an element of uncertainty. The negative outcomes are thus not a certainty, but rather a possibility. In very simple language, risk is thus the possibility of something undesirable occurring during the course of a set of actions.

A set of actions is, of course, an activity. We thus need to consider risk in the context of an activity. When performing any activity, a range of possible events may occur. Some of these outcomes may be positive, some may be neutral, and others might be negative. Risk, as alluded to by the preceding definition, is primarily interested in how often these negative events may occur. Two key elements are thus embedded in this definition.

1. **Outcomes**: What might happen during a given activity.
2. **Likelihood**: The probability (or possibility) of these outcomes.

These are thus the two principal dimensions of risk. Let's take a moment to address each of these elements in turn.

### 10.1.1 Determining Outcomes

Although the focus of this book is undeniably financial, risk is not restricted to financial markets. As a consequence, understanding and assessing risk goes beyond financial concepts. We can, in fact, learn much about the structure of risk by broadening our horizon of possible activities. Table 10.1 provides a list of *five* separate activities ranging from sky-diving to investing funds.

**Table 10.1** Risky activities

| Activity | Negative outcomes |
|---|---|
| Sky-diving | Faulty parachute, hit a power line |
| Mountain-climbing | Bad weather, faulty equipment |
| Bungee-jumping | Faulty cord, weight miscalculation |
| **Playing cards** | **Losing to your father-in-law** |
| Investing funds | $\downarrow$ in value of your investment |

Here is a list of diverse activities and a preliminary attempt to identify the set of possible risks that might occur when performing them.

Upon reflection, one can identify a number of possible negative outcomes for each of these activities. Identifying an exhaustive list of bad outcomes for an arbitrary activity, however, is often either extremely difficult or quite simply impossible. When climbing a mountain or sky-diving the precise nature and number of possible negative events that might occur can almost certainly not be completed catalogued.

This is a serious challenge. Effort is required to determine all of the possible things that can occur—as we'll see, this can be quite difficult. These should include negative outcomes (i.e., losses and injuries) and positive outcomes (i.e., gains and enjoyable experiences). The reality, however, is that it is often impossible, outside of stylized situations, to fully describe the exhaustive set of outcomes for a given activity. In the coming examples, we will explore the potential difficulties associated with identifying outcomes.

### 10.1.2 Assigning Probabilities

Knowledge of the possible outcomes associated with an activity is insufficient for risk assessment. Provided with a list of all of the horrible things that might happen during the course of any given activity would probably incite most of us to stay permanently within the confines of our own home. To avoid becoming shut off from society, however, we need a notion of how likely, or probable, these undesirable outcomes might be.

To sensibly assess the risk of an activity, therefore, we must assign a probability (i.e., likelihood) to each outcome. Table 10.2 returns to our group of possible activities and attempts to perform precisely this task.

How are these probabilities determined? In Table 10.2, the probability figures are basically guesses.[1] In reality, a sensible risk assessment requires a rather more objective approach for the determination of these probabilities.

**Table 10.2** Assigning probabilities

| Activity | Risk |
|---|---|
| Sky-diving | Faulty parachute (*5 %*), hit a power line (*1 %*) |
| Mountain-climbing | Bad weather (*20 %*), faulty equipment (2 %) |
| Bungee-jumping | Faulty cord (*1 %*), weight miscalculation (*3 %*) |
| **Playing cards** | **Losing to your father-in-law** (*80 %*) |
| Investing funds | ↓ in value of your investment (*50 %*) |

We return to our set of activities and attempt to assign a probability to each of these risks This can be very challenging.

[1]The exception is the 80 % probability of losing in cards to my father-in-law; this figure is derived from hard-earned experience!

Unfortunately, determining probabilities can be quite challenging. In some simple situations, one has enough structure to deduce the probabilities—this invariably occurs when faced with a finite and well-defined set of outcomes. In more complex settings, one must use historical data and statistical techniques to arrive at the probability associated with each outcome. In the most difficult cases, where no reasonable data is available, subjective reasoning must be employed. In the coming discussion, we will address a simple example, where the probabilities are known, and then generalize to a more complex setting, where they are not.

### 10.1.3 Getting to Risk

We've established that risk has *two* key dimensions: outcomes and likelihood. This perspective on risk provides a useful logical structure. To paraphrase, risk involves determination of the possible outcomes of any activity and assigning a probability to each of them. Given these outcomes and their associated likelihoods, one can then *combine* them in different ways to communicate risk. Figure 10.1 provides a schematic matrix of these two key dimensions. It provides a range of possible negative to positive outcomes along the vertical axis. Across the horizontal axis, we view the likelihood of these outcomes.

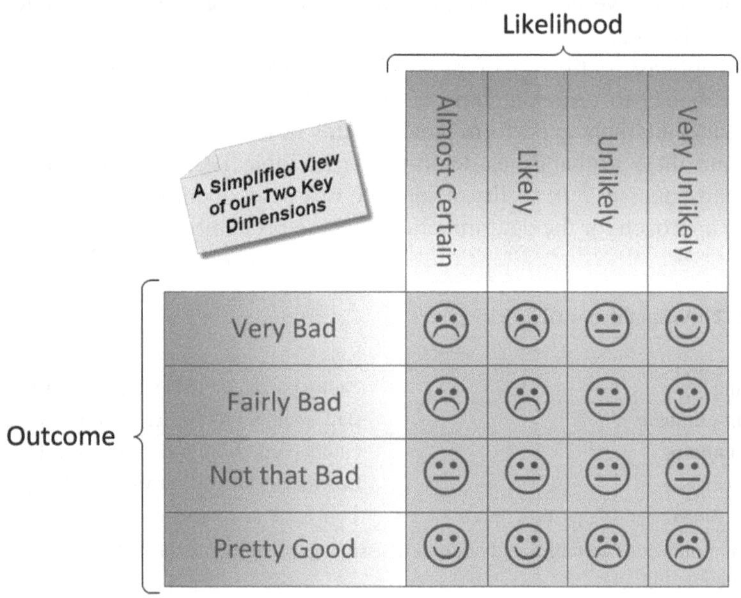


**Fig. 10.1** Outcomes and likelihood. This schematic provides some insight into the two key dimensions involved in assessing the risk of any activity. One needs to know what can happen (i.e., the outcomes) and their relative probability (i.e., likelihood)

Within the matrix in Fig. 10.1, we subjectively rank each combination of outcomes and likelihoods. Bad outcomes with large, or even moderate, likelihoods are naturally undesirable. The opposite is true for positive outcomes with large probabilities. Other combinations of outcomes and likelihoods are less easily classified.

It is tempting to conclude that an activity is risky when the possibility of bad or negative outcomes is high. While difficult to dispute, the statement lacks precision. Different people or institutions may have different tolerances for the probability of negative outcomes. They may also care about different types of negative outcomes more than others. A vanishing probability of an extremely negative outcome, for example, may be problematic for some, but perfectly acceptable for others.

This lack of clarity about what precisely constitutes acceptable and unacceptable levels of risk has led to multiple possible risk measures. Different measures of risk basically combine our two dimensions in various ways. Some measures focus on the average level of negative outcomes, whereas others focus on more extreme events. Some risk measures examine risk in a symmetric manner—considering both positive and negative outcomes—while others concentrate only on the negative outcomes. In short, different risk measures combine outcomes and likelihoods to answer different questions. Throughout our practical examples, we will explicitly establish the link between a number of common risk measures and their perspective on risk.

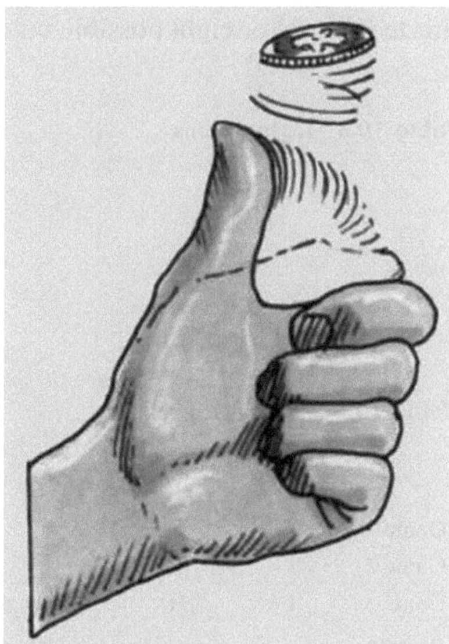

**Fig. 10.2** Old school. Despite the fact that the coin-toss example is perhaps the most old-fashioned example in statistics, it nevertheless holds a number of very useful insights

## 10.2 A Simple Example

Our first example involves a game of tossing coins (Fig. 10.2). This classic example from statistics, remains an excellent (but admittedly dry) way to introduce the rudiments of statistical analysis. Its simplicity is precisely what makes it so useful—both the set of outcomes and their associated likelihoods can be fully determined. This will prove a stark contrast to our more complex examples.

Imagine a game where you flip a *fair* coin with a colleague. If it's heads, you win $1 from your friend. If it's tails, then you pay your colleague $1. Since the coin is fair, the probability of either outcome is equal at 50 %. Table 10.3 highlights the outcomes, pay-outs, and associated probabilities.

We basically know everything about the set of outcomes and their associated probabilities. The expected pay-out is zero—in other words, on average, you should expect to break-even.[2] The worst case outcome is a loss of $1. Computing risk in this setting seems like overkill.

If you play this game *three* consecutive times, however, it becomes a bit more complicated. In this case, it seems legitimate to ask what type of risks are you assuming?

Since rather more can happen in this three-game setting, Table 10.4 laboriously works through all of the possible outcomes of our three consecutive coin tosses. If you play the game three times, there are *four* possible outcomes that can occur in *eight* different ways. More generally, with two outcomes per game and three games, this leads to $2^3$ or eight possible outcomes.

**Table 10.3** Tossing coins

| Outcome | Pay-out | Probability(%) |
|---|---|---|
| Heads | $1 | 50 |
| Tails | −$1 | 50 |

Here we summarize the possible outcomes, the pay-outs and the associated probabilities from our simple coin-tossing game.

**Table 10.4** Counting tosses

| | Possible outcomes | | | | | | | |
|---|---|---|---|---|---|---|---|---|
| | 1 | 2 | 3 | 4 | 5 | 6 | 7 | 8 |
| Game 1 | T | T | T | T | H | H | H | H |
| Game 2 | T | T | H | H | T | T | H | H |
| Game 3 | T | H | T | H | T | H | T | H |
| Pay-out | −$3 | −$1 | −$1 | *$1* | −$1 | *$1* | *$1* | $3 |

If you play the game three times, there are *four* possible pay-outs that can occur in *eight* different ways. This table counts them all.

[2]The computation for a single game is simply: $0.5 \cdot \$1 + 0.5 \cdot -\$1 = \$0$.

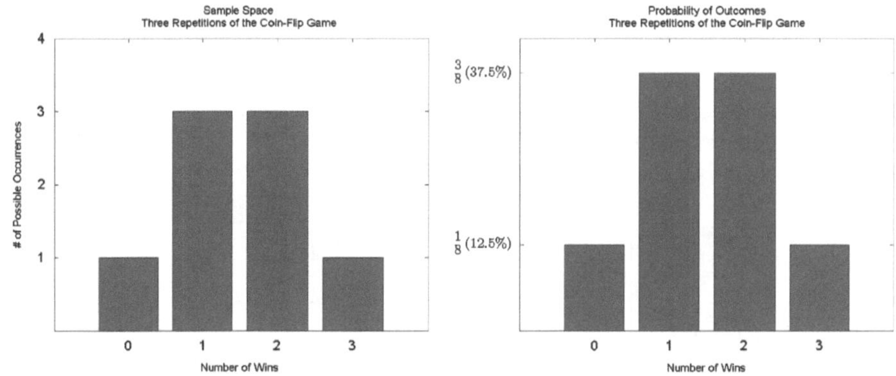


**Fig. 10.3** Coin-toss outcomes. The underlying figure illustrates the eight possible outcomes of three repetitions of our simple coin-toss game and their relative probabilities

**Table 10.5** Expected coin-toss return

| Number of wins | Probability(%) | P&L | Expected return |
|---|---|---|---|
| 0 | 12.5 | −\$3 | −37.5¢ |
| 1 | 37.5 | −\$1 | −37.5¢ |
| 2 | 37.5 | \$1 | 37.5¢ |
| 3 | 12.5 | \$3 | 37.5¢ |
| Total return | | | *\$0* |

Making use of the finite set of game outcomes, their pay-outs and the associated probabilities, we may compute the expected return for three repetitions.

This exceedingly well-defined set of outcomes and pay-outs—along with knowledge that the coin is fair—also permits us to precisely compute the associated probabilities. One may either gain \$1 or lose \$1 in three possible ways, respectively. Each pay-out thus has a $\frac{3}{8}$ or 37.5 % probability. There is only one way to lose \$3 and a single way to win \$3—the result is a $\frac{1}{8}$ or 12.5 % probability for each of these outcomes. Figure 10.3 outlines all of this information graphically.

Table 10.5 combines the outcomes and probabilities to explicitly compute the expected gain or loss associated with our three repetitions of the coin-toss game. We expect, on average, to break even and our intuition is justified. Before we move on to attempt to measure the risk associated with multiple repetitions of the game, it is useful to briefly reflect on what makes this simple game so special.

What do we know about this game? This simple example is an excellent opportunity to carefully note what information are available to us. We have *four* basic bits of information:

1. The *finite* set of outcomes are known in advance.
2. The outcome of each game cannot be known in advance—it is random.
3. Each game is *independent*.
4. The probability of each outcome is constant at 50 %.

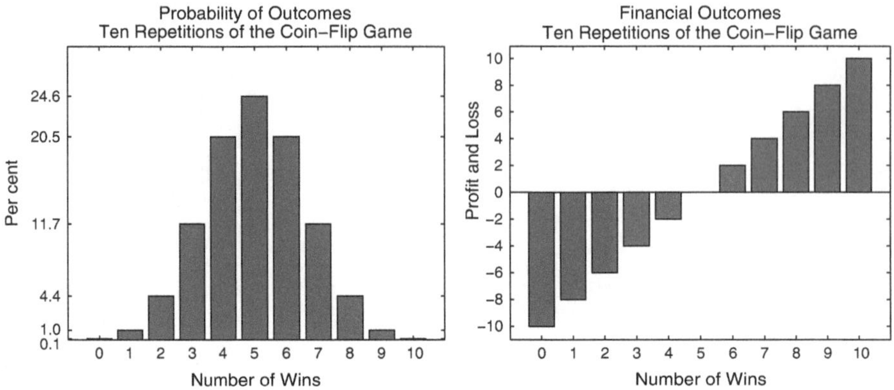


**Fig. 10.4** Ten repetitions. Ten repetitions of the coin-toss game yields 1,024 possibilities. The underlying figures outline the 11 possible pay-outs and their relative probabilities

The fact that, despite its inherent randomness, there is a well-defined set of finite outcomes is an enormous advantage. It is essentially this fact that permits such a simple characterization of our outcomes and likelihoods.

Perhaps this problem seems simple because we examine only three repetitions. Let's, therefore, look at 10 repetitions of the coin-toss game. Ten repetitions of our game yields $2^{10}$ or 1,024 possibilities. Figure 10.4 outlines the associated outcomes and pay-outs. This is clearly getting more difficult to handle, but, however numerous, there remains a finite number of outcomes. It still poses no problem.[3]

What is the risk associated with this game? Naturally, the answer depends on what you mean by risk. There are different ways to define uncertainty. Each measure of risk tries to answer a specific question. Each measure combines both outcomes and likelihood. Table 10.6 highlights *seven* possible measures of risk associated with ten repetitions of our game.

Table 10.6 highlights our risk measures both in terms of profit-and-loss (P&L) and number of wins. The expected pay-out is, therefore, zero, but the expected number of wins is 5. The volatility, or standard deviation, around our expected P&L is \$3.2. This represents the average uncertainty associated with ten repetitions of the game. Here we have two basic risk measures. We expect to break-even in the game, but also expect an average uncertainty of $\pm$\$3.2 in P&L.

What about more extreme notions of risk? The worst-case situation is a loss of \$10.[4] This is an extreme value, or minimum.[5] Two other classic risk measures also

[3]A sequence of coin tosses—each in itself a Bernoulli trial—is described by the binomial distribution. See Casella and Berger [4] for more on the binomial distribution. We suppress the notion of statistical distribution in this section, but will return to it explicitly in the following discussion.

[4]The probability of this outcome is very small at $\frac{1}{1024}$ or 0.1 %.

[5]Extreme-value theory is a complete area of study. See Embrechts et al. [8] for more information.

**Table 10.6** Measures of risk

| Measure | P&L | Wins | Question |
|---|---|---|---|
| Mean | \$0 | 5 | Average outcome? |
| Volatility | \$3.2 | 1.6 | Average uncertainty? |
| Minimum | −\$10 | 0 | *Absolute* worst case? |
| 99 % VaR | −\$8 | 1 | *Probability* based worst case? |
| 99 % Tail VaR | −\$8.2 | 0.9 | How bad is bad? |
| Skewness | \$0 | 0 | Symmetric? |
| Kurtosis | \$4.0 | 2.0 | Extreme values? |

This table displays a number of different risk measures associated with ten repetitions of the coin-toss game—Fig. 10.4 provides the possible pay-outs and their probabilities. Each risk measure attempts to answer a specific question.

attempt to assess extreme outcomes. Value-at-Risk (VaR) is a probabilistic measure of absolute risk that attempts to assess the worst-case loss with a given degree of confidence. We chose a confidence level of 99 % and computed its value as −\$8. The 99 % Tail VaR (or expected shortfall) is the expected outcome should you find yourself among the worst 1 % of observations. More simply, this risk measure answers the question: if things go very badly, how bad do they go on average. It takes a value of −\$8.2.

The final two measures are basically statistics. Skewness assesses the symmetry of the outcomes. If it takes a positive value, it implies that positive outcomes are more likely than their negative counterparts. A negative skewness thus implies that negative outcomes are more probable. Naturally, a risk manager would view negatively skewed activities as being riskier. In our coin-toss example, the skewness is identically zero implying, as evidenced by our graphics, that the outcomes are precisely symmetric.

Kurtosis is a measure of the probability of extreme outcomes. While it is essentially a technical statistical measure, it is useful to be aware that the kurtosis of a normal distribution is equal to 3. This is interesting because, given its popularity and ease of use, the normal distribution is a common point of comparison in risk analysis. The kurtosis of our coin-toss game is 2. This implies that the likelihood of extreme outcomes is smaller in this setting than would be suggested by a normal distribution.[6]

We have been able to gain significant insight into the risk of our coin-toss game through a careful classification of its outcomes and their associated probabilities. We now know, with quite a bit of accuracy, the average uncertainty, the probability of extreme outcomes, and the symmetry of risk associated with this activity. This is

[6]This is a familiar feature of the binomial distribution.

the essence of risk measurement.[7] Armed with this information, one may proceed to take a sensible decision as to whether or not to engage in it. In the next section, we will attempt to perform precisely the same analysis in a more complicated setting. We will see that a slight change in the underlying conditions associated with the activity dramatically increases the difficulty and accuracy of our analysis.

## 10.3 A More Complicated Example

Sadly, we are not in the business of flipping coins, but rather investing money. Consequently, it makes sense to consider an example that is closer to home. To do this, let's imagine a new game. We purchase an equally weighted portfolio of 2-year and 10-year US Treasury securities and hold them for one period. The question is simple: what is the risk associated with this game?

We've established that assessing the risk of any undertaking requires identifying its various outcomes and their associated probabilities. To do this, we have constructed Table 10.7 to succinctly summarize our US Treasury investment game. There is, however, a serious problem. We have no idea how to populate the entries in this table.

The reason is simple: the list of outcomes is vast. Table 10.7 is a huge simplification, but we still have no idea how to populate it. We have, quite simply, no clear *a priori* notion of the set of possible outcomes not to mention how much probability to assign to each of these (unknown) outcomes.

When unsure as to how to proceed, it is often useful to take a step back and place some structure on the problem. What do we know? What are the rules of this game? Using the same basic analysis as used in the analysis of the coin-toss problem, we may highlight the following points.

**Table 10.7**
Describing the second game

| Outcome | Pay-out | Probability |
|---|---|---|
| Big loss | ? | ? |
| Normal loss | ? | ? |
| Neither gain nor loss | ? | ? |
| Normal gain | ? | ? |
| Gain | ? | ? |

Attempting to describe the possible set of outcomes and probabilities associated with the second game is not so easy. We find that constructing something like Table 10.3 on page 302 is *not* possible.

[7]One can go much further in selecting a risk measure. Artzner et al. [1] posited a set of mathematical properties that each desirable risk measure should possess. Measures possessing all of these properties are termed coherent.

**Table 10.8** Comparing our two games

| Feature | Coin-toss | UST portfolio |
|---|---|---|
| Outcomes | Finite | Infinite |
| **Random** | **Yes** | **Yes** |
| Dependence between securities | *N/A* | Correlated |
| Dependence over time | Independent | Non-correlated |
| Probability | Known | Unknown |

This small table attempts to highlight the differences between our two examples: the coin-toss game and the two-bond portfolio investment.

1. The *infinite* set of outcomes can be approximated in advance.
2. The outcome of each game is random.
3. There are *dependencies* between repetitions of the game and the securities.
4. The probabilities of a gain or a loss in any given period are not known.

This simple list reveals some important points. Unlike the coin-toss example, there are an infinite number of possible return outcomes for our equally weighted 2- and 10-year US Treasury portfolio. This complicates matters considerably. The number of outcomes is unfortunately not the only difference with the previous simple example. Table 10.8 explicitly summarizes the differences between the coin-toss game and our US Treasury investment.

An infinite number of outcomes also requires the assignment of an infinite number of probabilities. Neither task seems particularly straightforward. Moreover, as if that was not enough, we cannot consider the two instruments in our US Treasury portfolio separately—their returns will not move together in lock-step, but they will certainly be related.[8]

Should we conclude that measuring risk on our investment portfolio is an impossible task? No, we should not. We should, however, appreciate that we are facing a rather difficult task. The natural starting point is the consideration of historical data. To understand the set of possible outcomes from our games, therefore, let us examine a set of historical outcomes for our 2-year and 10-year US Treasury yields. Figure 10.5 provides daily 2- and 10-year returns from January 1990 to November 2012. This amounts to a dataset of more than 5,700 returns for each instrument.

The investment returns for our US Treasury bonds have an infinite number of possible outcomes, but over the last 20 years, daily returns for each instrument have fallen in the range of $\pm$ 200 basis points. This is reassuring. The range of outcomes for 10-year bonds are significantly broader than for their 2-year equivalents. This

[8]With an investment portfolio, the returns also have a time dimension. Empirically, the returns from one time period to the next are generally uncorrelated. Consequently, we ignore this form of dependence, although it may, albeit weakly, actually exist.

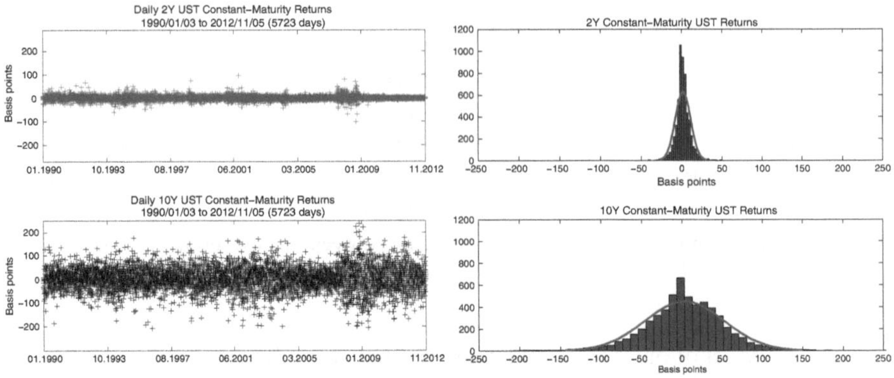


**Fig. 10.5** US treasury bond returns. The following figures provide two alternative perspectives of the possible daily return outcomes and their relative likelihood for 2- and 10-year US Treasury bond returns

is unsurprising given the relatively large duration and sensitivity to interest-rate movements associated with the 10-year sector.

How do we use this historical data? The empirical histograms illustrated in the right-hand graphic of Fig. 10.5 are probably the best place to start. The graphics summarize the relative frequency of different return outcomes—this is essentially a first step towards approximating the likelihood associated with our return outcomes. Both the 2- and 10-year bonds look to have about the same probability of positive and negative returns. Moreover, the most frequently occurring outcomes for both instruments lies around zero. Finally, there are dramatic differences in terms of extreme outcomes. The 10-year bond has exhibited numerous daily returns of $\pm$ 50 basis over the last 20 years, whereas this is a very rare occurrence for the 2-year bond.

This is promising, but we are principally interested in an equally weighted portfolio between these 2- and 10-year instruments. Figure 10.6 combines the data from Fig. 10.5 to provide the same perspective for an equally weighted portfolio of 2-year and 5-year US Treasury bonds.

The consequence of equally mixing the 2- and 10-year tenors is a set of returns that are more volatile than the 2-year bond, but more stable than the 10-year bond. The lower graphic in Fig. 10.6 is again an empirical summary of the relative frequency—which can be considered an estimate of the likelihood—for our investment portfolio. It is precisely this information that we will use to approximate the risk of this investment.

Using our historical dataset, we may now proceed to approximate risk. Table 10.9 summarizes the set of risk statistics introduced in our coin-toss example for the elements of our portfolio and our equally weighted portfolio.

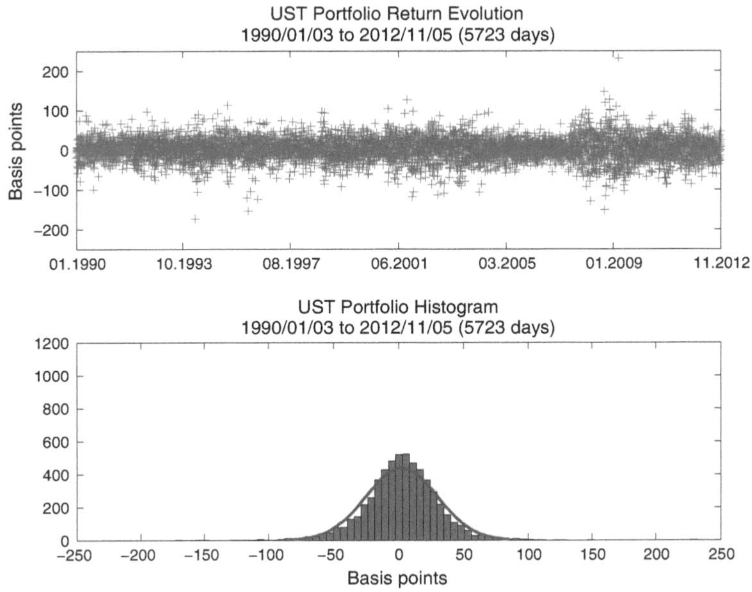


**Fig. 10.6** US treasury portfolio returns. Combining the information in Fig. 10.5, we provide some perspective on the returns of an equally weighted portfolio of 2- and 10-year US Treasury bonds

**Table 10.9** Historical risk measures

| Measure | 2Y | 10Y | **2+10Y** | Question |
|---|---|---|---|---|
| Mean | 1.7 | 2.9 | **2.3** | Average outcome? |
| Volatility | 9.5 | 49.0 | **27.7** | Average uncertainty? |
| Minimum | −101.4 | −285.7 | **−172.8** | *Absolute* worst case? |
| 99 % VaR | −23.4 | −129.0 | **−71.4** | *Probability* based worst case? |
| 99 % Tail VaR | −33.7 | −165.2 | **−92.2** | How bad is bad? |
| Skewness | 0.3 | −0.1 | **−0.1** | Symmetric? |
| Kurtosis | 13.9 | 5.8 | **5.7** | Extreme values? |

Using the historical data as a gauge of the possible outcomes and their associated probabilities, we can proceed to compute our set of risk measures. All values are in basis points.

The average daily return for our investment portfolio is about 2 basis points with a daily volatility of roughly 28 basis points. Without introducing any complex statistics, we should expect that most of the time daily returns will fall in a range of about ± 30 basis points. This is already a useful risk indicator.[9]

[9]None of the risk measures in Table 10.9, with the exception of the mean, is linear—that is, you cannot compute them as a simple weighted average of the underlying components. Try it as an exercise!

Moving beyond average levels of uncertainty, we should also consider the more extreme outcomes—these are, after all, the outcomes that concern us the most. The worst case outcome over the last 20 years was -173 basis points. Since this occurred once out of 5,723 daily observations, we may estimate its empirical probability as $\frac{1}{5{,}723}$ or 0.02 %. This may be a bit too extreme. The VaR and Tail VaR—again both measured at the 99 % level—are -71 and -92 basis points, respectively. Expected worst-case outcomes are thus significantly better than the absolute worst case.

The final two risk measures are kurtosis and skewness. Interestingly, our portfolio exhibits a slight negative skew—this implies that there is a marginally higher chance of negative returns. The kurtosis figure is roughly 6. Again, this is a bit tricky to interpret, but recall that the value for a normal distribution is about 3. Our kurtosis estimate suggests that the probability of extreme outcomes is significantly higher than what the normal distribution suggests. This should give us pause if, at some point, we opt to compute risk under the assumption of normally distributed returns.

Despite some initial concerns, we have actually succeeded in approximating the risk of our investment portfolio. There is no formal structure behind our computations. This is a shortcoming that we will address in the next section.

### 10.3.1 Enter the Distribution

We have extensively discussed the notions of outcomes and likelihoods. These two concepts form the backbone of statistical analysis and probability theory. As such, statistical techniques may be profitably used to assist us in the measurement of risk. In particular, a formal and extremely well researched mathematical relationship exists between the set of outcomes and their associated probabilities for a given random variable. A random variable can, quite loosely, be considered the outcome of a given activity. The set of possible outcomes and their relative probabilities are summarized by an object termed a statistical *distribution*. Figure 10.7 provides a graphical view of the cumulative distribution function and probability density function for two well-known distributions. Vastly more could be written about the notion of a distribution, but this will be left to the appropriate statistics textbooks.[10]

The important point is that, given our previous definition of risk, the statistical distribution is essentially tailor-made for risk measurement. The corollary is that since the study of statistical distributions is a technical subject, the study of risk is also unavoidably technical. Risk measurement is further complicated by the fact that one needs to keep economic objectives in mind and avoid focusing solely on the mathematical details. To this end, in the following discussion we will highlight a few important distributional assumptions made in risk measurement and, where possible, attempt to understand their reasonableness.

[10]The interested reader is referred to Appendix A or the following excellent statistical and probability theory references: Casella and Berger [4], Durrett [6] and Billingsley [3].

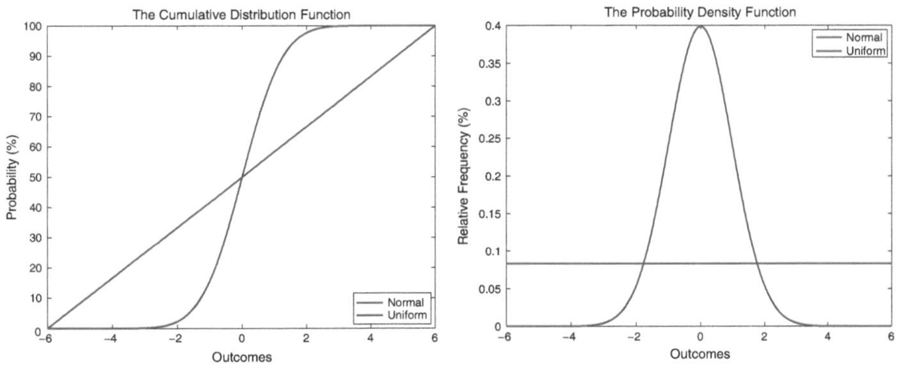


**Fig. 10.7** Sample distributions. The link between outcomes and probabilities for different random variables is described by a probability distribution. The cumulative distribution function and the probability density function are two alternative ways at examining the same distribution. They contain the same basic information

In risk measurement, two important assumptions are typically taken. These include:

1. the future will look like the past; and
2. returns follow a normal distribution.

Without the first assumption, one will not get far. Historical data provide us with the necessary input to both select an appropriate distribution and estimate its parameters. Reliance on historical data, however, is not without its dangers. The future need not look anything like the past.[11] Understanding the sensitivity of one's risk measures to the choice of distribution and their parameters is thus critical. This so-called sensitivity analysis will represent the remainder of discussion in this section.

The assumption of normally distributed returns is mostly a question of convenience. Normality can, and really should, be relaxed. There are many ways that this might be achieved. In the following discussion, however, we will focus on *two* key elements:

- getting away from normality; and
- the idea of dependence.

Let's address each in turn.

To motivate the importance of one's distributional choice, we begin with a simple experiment. Let us compare the true returns over our 22-year sample period with a

[11]This is the central thesis of Taleb [14], which offers a scathing review of risk-management practices.

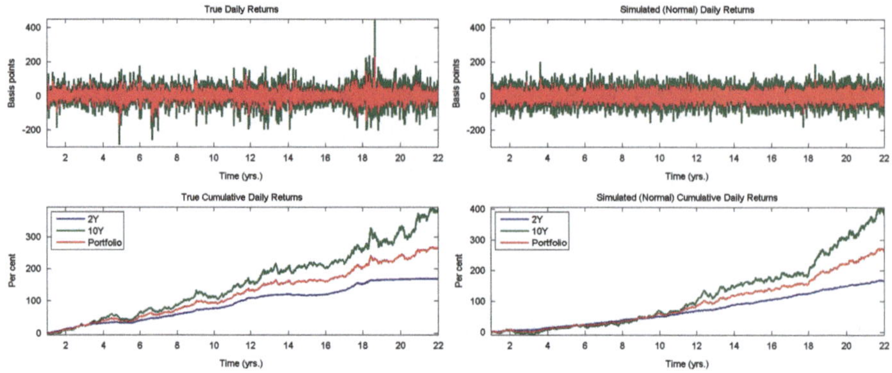


**Fig. 10.8** Assuming normal returns. We compare, over the same 22-year time interval, the true returns of our US Treasury portfolio to a simulated set of normally distributed returns. Observe the much greater regularity of normally distributed returns

set of normally distributed returns. This may appear a bit odd. How can we hope to observe a set of normally distributed returns? We cannot observe these returns, but we can simulate the returns from a normal distribution that shares the same parameters as our historical dataset. This is precisely what is done in Fig. 10.8

The two graphics in Fig. 10.8 contrast the true- and normally distributed returns. The regularity of the variation in daily returns—see the right-hand graphic—is striking. The true return observations exhibit varying levels of volatility.[12] The incidence of extreme positive or negative returns is also significantly higher than suggested by the normal distribution. Overall, therefore, the true distribution appears to have substantially more uncertainty than the simulated normal returns. The consequence is that the assumption of normally distributed returns for the purposes of risk management will, all else equal, likely lead to an underestimate of risk.

### 10.3.2 Relaxing Normality

If the assumption of normality leads to an underestimate of risk, then we should adopt another distribution. There are many ways that this might be accomplished and all of them are relatively complex. To keep our attention on the relevant conceptual ideas, we will abstract from these technical details—this does not imply, however, that they are unimportant.

---

[12]This is termed volatility clustering. Volatility behaviour is a broad area of financial research. See, for example, Engle [9].

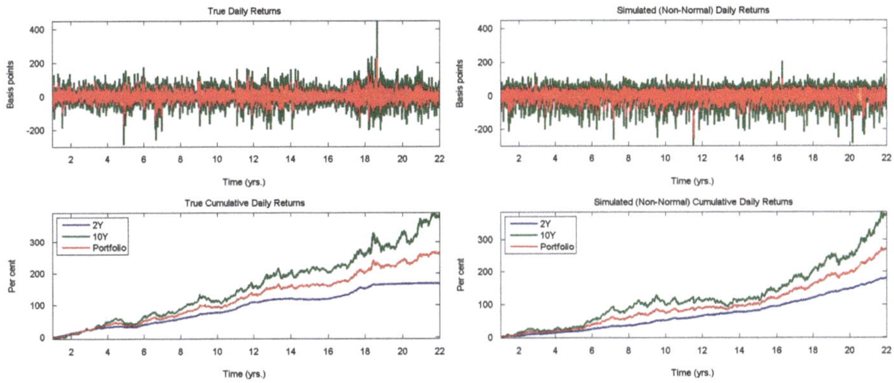


**Fig. 10.9** Non-normal returns. We compare, over the same 22-year time interval, the true returns of our US Treasury portfolio to a simulated set of non-normally distributed returns

Figure 10.9 compares the true returns over our data period with simulated *non-normal* returns.[13] The non-normally simulated returns have been draw from a distribution that was constructed with both excess kurtosis and a negative skew. In other words, the returns have been a greater probability of extreme and negative outcomes.

The non-normal return series looks significantly less regular than their normally distributed equivalents in Fig. 10.8. This is particularly true with regard to negative outcomes. The probability of negative outcomes appears significantly higher. The number of extreme negative outcomes is also correspondingly greater. These two effects are the impact of the negative skew element of our non-normal simulation. Visually, at least, non-normal return outcomes appear significantly different.

To gain some insight into the impact of different distributional assumptions on risk measurement, let's compute our risk statistics under the true, normal, and non-normal distributions. Table 10.10 illustrates the results of these computations.

The mean and standard deviation of our daily investment-portfolio returns are extremely similar. This is by construction. Each of the simulations forced the first two moments to be equal to the observed distribution.[14] The other measures, however, do not suffer from this constraint.

Our intuition from Figs. 10.8 and 10.9 appears to have been well founded. Across virtually all of the risk measures—save, of course, mean and volatility—the

[13]The non-normal distribution is constructed using the so-called Cornish–Fisher expansion. See Cornish and Fisher [5] or Maillard [12] for more detail.

[14]A *moment* measures the shape of a set of points—they are basically summary measures for a statistical distribution. The first moment (i.e., the mean) basically describes the shape of the set of average observations, while the second moment (very roughly) describes the average squared observations: this is often termed standard deviation or volatility. Third and fourth moments—also called skewness and kurtosis, respectively—examine cubic and quartic transformations of the observations.

**Table 10.10** Implications of distributional choice

| Measure | True | Normal | Non-normal |
|---|---|---|---|
| Mean | 2.3 | 2.3 | 2.5 |
| Volatility | 27.7 | 27.6 | 27.6 |
| Minimum | −172.8 | −94.3 | −219.2 |
| 99 % VaR | −71.4 | −61.8 | −81.7 |
| 99 % Tail VaR | −92.2 | −69.2 | −101.8 |
| **Skewness** | **−0.1** | **0.1** | **−0.9** |
| Kurtosis | 5.7 | 3.0 | 5.5 |

The underlying table illustrates the impact of one's choice of return distribution on our set of risk measures While numerous choices are possible we've indicated the true distribution, the normal distribution, and a possible non-normal transformation.

values computed under the assumption of a normal distribution suggest the lowest amount of risk. The negatively skewed non-normal distributional assumption not surprisingly leads to the highest risk estimates. By forcing a higher probability of negative and extreme outcomes, one naturally arrives at higher levels of risk.

The point of this analysis is not to establish that an assumption of normality is bad and that non-normality is good. The situation is rather more complex. The point of this analysis is to highlight the fact that there is no *single* correct value for risk. Different distributional assumptions will yield different approximations of risk. Since we face an infinity of possible outcomes and associated probabilities, we have no choice but to base our analysis on assumptions. It is thus of principal importance to understand the sensitivity of our risk measures to this distributional choice. In the next section, we expand this notion to consider alternative distributional parameters.

### 10.3.3 The Role of Dependence

The notion of dependence relates to the distributional relationship between two (or more) random variables. If one random variable is positive, is the other random variable also typically positive? This is called a positive relationship. In some cases, the relationship is negative. That is, one tends to be positive when the other is negative. In other cases, there is no discernible relationship between the two variables.

In a financial setting, the notion of dependence is extremely important, because it is intimately related to the notion of diversification. Diversification is the observation that holding a combination of securities typically leads to lower risk portfolios than holding a single security. Statistically, this stems from the fact that the returns of different securities do not move together perfectly. This acts as a bit of a hedge—when one security is down, the other will be down less or even up—and thus reduces risk. The bottom line is that the interaction, or dependence, between securities in a portfolio is important for risk. It thus requires deeper examination.

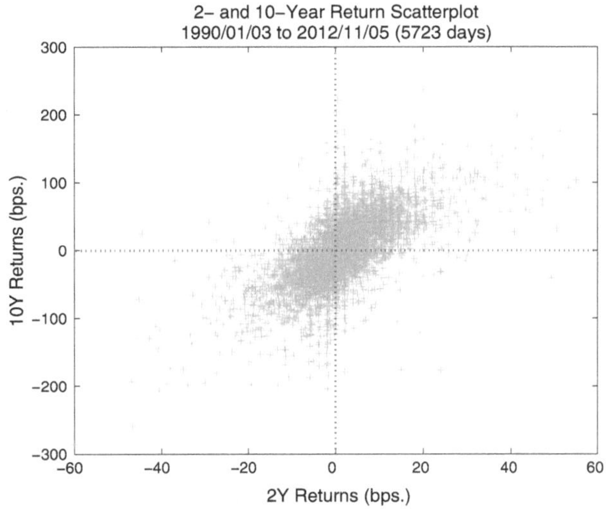


**Fig. 10.10** 2- and 10-year return dependence. This figure demonstrates a scatter-plot of daily 2- and 10-year US Treasury bond returns. Visually, this figure highlights the positive relationship between these two variables

Returning to our example, we may ask how the historical 2- and 10-year US Treasury bond returns move together? Figure 10.10 provides a scatter-plot of the two daily returns series suggesting they, in fact, move together quite strongly. When 2-year returns are positive, the associated 10-year return is also typically positive. The same applies for negative returns—this is evidenced by the majority of the points falling in the north–east and south–west quadrants of Fig. 10.10. The relationship is nonetheless imperfect. There are occasions when the 2-year bond has a negative return at the same time the 10-year return has positive return.[15] This suggests that there is some scope from diversification by a portfolio combining 2- and 10-year bonds.

One possible, and very popular, approach to the measurement of dependence between two random variables is the notion of correlation.[16] The historical correlation coefficient between the 2- and 10-year bonds in our dataset is slightly higher than 0.6—this implies a relatively strong level of positive dependence between 2- and 10-year yields. This correlation coefficient is a parameter of the joint distribution between 2- and 10-year bond returns. Since it is a parameter, it can be easily adjusted if we are willing to perform a few more simulations.

To visually see the implications of different choices of correlation coefficient, Fig. 10.11 highlights three different scatter-plots. To permit easy comparison, each of these returns is drawn from the non-normal distribution. Each of the graphics shows a simulation of 2- and 10-year bonds with a different correlation coefficient. To repeat, the distributions are the same, the only difference is the correlation coefficient.

[15]This is shown by the points in north–west and south–east quadrants of Fig. 10.10.

[16]Although we will make extensive use of it, correlation is not the only approach to measuring dependence. See Embrechts et al. [7] for an excellent discussion of the pros and cons of using correlation.

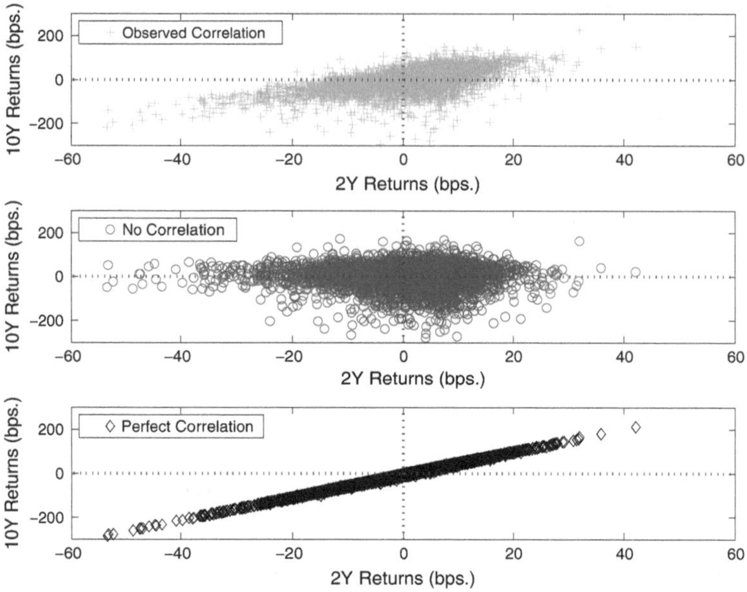


**Fig. 10.11** Implications of dependence. This figure demonstrates three different scatter-plots of daily simulated 2- and 10-year US Treasury bond returns for three distinct choices of correlation coefficient

The three choices of correlation coefficient are a bit extreme. We set it equal to the observed historical value of 0.6, a value of zero, and unity. Setting the correlation coefficient to zero is tantamount to assuming that 2- and 10-year US Treasury returns are independent. In other words, they are not related to one another. The middle graphic in Fig. 10.11 underscores this fact—the number of points in each quadrant is essentially equal.[17] This should give rise to significant opportunity for diversification.

A correlation coefficient of unity, conversely, assumes that they move together in perfect unison. Graphically, this amounts to a straight line through the origin. Under this assumption, there is basically no scope for diversification. This is nonetheless interesting because, while admittedly extreme, there is a tendency for return correlations among different assets to rise during times of crisis. Examining the impact of perfect correlation, therefore, provides some insight into the sensitivity of your portfolio to such crisis periods.

[17]There are actually more points in the two southern quadrants because of the negative skew in the simulated distribution. Without skewness, the points would be more equally spread throughout the four quadrants.

**Table 10.11** Implications of dependence

| Measure | $\rho = 0$ | $\rho = \mathbf{0.63}$ | $\rho = 1$ |
|---|---|---|---|
| Mean | 2.2 | **2.5** | 2.6 |
| Volatility | 25.3 | **27.6** | 30.1 |
| Minimum | −222.8 | **−219.2** | −364.8 |
| 99 % VaR | −80.1 | **−81.7** | −100.4 |
| 99 % Tail VaR | −109.5 | **−101.8** | −137.3 |
| Skewness | −1.4 | **−1.3** | −1.8 |
| Kurtosis | 8.1 | **7.0** | 11.9 |

This table illustrates the impact of the dependence between the instruments in one's return distribution on our set of risk measures. We compare the true, zero, and perfect correlation all in the context of the non-normal transformation described in Table 10.10.

We may now revisit in Table 10.11 our risk measure one final time under three possible dependence assumptions: observed, zero, and perfect correlation.[18]

Each of the risk measures is an approximately increasing function of the correlation—this is the diversification effect. An assumption of perfect correlation between our two risk factors increases the risk dramatically.[19] This stems from the fact that bad returns in one bond are not offset by better returns in the other instrument.

The key point is that one's assumptions about the dependence of returns between the various instruments in one's portfolio, much like the distributional choice, have important implications for risk. Coming back to our original definition, this amounts to describing the interaction of outcomes and likelihoods of two jointly performed activities. One cannot avoid making assumptions, but it is critical to test the sensitivity of your results to these choices.

## 10.4 A Specific Risk Measure

Having discussed risk in a generic sense and examined some key inputs into the computation, let us now turn to examine a specific risk measure—the tracking error. We will make extensive use of this measure, because it permits a detailed understanding of a portfolio's positioning relative to one's strategic benchmark and, as such, it is extremely useful within our portfolio-analytic framework. We will also examine this measure from backward- and forward-looking perspectives, which also provides some insight into the often-discussed, but sometimes misunderstood, concept of tracking error.

---

[18]Recall that $\rho$ is the Greek letter that is typically used by statisticians to designate the correlation coefficient.

[19]Since we are running simulations, there is still some noise in these computations. If we performed a very large number of simulations, however, we would expect risk to be an increasing function of correlation.

Tracking error, which as a concept comes from engineering, describes how well one system follows (i.e., tracks) another. Tracking, in the financial context, refers to return. In particular, it refers to active return. To ensure clarity, it is useful to carefully and clearly define the concept of active return. It is defined as follows:

$$\text{Active Return} = \text{Portfolio Return} - \text{Benchmark Return}, \tag{10.1}$$
$$R_{t,a} = R_{t,p} - R_{t,b}.$$

In a phrase, therefore, the objective of tracking error is to explain how closely portfolio returns follow (i.e., track) benchmark returns. If the active return is identically zero, for example, this would imply that the portfolio is perfectly tracking its strategic benchmark. Large values of active return conversely tend to suggest a weak relationship, or low level of tracking, between the portfolio and strategic benchmark.

As previously indicated, there are essentially two perspectives for thinking about tracking error—backward- and forward-looking. These are typically referred to as ex-post and ex-ante tracking error, respectively. The difference between ex-post and ex-ante tracking error relates to which direction you examine the time axis. Figure 10.12 provides a schematic representation of the time dimension to help us understand this idea.

If you stand at the current point in time, you may either look backwards towards realized active returns or forward to future active returns. Looking backwards is conceptually easier. Selecting a particular horizon, say 3 years, it is straightforward to examine active returns. This is basically historical data analysis. Looking forward is rather more challenging. We do not know future active returns and, as a result,

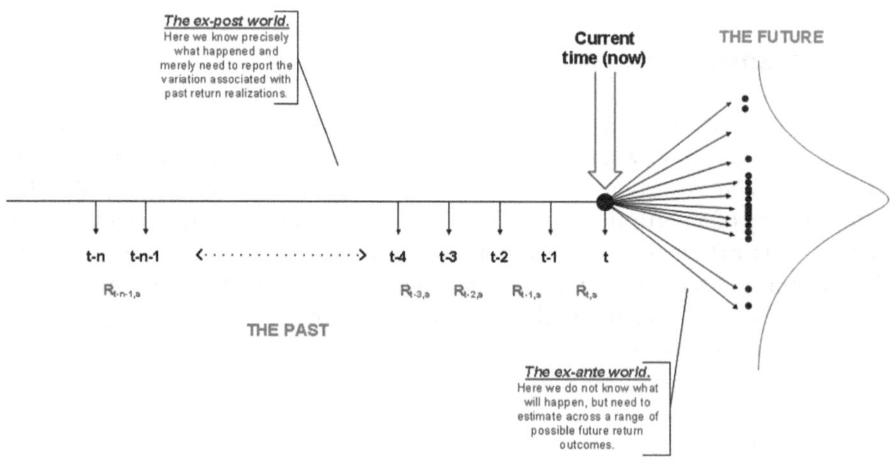


**Fig. 10.12** The time dimension. The difference between ex-post and ex-ante tracking error relates to which direction you examine the time axis

we find ourselves using the notions of outcome and likelihood to characterize the distribution of future active returns.

Both perspectives are completely legitimate, although forward-looking analysis is probably of more general interest. This is because, since the outcomes have not yet occurred, there is still time—should you not like what the measure is telling you—to react and manage one's portfolio. Nothing can be done about large negative active returns in the past, but a large ex-ante tracking error can be reduced by adjusting one's portfolio today. In summary, therefore, ex-post tracking error is more an historical summary statistic, whereas ex-ante tracking error is a true risk measure. Let's discuss each in a bit more detail.

### 10.4.1 Looking Backwards

This computation of ex-post tracking error is relatively straightforward. It is merely the standard deviation, or volatility, of the active return. Mathematically, it is defined as

$$\begin{aligned} \text{Ex-Post TE}(t-n,t) &= \sqrt{\text{var}(R_a)}, \\ &= \sqrt{\left(\frac{1}{n-2}\right)\sum_{i=0}^{n-2}(R_{t-i+1,a})^2}. \end{aligned} \tag{10.2}$$

It is a simple backward-looking measure of the dispersion of active returns. The metric used to determine the distance between portfolio and strategic benchmark returns is the standard deviation. Incidentally, a zero ex-post tracking error does not immediately imply that historical portfolio and strategic active returns are identical. If they always differed by a constant value, say $\gamma \in \mathbb{R}$, then there would be no variability in the active returns. The standard deviation of active returns, therefore, would be identically zero. Some claim this is a shortfall of the tracking-error measure, but the fact remains that whether the returns are equal or always differ by a constant, the portfolio is perfectly *tracking* the strategic benchmark.[20]

Figure 10.13 provides the return history—over the period of slightly more than 1 year—of a multi-currency portfolio relative to its benchmark. Visually, the portfolio appears to have quite closely tracked its strategic benchmark. The ex-post tracking error over this period, computed using Eq. (10.2), is about 12 basis points. This computation verifies the idea of a close link between the two portfolios. Note that, by convention, tracking error, whether ex-post or ex-ante, is reported as an annualized figure.

---

[20]Moreover, this is a very extreme example. It is probably rather difficult to actually construct a robust portfolio with a return that, under all market conditions, differs from the strategic benchmark by a constant amount.

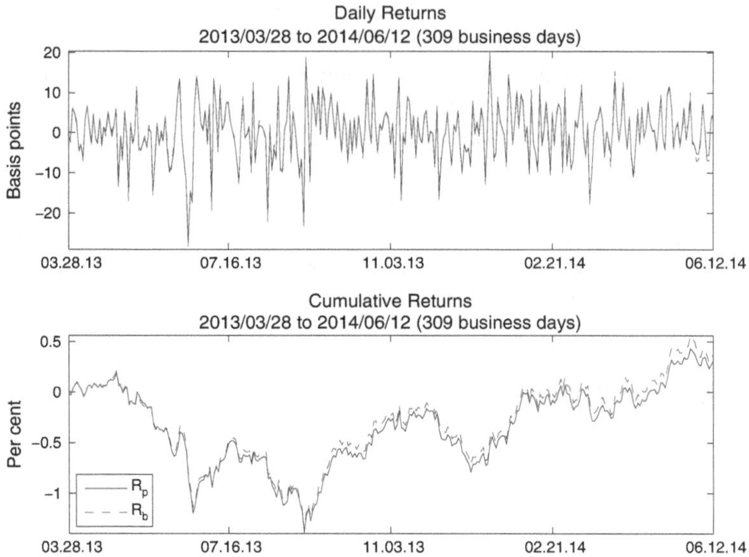


**Fig. 10.13** Ex-post TE. The ex-post TE over this period for this multi-currency portfolio is about 12 basis points. By convention, TE is reported as an annualized figure

**Table 10.12** Impact of time horizon

| Start date | End date | Period (years) | TE (bps) |
|---|---|---|---|
| Mar 31, 2009 | Jan 22, 2014 | 4.8 | 31.9 |
| Mar 31, 2010 | Jan 22, 2014 | 3.8 | 33.2 |
| Mar 31, 2011 | Jan 22, 2014 | 2.8 | 16.4 |
| Mar 30, 2012 | Jan 22, 2014 | 1.8 | 12.1 |
| Mar 28, 2013 | Jan 22, 2014 | 0.8 | 11.9 |

This table summarizes the portfolio's year-to-date ex-post TE over *five* alternative time horizons. Clearly, the choice of input data can make an important difference.

Although ex-post tracking error is simply computed, its value depends importantly on the selected dataset. As a backward-looking measure, ex-post tracking error is typically sensitive to the period over which it is computed. Table 10.12 summarizes our example portfolio's ex-post tracking-error over *five* alternative time horizons.

There are, at least, two elements contributing to the range of values in Table 10.12. First, market conditions are not constant over time. There will be times of higher and lower volatility and varying levels of correlation between market risk factors. Second, the portfolio positions also change over time. There will be periods, for example, when the portfolio manager is aggressively positioned relative to his or her strategic benchmark. In other periods, the portfolio manager will be conservatively, or neutrally, positioned. This analysis, therefore, clearly illustrates that, when markets and positions are not stable over time, the ex-post tracking-error will not be constant. To effectively use and interpret ex-post tracking error, it is important to appreciate this fact.

### 10.4.2 Looking Forward

Ex-ante tracking error is another story. The ex-ante tracking error is the standard deviation, or volatility, of the future expected active returns. The future dispersion of active returns, however, is *not* known. To compute it mathematically, we need to make use of a statistical framework. Ex-ante tracking error is defined as

$$\begin{aligned} \text{Ex-Ante TE}(t, t+1) &= \sqrt{\mathbb{E}\left((R_{t+1,a})^2\right)}, \\ &= \sqrt{\text{var}(R_{t+1,a})}, \end{aligned} \tag{10.3}$$

where $R_{t+1,a}$, the unknown future active return, is a random variable. We resolve the problem, therefore, by employing a statistical estimator for the variance of the random variable, $R_{t+1,a}$.

Working directly with expected active returns is cumbersome. Instead, one typically works with the underlying market-risk factors driving returns. Here enters our additive risk-factor decomposition developed in the previous chapters. In summary, the active return for a portfolio of $n$ instruments:

$$\begin{aligned} \sum_{i=1}^{n} \underbrace{\omega_{a_i} R_{t+1,a_i}}_{\substack{\text{Active return} \\ \text{of } i\text{th position}}} &\approx \sum_{i=1}^{n} \underbrace{\omega_{a_i}}_{\substack{i\text{th active} \\ \text{weight}}} \underbrace{g_i(\Delta f_{t,t+1,1}, \ldots, \Delta f_{t,t+1,\kappa})}_{\tilde{R}_{t+1,a_i}}, \\ &\approx \sum_{i=1}^{n} \omega_{a_i} \sum_{k=1}^{\kappa} \Phi_{t,i,k} \Delta f_{t,t+1,k}. \end{aligned} \tag{10.4}$$

where $\Phi_{t,i,k}$ and $\Delta f_{t,t+1,k}$ are the $k$th risk-factor sensitivity and $k$th risk-factor movement, respectively. The mapping $g$ links the return of a fixed-income instrument to its main risk factors. This is, in essence, our second-order Taylor series expansion of the bond-price function. It looks slightly different, because we have written it in a slightly different form. In the next chapter, we will examine how these computations are performed in a very detailed manner. The key insight, at this point, is that the active return can be written as a function of the active weights, the active exposures, and the changes in the risk factors. The random variables in our problem are now market risk factors and no longer active returns.

If we apply the variance operator to both sides of Eq. (10.4), switch to matrix algebra, and simplify, we will uncover a few interesting things. The result is

$$\begin{aligned}\sqrt{\operatorname{var}\left(\sum_{i=1}^{n}\omega_{a_i} R_{t+1,a_i}\right)} &\approx \sqrt{\operatorname{var}\left(\sum_{i=1}^{n}\omega_{a_i}\sum_{k=1}^{\kappa}\Phi_{t,i,k}\Delta f_{t+1,k}\right)}, \\ &\approx \sqrt{\operatorname{var}\left(\omega_a \Phi \Delta f_{t+1}\right)}, \\ &\approx \sqrt{(\omega_a \Phi)\operatorname{var}\left(\Delta f_{t+1}\right)(\omega_a \Phi)^T}.\end{aligned} \tag{10.5}$$

To summarize in words, $\omega_a$ is a vector of active weights, $\Phi$ is a matrix of risk-factor exposures, and $\operatorname{var}\Delta f_{t+1}$ is the covariance matrix of future market-risk factor movements. This is a rather compact form.

Risk measurement is about combining outcomes and likelihoods. Statistically, this amount to the use of probability distributions. Let's denote the distribution of future market risk-factor movements as $\Delta f_{t+1} \sim \mathscr{X}\left(0, \Omega_f\right)$. In this case, we have

$$\text{Ex-Ante TE}(t, t+1) = \sqrt{\operatorname{var}\left(R_{t+1,a}\right)} \approx \sqrt{\underbrace{(\omega_a \Phi)}_{\substack{\text{Risk-factor}\\ \text{weight}}}\ \tilde{\Omega}_f\ \underbrace{(\omega_a \Phi)^T}_{\substack{\text{Risk-factor}\\ \text{weight}}}}. \tag{10.6}$$

Ex-ante tracking error, as evidenced by Eq. (10.6), essentially amounts to combining a portfolio's active risk-factor weights and—an estimate of—the covariance matrix of future active returns, $\tilde{\Omega}_f$.[21] We will adopt this classical approach to the computation of the ex-ante tracking error.

What have we learned from this analysis? Equation (10.6) permits us to draw a number of important conclusions. In particular:

- Ex-ante TE is an approximation—we do not know the future, we merely seek to estimate it.
- Active weights ($\omega_a$) and exposures ($\Phi$) are known and fixed. We worked hard in previous chapters to construct these exposures and used them liberally in our performance attributions. Risk analysis will also make extensive use of these exposures.
- Our random variable of interest is *not* active returns ($R_{t+1,a}$), but rather market-risk factor changes ($\Delta f_{t+1}$). Uncertainty relates not to active returns, therefore, but to market-risk factor movements. This makes our performance analysis consistent with the risk dimension, since we consider the same set of risk factors.
- Uncertainty—dispersion and dependence—is summarized by the covariance matrix of market-risk factor movements ($\Omega_f$). The complexity of the ex-ante TE

[21] Much more will be said in the subsequent chapters on the construction, estimation, and analysis of this covariance matrix.

computation stems from estimation of $\Omega_f$ (i.e., $\tilde{\Omega}_f$). This will be an important actor in our subsequent risk development.

- Despite appearances to the contrary, this analysis has made no distributional assumptions. All we require is that our market risk factors follow a multivariate joint distribution, where the second moment (i.e., covariance matrix) exists. This is not a very strong assumption.

It also turns out that the ex-ante tracking is actually very closely related to the analytic definition of Value-at-Risk (VaR). Here is the definition of VaR in this context,

$$\text{VaR}(t, t+1, \alpha) \approx \mathcal{N}^{-1}(\alpha)\sqrt{(\omega_p \Phi)\, \tilde{\Omega}_f\, (\omega_p \Phi)^T}, \tag{10.7}$$

where $\alpha$ defines the level of confidence (i.e., 95 % or 99 %). The only difference is that the active risk-factor weights are replaced with their portfolio equivalents and the result is multiplied by a constant. Something important, however, has also happened. A distributional assumption has been made. The constant term stems from a normal distribution—in short, the VaR computation in Eq. (10.7) assumes the market risk factors follow a multivariate normal distribution.

Equation (10.7) is also interesting, because it shows that, using the same basic framework, we may compute both the ex-ante tracking error and the portfolio VaR. Both of these are interesting risk measures and, broadly speaking, they are consistent with one another. Quite simply, if you are in the business of computing tracking error then, at the cost of a distributional assumption, you are also completely capable of computing the analytic VaR of your portfolio.

To complete this section, let us now examine the ex-ante tracking-error computations for our same example portfolio. The results are found in Fig. 10.14. In contrast to the ex-post tracking error, which depend on a particular horizon, one

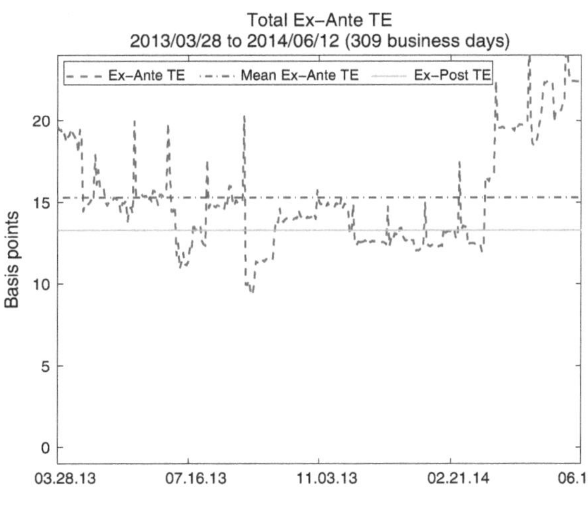


**Fig. 10.14** Ex-ante TE. The ex-ante tracking error, computed at each point over this time period is illustrated along with the average ex-ante value and the associated ex-post tracking error value

may compute the ex-ante tracking error on a daily basis. Each computation, and thus observation in Fig. 10.14, represents a separate forward-looking computation.

It is clearly not constant over time—the daily ex-ante tracking error over this period ranges from ten to 25 basis points. The average value is about 14 basis points, which compares to the average ex-post value, computed over the same period, of 12 basis points. At each point of time, the positions are fixed and the covariance matrix is updated. Thus, although the positions and our measure of market uncertainty change at each point in time, only the market element is truly random. This contrasts with the ex-post tracking error computation, where both positions and market conditions are stochastic. In summary, ex-post tracking error has two elements of uncertainty, whereas ex-ante risk has only one. The corollary is that ex-post tracking error, all else equal, will tend to be slightly larger than the ex-ante equivalent. While it does not appear to hold in this specific example, it is the general result.[22]

### 10.4.3 Comparing Forward- and Backward-Looking Perspectives

We've now looked independently at both notions of tracking error. It is now natural to ask the following question: should ex-post and ex-ante tracking, at least on average, be the same? Should they converge over long time periods? The short answer is: no, not exactly. The long answer is: it's a bit complicated. As we've seen, ex-ante and ex-post tracking error have quite different perspectives. There are *three* main sources of differences:

1. non-constant portfolio positions;
2. differing weighting schemes for volatilities and correlations between two computations; and
3. different time periods employed in computations.

One might cautiously conclude that for long time periods and fairly constant positions, the differences should not be significant. At any given instant, however, the differences can be quite important. Conceptually, we do expect some similarities. After all, we compute ex-ante tracking error to understand what might happen in the future. Ex-post tracking error examines what actually happened. If the two estimates, over long periods of time, were wildly different, then we might start to question the validity of our ex-ante estimates. Indeed, we dedicate a significant part of a future chapter to examining this question.[23]

Tracking error, as a risk measure, has a number of advantages. Both ex-post and ex-ante tracking error provide useful information about the portfolio's strategy. The majority of these benefits, however, relate to the ex-ante tracking error—as it is

[22]See Hwang and Satchell [10] for a more detailed discussion of this point.

[23]This notion is called backtesting and it is discussed in Chap. 12.

the true risk measure. Given a strategic benchmark, embedding an organization's investment preferences, ex-ante tracking error provides a measure of the *closeness* of the portfolio's investment policy relative to the strategic benchmark. This is a valuable tool for senior management, risk managers, and portfolio managers alike, because each of these stakeholders is interested in understanding to what degree the strategic vision is implemented. A second advantage of ex-ante tracking error is that, as a risk measure, it *jointly* considers all key market-risk factors. Thus, market risk arising from rate movements, credit-spread changes, inflation, and foreign-exchange volatility are combined into a single computation.[24] At the same time, there exists a methodology for decomposing either the ex-post or ex-ante tracking error into the contribution by individual risk factor.[25] This decomposition can be quite powerful, since it allows us to understand the relative mix of risks embedded in the overall risk figure. Finally, ex-ante tracking error can be usefully employed as a *yardstick*, or limiting device, for describing and measuring the permissible deviations from one's strategic benchmark. Again, most of these points apply to both ex-ante and ex-post tracking error, although the forward-looking ex-ante perspective is more useful.[26]

Naturally, ex-post and ex-ante tracking error are not the perfect risk measures and also have a number of limitations. Probably the most important criticism of tracking error, as a risk measure, is that it is *complex*. Its computation requires numerous approximations and estimations of a collection of slippery random variables. In the next two chapters, we will gain a significantly deeper understanding of this complexity. The second limitation of tracking error occurs by construction. It is a *relative*, not absolute, measure of risk. This implies that it provides essentially no information about the underlying magnitude of risk of the strategic benchmark—only one's position relative to it. In a related manner, it is also *symmetric*. A large tracking error may occur through sizable positions that either *reduce* or *increase* risk relative to the strategic benchmark—directionality of active risk is not provided. Additional information, above and beyond the tracking error, is required to appreciate the absolute level of portfolio risk and the nature of the active risks.[27] Finally, the tracking-error measure focuses solely on the *second moment* of active risks. Standard deviation is the key metric for measuring the distance between the portfolio and benchmark. While convenient and relatively robust, this choice nonetheless ignores potential asymmetries and tail behaviour in the joint active-risk return distribution.

---

[24] This has also traditionally been cited as one of the principal advantages of VaR. See Jorion [11] and Riskmetrics™[13].

[25] This follows from Euler's theorem for homogeneous functions and is discussed in detail in the next chapter.

[26] The ex-post tracking error provides useful information for assessing the historical closeness of the portfolio to its strategic benchmark. As a yardstick to measure permissible deviations, however, it is rather unhelpful. This is because, since it looks only backwards, no remedial action can be taken.

[27] Active exposures, as we have seen, can be helpful in this regard.

In summary, tracking error, although not without its limitations, is a useful risk measure. It serves, in this text, as the central portfolio-analytic risk measure. It will be regularly supplemented, however, by its cousin risk measure, VaR.

## 10.5 Using Tracking Error

To close out this chapter, we take a moment to consider another real-life example with a focus on how ex-ante tracking error might be employed. Figure 10.15 illustrates the daily ex-ante tracking error of an actual multi-currency, fixed-income portfolio over a 3-year period. This is useful information for a portfolio analyst. It tells us that the level of risk is not constant over time—it has ranged from 25 to 100 basis points over the sample period. Moreover, it can increase quite quickly from one period to the next.

While Fig. 10.15 is useful, we want more detail. We want to know something about the composition of the overall risk. Where is it coming from? Using a technique that will described thoroughly in the next chapter, we may decompose this risk into its underlying risk drivers and thus better understand what is going on. Figure 10.16 provides one possible decomposition of the overall ex-ante tracking error into three principal risk factors: treasury curves, credit spreads, and exchange rates.

Inspection of Fig. 10.16 reveals that the lion's share of the risk in this portfolio has historically come from the various treasury curves, but that recently the proportion of credit-spread risk has risen dramatically. Exchange-rate risk is apparently *not* an important risk element for this portfolio, but it does occasionally arise and persist for a number of periods.

We can, of course, go even further and drill down into the most important element of overall risk: the contribution of the various treasury curves. Figure 10.17

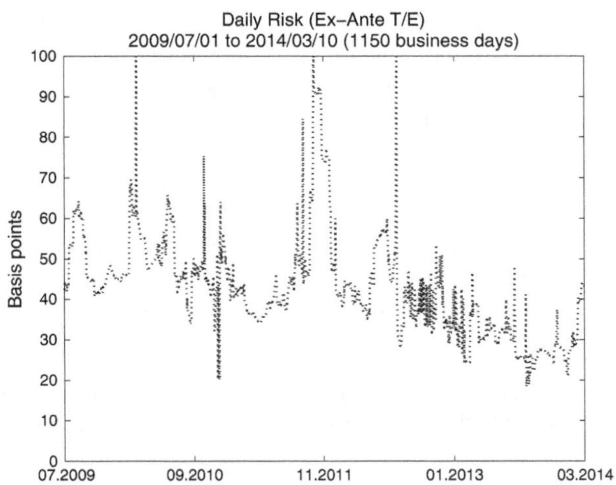


**Fig. 10.15** Ex-ante tracking error. Here we see the daily evolution of the ex-ante tracking error for a multiple-asset, multiple-currency portfolio

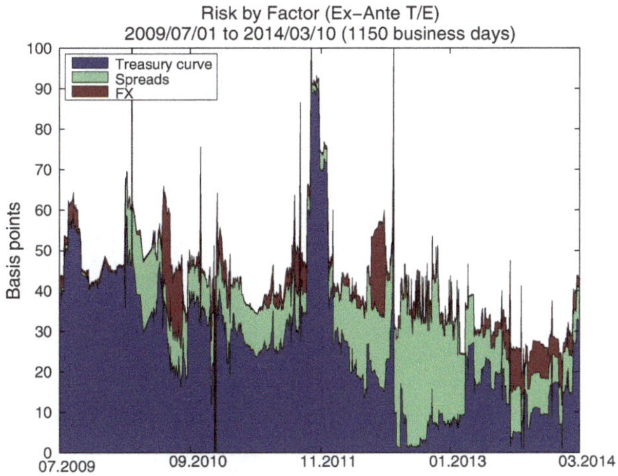


**Fig. 10.16** Tracking-error by risk factor. This figure decomposes the ex-ante tracking error into the contribution by the individual risk factors: treasury curve, credit spreads, and foreign-exchange rates

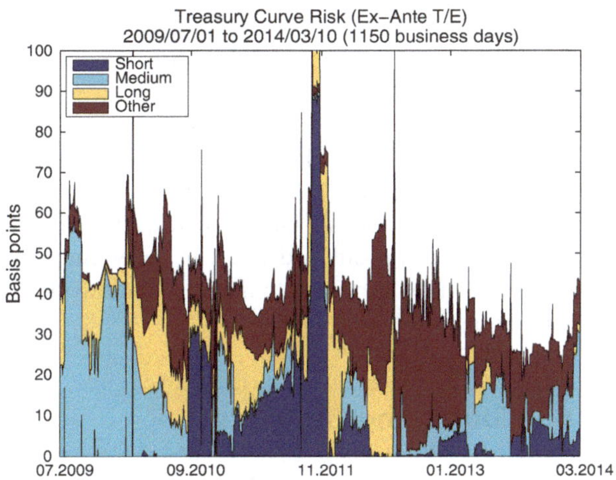


**Fig. 10.17** Drilling into curve risk. This figure breaks down the treasury-curve risk into the short-, medium, and long-term elements of the curve. The remaining risk factors—credit spreads and exchange rates—are combined into the other category

groups the credit-spread and exchange-rate risk together and further decomposes the treasury curve risk into three sub-categories: short, medium, and long tenors. No clear trend appears. There have been periods where risk arises predominately from each of the curve sectors. There was, in particular, a period in late 2011 when roughly 90 basis points of active risk was consumed by short-tenor treasury-curve risk. Such information is extremely valuable for a portfolio manager.

At the risk of testing the reader's patience, we may drill down further from yet another perspective: credit-spread risk. Figure 10.18 breaks down the credit-spread risk into agency and supranational, core-European government, and peripheral-European government categories. Modest amounts of core- and peripheral-European sovereign risk have been undertaken, but the majority of the credit-risk exposure has come through the agency and supranational sectors. This

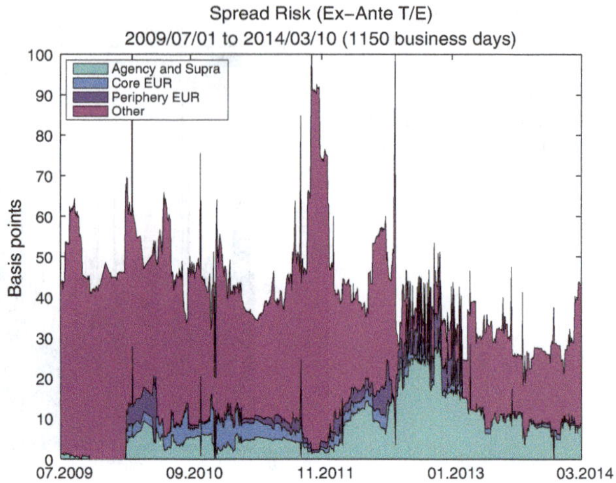


**Fig. 10.18** Drilling into spread risk. This figure breaks down the credit-spread risk into Agency and Supranational, core European government, and peripheral European government categories. The remaining risk factors—treasury curves and exchange rates—are combined into the other category

is particularly true in recent periods, where credit-spread risk has been a dominant contributor to overall risk.

This final section has essentially been an appetizer for the next chapter, where detailed mathematical explanations of risk computations will be provided. To accomplish this, we will not only employ the rudiments of risk measurement—outcomes and likelihoods—but we will also need to use the additive risk-factor-based return decomposition derived in early chapters. The combination of these elements will allow us to create the risk framework used to generate the insightful graphics presented in this section.

## 10.6 Concluding Thoughts

Risk, as the reader now hopefully appreciates, is the most complex of our three portfolio-analytic elements. Risk associated with any activity has two principal dimensions: the set of outcomes associated with that activity and their corresponding probabilities.[28] In the context of a simple coin-toss example, we were able to use these notions to define a broad set of risk measures. As we moved to a more complex example, we found a significant stumbling block in the form of an infinite number of possible outcomes. Statistical analysis came to our rescue and we found that the technical link between outcomes and likelihoods is an object termed the statistical distribution. The distribution is tailor-made for risk analysis. As such, much of risk analysis is basically applied statistics. The job of the analyst, however, is to ensure that statistics remains a tool for assisting in the assessment of the risk in one's portfolio. The statistics, however important, should *not* take

[28]An interesting introduction to the foundations of risk analysis is found in Berstein [2].

the forefront. Sensitivity analysis—which we saw in the form of consideration of different distributional assumptions and parameters—can be a powerful ally in managing this trade-off between statistical details and economic realism.

## References


1. P. Artzner, F. Delbaen, J.M. Eber, D. Heath, Coherent measures of risk. Math. Finance **9**(3), 203–228 (2001)
2. P.L. Bernstein, *Against the Gods: The Remarkable Story of Risk* (Wiley, New York, 1996)
3. P. Billingsley, *Probability and Measure*, 3rd edn. (Wiley, New York, 1995)
4. G. Casella, R.L. Berger, *Statistical Inference* (Duxbury Press, Belmont, 1990)
5. E.A. Cornish, R.A. Fisher, Moments and cumulants in the specification of distributions. Rev. Inst. Int. Stat. **5**(4), 307–320 (1938)
6. R. Durrett, *Probability: Theory and Examples*, 2nd edn. (Duxbury Press, Belmont, 1996)
7. P. Embrechts, A. McNeil, D. Straumann, Correlation and dependence in risk management: properties and pitfalls. ETH Working Paper, 1999
8. P. Embrechts, S.I. Resnick, G. Samorodnitsky, *Extreme Value Theory as a Risk Management Tool* (Cornell University, Rhodes Hall, Fall 1996)
9. R. Engle, Dynamic conditional correlation: a simple class of multivariate generalized autoregressive conditional heteroskedasticity models. J. Bus. Econ. Stat. **20**(3), 339–350 (2002)
10. S. Hwang, S. Satchell, Tracking error: ex-ante versus ex-post measures. Financial Economics Research Centre Working Paper WP01-15, 2001
11. P. Jorion, *Value at Risk: The New Benchmark for Managing Financial Risk* (McGraw-Hill, New York, 2006)
12. D. Maillard, A user's guide to the cornish fisher expansion. Conservatoire national des arts et métiers, 2012
13. J.P. Morgan/Reuters, *RiskMetrics™ — Technical Document* (Morgan Guaranty Trust Company, New York, 1996)
14. N.N. Taleb, *The Black Swan: The Impact of the Highly Improbable* (Random House, New York, 2007)

# 11 Portfolio Risk

*Prediction is very difficult, especially if it's about the future.*

Niels Bohr

Computing the risk of a portfolio, as we saw in the previous chapter, is an exercise in prediction. It is not prediction in the typical sense as we are *not* trying to predict the future return of our portfolio at some future point in time, but rather we are trying to predict the range of possible future portfolio-return outcomes. In other words, computing a portfolio's risk involves predicting the statistical distribution of future return outcomes.

Such an exercise is both difficult and complicated. It is difficult, because generally speaking, prediction is hard. The only really useful information available to form our distributional predictions is past data.[1] While past data is helpful, it is always somewhat dangerous to expect the future to look like the past. It is complicated, because the returns of fixed-income portfolios depend on a wide range of risk factors. Predicting fixed-income portfolio returns amounts to prediction of the *joint* evolution of these risk factors.

Strategic asset allocation is also concerned with predicting the distribution of future portfolio returns in an effort to help an institution, or individual investor, select the optimal security mix for its portfolio. An important difference between the computation of portfolio risk and strategic asset allocation is the time horizon. Portfolio-risk calculations are typically performed on a daily basis and typically look at possible movements over the coming days, weeks, or month. Strategic asset allocation is performed relatively infrequently—perhaps once every 1–3 years—and generally looks numerous years into the future.

[1]Computing the risk of a fixed-income portfolio involves predicting the range of possible risk-factor outcomes, but these predictions need to be based on something. In practice, these turn out to be, in large part, formulated using past data as a guide.


D.J. Bolder, *Fixed-Income Portfolio Analytics*,
DOI 10.1007/978-3-319-12667-8_11

These seemingly small differences have two important implications that make the computation of portfolio risk somewhat easier relative to strategic asset allocation. The first implication relates to expected returns. Typically, when computing portfolio risk, it is assumed that the expected return of each fixed-income instrument, and hence the portfolio, is zero. The underlying idea is that the average return is close to zero over short time periods and that, in any event, a risk manager is not trying to predict future returns, but instead focuses on expected future volatility.

Typical practice in risk computations is to focus solely on the uncertainty of a portfolio's return. Such an assumption is impossible in the performance of strategic asset allocation because the time horizon is much longer and consequently the return is necessarily not zero. Moreover, the expected return is a critical input for deciding on the relative desirability of various asset types. Assuming it to be zero would basically defeat the principal purpose of the exercise.

The second implication relates to precision. We should expect a higher degree of accuracy with short-term portfolio risk computations relative to longer-term strategic asset allocation computations. The basic reasoning is that it is somewhat easier, albeit still quite difficult, to predict the range of future outcomes for the coming few weeks or month, than it is to predict the range of outcomes that may occur in several year's time.

Prediction, by definition, is thus related to time. Figure 11.1 provides a schematic view of how one can view the time dimension in risk computations. Standing at the current point in time, you may look backwards and ask:

- What was the risk of my portfolio over the last month?
- What has been the volatility of my portfolio's monthly return over the last 3 years?

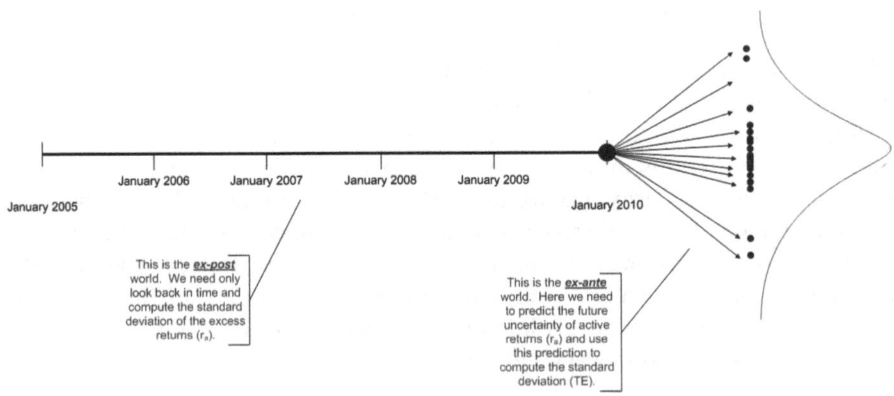


**Fig. 11.1** Ex-post versus ex-ante perspective. This schematic describes the difference between an ex-post (backward-looking) and ex-ante (forward-looking) view of the world

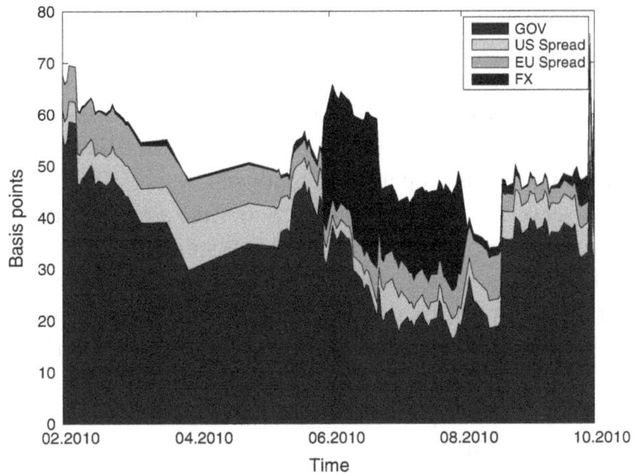


**Fig. 11.2** A sample tracking error history. This figure outlines the evolution of the ex-ante tracking error for a multicurrency fixed-income portfolio over the course of a roughly 9-month period. Observe that it decomposes the risk into the contribution from *four* different risk factors

These are both backward-looking, or ex-post, views of the world. While interesting questions, the focus of this chapter is the forward-looking, or ex-ante view, of the world. An ex-ante view implies that one's looks to the next day, week, or month and asks: what could happen to the return of my portfolio? Correspondingly, this chapter centres around *three* principal questions:

- How much risk do I have in my portfolio?
- How much risk arises from my portfolio relative to my benchmark?
- Where does this risk come from?

The answer to each of these questions may be estimated using quantitative techniques. While there are a variety of possible ways to answer these questions, this chapter will walk through *one* possible approach with some directions as to how one might go in other directions.

Before jumping into the mathematical details, let's make our objectives a bit more concrete by looking at an example. Figure 11.2 outlines, for a sample multiple currency and multiple asset-class portfolio, the evolution of the daily ex-ante tracking error over the 9-month period from February to October 2010. It demonstrates not only the overall level of daily risk for each business date in this period, but also the relative risk contribution by three main risk factors.[2] Neither the level of risk nor the contribution of government bond yields, credit spreads, and foreign-exchange rates are constant over this period. This useful and illuminating graphic can form an important element of one's daily portfolio oversight.

Our principle objective in this chapter is to demonstrate how such a graph might be constructed. Our analysis is not restricted to tracking error. Different

[2]This breakdown of risk by the contribution from each risk factor is conceptually analogous to performance attribution—the allocation of return to different risk factors.

perspectives lead to different risk measures. Thinking about the return of one's portfolio relative to its benchmark leads one to the concept of ex-ante tracking error, while consideration of the *worst-case* outcome for one's portfolio return brings up the idea of Value-at-Risk (VaR). Both of these ideas involve looking into the future and predicting the span of portfolio-return outcomes that might occur over the course of one's period of analysis.

A secondary objective is to encourage a rather cautious view of all risk measures. Despite the appearance of accuracy stemming from generous use of mathematics and statistics, risk measurement remains a relatively subjective area of analysis. Another way to see this is to realize that prediction involves assumptions that are, by definition, subjective. Some key examples include:

- there are multiple ways to perform such an analysis and Fig. 11.2 represents only one possibility;
- the computations make use of historical data and use of a different data frequency (i.e., daily, monthly, or annually) or data sample period (i.e., the last 2-, 5-, or 10-years of data) is entirely possible; and
- the computations place a certain amount of weight on each historical data point—other choices are naturally possible.

Given the range of possible choices involved in this analysis, it should come as no surprise that the risk figures presented in Fig. 11.2 *cannot* represent a definitive picture of this portfolio's forward-looking risk over this period. Instead, it is an approximation.[3]

## 11.1 The Punchline

In the following discussion, we will perform a rather wide range of computations to describe *two* main risk measures: the ex-ante tracking error and the Value-at-Risk (VaR). To support these calculations, we invoke a number of mathematical and statistical results. Without a clear idea of where we are going, this may be confusing and difficult to follow. Figure 11.1 provides a graphical preview of what we are trying to accomplish, but we will also offer a brief insight into the final risk computations to help ease the computational burden in the coming pages.

Ex-ante tracking error, a measure of relative risk, is the standard deviation of your active positions relative to a benchmark. It is defined as,

$$\begin{aligned} \text{TE} &= f(\text{Active weights, Market uncertainty}), \\ &\triangleq \sqrt{\omega_a^T \Omega_R \omega_a}. \end{aligned} \tag{11.1}$$

[3] In the next chapter, we will explicitly examine the subjective nature of our risk computations and present a possible framework for assessing their accuracy.

Forward-looking tracking error basically boils down to the interplay between two elements. It is determined by your active positions—that is, over- and underweight of individual fixed-income positions relative to the benchmark—which are defined by $\omega_a$. It also depends upon your assessment of market uncertainty. In our framework, market uncertainty is summarized in the covariance matrix of security returns, denoted as $\Omega_R$.[4] Equation (11.1) indicates that the tracking error has a relatively simple quadratic form. It is precisely this form that we will exploit later to perform the risk decomposition.

VaR is a probabilistic measure of absolute risk that attempts to assess, with a given degree of confidence, the worst-case loss of your portfolio over some time interval. It is written mathematically as,

$$\begin{aligned} \text{VaR} &= g(\text{Portfolio weights, Market uncertainty}), \\ &\triangleq \underbrace{\mathcal{N}^{-1}(\alpha)}_{\substack{\text{Confidence} \\ \text{level} \\ (\text{i.e., } \alpha = 0.95)}} \sqrt{\omega_p^T \Omega \omega_p}. \end{aligned} \tag{11.2}$$

Again, the VaR risk measure depends on one's assessment of market uncertainty—as proxied by the covariance matrix of security returns, $\Omega_R$—and a set of weights. In contrast to the ex-ante tracking error, one does not use active weights, but rather the portfolio weights.

There are a number of strong similarities between the ex-ante tracking error and VaR. Both have a quadratic form in a set of weights and both make use of a covariance matrix. The key difference with the VaR computation is the extra multiplicative term, $\mathcal{N}^{-1}(\alpha)$, at the beginning of Eq. (11.2). This extra term essentially specifies the level of confidence in one's computation.

Both ex-ante tracking error and VaR capture market uncertainty through the covariance matrix, $\Omega_R$. In the approach presented in this chapter, $\Omega_R$ is the *heart* of risk. The multivariate distribution of our portfolio's returns, $r$, is defined as,

$$r \sim \mathcal{X}\left(\vec{0}, \Omega_R, \cdots\right). \tag{11.3}$$

That is, the portfolio's returns follow the distribution $\mathcal{X}(\cdots)$ with a zero expected return, $\vec{0}$, and covariance matrix of $\Omega_R$.[5]

---

[4]Much more will be said, later in the chapter, about this matrix. For the moment, it suffices to treat it as our assessment of market uncertainty.

[5]Note we have not automatically assumed a normal distribution. The ex-ante tracking error is merely the standard deviation of the distance between portfolio and benchmark returns—no mention of the underlying distribution is either made or required. The VaR computation, in Eq. (11.2), in contrast, requires a distributional assumption. In this chapter, we will assume Gaussianity, although this will be partially relaxed in the subsequent chapter.

While Eq. (11.3) clarifies matters somewhat, it still remains a rather unsatisfying definition as it leaves a number of questions unanswered. It does not tell us, for example, if $\Omega_R$ must be defined in terms of security return uncertainty or if it can be adjusted to say something about the uncertainty of the portfolio's market risk factors. It also does not explain to us how such a quantity is estimated or, indeed, if it even needs to be estimated at all. Each of these issues will be addressed in detail in this chapter.

The principle objective of this chapter is thus to explain how we arrive at the formulae, in Eqs. (11.1) and (11.2), for ex-ante tracking error and VaR. We also dedicate substantial time to the discussion of the covariance matrix, $\Omega_R$. We will make extensive use of the concepts introduced in previous chapters to establish a link between the exposure of one's portfolio and strategic benchmark to market movements and thereby set the stage for the computation of risk. Our analysis will include a framework for breaking down, from the individual security level, the individual contributions of different risk factors to our overall risk measures. As usual, we explore all of the computations with the liberal use of a practical example.

## 11.2 Getting Started

To create an example, we need to define a set of fixed-income instruments. We have selected three plain vanilla US Treasury bonds with 2-, 5-, and 10-year tenors, respectively. These instruments can be placed in either our portfolio or strategic benchmark. Three bonds, we realize, do *not* represent a very realistic portfolio. They are sufficient, however, to create a non-trivial setting and thereby reasonably animate our discussion. Table 11.1 outlines some summary statistics for these three US Treasury (UST) treasury bonds. This analysis is performed as at November 2009.

There is nothing particularly special about these three bonds other than the fact that they loosely cover the 2- to 10-year spectrum of the US yield curve. A real portfolio and benchmark, of course, would have many more instruments. We can and will, however, be able to illustrate all of the basic ideas associated with

**Table 11.1** A sample portfolio

| ISIN | Maturity | Coupon | Yield | Duration |
|---|---|---|---|---|
| US912828JU50 (2Y) | Nov 15, 2011 | 1.75 % | 0.93 | 1.98 |
| US912828DC17 (5Y) | Nov 15, 2014 | 4.25 % | 2.35 | 4.47 |
| US912810ED64 (10Y) | Aug 15, 2019 | 8.125 % | 3.56 | 7.18 |

To demonstrate the basic concepts in the computation of portfolio risk, we will use a sample portfolio comprised of the underlying three UST bonds at 2-, 5-, and 10-year tenors; these bonds will be found in both the portfolio and benchmark.

risk computation within the context of this small, compact portfolio and strategic benchmark.[6]

### 11.2.1 Portfolio Weights

The first step in the computation of one's risk is to provide a mathematical description of one's portfolio and strategic benchmark. Although this may sound daunting, it basically amounts to the determination of the portfolio and benchmark weights. While conceptually simple, there are nonetheless a few tricky practical points to be considered.

Let $n$ denote the total number of instruments in *both* the portfolio and the benchmark. We must jointly consider the instruments in both the portfolio and the benchmark. This is a key point as we need to determine the risk of all instruments. There are *three* possible cases for a given fixed-income instrument: it may be found in

1. the benchmark, but not the portfolio;
2. the portfolio, but not the benchmark;
3. both the benchmark and the portfolio.

All three cases occur commonly in portfolio management, although given the typically large size of benchmarks relative to portfolios, the first case occurs with the highest frequency.

With this in mind, we proceed to define the portfolio weights, $\omega_p$. It is an $n$-dimensional vector, $\omega_p \in \mathbb{R}^{n\times 1}$, representing the proportional market weight of each of the $n$ instruments held in the portfolio. It has the following form,

$$\omega_p = \begin{bmatrix} \omega_{1,p} \\ \vdots \\ \omega_{n,p} \end{bmatrix}, \tag{11.4}$$

where,

$$\sum_{k=1}^{n} \omega_{k,p} = 1. \tag{11.5}$$

In a more realistic setting, we would expect to see many of the entries of $\omega_p$ equal to zero. This is because there are often hundreds of instruments in a fixed-income

[6]Three bonds permit us, more specifically, to explicitly write out all of the computations without confusing the reader with enormous tables of numbers.

benchmark, but only 15–30 positions in a typical fixed-income portfolio. All of the benchmark bonds not held in the portfolio, therefore, will have a zero weight in $\omega_p$.[7]

In a conceptually similar manner, we let $\omega_b \in \mathbb{R}^{n\times 1}$ represent the proportional market weight of each of the $n$ instruments held in the benchmark as,

$$\omega_b = \begin{bmatrix} \omega_{1,b} \\ \vdots \\ \omega_{n,b} \end{bmatrix} \tag{11.7}$$

where,

$$\sum_{k=1}^{n} \omega_{k,b} = 1. \tag{11.8}$$

Since the benchmark is generally comprised of a large number of bonds, we should expect to see a large number of relatively small weights. We only expect to see zeros for positions in the portfolio—such as derivatives, positions in ex-benchmark bonds, and perhaps cash—that are not present in the benchmark.

Armed with this parsimonious description of our portfolio and benchmark, we may proceed a step further to describe the *active* positions in our benchmark. An active position is defined as a position held in the portfolio that differs from the benchmark. If the proportion of a given bond held in the portfolio is exactly identical to the proportion in the benchmark, then a position is flat, inactive, or neutral.

---

[7]If one introduces derivative positions, such as futures, computation of the market proportional weight becomes tricky, because futures typically have market values very close to zero implying a zero portfolio weight. In this case, it is common, for derivatives, to use the bond-equivalent exposure of the futures contract in the numerator, but not the denominator. If one replaces the futures exposure in both the numerator and denominator of the market-weight computation, one ignores the leverage effect associated with the use of futures. The leverage effect occurs because, with bond or rate futures, one can increase or decrease one's portfolio duration (or alter key-rate durations) without any cash investment. One reasonable approach for computation of weights in a portfolio with bond futures, therefore, is to set;

$$\omega_{i,p} = \frac{\text{Exposure}_i}{\sum_{k=1}^{n} \text{Market Value}_k}. \tag{11.6}$$

In this case, the leverage effect is captured, but it is no longer the case that the portfolio weights sum to one as indicated in Eq. (11.5). This solution would fail in, an admittedly extreme, situation where the entire portfolio was comprised of futures.

These active weights, denoted as $\omega_a \in \mathbb{R}^{n \times 1}$, are described mathematically as the difference between the portfolio and benchmark weights,

$$\begin{aligned} \omega_a &= \omega_p - \omega_b, \\ &= \begin{bmatrix} \omega_{1,p} \\ \vdots \\ \omega_{n,p} \end{bmatrix} - \begin{bmatrix} \omega_{1,b} \\ \vdots \\ \omega_{n,b} \end{bmatrix}. \end{aligned} \tag{11.9}$$

This vector computation is precisely why we need $\omega_p$ and $\omega_b$ to have the same dimensions. Let's think a bit about Eq. (11.9). If one's portfolio perfectly replicated the benchmark, the active weight vector would be full of zeros. This would imply, quite reasonably, that the portfolio is absent of active risk. This does *not* imply, however, that the portfolio is absent of risk, but rather that it has the same inherent risk as the underlying strategic benchmark.

Returning to our simple example, we need to define the portfolio and benchmark weights. To illustrate a number of different outcomes, we have opted for a single strategic benchmark, but *three* alternative portfolio implementations. Table 11.2 highlights the portfolio, benchmark, and active weights for our simple example.

Using Eq. (11.9), it is straightforward to compute the active weights for each portfolio. For the first portfolio, the active weights are,

$$\begin{aligned} \omega_a &= \omega_p - \omega_b, \\ &= \begin{bmatrix} \omega_{1,p} \\ \omega_{2,p} \\ \omega_{3,p} \end{bmatrix} - \begin{bmatrix} \omega_{1,b} \\ \omega_{2,b} \\ \omega_{3,b} \end{bmatrix}, \\ &= \begin{bmatrix} 0.25 \\ 0.25 \\ 0.50 \end{bmatrix} - \begin{bmatrix} 0.50 \\ 0.00 \\ 0.50 \end{bmatrix} = \begin{bmatrix} -0.25 \\ 0.25 \\ 0.00 \end{bmatrix}. \end{aligned} \tag{11.10}$$

The weights for each of the three sample portfolios summarized in Table 11.2, are provided in Table 11.3. The reader is invited to predict which portfolio has the

**Table 11.2** Portfolio and benchmark weights

| ISIN | Benchmark ($\omega_b$) | Portfolio ($w_p$) 1 | 2 | 3 |
|---|---|---|---|---|
| US912828JU50 (2Y) | 0.50 | 0.25 | 0.00 | 0.40 |
| US912828DC17 (5Y) | 0.00 | 0.25 | 0.50 | 0.00 |
| US912810ED64 (10Y) | 0.50 | 0.50 | 0.50 | 0.60 |

This table summarizes the weights of the UST bonds, summarized in Table 11.1, found in the benchmark and the three sample portfolios.

**Table 11.3** Active weights

| ISIN | Benchmark ($\omega_a$) | Portfolio ($\omega_a$) 1 | 2 | 3 |
|---|---|---|---|---|
| US912828JU50 (2Y) | 0.00 | −0.25 | −0.50 | −0.10 |
| US912828DC17 (5Y) | 0.00 | 0.25 | 0.50 | 0.00 |
| US912810ED64 (10Y) | 0.00 | 0.00 | 0.00 | 0.10 |

This table uses the portfolio and benchmark weights, summarized in Table 11.2, to compute the active weights for our three sample portfolios.

most relative and absolute risk—the answer will be provided through the course of the chapter.

### 11.2.2 Incorporating Risk-Factor Exposures

The second step in deriving our risk measures is to determine the exposure of one's portfolio to a desired set of fixed-income risk factors. Although this is a fairly challenging task on its own, the heavy lifting for the link between return and risk factors was already done in previous chapters. Using Taylor's theorem, we found that we could approximate the return of a fixed-income security with the following additive risk-factor-based decomposition:

$$r \approx \overbrace{\underbrace{y\Delta t}_{\text{Carry return}} + \mathbb{I}_{\text{ILB}} \underbrace{\pi(t,t+1)}_{\text{Inflation carry}} - \underbrace{\sum_{i=1}^{\nu} \kappa_i \Delta y_{i,\text{TRE}}}_{\text{Curve return}} - \underbrace{D_S \Delta s^{\text{OAS}}}_{\text{Credit return}}}^{\text{Local currency return}} + \underbrace{\sum_{i=1}^{\alpha} \mathbb{I}_{FX_i} \left( \frac{E_{i,1} - E_{i,0}}{E_{i,0}} \right)}_{\text{FX return}}, \tag{11.11}$$

where the exposure to each of the $\alpha$ fx-rates is an indicator variable,

$$\mathbb{I}_{FX_i} = \begin{cases} 0 : \text{Not exposed to currency } i \\ 1 : \text{Exposed to currency } i \end{cases}, \tag{11.12}$$

that takes a value of 1, if exposed to currency $i$ and zero, if not. The exposure to inflation is conceptually similar to currency exposure and given as,

$$\mathbb{I}_{ILB} = \begin{cases} 0 : \text{A nominal bond} \\ 1 : \text{An ILB} \end{cases}. \tag{11.13}$$

Despite the usefulness of our additive decomposition, it needs a more convenient form for the computation of portfolio risk. We will be performing computations with large numbers of instruments and using an expression like Eq. (11.11) would make both the computation and exposition quite clumsy. For this reason, we transform Eq. (11.11) into matrix notation as follows,

$$r \approx \underbrace{\begin{bmatrix} y & \mathbb{I}_{ILB} & -\kappa_1 & \cdots & -\kappa_\nu & -D_S & \mathbb{I}_{FX_1} & \cdots & \mathbb{I}_{FX_\alpha} \end{bmatrix}}_{\text{Instrument exposures}} \underbrace{\begin{bmatrix} \Delta t \\ \pi(t, t+1) \\ \Delta y_{1,\text{TRE}} \\ \vdots \\ \Delta y_{\nu,\text{TRE}} \\ \Delta s^{\text{OAS}} \\ \frac{E_{1,1} - E_{1,0}}{E_{1,0}} \\ \vdots \\ \frac{E_{\alpha,1} - E_{\alpha,0}}{E_{\alpha,0}} \end{bmatrix}}_{\substack{\text{Changes in} \\ \text{market factors}}} \tag{11.14}$$

This is a convenient representation, because it separates the portfolio risk-factor exposures, or sensitivities, from the changes in the market risk factors. Why is this separation useful? At a given instant in time, the portfolio exposures are not expected to change. We can think, therefore, of the first vector in Eq. (11.14) as being fixed or static. This is only half the story. Computing risk involves understanding how the portfolio, at a given point in time, is impacted by different market scenarios—whether historical, simulated, or just imagined. These different market scenarios are summarized by the second vector in Eq. (11.14). It represents market risk-factor movements and is expected to change; it is, therefore, variable or dynamic. Equation (11.14), therefore, separates the static and dynamic elements from our additive return decomposition.

The first vector in Eq. (11.14) summarizes the portfolio exposures for a single fixed-income instrument. Having moved into a matrix setting, there is nothing stopping us from collecting the exposures for multiple instruments into a single matrix. For a general situation with a portfolio and benchmark comprised of $n$ instruments, the portfolio exposures can be collected as follows,

$$\begin{bmatrix} \cancel{y_1} & \cancel{\mathbb{I}_{ILB_1}} & -\kappa_{1,1} & \cdots & -\kappa_{1,\nu} & -D_{1,S} & \mathbb{I}_{FX_{1,1}} & \cdots & \mathbb{I}_{FX_{1,\alpha}} \\ \vdots & \vdots & & \ddots & \vdots & \vdots & \vdots & \ddots & \vdots \\ \cancel{y_n} & \cancel{\mathbb{I}_{ILB_n}} & -\kappa_{n,1} & \cdots & -\kappa_{n,\nu} & -D_{n,S} & \mathbb{I}_{FX_{n,1}} & \cdots & \mathbb{I}_{FX_{n,\alpha}} \end{bmatrix} \tag{11.15}$$

collecting the carry, curve, credit, and foreign-exchange factors together.

We now introduce an important assumption. Our risk measures, whether VaR or ex-ante tracking error, are computed under the assumption that market movements impact only the current positions. In other words, the portfolio is assumed to remain constant as time moves forward. This implies that the carry contribution, $y\Delta t$ does *not* contribute to the return volatility of market movements and is, consequently suppressed.[8]

Collecting all portfolio and benchmark exposures into a single matrix is a natural next step. We denote this object as $K$ and term it the *exposure* matrix.[9] It has the following form:

$$K = \begin{bmatrix} -\kappa_{1,1} & \cdots & -\kappa_{1,\nu} & -D_{1,S} & \mathbb{I}_{FX_{1,1}} & \cdots & \mathbb{I}_{FX_{1,\alpha}} \\ \ddots & \vdots & \vdots & \vdots & & \ddots & \vdots \\ -\kappa_{n,1} & \cdots & -\kappa_{n,\nu} & -D_{n,S} & \mathbb{I}_{FX_{n,1}} & \cdots & \mathbb{I}_{FX_{n,\alpha}} \end{bmatrix}, \tag{11.16}$$

where $K \in \mathbb{R}^{n \times m}$, $n$ is the number of instruments and $m$ denotes the number of market risk factors.

$K$ is a convenient mathematical object. Each row represents the collection of exposures for each individual instrument in our portfolio and benchmark. Each column, conversely, represents the exposures for all of the securities in our portfolio and benchmark to a single market risk factor. In short, $K$ summarizes the market-risk characteristics of our portfolio and strategic benchmark. Even better, given any set of changes in the factors, we can use the exposure matrix, $K$, to compute the return for each individual security and each individual risk factor.

Equation (11.15) provides us with all that is required to compute the exposure matrix for our simple example. Our three-bond exposure matrix is populated as,

$$\begin{aligned} K &= \begin{bmatrix} -\kappa_{1,6M} & -\kappa_{1,2Y} & -\kappa_{1,5Y} & -\kappa_{1,7Y} & -\kappa_{1,10Y} & -D_{1,S} & \mathbb{I}_{FX_1} \\ -\kappa_{2,6M} & -\kappa_{2,2Y} & -\kappa_{2,5Y} & -\kappa_{2,7Y} & -\kappa_{2,10Y} & -D_{2,S} & \mathbb{I}_{FX_2} \\ -\kappa_{3,6M} & -\kappa_{3,2Y} & -\kappa_{3,5Y} & -\kappa_{3,7Y} & -\kappa_{3,10Y} & -D_{3,S} & \mathbb{I}_{FX_3} \end{bmatrix}, \\ &= \begin{bmatrix} -0.014 & -1.943 & -0.019 & 0 & 0 & 0 & 0 \\ -0.032 & -0.021 & -4.169 & -0.060 & 0 & 0 & 0 \\ -0.048 & -0.311 & -0.060 & -1.182 & -5.038 & 0 & 0 \end{bmatrix}. \end{aligned} \tag{11.17}$$

There are, in our simple example, seven possible risk factors: that is, $K \in \mathbb{R}^{3 \times 7}$. As these are US Treasury bonds, there is no spread duration, *no* foreign-exchange risk, and no exposure to inflation. Although it is relatively simple, the exposition nonetheless demonstrates clearly how to incorporate more complex securities into this framework.

[8]Another way to see this is to realize that the carry is computed as a function of the current yield and the change in time. The current yield is known today and, as such, exhibits no future variability. It is not a random variable. Quite simply, the carry return is constant and can be removed from our calculations.

[9]Since we, at least in this text, use the term exposure and sensitivity interchangeably, we may also call this the sensitivity matrix.

### 11.2.3 Handling Market Movements

The key element of risk is predicting the range of future market movements: more concretely, this is the second vector in Eq. (11.14). Before we can appropriately discuss this vector, we need to clarify some important statistical concepts and provide the appropriate notation. We use two related, but subtly different, versions of market movements throughout the upcoming text:

1. $M \in \mathbb{R}^{m \times 1}$ is a *random vector* of market movements. Its distribution is unknown.
2. $\hat{M}_t \in \mathbb{R}^{m \times 1}$ is a single *realization* (or observation) of the market movements from time $t-1$ to $t$.

$M$ is a statical object termed a random variable. We do not, nor will we ever, know its true statistical distribution. $\hat{M}_t$ is an observed value of $M$—many observations of $\hat{M}_t$ form a historical dataset. We can use these historical observations to empirically *estimate* the distribution $M$.

To be more concrete, we may assume that $M \sim \mathcal{N}(\mu, \Omega)$. This is not a problem, but we need to be aware that neither $\mu$ nor $\Omega$ will ever be known with certainty.[10] They must estimated. In practice, we employ the estimators $\tilde{\mu}$ and $\tilde{\Omega}$ to estimate $\mu$ and $\Omega$.[11] The historical observations may be employed to form these estimates. We will present a few alternative estimators for the covariance matrix of $M$. In summary, it is important to understand and appreciate the difference between the random variable, the observed realizations, and the estimator.

To compute the risk of a portfolio, we need to look at many sets of market movements. Matrix notation is helpful in this regard, because we can easily generalize the single vector of market movements in Eq. (11.14) into a full collection of market movements within a single matrix. We call this the *market-movement matrix* and it has the following form,

$$\hat{M} = \left[\hat{M}_1 \ \hat{M}_2 \cdots \hat{M}_T\right] = \begin{bmatrix} \Delta y_{1,1}^{\text{TRE}} & \Delta y_{1,2}^{\text{TRE}} & \cdots & \Delta y_{1,T}^{\text{TRE}} \\ \vdots & \vdots & \ddots & \vdots \\ \Delta y_{\nu,1}^{\text{TRE}} & \Delta y_{\nu,2}^{\text{TRE}} & \cdots & \Delta y_{\nu,T}^{\text{TRE}} \\ \Delta s_1^{\text{OAS}} & \Delta s_2^{\text{OAS}} & \cdots & \Delta s_m^{\text{OAS}} \\ \frac{E_1 - E_0}{E_0} & \frac{E_2 - E_1}{E_1} & \cdots & \frac{E_T - E_{T-1}}{E_{T-1}} \end{bmatrix}, \tag{11.18}$$

where $\hat{M} \in \mathbb{R}^{m \times T}$. This is basically our historical dataset.

$M$ is thus a parsimonious representation of a collection of realized market risk factor movements. Each column of $M$ represents the collection of risk-factor changes over a given time interval. Each column is thus a scenario: either historical

[10] We assume that $\mu \equiv \vec{0}$, but this is also a form of estimation.

[11] In statistics, an *estimator* is an algorithm or rule for computing an estimate of some aspect of a distribution from a historical dataset. See Casella and Berger [3].

realization, a simulated outcome, or a stress test. Each row, in contrast, represents the distribution of possible movements across all scenarios for a given risk factor.

Unlike the portfolio exposures, which given your strategic benchmark and portfolio holdings are uncontroversially computed, it is not immediately obvious how to populate the market-movement matrix, $\hat{M}$. This is the input for the computation of the uncertainty in market risk factors. One can use historical, simulated, or completed invented data. To be fair, data simulated from a model is, in many ways, very similar to historical data. The reason is that the model parameters must be estimated using historical data. In brief, to understand the range of risk-factor outcomes and their relative likelihood, as we saw in the previous chapter, it is almost inevitable that you need to use historical data, in one way or another, to populate the market-movement matrix.

Like most things in life, the best approach is relatively balanced. Historical data is a natural starting point. A good data sample provides some insight into our risk-factor dynamics. Concerns about the use of backward-looking data can then be, at least partially, assuaged with sensitivity analysis and stress-testing.

In our example, the data requirement is rather light since we only need UST yield curves. Figure 11.3 provides a graphical illustration of monthly UST yield curves—estimated using the techniques described in earlier chapters—from January 1997 to October 2009. This 13-year period covers a fairly wide range of events, but also encapsulates a period of relatively consistent monetary and fiscal policy in the United States.

Figure 11.4 shows the collection of UST yield curves and the monthly changes in these yield curves over the 13-year period. The lower graphic in Fig. 11.4 is the ultimate input into our market-change matrix, $\hat{M}$, described in Eq. (11.18). The mean yield-curve change across all tenors seems to be centred around zero, with an average dispersion between 25 and 50 basis points relatively evenly spread in both directions around zero. The most extreme movements appear to be on the order of

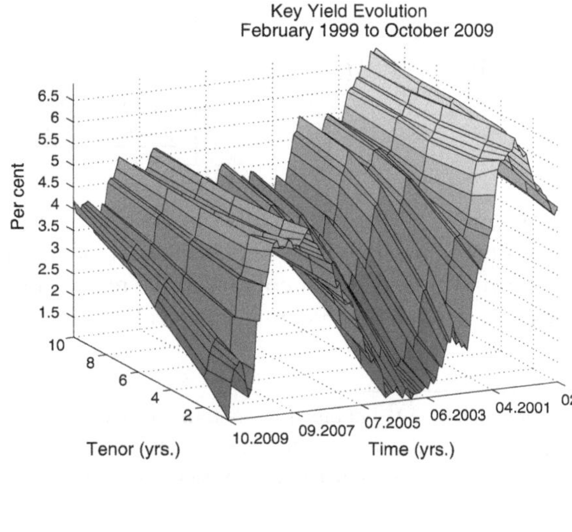


**Fig. 11.3** Historical UST curves. This figure summarizes the input UST curve data used in our simple example—these are the UST curves from January 1997 to October 2009. Typically, one would expect to have a wider range of risk factors, but for our simple UST-bond example, the only source of risk is the UST curve

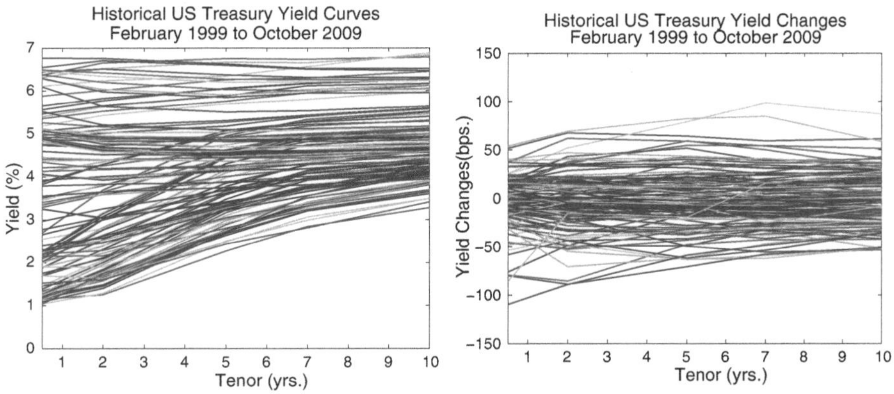


**Fig. 11.4** Another view of our input data. This figure illustrates the input UST yield curves from two additional perspectives. The upper graphic outlines all of the monthly yield curves from January 1997 to October 2009, while the lower graphic summarizes the monthly changes in curves, or first differences, across the period

$\pm$100 basis points—the changes, of course, do not always impact all tenors across the yield curve in an identical fashion.

The distribution of *changes* in financial-market factors typically exhibits greater stability than the distribution of their *levels*. The upper graphic in Fig. 11.4 proves some evidence for this claim, by illustrating the associated UST yield-curve levels—it ranges from flat yields curves at the 7 % level to yield curves with a slope of almost 300 basis points and situations of short rates at close to zero.[12]

Figure 11.5 takes the results in Figs. 11.3 and 11.4 one step further by providing a view of the empirical distributions of the monthly differences at *four* key tenors along the UST yield curve—the 6-month, 2-, 5-, and 10-year sectors. Each histogram has a normal distribution—computed using the sample mean and volatility—superimposed on top of the data. Yield-curve movements are *not* strictly normally distributed across any of these four key-rate tenors. In all cases, there is a greater probability of extreme events (i.e., as described by the tails of the distribution) and a correspondingly tighter concentration in the central part of the distribution.[13] The empirical distributions in Fig. 11.5 are not so far from normality that they cannot, at least, be loosely approximated with the Gaussian distribution.

[12]The distribution of changes in market risk factors is not always stable, but we are on somewhat safer statistical ground when working with differences—or rather changes—in market factor rather than their levels.

[13]The technical term for this type of distribution is *leptokurtotic*, which is often called fat tails. In fact, the Greek translation is closer to *slender waist*. The two notions are essentially equivalent for a symmetric distribution. A tighter concentration of observations around the mean is typically accompanied by a larger number of observations further from the mean in the tails of the distribution.

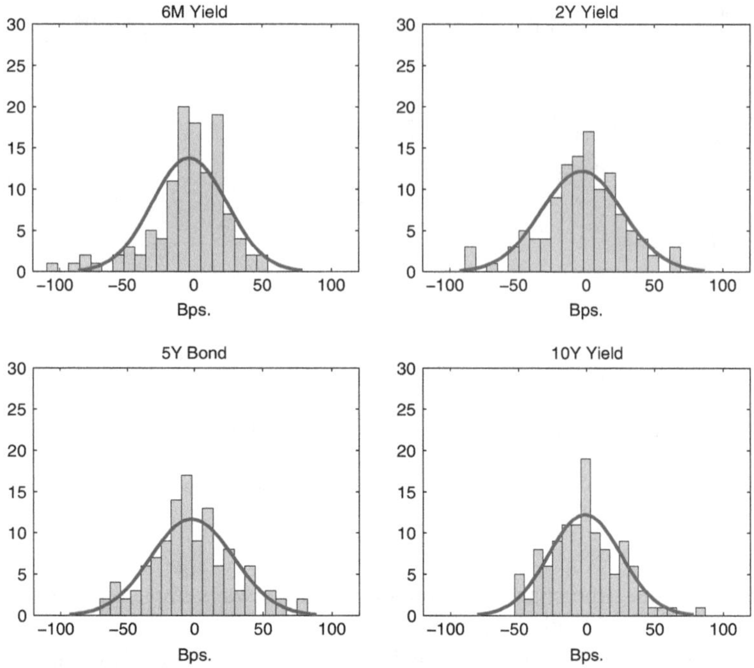


**Fig. 11.5** Empirical key-rate distributions. This figure provides four histograms describing the distribution of yield changes at the 6-month, 2-year, 5-year, and 10-year yield tenors over the period from January 1997 to October 2009

Moreover, this assumption is only required for VaR computations. This assumption will be relaxed in the next chapter.

### 11.2.4 Computing Return Distributions

We transformed our additive return decomposition of our portfolio and strategic benchmark security returns into the product of two components: an exposure matrix ($K$) and a market-movement matrix ($\hat{M}$). With these two matrices, we can easily proceed to compute the return for each security across each of the $T$ scenarios in the market-movement matrix. We need only take the product of $K$ and $\hat{M}$ as follows

$$R = K\hat{M}, \tag{11.19}$$

where $R \in \mathbb{R}^{n \times T}$. Let's call $R$ the *security return matrix*. For each of the $n$ securities in either the portfolio, the strategic benchmark, or both, we have the security returns for each of the $T$ scenarios. Each row of $R$ describes the return for an individual security across all market risk-factor scenarios. Each column, conversely, describes the return of all instruments in the portfolio and benchmark for a single market scenario.

The security return matrix is populated with the *raw* security returns. What this means is that the returns are computed with no notion of the relative size of the position in one's portfolio or benchmark. To use them for the computation of risk, it is necessary to weight these returns in such a way that the actual strategic benchmark and portfolio positions are appropriately reflected. This weighting is easily accomplished using the weighting vectors, $\omega_p$, $\omega_b$, and $\omega_a$ defined in Eqs. (11.4)–(11.9). More simply,

$$R_x = \omega_x^T R, \tag{11.20}$$

for $x = p, b$, and $a$ or, in words, the portfolio, benchmark, and active returns, respectively.

Figure 11.6 outlines, using the preceding logic, for the UST bonds in our simple example, the empirical distribution of returns. It highlights each row of the security return matrix for the 2-year, 5-year, and 10-year UST bonds across all of the market movements that occurred in our 13-year data sample. It also includes, in the bottom right-hand quadrant, the distribution of benchmark returns.

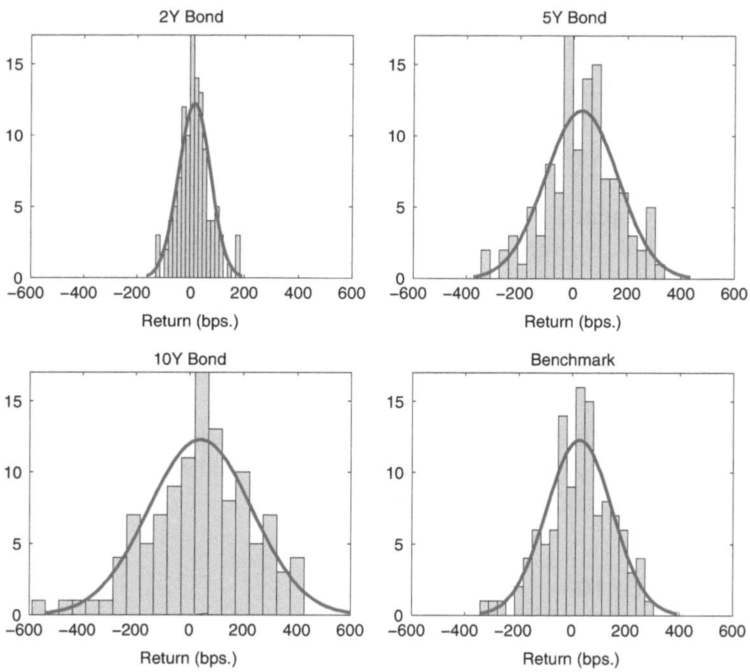


**Fig. 11.6** Individual security returns. This figure outlines the corresponding return distribution for the 2-year, 5-year, 10-year, and benchmark-bond portfolios associated with the yield change outlined in Fig. 11.5

There are three principal points to draw from Fig. 11.6. First, all of the distributions appear to be approximately centred around zero. This provides some justification for our assumption of zero expected return in our risk computations. Second, as the tenor of the instrument increases, the dispersion of the returns increases. This is natural as, with relatively similar yields changes across the curve, the return movements are principally driven by the exposure of the instrument to yield movements. The longer the tenor of a bond, generally speaking, the greater its modified duration and consequently the greater its sensitivity to yield-curve movements. Third, these distributions do *not* appear to be normally distributed, although the deviations do not appear overly dramatic.

## 11.3 Understanding and Exploring $\Omega_R$

Returns may, given a set of market risk-factor exposures, be computed for a wide range of market outcomes. The next step is capturing the uncertainty associated with our risk factors, which brings us to the covariance matrix, $\Omega_R$. Indeed, we now have everything needed for its estimation. It is defined, quite simply, as

$$\Omega_R = \text{cov}(R). \tag{11.21}$$

This $n \times n$ matrix summarizes the dispersion of return of the instruments found in our portfolio and strategic benchmark. The definition in Eq. (11.21) is nevertheless somewhat lacking. A number of aspects of the covariance matrix need to be explored before we can begin confidently using it for the computation of our portfolio's risk position. Specifically, we need to consider:

- How is $\Omega_R$ computed?
- What is the link between the dispersion in market risk factors and instrument returns?
- How can we ensure that our estimate of $\Omega_R$ takes into account recent market movements?
- How might $\Omega_R$ be used to simulate future market movements?

In this section, we address each of these questions in turn. Much, if not all, of the material discussed is quite technical in nature, but is nonetheless critically important to a thorough understanding of risk computations.

### 11.3.1 Variance 101

To really understand the covariance matrix, the best place to start is with the notion of variance. Variance is a standard statistical measure of dispersion. Given two random variables, $X_1$ and $X_2$, the variance of their sum is written as,

$$\text{var}(X_1 + X_2) = \text{var}(X_1) + \text{var}(X_2) + 2\text{cov}(X_1, X_2). \tag{11.22}$$

The variance is thus comprised of a contribution from the first random variable, $X_1$, the second random variable, $X_2$, and the interaction between these two variables, described by the covariance $\text{cov}(X_1, X_2)$.[14] The variance of multiple random variables, therefore, jointly characterizes dispersion and pairwise dependence of each random variable.

If, as is the case in our situation, we wish to compute the weighted variance of two random variables, $X_1$ and $X_2$ with weights $\omega_1$ and $\omega_2$, respectively, then the variance is described as

$$\text{var}(\omega_1 X_1 + \omega_2 X_2) = \omega_1^2 \text{var}(X_1) + \omega_2^2 \text{var}(X_2) + 2\omega_1\omega_2 \text{cov}(X_1, X_2). \quad (11.24)$$

While perfectly correct from a technical perspective, working with Eq. (11.24) is practically challenging. This becomes particularly evident when we generalize to an $n$-dimensional setting. Imagine, for example, that we wish to compute the weighted variance of the sum of random variables $\{X_1, \ldots, X_n\}$ with a set of weights $\{\omega_1, \ldots, \omega_n\}$. The resulting expression is,

$$\text{var}\left(\sum_{i=1}^{n} \omega_i X_i\right) = \sum_{i=1}^{n}\sum_{j=1}^{n} w_i w_j \text{cov}(X_i, X_j). \quad (11.25)$$

This resulting double sum is unwieldy.[15] There are only $n$ independent variance terms in this sum with one for each random variable, but there are unfortunately $\frac{(n-1)\cdot n}{2}$ covariance terms. If, therefore, you have 25 securities in your portfolio and benchmark, which would be on the low side, you could expect to have 25 variance terms and a dismayingly large 300 covariance terms.

The double sum in Eq. (11.25) is inconvenient and, to make progress on a tractable risk computation, it must be replaced. Once again, matrix algebra comes to the rescue. The covariance matrix enormously simplifies the computation of the

[14] A simple way to remember, and conceptualize, Eq. (11.22), is to consider the square of the sum of two ordinary variables, $x_1$ and $x_2$,

$$(x_1 + x_2)^2 = x_1^2 + x_2^2 + 2x_1x_2. \quad (11.23)$$

This quadratic equation is the deterministic equivalent of variance. As is the case with Eq. (11.22), each variable $x_1$ and $x_2$ contributes independently to the result of Eq. (11.23), while an interaction term, $2x_1x_2$, is included. Covariance is the stochastic analogue of this interaction term.

[15] Recall that the covariance of a random variable with itself is merely its variance. That is,

$$\text{cov}(X_i, X_i) = \text{var}(X_i), \quad (11.26)$$

for all choices of $i$.

weighted variance of an arbitrary number of random variables. For *three* random variables $X_1$, $X_2$, and $X_3$, the covariance matrix is defined as,

$$\Omega_{X_1,X_2,X_3} = \begin{bmatrix} \text{var}(X_1) & \text{cov}(X_1, X_2) & \text{cov}(X_1, X_3) \\ \text{cov}(X_1, X_2) & \text{var}(X_2) & \text{cov}(X_2, X_3) \\ \text{cov}(X_1, X_3) & \text{cov}(X_2, X_3) & \text{var}(X_3) \end{bmatrix}. \tag{11.27}$$

This matrix holds, in a organized manner, all of the variance and covariance information regarding our three random variables.[16] The elements along the main diagonal are the positive variance terms, while the off-diagonal elements in the matrix are the covariance terms. These off-diagonal terms may be either positive or negative. The covariance matrix is also, by construction, symmetric.[17] This technical feature turns out to be advantageous when we seek to use the covariance matrix for the simulation of future security returns and risk factors.

Using $\Omega_{X_1,X_2,X_3}$ we can compute the weighted variance of the three random variables with weights, $\{\omega_i, i = 1, .., 3\}$ by

$$\begin{aligned} \text{var}\left(\sum_{i=1}^{3} \omega_i X_i\right) &= \underbrace{\begin{bmatrix}\omega_1 & \omega_2 & \omega_3\end{bmatrix}}_{\omega} \underbrace{\begin{bmatrix} \text{var}(X_1) & \text{cov}(X_1, X_2) & \text{cov}(X_1, X_3) \\ \text{cov}(X_1, X_2) & \text{var}(X_2) & \text{cov}(X_2, X_3) \\ \text{cov}(X_1, X_3) & \text{cov}(X_2, X_3) & \text{var}(X_3) \end{bmatrix}}_{\Omega_{X_1,X_2,X_3}} \underbrace{\begin{bmatrix}\omega_1 \\ \omega_2 \\ \omega_3\end{bmatrix}}_{\omega}, \\ &= \omega^T \Omega_{X_1,X_2,X_3} \omega. \end{aligned} \tag{11.28}$$

This clearly shows the quadratic form of the variance function and generalizes easily to an $n$-dimensional setting. With $n$ random variables, $\{X_i, i = 1, \ldots, n\}$ and $n$ corresponding weights summarized in the weighting vector,

$$\omega^T = \begin{bmatrix}\omega_1 & \ldots & \omega_n\end{bmatrix} \tag{11.29}$$

we have

$$\text{var}\left(\sum_{i=1}^{n} \omega_i X_i\right) = \omega^T \begin{bmatrix} \text{var}(X_1) & \cdots & \text{cov}(X_1, X_n) \\ \vdots & \ddots & \vdots \\ \text{cov}(X_1, X_n) & \cdots & \text{var}(X_n) \end{bmatrix} \omega. \tag{11.30}$$

This elegant matrix representation of variance dramatically simplifies our computations.

[16]It is rather more parsimonious, and common, to show two-variable examples of a covariance matrix. In our view, however, the two variable example is somewhat overly simple and does not provide an adequate feeling of its form.

[17]This means the elements on either side of the diagonal are the mirror reflection of one another. Technically, a matrix $S$ is symmetric if $S = S^T$.

### 11.3.2 Linking Covariance and Correlation

The covariance matrix nevertheless has an important disadvantage. The variance and covariance terms are squared units of the underlying random variables, which renders the values difficult, if not impossible, to interpret. There is thankfully a decomposition of the covariance matrix that can be helpful in understanding the individual entries. In particular, one can write any covariance matrix as,

$$
\begin{aligned}
\Omega &= \begin{bmatrix} \mathrm{var}(X_1) & \mathrm{cov}(X_1, X_2) & \mathrm{cov}(X_1, X_3) \\ \mathrm{cov}(X_1, X_2) & \mathrm{var}(X_2) & \mathrm{cov}(X_2, X_3) \\ \mathrm{cov}(X_1, X_3) & \mathrm{cov}(X_2, X_3) & \mathrm{var}(X_3) \end{bmatrix}, \\
&= \begin{bmatrix} \sigma^2(X_1) & \sigma(X_1)\sigma(X_2)\rho(X_1, X_2) & \sigma(X_1)\sigma(X_3)\rho(X_1, X_3) \\ \sigma(X_1)\sigma(X_2)\rho(X_1, X_2) & \sigma^2(X_2) & \sigma(X_2)\sigma(X_3)\rho(X_2, X_3) \\ \sigma(X_1)\sigma(X_3)\rho(X_1, X_3) & \sigma(X_2)\sigma(X_3)\rho(X_2, X_3) & \sigma^2(X_3) \end{bmatrix},
\end{aligned}
\tag{11.31}
$$

where $\rho(X_i, X_j)$ denotes the correlation coefficient between the random variables, $X_i$ and $X_j$.[18]

At first this does *not* seem to dramatically improve the situation, but with this structure, our covariance matrix can be further decomposed as follows,

$$
\begin{aligned}
\Omega &= \underbrace{\begin{bmatrix} \sigma(X_1) & 0 & 0 \\ 0 & \sigma(X_2) & 0 \\ 0 & 0 & \sigma(X_3) \end{bmatrix}}_{\substack{\text{Volatility} \\ \text{matrix, } V}} \underbrace{\begin{bmatrix} 1 & \rho(X_1, X_2) & \rho(X_1, X_3) \\ \rho(X_1, X_2) & 1 & \rho(X_2, X_3) \\ \rho(X_1, X_3) & \rho(X_2, X_3) & 1 \end{bmatrix}}_{\substack{\text{Correlation} \\ \text{matrix, } C}} \underbrace{\begin{bmatrix} \sigma(X_1) & 0 & 0 \\ 0 & \sigma(X_2) & 0 \\ 0 & 0 & \sigma(X_3) \end{bmatrix}}_{\substack{\text{Volatility} \\ \text{matrix, } V}}, \\
&= VCV.
\end{aligned}
\tag{11.33}
$$

This is progress. In Eq. (11.33), we have broken down our covariance matrix into the product of two matrices: a diagonal volatility matrix and correlation matrix.

Although the correlation matrix summarizes the same covariance information between our random variables, it permits a much greater ease of interpretation.

[18] We have merely used the identity,

$$
\rho(X, Y) = \frac{\mathrm{cov}(X, Y)}{\sigma(X)\sigma(Y)},
\tag{11.32}
$$

to replace the covariance terms with the product of the correlation coefficient and the volatilities of the random variables.

**Table 11.4** A correlation matrix

| Tenor | 6M | 2Y | 5Y | 7Y | 10Y | 15Y | 20Y | 30Y |
|---|---|---|---|---|---|---|---|---|
| 6M | 1.00 | 0.88 | 0.75 | 0.66 | 0.59 | 0.47 | 0.46 | 0.47 |
| 2Y | 0.88 | 1.00 | 0.93 | 0.86 | 0.80 | 0.72 | 0.69 | 0.65 |
| 5Y | 0.75 | 0.93 | 1.00 | 0.97 | 0.92 | 0.85 | 0.84 | 0.81 |
| 7Y | 0.66 | 0.86 | 0.97 | 1.00 | 0.97 | 0.92 | 0.91 | 0.87 |
| 10Y | 0.59 | 0.80 | 0.92 | 0.97 | 1.00 | 0.95 | 0.94 | 0.89 |
| 15Y | 0.47 | 0.72 | 0.85 | 0.92 | 0.95 | 1.00 | 0.99 | 0.94 |
| 20Y | 0.46 | 0.69 | 0.84 | 0.91 | 0.94 | 0.99 | 1.00 | 0.96 |
| 30Y | 0.47 | 0.65 | 0.81 | 0.87 | 0.89 | 0.94 | 0.96 | 1.00 |

This table summarizes the correlation structure of the UST yield curves as illustrated in Fig. 11.3.

Returning to our specific example, Table 11.4 summarizes the actual correlation matrix for the historical UST yield-curve data described in Figs. 11.3 and 11.4.[19]

A correlation coefficient can only take values ranging from −1 to 1. A value of unity implies that the two random variables move perfectly in unison. All of the diagonal values in the correlation matrix are thus equal to one; these points represent the correlation of each random variable with itself. The off-diagonal elements represent the correlation coefficients between the random variables. In this way, the dispersion (i.e., variance) and dependence (i.e., covariance) elements of our covariance matrix are conveniently separated into *two* distinct parts.

For the changes in UST yields in our dataset, the correlation between adjacent points on the yield curve is typically quite high. The correlation between 2- and 5-year yield movements, for example, is 0.93 implying that movements in this part of the yield curve are very highly correlated. Movements in more distant yield points, conversely, appear to be somewhat less highly correlated although all values in this correlation matrix are positive. Correlation between 6-month and 30-year yield movements, for example, is still 0.47. This relatively strong positive correlation with particularly high correlation among adjacent yield-curve sectors is a common feature among yield curves in virtually all currencies.

Although the correlation matrix is certainly an improvement over the virtually incomprehensible entries in the covariance matrix, we can actually take our analysis of the correlation matrix a step further. Figure 11.7 uses what is termed a contour plot to graphically describe the correlation matrix summarized in Table 11.4.

The contour plot in Fig. 11.7 permits one, at a glance, to understand the correlation between one's risk factors. It does, however, require a bit of practice. The colours describe the degree of correlation between, in our specific case, yield changes for eight different nodes in the UST curve. The warmer, red and orange, colours indicate relatively high correlation in the range of 0.8–1. The cooler colours, yellow and green, describe somewhat lower correlation in area of 0.65–0.80. Finally,

[19]More specifically, it summarizes the correlation structure of the monthly differences of eight key points in the UST yield curve.

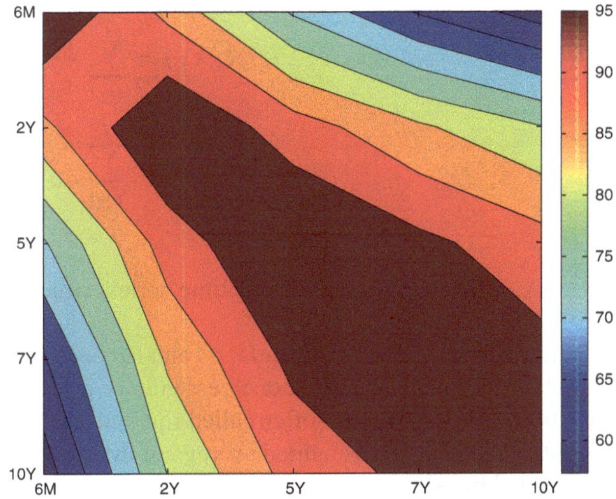


**Fig. 11.7** A correlation matrix, graphically. This table provides a heat map, or contour plot, of the correlation matrix in Table 11.4. The warmer the colour, the larger the correlation

correlation coefficients of less than about 0.65 are described by cold colours represented with various shades of blue.[20] For small numbers of risk factors, this may appear somewhat excessive. For larger numbers of risk factors, however, it may prove useful for analysts and be particularly helpful in communicating the correlation structure of one's risk factors to senior management.

### 11.3.3 Classic and Alternative Estimators of $\Omega_R$

Let us now turn our attention to the computation of the individual entries in the covariance matrix. Again this is conceptually straightforward, but a bit challenging in practice. This is because the data used to populate these entries stems from historical time series. Standard computations of variance and covariance make assumptions about the relative importance of each individual observation. Alternative assumptions are not only possible but, in many cases, desirable. A closer look is therefore warranted.

To do this, we need to carefully return to the definition of variance. Previously, we discussed $n$ distinct random variables, $X_1, \ldots, X_n$. For a time series, since we introduce the time dimension, we consider $T$ separate realizations of a single random variable, $X$, defined as $\{X_t, t = 1, \ldots, T\}$. Each observation is indexed to a separate point in time. With this in mind, the classical variance estimator for our time series is given as,

[20]Given the association of the colours with temperatures, the graphic in Fig. 11.7 is often also termed a *heat map*.

$$\tilde{\text{var}}\,(X\,|\,X_1,\ldots,X_T) = \sum_{t=1}^{T} \frac{\left(X_t - \bar{X}\right)^2}{T-1}, \\ = \sum_{t=1}^{T} \underbrace{\theta_t}_{\frac{1}{T-1}} \underbrace{X_t^2}_{\substack{\text{Squared}\\ \text{error}}}, \tag{11.34}$$

where $\bar{X}$ is the mean value of our time series, which we assume to be identically zero.[21]

While this is a familiar formula, it can also be seen from a different perspective. The classical variance estimator is essentially a weighted average of the squared deviations from the mean—often called squared errors. Moreover, and this is the key insight, it is an *equally* weighted average of these squared errors—each observation is weighted by $\frac{1}{T-1}$.

This equal-weight assumption explicitly assumes that each realization of $X$ is identically and independently distributed.[22] This may not always be the best approach. This is particularly true with financial time series where the distributional parameters often vary over time. As such, more recent observations may be more important in estimating the *current* variance and covariance of one's risk factors than observations from several years ago.

Figure 11.8 provides some motivation for this idea. It describes the evolution over our sample period of the credit spread between 10-year Austrian and German government bonds. Changes in this credit spread do not appear to be constant over time. There appear to be periods of relative calm interrupted by episodes of increased uncertainty with spreads exhibiting larger upward and downward movements—this phenomenon is termed volatility clustering. One may thus wish to place greater weight on more recent observations to improve one's assessment of the *current* uncertainty in one's risk factors.

An entire literature in the field of econometrics attempts to capture the time-varying nature of the financial-variable volatility. Well-known models such as ARCH and GARCH are examples of some of the approaches proposed in this literature.[23] While these are flexible and sensible models of time-varying variance, they are computationally involved and require a substantial amount of experience for their efficient implementation. There is, however, a simpler alternative method that provides a quite reasonable approach to these more complex econometric

[21]Our notation $\tilde{\text{var}}\,(X\,|\,X_1,\ldots,X_T)$ implies an estimate of the var$(X)$ conditional upon the data $X_1,\ldots,X_T$. It is unconditional with respect to time.

[22]This is often referred to as the *i.i.d.* assumption.

[23]Autoregressive conditional heteroskedasticity (ARCH) and generalized ARCH (GARCH) models are basically autoregressive models operating on the squared-error term—there are many different variations of these approaches of varying degrees of complexity. For more detailed information on these models, a good start is Hamilton [7] or Engle [5].

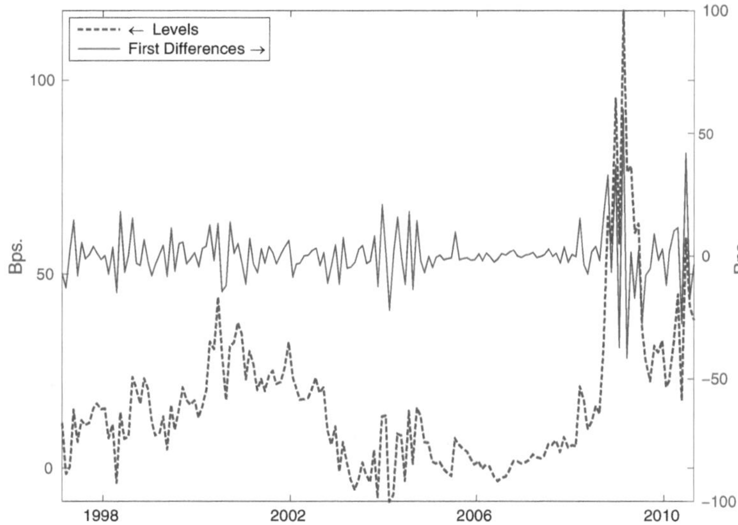


**Fig. 11.8** Volatility clustering. The underlying graph shows the evolution of the spread of 10-year Austrian government bonds to 10-year German Bunds. Note the distinct periods of heightened volatility; this is known as volatility clustering

methods.[24] The principal idea behind this approach is to carefully adjust the weights in Eq. (11.34) to increase the importance of more recent observations.

To implement Exponential weighting, one replaces the equal weights $\left(\frac{1}{T-1}\right)$ with a more general weight, $\theta_t$, as follows

$$
\begin{aligned}
\tilde{\text{var}}\left(X_{\mathbf{T+1}} \mid X_1, \ldots, X_T\right) &= \sum_{t=1}^{T} \theta_t X_t^2, \\
&= \sum_{t=0}^{T-1} \underbrace{(1-\lambda)\lambda^t}_{\theta_t} \underbrace{X_{T-t}^2}_{\text{Squared error}}, \qquad (11.35)
\end{aligned}
$$

where $\lambda \in (0, 1)$ is termed the smoothing parameter. Exponential weighting Observe that this is a conditional variance estimator—it is *conditional* on when the realization occurs over time.[25]

The conditional covariance estimator of two random variables $X_1$ and $X_2$ with time series data, $\{X_{1,t}, t = 1, \ldots, T\}$ and $\{X_{2,t}, t = 1, \ldots, T\}$, computed using

---

[24]This clever approach was first suggested by RiskMetrics™[9].

[25]The notation has slightly changed. $\tilde{\text{var}}\left(X_{T+1} \mid X_1, \ldots, X_T\right)$ is still an estimate of var$(X)$ conditional on the data $X_1, \ldots, X_T$. The difference is that it is an estimate of the variance for the next period, $T + 1$.

exponential weighting is correspondingly given as,

$$\begin{aligned}\tilde{\text{cov}}\left(X_{1,T+1}, X_{2,T+1} \mid X_{i,1}, \ldots, X_{i,T} \text{ for } i = 1,2\right) &= \sum_{t=1}^{T} \underbrace{\theta_t}_{\frac{1}{T-2}} \underbrace{X_{1,t} \cdot X_{2,t}}_{\substack{\text{Interaction}\\\text{term}}},\\ &= \sum_{t=0}^{T-1} \underbrace{(1-\lambda)\lambda^t}_{\omega_t} X_{1,T-t} \cdot X_{2,T-t},\end{aligned} \tag{11.36}$$

where, based upon our previous assumption, the mean of the two random variables is assumed to be zero.

Are these sensible weights? If we sum them, we arrive at a geometric sum that can be dramatically simplified.

$$\begin{aligned}(1-\lambda) \underbrace{\sum_{t=0}^{T-1} \lambda^t}_{\substack{\text{Geometric}\\\text{series}}} &= \cancel{(1-\lambda)} \left(\frac{1-\lambda^T}{\cancel{1-\lambda}}\right),\\ &= 1-\lambda^T.\end{aligned} \tag{11.37}$$

For permissible values of the smoothing parameter, $\lambda \in (0,1)$, we have a geometric series. For relatively large values of $T$, the weights are approximately equal to unity. If one has 120 months of data and $\lambda = 0.95$, then the sum of the weights is equal to $1 - 0.95^{120}$, which is equal to 0.9979. As the sample size tends to infinity, the sum of the weights tends to one.[26]

Let us now examine the implications of different choices of $\lambda$. Figure 11.9 highlights the relative weight on each of the observations in our data sample associated with the standard variance computation (i.e., equal weights) and exponential weighting associated with *three* different possible choices for the parameter, $\lambda$. As the value of $\lambda$ approaches unity, the weighting function approaches equal weights.

[26]This is easy to see by starting with Eq. (11.37), increasing the sample size to infinity, and recalling the properties of geometric series,

$$\begin{aligned}(1-\lambda) \underbrace{\sum_{t=0}^{\infty} \lambda^t}_{\substack{\text{Geometric}\\\text{series}}} &= (1-\lambda) \left(\frac{1}{1-\lambda}\right),\\ &= 1.\end{aligned}$$

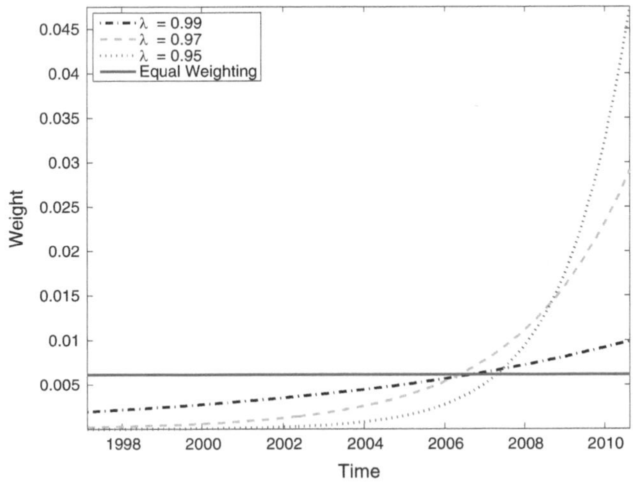


**Fig. 11.9** Weighting functions. This figure demonstrates the impact of different values for the weighting parameter, $\lambda$, in an exponential weighting scheme for one's covariance matrix

When $\lambda$ is set to a value of 0.95 with monthly data, there is virtually no weight at all on the first half of the sample and quite substantial weights on observations occurring over the last year.

The basic idea behind the use of exponential weighting is that the amount of importance placed upon each observation declines exponentially as we move from the current point in time.[27] The underlying shaded box provides, for the interested reader, a bit more insight into the derivation of the exponential-weighting formula.[28]

Equation (11.35) provides the formula for the exponentially weighted moving average (EWMA), but it does not provide any insight into its origins. The formal definition of an EWMA is as follows,

$$A_t = (1-\lambda)X_t + \lambda A_{t-1}, \tag{11.38}$$

where $A_t$ is the EWMA at time $t$. It is a recursively defined convex combination of the current observation and the EWMA from the previous point in time, $t-1$. At first glance, this does not overly resemble the expression

(continued)

[27]For this reason, $\lambda$ is also termed the *decay factor*.

[28]This description of exponential weighting is also something of an approximation, which is conceptually quite useful. The next chapter provides a few small, but necessary, precisions.

in Eq. (11.35). If we start from $A_1$ and assume that $A_0 = (1-\lambda)X_0$, we can work forward to derive the formula in Eq. (11.38) as,

$$\begin{aligned} A_1 &= (1-\lambda)X_1 + \lambda A_0, \\ &= (1-\lambda)X_1 + \lambda(1-\lambda)X_0. \end{aligned} \tag{11.39}$$

Using this expression and moving on to $A_2$, we have

$$\begin{aligned} A_2 &= (1-\lambda)X_2 + \lambda A_1, \\ &= (1-\lambda)X_2 + \lambda\left(\underbrace{(1-\lambda)X_1 + \lambda(1-\lambda)X_0}_{\text{Equation (11.39)}}\right), \\ &= (1-\lambda)X_2 + (1-\lambda)\lambda X_1 + (1-\lambda)\lambda^2 X_0, \\ &= (1-\lambda)\left(\underbrace{\lambda^0}_{=1} X_2 + \lambda^1 X_2 + \lambda^2 X_0\right), \\ &= (1-\lambda)\sum_{t=0}^{2} \lambda^t X_{3-t}. \end{aligned} \tag{11.40}$$

Thus, it is not difficult to generalize this formula to $T$ observations as,

$$A_T = (1-\lambda)\sum_{t=0}^{T-1} \lambda^t X_{T-t}. \tag{11.41}$$

Thus, we see clearly how the definition in Eq. (11.39) permits us to derive the weighting formula in Eq. (11.35). The identical logic applies to the computation of covariance.

Is such an approach really necessary? The answer: it depends.[29] In Fig. 11.10, we observe the evolution of the 10-year German Bund yield. Monthly changes over this period appear to be centred around zero and rarely exceed an increase or decrease of 40 basis points in any given month. The squared errors also appear to be relatively flat throughout the period, albeit with a slight increase in volatility towards the end of the sample. The bottom graphic in Fig. 11.10 overlays the exponential weighting function, with a decay factor of $\lambda = 0.95$, onto the squared deviations from the

[29] An interesting comparison is found in Alexander and Leigh [2].

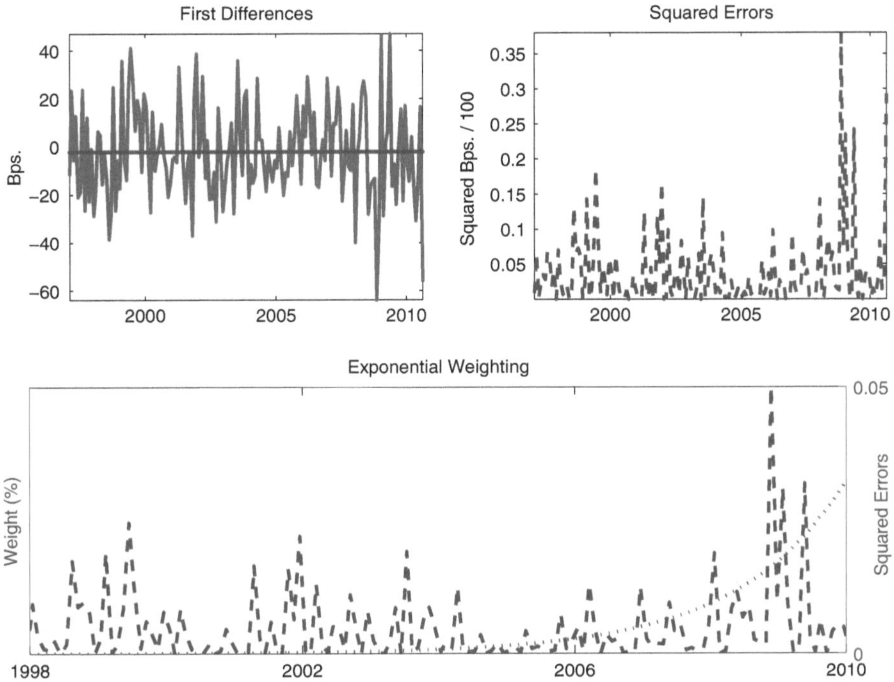


**Fig. 11.10** An almost I.I.D. example. This figure highlights the first difference of 10-year Bund yields, their associated squared errors and the weights associated with an exponentially weighted scheme

mean. The exponentially weighted approach clearly places a greater emphasis on the more recent, volatile, period. What are the results? The monthly volatility, estimated using the standard equal-weighted approach in Eq. (11.33), is 19 basis points. Using the weighting function shown in Fig. 11.10 one arrives at a value of 23 basis points. In this case, it appears that 10-year German Bund yield changes exhibit fairly stable levels of volatility over our sample period.

Figure 11.11 provides a counterexample. It outlines the credit spread between 10-year Greek and German government bond yields. For much of the period, this credit spread was less than 100 basis points with modest volatility. In 2009 and 2010, however, the level of the spread spiked dramatically with a corresponding increase in volatility. Computing the volatility using the standard equally weighted approach generates a monthly spread volatility of 53 basis points. Using an exponential weighting scheme, with $\lambda = 0.95$ as before, it more than doubles to 128 basis points. In this case, exponential weighting makes a significant difference.

For some risk factors, the choice of weighting scheme appears to make relatively little difference, whereas for others it generates important differences. It is, of course, up to the individual analyst to decide whether to use equal, exponential, or some more complicated form of weighting such as ARCH or GARCH for the

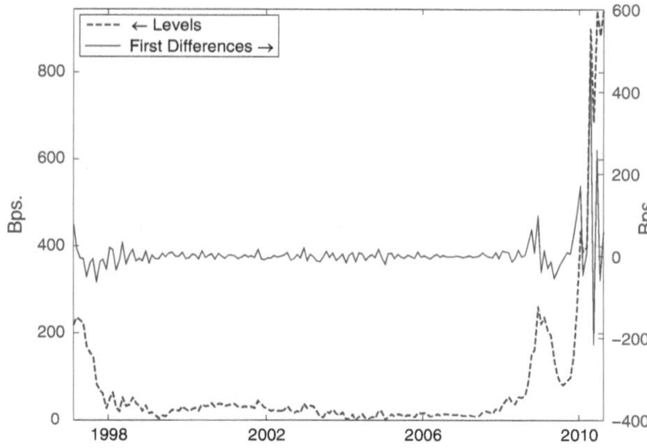


**Fig. 11.11** A non-I.I.D. example. This figure illustrates the squared errors and the exponential weights for 10-year Greek government bond yields over the period

computation of the individual elements in her covariance matrix. We highlighted the exponential weighting scheme in this section to raise the issue and provide a relatively straightforward alternative to the equal weighting inherent in standard computations. It will be considered in more detail—with a particular focus on its impact on our risk measures—in the next chapter.

### 11.3.4 Simulating Random Realizations

As a final point in this section, we seek to provide some insight into how one may simulate risk-factor outcomes from one's covariance matrix. There are many sophisticated ways to simulate risk factors. The approach that we will discuss is just one possibility and we make no claim that it is the best of many possible methods.[30] It is, however, relatively easy to perform and makes use of the covariance matrix that one must anyway construct for risk computations.

The idea is based on some standard properties of the Gaussian distribution. We begin with a single random variable $X$ that is normally distributed as follows,

$$X \sim \mathcal{N}\left(a, b^2\right), \tag{11.42}$$

where $a, b \in \mathbb{R}$. $X$ has a mean of $a$ and a volatility of $b$. We then require a second standard normally distributed random variable, $Z$, where[31]

$$Z \sim \mathcal{N}(0, 1) \tag{11.43}$$

[30]See Fishman [6] for an excellent general treatment of simulation methods.

[31]The so-called standard normal variate has a mean of zero and a variance of one.

We can always construct $X$ using $Z$ as well as the parameters of $X$—$a$ and $b$. The construction follows this simple formula,

$$X = a + \sqrt{b^2} Z. \tag{11.44}$$

Why is this useful? Equation (11.44) can be exploited to generate a large number of random outcomes for $X$. Generation of standard normal variates is straightforward with most software packages. One need only generate standard normal variates, $\{Z_i, i = 1, \ldots, n\}$ and then apply Eq. (11.44).

We want to make use of the relationship described in Eq. (11.44), but there is a problem. We are not dealing with a univariate random variable, but rather the joint distribution of an arbitrary number, say $n$ of normally distributed risk factors. Let us try to apply Eq. (11.44) where our random variable, $\vec{X}$, is now vector-valued as

$$\vec{X} \sim \mathcal{N}\left(\vec{a}, B\right), \tag{11.45}$$

where $\vec{a} \in \mathbb{R}^{n\times 1}$ and $B \in \mathbb{R}^{n\times n}$.[32] The required standard normal variable, $\vec{Z}$, is also vector-valued and is distributed as,

$$\vec{Z} \sim \mathcal{N}\left(\vec{0}, I\right). \tag{11.46}$$

For $\vec{Z}$ the mean is a vector of zeroes, whereas the covariance matrix is merely an identity matrix of appropriate dimension.

Naively applying Eq. (11.44) with the definitions in Eqs. (11.45) and (11.46) yields the following result,

$$\begin{bmatrix} X_1 \\ \vdots \\ X_n \end{bmatrix} ?=? \begin{bmatrix} a_1 \\ \vdots \\ a_n \end{bmatrix} + \sqrt{\begin{bmatrix} b_{1,1} & \cdots & b1,n \\ \vdots & \ddots & \vdots \\ bn,1 & \cdots & b_{n,n} \end{bmatrix}} \begin{bmatrix} Z_1 \\ \vdots \\ Z_n \end{bmatrix},$$
$$X ?=? a + \sqrt{B} Z. \tag{11.47}$$

This may appear to be a legitimate application of Eq. (11.44), but applying the square-root operator to a matrix is *not* generally possible. Indeed, it is not immediately clear what it would mean. Under certain conditions, however, one may perform a computation on a matrix that, in a number of respects, is the conceptually equivalent of the square root of a number.

[32]The mean of $\vec{X}$ is a vector of values, $\vec{a}$, and its variance is described by the covariance matrix, $B$.

The question is, what does it mean to compute,

$$\sqrt{\begin{bmatrix} b_{1,1} & \cdots & b1,n \\ \vdots & \ddots & \vdots \\ bn,1 & \cdots & b_{n,n} \end{bmatrix}}? \tag{11.48}$$

A good starting point would be to find a matrix, $B^*$ whose product with itself (i.e., something like $B^*B^*$) yields $B$. While this requires some rather advanced results in matrix algebra and a technique called the Cholesky decomposition, we will conceptualize the approach through the notion of the square root.[33] One cannot, for example, compute the square root of any number and expect to get a sensible number for use in one's risk calculations.[34] Similarly, a matrix must be positive for us to apply a pseudo square-root operator. Of course, the notion of positiveness is also somewhat more involved for matrices. A matrix must be, in the language of linear algebra, *positive definite*.[35] If the matrix is also symmetric and has all real-valued entries, then there exists a unique matrix, $U$, such that,

$$U^T U = B. \tag{11.50}$$

This $U$ is an upper-triangular matrix termed the Cholesky decomposition of $B$. A covariance matrix meets all three of these criteria—positive definiteness, symmetry, and real-valued entries—and, thus, such a matrix, $U$, exists for all covariance matrices.

We need only return to Eq. (11.47) and replace the vaguely defined and technically incorrect notion of *square root of B* with the Cholesky decomposition as,

$$X = a + U^T Z. \tag{11.51}$$

This may be employed, in a conceptually identical fashion to Eq. (11.44), to move from standard multivariate normal variates to *correlated* normal variables.

We may now return to our example and use this result to generate different instrument returns and risk-factor outcomes. We merely generate a collection of $n$

---

[33]For a more detailed description of the Cholesky decomposition, see Press et al. [10].

[34]The square root of any negative number, for example, has a solution in the complex domain. Complex numbers, while common in, say, electrical engineering, are quite a rarity in asset-management risk reports.

[35]A positive-definite matrix is the matrix analogue of a positive real number. Technically, a symmetric real-valued matrix $B \in \mathbb{R}^{n \times n}$ is positive definite if,

$$xBx^T > 0, \tag{11.49}$$

for all real-valued vectors $x \in \mathbb{R}^{1 \times n}$. That is, any quadratic function of $B$ generates a positive value. This is exactly the case for a covariance matrix as the variance of any combination of one's instruments must be a strictly positive number.

standard random variables, $\vec{Z}$, distributed according to Eq. (11.46). Then we define,

$$\tilde{R} = \vec{0} + U_{\Omega_R}^T \vec{Z}, \tag{11.52}$$

where $U_{\Omega_R}$ is the Cholesky of the covariance of the security-return matrix, $R$. The consequence of the calculation in Eq. (11.52) is that,[36]

$$\tilde{R} \sim \mathcal{N}\left(\vec{0}, \Omega_R\right) \tag{11.53}$$

$\tilde{R}$ is a simulated set of returns (or risk factors) from our covariance matrix, $\Omega$. One can easily generate an arbitrary number of simulated security return vectors by merely repeatedly drawing standard normal vectors, $\vec{Z}$, and applying Eq. (11.52).

We may also want to simulate different risk-factor outcomes. To do this, we need to recall a simple property of variance. If $a \in \mathbb{R}$ is a scalar and $X$ is a univariate random variable, then

$$\text{var}(aX) = a^2\text{var}(X). \tag{11.54}$$

If $\vec{a} \in \mathbb{R}^{n\times 1}$ and $\vec{X}$ is an $n$-dimensional random vector, then

$$\text{var}\left(\vec{a}\vec{X}\right) = a^T\text{var}(X)a. \tag{11.55}$$

This fact is extremely useful when one recalls that the security return matrix, $R$, is the product of the exposure matrix, $K$, and the market movement matrix, $M$. The exposure matrix is constant, whereas the market-movement matrix is a random variable. Thus,

$$\begin{aligned} \text{var}(R) &= \text{var}(KM), \\ &= K^T \underbrace{\text{var}(M)}_{\Omega_M} K, \end{aligned} \tag{11.56}$$

where $\text{var}(M) = \Omega_M$ is the covariance matrix of market movements. Equation (11.56) represents the link between the covariance matrix of security returns and the market risk factors. In short, the covariance of security returns, $\Omega_R$, and the covariance matrix of the market risk factors, $\Omega_M$, are linked by the exposure matrix, $K$.

---

[36]Recall that we assume a zero-expected return for the securities in the portfolio and benchmark, because we are looking at a small time step.

To simulate market risk-factor outcomes, we employ Eq. (11.56) and the previously described approach as,

$$\hat{M}_{\text{SIM}} = \vec{0} + U_{\tilde{\Omega}_M}^T \vec{Z}, \tag{11.57}$$

where $\tilde{\Omega}_M$ is the estimated covariance of the market-movement matrix and $U_{\tilde{\Omega}_M}$ is the associated Cholesky decomposition of this matrix. $\hat{M}_{\text{SIM}}$ is thus a collection of simulated market risk-factor realizations consistent with the estimated covariance matrix of $M$. The consequence of the calculation in Eq. (11.52) is that,

$$\hat{M}_{\text{SIM}} \sim \mathcal{N}\left(\vec{0}, \tilde{\Omega}_M\right). \tag{11.58}$$

This demonstrates how one may simulate both individual security returns and market-risk factor outcomes using the appropriate covariance matrix.

This was a substantial amount of mathematical effort. It is nonetheless a useful way to simulate risk-factor outcomes as it allows one to handle a large number of diverse risk factors. We may now examine the results for the input data used in our simple example. Figure 11.12 outlines the results of the use of Eq. (11.57) to simulate a variety of UST yield-curve outcomes. The upper graphic in Fig. 11.12 illustrates the observed monthly UST yield-curve changes over our sample period. The lower graphic, conversely, provides a view of 1,000 simulated UST yield-curve changes. The results are fairly similar as the simulation approach is based on the estimated covariance matrix of the underlying data. The simulation approach does not generate as many extreme values, however, since it is based on the assumption of normality of return and market-risk factor movements. The following shaded box illustrates another possible approach to simulating market risk factors using the ideas from Chap. 6.

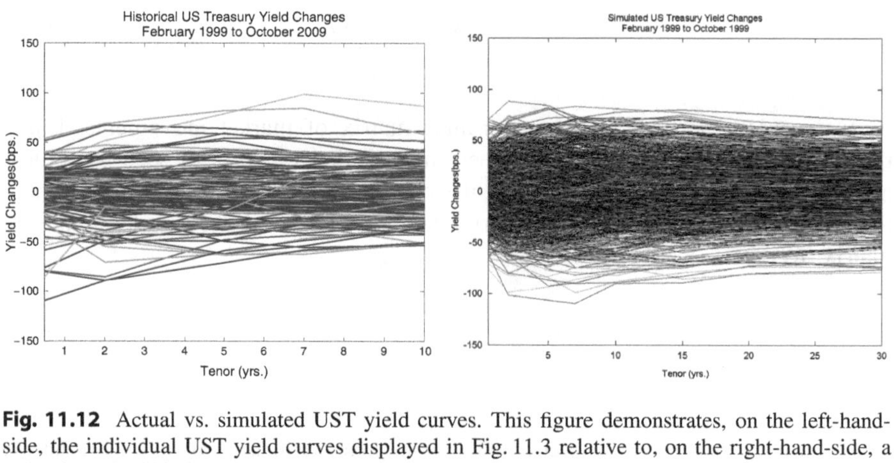


**Fig. 11.12** Actual vs. simulated UST yield curves. This figure demonstrates, on the left-hand-side, the individual UST yield curves displayed in Fig. 11.3 relative to, on the right-hand-side, a collection of 1,000 simulated curves use the Cholesky decomposition described in Eq. (11.57)

Simulation of future risk-factor outcomes are not restricted to the approach described in the preceding text. Indeed, a wide range of sensible simulation-based models exist. To give the reader a sense of what is possible, we will show how we might employ the techniques described in Chap. 6 to simulate future yield-curve outcomes. In particular, we will make use of the extended Nelson–Siegel model as suggested by Diebold and Li [4].

The first step is to estimate, using a reasonably sized dataset, the following VAR model for the Nelson–Siegel risk-factor dynamics,

$$\begin{bmatrix} a_t \\ b_t \\ c_t \end{bmatrix} = \begin{bmatrix} h_0 \\ h_1 \\ h_2 \end{bmatrix} + \begin{bmatrix} F_{1,1} & \cdots & F_{1,3} \\ \vdots & \ddots & \vdots \\ F_{3,1} & \cdots & F_{3,3} \end{bmatrix} \begin{bmatrix} a_{t-1} \\ b_{t-1} \\ c_{t-1} \end{bmatrix} + \begin{bmatrix} \epsilon_{a,t} \\ \epsilon_{b,t} \\ \epsilon_{c,t} \end{bmatrix},$$

$$X_t = H + FX_{t-1} + \epsilon_t. \tag{11.59}$$

where $a_t, b_t$, and $c_t$ are the level, slope, and curvature factors for $t = 0, \ldots, T$. The conditional factor distribution is Gaussian as follows,

$$X_t \mid X_{t-1} \sim \mathcal{N}\left(H + FX_{t-1}, \text{var}(\epsilon_t)\right). \tag{11.60}$$

To actually estimate it, we need to stack up our $T + 1$ observations,

$$\begin{bmatrix} a_1 & b_1 & c_1 \\ \vdots & \vdots & \vdots \\ a_T & b_T & c_T \end{bmatrix} = \begin{bmatrix} 1 & a_0 & b_0 & c_0 \\ \vdots & \vdots & \vdots & \vdots \\ 1 & a_{T-1} & b_{T-1} & c_{T-1} \end{bmatrix} \begin{bmatrix} h_1 & h_2 & h_3 \\ F_{1,1} & F_{1,2} & F_{1,3} \\ F_{2,1} & F_{2,2} & F_{2,3} \\ F_{3,1} & F_{3,2} & F_{3,3} \end{bmatrix} + \begin{bmatrix} \epsilon_{a,1} & \epsilon_{b,1} & \epsilon_{c,1} \\ \vdots & \vdots & \vdots \\ \epsilon_{a,T} & \epsilon_{b,T} & \epsilon_{c,T} \end{bmatrix},$$

$$\begin{bmatrix} X_1^T \\ \vdots \\ X_T^T \end{bmatrix} = \begin{bmatrix} 1 & X_0^T \\ \vdots & \vdots \\ 1 & X_{T-1}^T \end{bmatrix} \begin{bmatrix} H^T \\ F^T \end{bmatrix} + \begin{bmatrix} \epsilon_1^T \\ \vdots \\ \epsilon_T^T \end{bmatrix},$$

$$Y = X\beta + \epsilon.$$

One selects choice of parameters, $\hat{\beta}$, that try to make $\epsilon = Y - X\beta$ as small as possible. The solution is $\hat{\beta} = (X^T X)^{-1} X^T Y$. Most importantly, for our simulations, the estimated covariance matrix is

$$\boxed{\hat{\text{var}}(\epsilon_t)} = \hat{\Omega} = \frac{1}{n-k-1}\left(Y - X\hat{\beta}\right)^T \left(Y - X\hat{\beta}\right). \tag{11.61}$$

(continued)

Since, $\hat{\Omega}$ is a real-valued, positive-definite, symmetric matrix, it has a Cholesky decomposition of the form, $U^T U = \hat{\Omega}$.

Given that $X_{t+1} \sim \mathcal{N}(H + FX_t, \Omega)$ and defining $\nu_{t+1} \sim \mathcal{N}(0, I)$ as a three-dimensional standard normal variate, then

$$\tilde{\epsilon}_{t+1} = \vec{0} + U^T \nu_{t+1} = U^T \nu_{t+1}.$$

$\tilde{\epsilon}_t$ is thus a simulated set of multivariate normally distributed variables with a mean of zero and the covariance matrix, $\Omega$. With the previous results, we can easily simulate a set of vector outcomes,

$$X_{t+1,i} = H + FX_t + \underbrace{U^T \nu_{t+1,i}}_{\tilde{\epsilon}_{t+1,i}}.$$

for $i = 1, \ldots, N$ simulations. These $N$ simulated level, slope, and curvature factors are then transformed into yield-curve using the

$$\begin{aligned} y(t+1, T, i) &= a_{t+1,i} + b_{t+1,i}\left(\frac{1 - e^{-\lambda(T-t)}}{\lambda(T-t)}\right) \\ &+ c_{t+1,i}\left(\frac{1 - e^{-\lambda(T-t)}}{\lambda(T-t)} - e^{-\lambda(T-t)}\right), \end{aligned} \tag{11.62}$$

for $i = 1, \ldots, N$. This provides $N$ yield-curve simulations for time, $t+1$. The desired yield changes are extracted and put into the market-movement matrix, $\tilde{M}$, for one's risk computations. Conceptually, this is similar to the previously described approach, although in this setting, the predictive structure of the yield-curve model (i.e., non-zero mean) is exploited.

## 11.4 The Final Results

We are finally in a position to return to the risk formulae introduced at the beginning of this chapter. Recall that the tracking error is simply,

$$\begin{aligned} \text{Tracking error} = \sigma_a &= \sqrt{\text{var}(R_a)}, \\ &= \sqrt{\omega_a^T \underbrace{\text{var}(R)}_{\Omega_R} \omega_a}, \end{aligned} \tag{11.63}$$

which may be expanded to,

$$\text{Tracking error} = \sqrt{(\omega_p - \omega_b)^T K^T \underbrace{\text{var}(M)}_{\Omega_M} K(\omega_p - \omega_b)}. \quad (11.64)$$

The tracking error is a function of the uncertainty regarding security returns and the active portfolio weights.[37] Each of these individual components should now be very familiar. The covariance matrix takes into account the portfolio exposures and the market movements over our data sample, while the active weights are merely the difference between our portfolio positions and the portfolio's benchmark.

Figure 11.13 demonstrates the active portfolio returns using *historical* market movements. Each graph is computed by transforming the raw security returns, in $R$, with the portfolio weight vector, $\omega_a$, into the portfolio returns—specifically, $\omega_a R$. The second portfolio exhibits the largest dispersion in active returns with observations ranging from ±400 basis points. The first portfolio demonstrates about half the range of active returns, while the third portfolio appears to have the closest returns to the benchmark. None of these empirical active-return distributions appears to be strictly Gaussian.

Figure 11.14 demonstrates the active portfolio returns using *simulated* market movements. Raw simulated security returns, in $\tilde{R}$, are transformed with the portfolio weight vector, $\omega_a$, into the portfolio returns. The relative dispersion of the *three* sample portfolios has not changed, but given the much larger number of simulated observations, the distributions are significantly more filled out than in Fig. 11.13. Moreover, since the very computation of the simulated returns assumes normality, it should be no surprise that the active-return distributions appear almost perfectly normally distributed.

Table 11.5 outlines, for our simple example, the actual tracking error and VaR for each of our three sample portfolios. As hinted by Figs. 11.13 and 11.14, the second portfolio exhibits the largest ex-ante tracking error with a value of almost 150 basis points. The simulated value is slightly higher than the computation using the historical data—this is because the tails of the distribution have been filled out

[37]To be very precise, given that we do *not* know the covariance of market movements, but instead have to estimate it, we should write Eq. (11.63) as,

$$\begin{aligned} \hat{\sigma}_a &\approx \sqrt{\omega_a^T \underbrace{\tilde{\text{var}}(R)}_{\tilde{\Omega}_R} \omega_a}, \\ &= \sqrt{(\omega_p - \omega_b)^T K^T \underbrace{\tilde{\text{var}}(M)}_{\tilde{\Omega}_M} K(\omega_p - \omega_b)}. \end{aligned} \quad (11.65)$$

To keep the notational burden to a minimum, we will suppress the estimated values. The reader, however, should always be aware that we are working with covariance estimators and that our tracking-error and VaR estimates are only as good as these estimates.

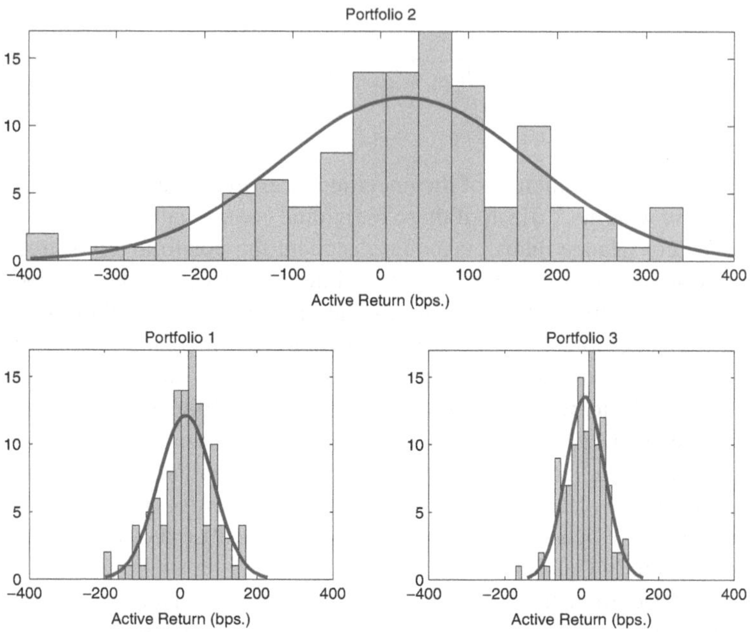


**Fig. 11.13** Empirical portfolio return distributions. This figure outlines histograms for the three portfolios described in Table 11.2 using the historical yields curves described in Fig. 11.12

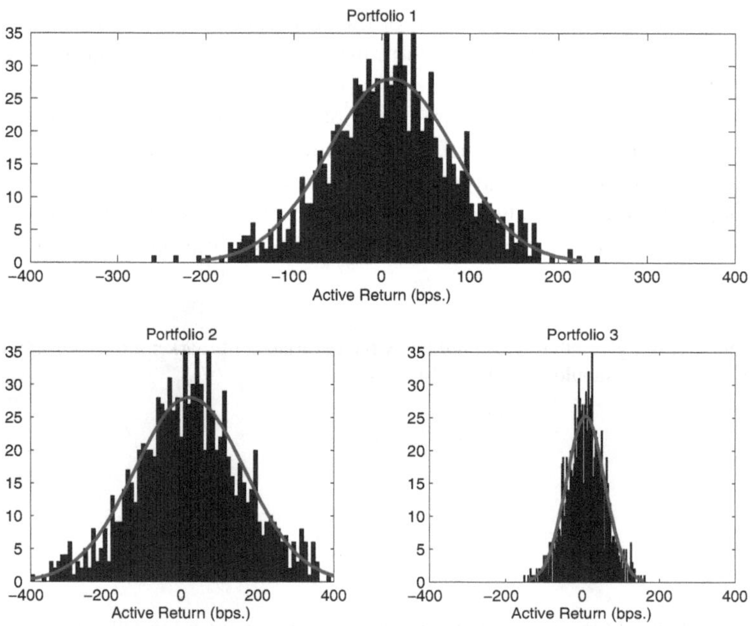


**Fig. 11.14** Simulated portfolio return distributions. This figure outlines histograms for the three portfolios described in Table 11.2 using the simulated yields curves described in Fig. 11.12

**Table 11.5** Ex-ante tracking error and value-at-risk

| Statistic | Benchmark ($\omega_b$) | Portfolio ($w_p$) 1 | 2 | 3 |
|---|---|---|---|---|
| US912828JU50 (2Y) | 0.50 | 0.25 | 0.00 | 0.40 |
| US912828DC17 (5Y) | 0.00 | 0.25 | 0.50 | 0.00 |
| US912810ED64 (10Y) | 0.50 | 0.50 | 0.50 | 0.60 |
| Tracking Error $\left(\sqrt{\omega_a^T \Omega_R \omega_a}\right)$ | | | | |
| Simulated | n/a | 73.7 | 147.4 | 52.5 |
| Historical | n/a | 71.1 | 142.1 | 50.2 |
| VaR $\left(\mathcal{N}^{-1}(0.95)\sqrt{\omega_p^T \Omega_R \omega_p}\right)$ | | | | |
| Simulated | n/a | 848.1 | 964.8 | 815.0 |
| Historical | n/a | 805.9 | 918.3 | 773.2 |

This table outlines the results of the historical and simulated ex-ante tracking error and VaR figures for our simple example.

somewhat by the larger number of distributions.[38] The second portfolio, with its concentration in the 5- and 10-year sectors, also has the highest VaR value at more than 9 %. Again, the simulated VaR is slightly larger than its historical counterpart.

The third portfolio, which quite closely matches the benchmark, has the smallest tracking error at about 50 basis points. It also has the lowest absolute risk, as measured by a 95 % VaR of around 8 %. This result is due to the highest concentration among all three portfolios in the 2-year sector—the instrument with the lowest level of risk. The first portfolio lies between the second and third portfolios in terms of both relative and absolute risk with a tracking error and 95 % VaR of around 75 basis points and 8.5 %, respectively. In summary, while they require significant data and a number of assumptions, the ultimate risk computations are not that complex.

## 11.5 Attributing Risk

We have invested substantial time and effort in approximating the risk of our three sample portfolios. The figures in Table 11.5 are quite helpful in determining how much risk a given portfolio is taking relative to its benchmark or on an absolute basis. We can make useful statements about how one portfolio is riskier, or less risky, than another. This analysis is silent, however, with respect to the source of this risk. Analogous to the performance setting, total risk is like total return. It is useful, but we want more. It would be enormously helpful, for example, to be able

[38]This is not a general result, but instead stems from the fact that the empirical distributions are, in this particular case, not extremely far from the normal distribution. See Abken [1] for more detail on assessing VaR simulation approaches.

to determine how much of the portfolio's relative or absolute risk stems from an investment, or failure to invest, in a given instrument or a particular risk factor. What we really desire is a decomposition of our risk measures by individual risk factor.

Such a decomposition is possible in this framework. It requires, nevertheless, further investment in some mathematical structure. It relies on the use of some basic results in matrix algebra and a mathematical property of our risk measures. We begin by computing the sensitivity of our risk measure to a small change in the portfolio weights. We use the tracking error computation as our example, although the computations are basically identical for the portfolio's variance, or VaR for that matter.

Such a sensitivity requires, of course, the computation of a partial derivative. Specifically, we seek to compute the partial derivative of the tracking error with respect to its active weights or $\frac{\partial \sigma_a}{\partial \omega_a}$. Direct computation, however, is a bit messy because of the square-root term. To circumvent this problem, we employ the chain rule to the ex-ante tracking *variance* and re-arrange as follows,

$$\begin{aligned}\frac{\partial \sigma_a^2}{\partial \omega_a} &= \underbrace{2\sigma_a \frac{\partial \sigma_a}{\partial \omega_a}}_{\substack{\text{By the}\\ \text{chain rule}}},\\ \frac{\partial \sigma_a}{\partial \omega_a} &= \frac{1}{2\sigma_a}\frac{\partial \sigma_a^2}{\partial \omega_a}.\end{aligned} \tag{11.66}$$

This is essentially a trick. The result is that we now have isolated the partial derivative of tracking error in terms of the tracking variance, which, due to its quadratic form, is significantly easier to differentiate.

Equation (11.66) is quickly resolved as,

$$\begin{aligned}\frac{\partial \sigma_a}{\partial \omega_a} &= \frac{1}{2\sigma_a}\frac{\partial \left(\omega_a^T \Omega \omega_a\right)}{\partial \omega_a},\\ &= \frac{1}{2\sigma_a} 2\Omega w_a,\\ &= \sigma_a^{-1}\Omega w_a.\end{aligned} \tag{11.67}$$

This result, which is itself an $n$-dimensional vector with one partial derivative for each of the active weights in $\omega_a$, is termed the *marginal security* tracking-error vector. The $i$th element of this vector is the contribution to tracking error of a small increase in the weight of the $i$th security.

This is a nice concept, but the marginal tracking error is a bit difficult to interpret. The sum of the marginal tracking errors, for example, has no real meaning since one cannot increase the weight on one security without simultaneously reducing the weight for another. Moreover, the sum of the marginal tracking errors has no clear relationship to the total tracking error. Finally, the marginal tracking error may be either negative or positive.[39]

While the marginal security tracking error tells us the tracking error's sensitivity to a change in active weights, we are looking for the contribution of a given security to the tracking error. This is easily resolved. If we multiply the marginal security tracking error with the portfolio weights, we have

$$\omega_a^T \frac{\partial \sigma_a}{\partial \omega_a} = \sum_{i=1}^{n} \omega_{a,i} \left. \frac{\partial \sigma_a}{\partial \omega_a} \right|_i , \tag{11.68}$$

where each individual term $\omega_{a,i} \left. \frac{\partial \sigma_a}{\partial \omega_a} \right|_i$ is the contribution of the $i$th security to the tracking error. The sum of these contributions is equal to the tracking error. This is not immediately obvious, but is shown easily as,

$$\begin{aligned} \omega_a^T \frac{\partial \sigma_a}{\partial \omega_a} &= \omega_a^T \underbrace{\sigma_a^{-1} \Omega w_a}_{\substack{\text{Equation} \\ (11.67)}}, \\ &= \frac{\overbrace{\omega_a^T \Omega w_a}^{\sigma_a^2}}{\sigma_a}, \\ &= \sigma_a. \end{aligned} \tag{11.69}$$

The product of each active weight with the marginal security tracking error, therefore, describes the contribution of each security to the overall active risk.

Let us define the tracking error contribution as,

$$\text{Tracking-error contribution of } i\text{th security} = \omega_{a,i} \left. \frac{\partial \sigma_a}{\partial \omega_a} \right|_i , \tag{11.70}$$

for $i = 1, \ldots, n$. This is precisely what we were trying to achieve: a security by security breakdown of the tracking error. Again, this value can be either positive or negative. In some cases, adding an instrument to one's portfolio adds risk by bringing it further from the benchmark, whereas in other cases, such an addition decreases risk by approaching one's portfolio to the benchmark.[40]

[39]This is, however, quite natural. Changes in some positions will increase risk, while changes in others will act to reduce it.

[40]Since both active weights and marginal security tracking error may both be either negative or positive, there is a fairly rich range of possible outcomes.

We have been fairly relaxed about the derivation of this result. It turns out that it can be established more formally. The interested reader is directed to the underlying shaded box for a more detailed, although still not completely rigorous, discussion.

The curious reader may be wondering about the robustness of the result in Eqs. (11.68)–(11.70). How can one be certain that the tracking error can be written as the sum of active weights and the partial derivatives with respect to these weights? The robustness of the result stems from a mathematical result: Euler's theorem for homogeneous functions. It—presented here very briefly without proof—holds that:

**Theorem 11.1** *A continuous, differentiable, kth degree homogeneous function* $f : \mathbb{R}^n_+ \rightarrow \mathbb{R}$ *has the property,*

$$f(\lambda x) = \lambda^k f(x), \tag{11.71}$$

*for all* $x \in \mathbb{R}^n_+$ *and* $\lambda > 0$ *if and only if,*

$$kf(x) = \sum_{i=1}^{n} x_i \frac{\partial f(x)}{\partial x_i}. \tag{11.72}$$

This interesting result holds that one can decompose a homogeneous function as a kind of weighted sum of its partial derivatives. Its applicability, however, depends upon the homogeneity of one's function. The first order of business, therefore, is to examine our risk measure to see if this is true. Let's look at the ex-ante tracking error as a function of a set of weights,

$$f(\omega_a) = \sqrt{\omega_a^T \Omega_R \omega_a}. \tag{11.73}$$

If we multiply the argument of our tracking error function by $\lambda \in \mathbb{R}$, we may simplify as follows,

$$\begin{aligned}
\underbrace{f(\lambda\omega_a)}_{\sigma_a} &= \sqrt{(\lambda\omega_a)^T \Omega_R (\lambda\omega_a)}, \\
&= \sqrt{\lambda^2 \omega_a^T \Omega_R \omega_a}, \\
&= \lambda\sqrt{\omega_a^T \Omega_R \omega_a}, \\
&= \lambda f(\omega_a).
\end{aligned} \tag{11.74}$$

(continued)

Ex-ante tracking error is a first-degree homogeneous function—indeed, this is a consequence of the fact that standard deviation is a first degree homogeneous function.

Applying Euler's theorem, with $k = 1$, we have

$$f(\omega_a) = \sum_{i=1}^{n} \omega_{i,a} \frac{\partial f(\omega_a)}{\partial \omega_{i,a}},$$

$$\sigma_a = \sum_{i=1}^{n} \omega_{i,a} \frac{\partial \sigma_a}{\partial \omega_{i,a}}, \tag{11.75}$$

which is precisely the result derived in Eqs. (11.68)–(11.70). One may either derive this result by manipulating $\sigma_a$ or applying Euler's theorem for homogeneous functions. The latter approach provides us with a bit more comfort in the robustness of our result.

There is no real difference with the VaR decomposition.[41] One merely needs to compute the sensitivity of the portfolio variance for a change in the portfolio security weights. Once again, this is a partial derivative. Using the same trick—and noting that the coefficient, $\mathcal{N}^{-1}(\alpha)$, is a constant and has no affect on the computation—the marginal security portfolio variance is,

$$\frac{\partial \sigma_p}{\partial \omega_p} = \sigma_p^{-1} \Omega w_p. \tag{11.76}$$

The corresponding contribution to VaR

$$\text{VaR Contribution of } i\text{th security} = \mathcal{N}^{-1}(\alpha) \cdot \omega_{p,i} \left. \frac{\partial \sigma_p}{\partial \omega_p} \right|_i, \tag{11.77}$$

for your choice of $\alpha$ and $i = 1, \ldots, n$.

The enormous advantage of the decomposition of the tracking-error or VaR contribution by individual security is that *no* additional information is required. It can be determined entirely using the covariance matrix and the appropriate weighting vector.

Returning to our example and employing Eqs. (11.69) and (11.77), we may now determine the risk contributions for our three sample portfolios. Table 11.6 breaks down the contribution from each security to the tracking error and VaR. The

[41] This is because both risk measures are functions of the standard-deviation measure, which is a first-degree homogeneous function.

**Table 11.6** Instrument attribution

| Bond | Portfolio 1 | | | Portfolio 2 | | | Portfolio 3 | | |
|---|---|---|---|---|---|---|---|---|---|
| | $\omega_a$ | Ctn | VaR Ctn | TE $\omega_a$ | Ctn | VaR Ctn | TE $\omega_a$ | VaR Ctn | TE Ctn |
| 2Y | −0.25 | 76.0 | *−41.5* | −0.5 | 376.7 | *−82.9* | −0.1 | 120.6 | *−14.7* |
| 5Y | 0.25 | 187.8 | *112.5* | 0.5 | 541.6 | *225.1* | 0 | 0 | *0* |
| 10Y | 0 | 542.2 | *0* | 0 | 0 | *0* | 0.1 | 652.6 | *64.9* |
| Total | 0 | 805.9 | *71.1* | 0 | 918.3 | *142.1* | 0 | 773.2 | *50.2* |

This table describes, for our benchmark and three sample portfolios, the results of the instrument-level risk decomposition of both ex-ante tracking error and 95 % VaR.

**Table 11.7** T/E sign

| Marginal tracking error | Active weight position: Positive | Active weight position: Negative |
|---|---|---|
| Positive | *+ve TE contribution* | *−ve TE contribution* |
| Negative | *−ve TE contribution* | *+ve TE contribution* |

This table describes the four possible cases that can contribute to a positive or a negative tracking error contribution. It depends entirely on the respective signs of the active weights and the marginal tracking-error values.

total tracking error and VaR figures exactly coincide to the computations found in Table 11.5. In some cases, the tracking-error contribution from a given security is negative. The 2-year bond position in each of the three sample portfolios has, in fact, a negative tracking-error contribution.

The sign of the tracking-error contribution will depend on the sign of the active weight, which may be positive or negative, and the sign of the marginal tracking-error value. The marginal tracking-error may also be either negative or positive indicating that some positions can marginally increase in risk whereas others lead to marginal risk reductions. As both active weights and marginal tracking errors may be either positive or negative, we may observe a range of positive and negative tracking error contributions despite the fact that the total tracking error must be positive.

Table 11.7 indicates the four possible cases that can contribute to a positive or a negative tracking error contribution. In our simple example, given that we have negative weights for all 2-year positions and negative tracking-error contributions, we can safely state that the marginal tracking-error contribution for the 2-year bond in each sample portfolio is positive. The positive active weights and positive tracking-error contributions for all other 5- and 10-year bond positions lead us to conclude that the marginal-tracking error values are positive for these instruments as well.

While interesting, risk attribution at the individual security level is cumbersome in practice. This is because, between the portfolio and strategic benchmark, there are often hundreds of securities making the results hard to examine and interpret.[42] This can be mitigated by computing the risk contribution associated with each individual risk factor—this typically leads to a more manageable and easily interpretable set of risk factors. For our example these risk factors are a subset of key rates along the UST curve from 6-months to 10 years—to the overall risk position of our portfolio. Again, we'll focus on the tracking-error case with the knowledge that the VaR computation is conceptually identical. This requires a bit of tedious algebra on the tracking-error expression to compute the active risk-factor positions, but the basic approach is conceptually identical to the security attribution.

We again start with the active variance, $\sigma_a^2$ and make use the basic results in Eqs. (11.54) and (11.55) as follows,

$$\begin{aligned}
\sigma_a^2 &= \omega_a^T \Omega \omega_a, \\
&= \omega_a^T \text{var}(R) \omega_a, \\
&= \omega_a^T \text{var}(KM) \omega_a, \\
&= \underbrace{\omega_a^T K^T \text{var}(M) K \omega_a}_{\substack{\text{From Eq.} \\ (11.55)}}, \\
&= \underbrace{(K\omega_a)^T \text{var}(M) (K\omega_a)}_{\substack{\text{Recall that} \\ A^T B^T = (AB)^T}}, \\
&= \xi_a^T \text{var}(M) \xi_a, \\
&= \xi_a^T \Omega_M \xi_a.
\end{aligned} \tag{11.78}$$

The consequence is that we have rewritten the tracking error as a function of the active risk-factor positions, $\xi_a = K(\omega_p - \omega_b)$, and the covariance matrix of market movements, $\Omega_M$. Simply put, $\xi_a$ is the vector of active risk positions.

We may now employ exactly the same ideas and techniques to arrive at a second possible risk attribution based on the individual risk factors. The sensitivity of the tracking error to a change in the active factor positions is given as,

$$\begin{aligned}
\frac{\partial \sigma_a}{\partial \xi_a} &= \frac{1}{2\sigma_a} \frac{\partial \sigma_a^2}{\partial \xi_a}, \\
&= \frac{1}{2\sigma_a} \frac{\partial \left(\xi_a^T \Omega_M \xi_a\right)}{\partial \xi_a},
\end{aligned}$$

[42]One useful idea is to look at only those instruments with the top five or ten contributions to overall risk.

$$
\begin{aligned}
&= \frac{1}{2\sigma_a} 2\Omega_M \xi_a, \\
&= \sigma_a^{-1} \Omega_M \xi_a. \qquad (11.79)
\end{aligned}
$$

We term this the *marginal factor* tracking error vector and, as before, in Eq. (11.69) we have that

$$
\text{TE contribution of } i\text{th risk factor} = \xi_{a,i} \left. \frac{\partial \sigma_a}{\partial \xi_a} \right|_i, \qquad (11.80)
$$

for $i = 1, \ldots, m$. Thus, with relatively modest additional effort, we have derived a factor-based breakdown of the tracking error.

Using Eq. (11.80), we proceed in Table 11.8 to break down the contribution from each *factor* to the tracking error (and VaR). This will be our final look at our practical example.

Not surprisingly, most of the active risk contribution stems from exposure to longer-term yields. Once again, nothing keeps the ex-ante tracking error contribution from an individual factor from being a negative number. The 2-year sector appears to contribute negatively to the overall relative risk across all portfolios.[43] In this simple example, the risk attribution may appear somewhat superfluous. For a larger portfolio with exposure to multiple currencies, various credit sectors, and inflation-linked bonds such a risk attribution breakdown can prove extremely useful—both on a individual security and risk-factor level.

The focus of this chapter has been on ex-ante risk measures, but it is equally possible—using the same basic technique—to perform a decomposition of ex-post risk measures. The underlying shaded box provides a detailed description of how this might be accomplished for ex-post tracking error.

**Table 11.8** Factor attribution

| | Portfolio 1 | | | Portfolio 2 | | | Portfolio 3 | | |
|---|---|---|---|---|---|---|---|---|---|
| Factor | $\xi_a$ | VaR Ctn | TE Ctn | $\xi_a$ | VaR Ctn | TE Ctn | $\xi_a$ | VaR Ctn | TE Ctn |
| 6M | 0 | 3.9 | *0.3* | −0.01 | 4.3 | *0.5* | −0.1 | 3.7 | *0.2* |
| 2Y | 0.43 | 106.7 | *−36.5* | 0.87 | 39.5 | *−72.9* | 0.16 | 147.3 | *−12.2* |
| 5Y | −1.04 | 227.7 | *105.8* | −2.07 | 405.0 | *211.7* | −0.06 | 61.2 | *5.4* |
| 7Y | −0.01 | 97.2 | *1.42* | −0.03 | 99.7 | *2.8* | −0.12 | 113.6 | *11.1* |
| 10Y | 0 | 370.3 | *0* | 0 | 369.8 | *0* | −0.50 | 447.6 | *45.7* |
| Total | n/a | 805.9 | *71.1* | n/a | 918.3 | *142.1* | n/a | 773.2 | *50.2* |

This table describes, for our benchmark and three sample portfolios, the results of the risk decomposition, at the risk-factor level, of both ex-ante tracking error and VaR.

[43]The logic outlined in Table 11.7 applies equally well to the results in Table 11.8.

This chapter has focused on ex-ante risk. Ex-post risk computation and attribution is also interesting and possible. To see how this works, let us define the active return of a portfolio as follows,

$$r_{a,t} = r_{p,t} - r_{b,t}, \tag{11.81}$$

where, for a given point in time $t$, $r_p$ and $r_b$ denote the portfolio and benchmark returns, respectively. This is merely the difference between the portfolio and benchmark return. Now imagine that we have portfolio values ranging backwards $n + 1$ periods from the current point in time, $T$. With this information, we may compute $n$ active returns. The ex-post tracking error is merely the standard deviation of these $n$ active returns. Mathematically, we have

$$\begin{aligned} \text{TE}(T-n,T) &= \sqrt{\text{var}(r_a)}, \\ &= \sqrt{\left(\frac{1}{n-1}\right)\sum_{i=0}^{n-1}(r_{a,T-i}-\hat{r}_a)^2}, \end{aligned} \tag{11.82}$$

where $\hat{r}_a = \frac{1}{n}\sum_{i=0}^{n-1} r_{a,T-i}$ is the average active return over the sample period.

Now, imagine that we have decomposed active return into $m$ different return categories—these might be risk factors, currencies, countries, or types of issuer. We write the active returns as,

$$r_{a,t} = x_{1,t} + \cdots + x_{m,t} \tag{11.83}$$

for $t = T-n, \ldots, T$. We wish to attribute the ex-post risk along the same set of categories. To do this, we begin by examining its impact on the ex-post tracking error computation,

$$\begin{aligned} \text{TE}(T-n,T) &= \sqrt{\text{var}(r_a)}, \\ &= \sqrt{\text{var}(x_1+\cdots+x_m)}, \\ &= \sqrt{\sum_{k=1}^{m}\text{var}(x_k) + 2\sum_{k=1}^{m}\sum_{j=1}^{m}\text{cov}(x_k,x_j)}, \end{aligned}$$

(continued)

$$
= \sqrt{\begin{array}{l} \underbrace{\sum_{k=1}^{m}\left(\frac{1}{n-1}\right)\sum_{i=0}^{n-1}(x_{k,T-i}-\hat{x}_k)^2}_{\text{var}(x_k)} \\ +2\sum_{k=1}^{m}\sum_{j=1}^{m}\underbrace{\left(\frac{1}{n-2}\right)\sum_{i=0}^{n-1}(x_{k,T-i}-\hat{x}_k)\left(x_{j,T-i}-\hat{x}_j\right)}_{\text{cov}(x_k,x_j)} \end{array}} \tag{11.84}
$$

This is a fairly ugly and generally unwieldy expression. The situation can be dramatically improved if we move from summation to matrix notation.

$$
\begin{aligned}
\text{TE}(T-n,T) &= \sqrt{\sum_{k=1}^{m}\text{var}(x_k) + 2\sum_{k=1}^{m}\sum_{j=1}^{m}\text{cov}(x_k,x_j)}, \\
\sqrt{\text{var}(r_a)} &= \sqrt{[1 \cdots 1]\begin{bmatrix} \text{var}(x_1) & \text{cov}(x_1,x_2) & \cdots & \text{cov}(x_1,x_m) \\ \text{cov}(x_1,x_2) & \text{var}(x_2) & \ddots & \text{cov}(x_2,x_m) \\ \vdots & \vdots & \ddots & \vdots \\ \text{cov}(x_1,x_m) & \text{var}(x_2,x_m) & \cdots & \text{var}(x_m) \end{bmatrix}\begin{bmatrix}1\\ \vdots \\ 1\end{bmatrix}}, \\
\sqrt{\text{var}(x_1+\cdots+x_m)} &= \sqrt{\mathbb{I}^T\Omega_x\mathbb{I}},
\end{aligned} \tag{11.85}
$$

where $\mathbb{I}^T \in \mathbb{R}^{1\times m}$ is a vector of ones and $\Omega_x \in \mathbb{R}^{m\times m}$ is the covariance matrix of our return decomposition.

To decompose the ex-post tracking error, we need to compute the marginal tracking error associated with each of the individual risk factors, $x_1,\ldots,x_m$. More specifically, we require

$$
\frac{\partial \overbrace{\sqrt{\text{var}(x_1+\cdots+x_m)}}^{\text{TE}(T-n,t)}}{\partial x_k} \text{ for } k=1,\ldots,m. \tag{11.86}
$$

As we can see from Eq. (11.85), each of these partial derivatives is a complex function of the variance of $x_k$ and the covariance of $x_k$ with all other $x'$s ranging from $x_1$ to $x_m$. This could potentially get quite ugly. Fortunately, we

(continued)

have a very elegant representation of the ex-post tracking error in Eq. (11.85). Indeed, Eq. (11.86) may be re-written much more compactly as follows:

$$\frac{\partial\sqrt{\mathbb{I}^T \Omega_x \mathbb{I}}}{\partial \mathbb{I}}. \tag{11.87}$$

The result of this computation is a gradient vector in $\mathbb{R}^{m\times 1}$, where each element represents the appropriate partial derivative of our ex-post tracking error to each risk factor, $x_1, \ldots, x_k$.

It can be resolved as follows:

$$\begin{aligned}\frac{\partial\sqrt{\mathbb{I}^T \Omega_x \mathbb{I}}}{\partial \mathbb{I}} &= \frac{1}{2\sqrt{\mathbb{I}^T \Omega_x \mathbb{I}}}\frac{\partial\left(\mathbb{I}^T \Omega_x \mathbb{I}\right)}{\partial \mathbb{I}},\\ &= \frac{1}{\cancel{2}\sqrt{\mathbb{I}^T \Omega_x \mathbb{I}}}\cancel{2}\Omega_x \mathbb{I},\\ &= \frac{\Omega_x \mathbb{I}}{\sqrt{\mathrm{var}(x_1, \ldots, x_m)}}. \end{aligned}\tag{11.88}$$

The element by element product of $\mathbb{I}$ with Eq. (11.88) provides the ex-post tracking contribution from each of the individual risk factors, $x_1, \ldots, x_m$.[44]

In short, the same basic risk attribution based upon Euler's theorem for homogeneous functions is also possible in the ex-post setting. The only difference is that the weights on the individual risk factors are all equal to unity; the risk factors themselves must come from a separate performance attribution.

This has been a extremely detailed chapter with a substantial number of definitions and derivations. To thoroughly explain how one computes risk exposures, it is inevitable. Nevertheless, the cast of characters is indeed so large that it likely feels at times like a nineteenth century Russian novel by Tolstoy or Dostoevsky. Table 11.9 tries to help somewhat by providing a summary of the various vectors and matrices employed in our risk computations and their dimensions. One can think of Table 11.9 as the cast of characters employed for the computation of risk.

[44]This is easily verified: $\mathbb{I}^T \frac{\Omega_x \mathbb{I}}{\sqrt{\mathrm{var}(x_1, \ldots, x_m)}} = \sqrt{\mathbb{I}^T \Omega_x \mathbb{I}} = \sqrt{\mathrm{var}(r_a)} = \mathrm{TE}(T-n, T)$.

**Table 11.9** Risk notation

| Value | Description | Dimension | Definition |
|---|---|---|---|
| $\omega_p, \omega_b$ | Weights | $n \times 1$ | Data: user input |
| $\omega_a$ | Active weights | $n \times 1$ | $\omega_p - \omega_b$ |
| $K$ | Exposure matrix | $n \times m$ | Data: user input |
| $\xi_a$ | Factor weights | $m \times 1$ | $K^T \omega_a$ |
| $M$ | Market-movement vector | $m \times 1$ | Random variable |
| $\hat{M}_t$ | Realization of $M$ at $t$ | $m \times 1$ | Data: user input |
| $\Omega_M$ | Factor covariance matrix | $m \times m$ | $\text{var}(M)$ |
| $\tilde{\Omega}_M$ | Classical estimator of $\Omega_M$ | $m \times m$ | $\tilde{\text{var}}(M \mid \hat{M}_1, \ldots, \hat{M}_T)$ |
| $\tilde{\Omega}_{M_{T+1}}$ | Modified estimator of $\Omega_M$ | $m \times m$ | $\tilde{\text{var}}(M_{T+1} \mid \hat{M}_1, \ldots, \hat{M}_T)$ |
| $R$ | Security return vector | $n \times 1$ | Random variable |
| $\hat{R}_t$ | Realization of $R$ at $t$ | $n \times m$ | $\approx K\hat{M}$ |
| $\Omega_R$ | Return covariance matrix | $n \times n$ | $\text{var}(R)$ |
| $\tilde{\Omega}_R$ | Classical estimator of $\Omega_R$ | $n \times n$ | $K^T \tilde{\text{var}}(M \mid \hat{M}_1, \ldots, \hat{M}_T) K$ |
| $\tilde{\Omega}_{R_{T+1}}$ | Modified estimator of $\Omega_R$ | $n \times n$ | $K^T \tilde{\text{var}}(M_{T+1} \mid \hat{M}_1, \ldots, \hat{M}_T) K$ |
| $\frac{\partial \sigma_a}{\partial \omega_a}$ | Marginal security TE | $n \times 1$ | $\sigma_a^{-1} \Omega_R \omega_a$ |
| $\frac{\partial \sigma_a}{\partial \xi_a}$ | Marginal factor TE | $m \times 1$ | $\sigma_a^{-1} \Omega_M \xi_a$ |

This table summarizes the key notation used in this chapter along with a short definition and, where applicable, its dimensions.

## 11.6 Concluding Thoughts

This chapter centred around *two* principal questions. How much risk is there in my portfolio and where does it come from? The answer: it depends. It depends on how risk is defined and quantified. It depends on one's assumptions and one's estimation techniques. Finally, it depends on whether you consider risk from a relative or an absolute perspective. This chapter provided an analytical approach based on portfolio exposures and a covariance matrix estimated from historical monthly data. This is *not* the only way to approximate one's risk positions. It is *one* relatively straightforward approach that allows one to answer, in a sensible and structured manner, the previously posed questions.[45]

The presented approach is not without some disadvantages. Perhaps the greatest disadvantage is its explicit or implicit assumption of a normally distributed market risk-factor movements.[46] The true distribution of risk-factor changes typically exhibit fatter tails than that predicted by a Gaussian distribution. An assumption of normality, therefore, will tend to understate the actual risk in one's portfolio.

[45]For a more comprehensive look at the VaR measure, for example, and some alternative approaches for its computation, see Jorion [8].

[46]Ex-ante tracking error does not explicitly assume normality, but its restriction to the second moment of the return distribution may be viewed as an implicit assumption of normality. At the very least, it ignores the higher moments of the return distribution.

One should nonetheless exercise a bit of caution in jumping to more complex risk models. This approach makes a number of simplifying assumptions, but with a bit of effort it is readily understood and, more importantly, relatively easily communicated to consumers of risk information. We should not understate the value of simple, easy-to-explain models. They can be very useful as long as we understand their weaknesses and do not treat the results as truth. Indeed, assessing the reliability and accuracy of our risk estimates—as well as their key associated approximations and assumptions—is the topic of the next chapter.

## References


1. P.A. Abken, *An Empirical Evaluation of Value at Risk by Scenario Simulation* (Comptroller of the Currency, Washington, DC, 2000)
2. C.O. Alexander, C.T. Leigh, On the covariance matrices used in value at risk models. J. Deriv. **4**, 50–62 (1997)
3. G. Casella, R.L. Berger, *Statistical Inference* (Duxbury Press, Belmont, 1990)
4. F.X. Diebold, C. Li, Forecasting the term structure of government bond yields. University of Pennsylvania Working Paper (2003)
5. R. Engle, Dynamic conditional correlation: a simple class of multivariate generalized autoregressive conditional heteroskedasticity models. J. Bus. Econ. Stat. **20**(3), 339–350 (2002)
6. G.S. Fishman, *Monte Carlo: Concepts, Algorithms, and Applications*. Springer Series in Operations Research (Springer, New York, 1995)
7. J.D. Hamilton, *Time Series Analysis*, Chapter 22 (Princeton University Press, Princeton, 1994)
8. P. Jorion, *Value at Risk: The New Benchmark for Managing Financial Risk* (McGraw-Hill Ryerson Limited, New York, 2006)
9. J.P. Morgan/Reuters, *RiskMetrics™ — Technical Document* (Morgan Guaranty Trust Company, New York, 1996)
10. W.H. Press, S.A. Teukolsky, W.T. Vetterling, B.P. Flannery, *Numerical Recipes in C: The Art of Scientific Computing*, 2nd edn. (Cambridge University Press, Cambridge, 1992)

# Exploring Uncertainty in Risk Measurement 12

Доверяй, но проверяй*(Trust, but verify)*

Russian proverb

This the final chapter dedicated to the third dimension of portfolio analytics: risk. As demonstrated in Chap. 10, assessing risk requires a careful examination of the possible outcomes related to a given activity and their associated likelihoods. Chapter 11 then proceeded—employing precisely these outcomes, likelihoods and a liberal dose of statistics—to work through the details required to actually produce different risk measures. While these results are useful and essential for understanding the risk dimension of fixed-income portfolios, stopping at this point would leave this task incomplete and, quite possibly, be somewhat dishonest. Why? The reason is risk is complicated. Unlike overall performance figures—which excepting some uncertainty related to external cash-flows are unequivocal values—there is *no* single, correct risk value. We may, therefore, unequivocally compare our daily performance attributions to daily returns. Risk is different. Were you to ask five experienced risk managers to compute a given risk measure, you would quite likely receive five slightly different answers.[1] Moreover, none of them could be considered wrong. The differences would stem from varying assumptions and inputs. This is unsettling and suggests there is a need to test our risk estimates. This topic, quite simply, deserves further attention.

Risk measurement is *not* an exact science. This is just part of life in the world of risk measurement and it is not necessarily problematic. Nevertheless, this fact needs to be both appreciated and well understood. We stress that it is critical to explore and systematically gain knowledge of those elements that have an important impact on our risk measures. In other words, we require techniques to ensure that our risk

[1]Indeed, you may receive more than five responses. It is not impossible that one, or more, risk managers may provide multiple estimates.



D.J. Bolder, *Fixed-Income Portfolio Analytics*,
DOI 10.1007/978-3-319-12667-8_12

measures are forecasting true uncertainty over time. The principal objective of this chapter, therefore, is to consider a few possible tools that might profitably be used for this purpose. In doing so, we hope to better trust our risk computations and, perhaps even more importantly, to gain some intuition as to when they might fail us.

There are two basic tools in our toolbox that can help us in this regard: sensitivity analysis and backtesting. Sensitivity analysis is, as the name suggests, an organized analysis of those inputs and assumptions that are most important for our risk measure outputs. Backtesting is a comparison of our risk estimates to actual return outcomes with an eye towards judging their success in estimating the true risk.

Risk estimation is basically a type of forecast—in particular, we are essentially forecasting the future return distribution of the fixed-income assets in our portfolios and strategic benchmarks. Sensitivity analysis asks "what is important in determining the final forecast?" Backtesting, in contrast, asks "are these forecasts any good?" These are two sides of the same coin. Sensitivity analysis highlights the key levers in producing sensible forecasts and backtesting feeds back to verify that the forecasts are consistent with reality. Successfully performing both can dramatically improve your capacity to estimate and manage risk.

In the following pages, we will examine sensitivity analysis and backtesting techniques in turn. It should nonetheless be noted that an entire book could be dedicated to this topic. This implies that although we will try to thoroughly treat the topic, we will nonetheless *not* cover all possible techniques and approaches. Our hope, however, is that this will represent a solid starting point. As usual, we will attempt to make this, at times, rather complex analysis more concrete by demonstrating the main ideas in the context of a collection of practical examples.

## 12.1 Sensitivity Analysis

Sensitivity analysis is a common practice in all areas of practical quantitative analysis. In operations research, engineering, and management science it is used extensively. The scope of sensitivity analysis is extremely broad. Nevertheless, it can be summarized by one simple question: "what if?" *What if* this parameter was bigger or smaller? *What if* we used a different data set? *What if* we added (or eliminated) this aspect of our model? *What if* we relax this assumption? Answers to these questions can reveal an enormous amount about the robustness of our analysis. As a general rule, good solutions to quantitative problems are robust solutions. A solution that changes dramatically in the face of a small change in an input is probably not a very desirable choice.

In a portfolio-analytics setting, risk measurement typically focuses on ex-ante tracking error and portfolio VaR. Our task is thus to seek those assumptions and inputs that have the largest impact on these two measures. Quite simply we seek the most interesting "what-if" questions. Not surprisingly, based upon the discussion from previous chapters, all of the "what-if" questions surround the covariance matrix of risk-factor movements, $\Omega_M$. This is the central input in our risk estimates.

To this end, we propose the following *three* principal "what-if" questions in our sensitivity analysis of the covariance matrix:

1. *What if* we change the historical observations used in the estimation of $\Omega_M$ from a daily to a monthly frequency?
2. *What if* we alter the parameter employed in our weighting of this historical data and/or *what if* we change the overall sample size used in our estimations? As we will see, these two questions are related.
3. *What if* we alter the dependence structure between the risk-factor movements?

This represents a fairly full program of analysis. The first question looks at what type of data we should use. The second asks how much importance we should attribute to each of these datapoints. The final question looks at the interactions between our data elements—should we accept the statistical estimates or consider alternatives. In some respects, this is the truest type of sensitivity analysis, because it questions the very foundation of our prediction model. We will consider each element consecutively in the context of a numerical example—highlighting the interesting points along the way—and then examine the relative importance of these different dimensions.

### 12.1.1 Setting the Stage

To appropriately analyse our *three* "what-if" questions, we will need a rather sizeable dataset. With this dataset we can change different elements, estimate the corresponding risk, and then see what happens. We elected to use the roughly $11\frac{1}{2}$-period from January 2003 to April 2014 and focus on the European sovereign market. Naturally, we need to examine this from both a daily and monthly perspective. Figure 12.1 illustrates the evolution of *four* German-Bund yields and two interesting credit spreads—France and Italy versus German Bunds—over the full period.

What might we conclude from this data? A few interesting points arise. There does not appear to be any substantial difference between monthly and daily data—the overall story is the same, although the movements at the monthly frequency are naturally smoother. Both credit spreads appear quite calm until early 2008, when they both started to widen. The Italian spread virtually exploded in 2010 at the onset of the European sovereign debt crisis. Finally, although the level of the various Bund yields has evolved over the full period, the volatility looks, at least visually, to have been fairly stable. Overall, however, there looks to be quite a bit of action in this dataset, which should make for an interesting analysis.

Armed with the risk-factor input data, the next order of business is to build a few portfolios. Using the framework built in the previous chapter, we will construct a strategic benchmark and *three* alternative portfolios. Again, it bears repeating that these are extremely simple portfolios. This is admittedly somewhat unrealistic, but

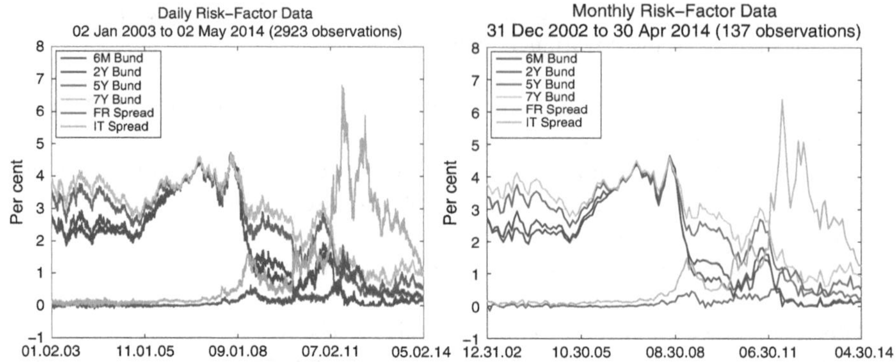


**Fig. 12.1** Risk factor data. Over the roughly $11\frac{1}{2}$-period from January 2003 to April 2014, we examine six risk-factors from the European sovereign market at two alternative data frequencies: monthly and daily

**Table 12.1** Three simple portfolios

| Instrument | Country code | Benchmark ($\omega_b$) | Portfolio ($w_p$) 1 | 2 | 3 |
|---|---|---|---|---|---|
| Cash | EUR | – | 20 % | – | – |
| 2Y | DE | 50 % | 40 % | 30 % | 40 % |
| 2Y | FR | – | – | 10 % | – |
| 2Y | IT | – | – | 10 % | 10 % |
| 5Y | DE | 50 % | 40 % | 40 % | 40 % |
| 5Y | FR | – | – | 10 % | – |
| 5Y | IT | – | – | – | 10 % |

The following portfolios are comprised of two tenors—2- and 5-years—and three sovereign issuers: Germany, France and Italy The benchmark is comprised entirely of German Bunds.

it provides the very important advantage of permitting us to explicitly write out all of the computations and transparently examine each of the important elements.

This point notwithstanding, Table 12.1 provides us with rather more interesting portfolios than in the previous chapter. The strategic benchmark is merely a 50/50 mix of 2- and 5-year German Bunds. The first portfolio is defensively positioned, and as such not terribly interesting, with underallocations to Bunds and sizeable cash position. The second and third portfolios, however, have active positions in French and Italian treasury securities. This brings two new—and potentially quite interesting—risk factors into play.

Active weights, $\omega_a = \omega_p - \omega_b$, are easily defined from Table 12.1 and thus not explicitly shown. Active risk-factor exposures, however, are less obvious. To better understand these, we must first use the risk-factor exposures to construct our exposure matrix, $K$. We have a total universe—within the strategic benchmark and our three portfolios—of seven separate securities.[2] We may consequently populate our exposure matrix with the key-rate and spread durations as,

$$K = \begin{bmatrix} -\kappa_{1,6M} & -\kappa_{1,2Y} & -\kappa_{1,5Y} & -\kappa_{1,7Y} & -D_{1,\text{FR}} & -D_{1,\text{IT}} \\ \vdots & \vdots & \vdots & \vdots & \vdots & \vdots \\ -\kappa_{7,6M} & -\kappa_{7,2Y} & -\kappa_{7,5Y} & -\kappa_{7,7Y} & -D_{7,\text{FR}} & -D_{7,\text{IT}} \end{bmatrix}, \tag{12.1}$$

$$= \begin{bmatrix} 0.00 & 0.00 & 0.00 & 0.00 & 0.00 & 0.00 \\ -0.01 & -1.95 & -0.01 & 0.00 & 0.00 & 0.00 \\ -0.01 & -1.85 & -0.15 & 0.00 & -2.01 & 0.00 \\ -0.01 & -1.95 & -0.01 & 0.00 & 0.00 & -1.97 \\ -0.03 & -0.02 & -4.15 & -0.06 & 0.00 & 0.00 \\ -0.02 & -0.02 & -4.25 & -0.16 & -4.45 & 0.00 \\ -0.03 & -0.02 & -4.15 & -0.06 & 0.00 & -4.26 \end{bmatrix},$$

where $K \in \mathbb{R}^{n \times m}$, where $n$ is the number of securities (i.e., 7) and $m$ is the number of market risk factors (i.e., 6). In other words, each row represents a security, whereas each column denotes a risk factor.[3] This object—as the personification of the additive linear mapping between returns and risk factors developed in Chap. 3—tells us everything we need to approximate security returns for a given set of risk-factor movements.

We also have everything that we require to compute the risk-factor exposures. Recall, from the previous chapter, that our active risk factors are simply a function of the active weights and the exposure matrix,

$$\xi_a = K^T \omega_a. \tag{12.2}$$

This essential aspect of the risk computation provides us with important insight into the relative weighting of the volatility and correlation estimates embedded into the covariance matrix $\Omega_M$. Using Table 12.1 and the exposure matrix embedded in Eq. (12.1), we may compute the individual elements of the active risk-factor exposure vector, $\xi_a$. The results are illustrated in Table 12.2.

What might we conclude? Portfolio #1 is, as expected, solely exposed—in a defensive way relative to the strategic benchmark—to the Bund curve, Portfolio #2 is exposed to all risk factors, while Portfolio #3 has only a single active risk-factor exposure—the Italian-Bund spread. This should have important consequences for our risk computations and the associated sensitivity analysis.

---

[2] Cash, while admittedly somewhat dull, is also a security.

[3] Given the absence of non-EUR denominated securities, and because the base currency of this investor is assumed to be EUR, we have dispensed with the foreign-exchange risk factor.

**Table 12.2** Active risk

| Risk factor | $\xi_{a,1}$ | $\xi_{a,2}$ | $\xi_{a,3}$ |
|---|---|---|---|
| 6M | 0.002 | 0.001 | 0.000 |
| 2Y | 0.099 | 0.010 | 0.000 |
| 5Y | 0.208 | −0.024 | 0.000 |
| 7Y | 0.003 | −0.010 | 0.000 |
| FR | 0.000 | −0.646 | 0.000 |
| IT | 0.000 | −0.197 | −0.623 |

Here we see the active-risk exposures associated with each of the portfolios—these are merely a function of the active weights and the sensitivity matrix—see Eq. (12.2). Note that Portfolio #3 has exposure to only a single risk factor.

**Table 12.3** Risk-factor volatilities

| Country code | Risk factor | Daily volatility | | Monthly volatility | |
|---|---|---|---|---|---|
| | | **Daily data** | Monthly data | Daily data | **Monthly data** |
| DE | 6M | **4.9** | 6.0 | 22.3 | **20.8** |
| DE | 2Y | **4.9** | 6.1 | 22.4 | **21.3** |
| DE | 5Y | **5.5** | 6.4 | 25.3 | **22.1** |
| DE | 7Y | **5.5** | 6.3 | 25.3 | **21.8** |
| FR | Spread | **4.9** | 2.8 | 22.3 | **9.6** |
| IT | Spread | **9.7** | 11.3 | 44.6 | **39.1** |

This table summarizes, for both frequencies, the equally weighted risk-factor volatility estimates. Note that scaling is required—either up or down—to permit a direct comparison. The results are quite enlightening.

### 12.1.2 The Data Frequency

We've introduced our risk-factor data and constructed a few portfolios. The next order of business, of course, is to estimate a covariance matrix and compute our risk measures. Here we face a few important decisions. The first decision is what data frequency should we employ? Does it make a difference? Yes, in fact, it often does. Different frequencies may give quite different estimates of volatility. To see this, let us examine Table 12.3, which provides, for each risk factor, an equally weighted estimate of volatility using all of the data in our sample dataset.

At first glance, it is difficult to compare the figures. Indeed, the only *true* values are the columns in bold. These represent the daily volatility estimated with daily data and the monthly volatility estimated with monthly data. Other than the rather obvious statement that monthly volatilities are, virtually by definition, larger than daily volatilities, we cannot say much. The only way to directly compare these estimates is to put them on an equal footing. This requires scaling *up* daily volatility to a monthly frequency or scaling *down* monthly volatility to a daily frequency. Exactly this computation is performed in the non-bold columns of Table 12.3.

Before interpreting the results of Table 12.3, it is worth briefly investigating whether the scaling of volatilities is a sensible practice. To answer this question, it is necessary to understand precisely how scaling is done and, most importantly, under what assumptions.

The computation is quite simple, but it may only be sensibly performed if some reasonably strong assumptions hold true. How is it done and what must be assumed? Imagine that a collection of random returns, $r_1, \ldots, r_T$ are independent and identically distributed. In other words, each $r_t$ is drawn from,

$$r_t \sim \mathscr{X}\left(\mu_r, \sigma_r^2\right) \text{ for all } t, \tag{12.3}$$

where $\mathscr{X}(\cdot, \ldots)$ is some generic probability distribution. We are interested in the distribution of the sum of these random returns.[4] This sum is distributed as follows,

$$\sum_{t=1}^{T} r_t \sim \mathscr{X}\left(\sum_{t=1}^{T} \mu, \sum_{t=1}^{T} \sigma_r^2\right) = \mathscr{X}(T\mu_r, T\sigma_r^2). \tag{12.4}$$

Given that we have independence, the variance of the sum is merely the sum of the variances. Consequently, the volatility is $\sqrt{T}\sigma_r$.

Assuming that returns are i.i.d., therefore, one might transform daily volatility into a monthly frequency merely by multiplying the daily estimate by $\sqrt{21}$.[5] This is the origin of what is termed the *square-root* rule for scaling volatilities. This rule has its roots in mathematical finance and represents the working approximation employed by market participants.[6] Alternatives are formally considered by Diebold et al. [6], although it should be noted that the scale-root rule is extensively used in practice and we will make no exception during our analysis.[7]

The obvious question is "are risk-factor movements independently and identically distributed?" As usual, the answer is: it depends. Figure 12.2 outlines the evolution of the risk-factor changes introduced in Fig. 12.1. For some of the risk factors—particularly the underlying treasury yields—the i.i.d. assumption appears fairly harmless. In all cases, it would seem that assuming a mean of zero is quite reasonable. What is clear, however, is that for some of the risk factors—most notably the spreads—the dispersion is certainly *not* constant over time and, consequently, it appears that we may soundly reject the i.i.d. hypothesis. In conclusion, although the square-root rule is used extensively in practice, it is *not* on the strongest theoretical ground.

---

[4]While the simple sum of a set of returns is not the same as the geometric sum, it is typically fairly close.

[5]Moving from monthly to daily, the factor becomes $\sqrt{\frac{1}{12}}$.

[6]In mathematical finance, when working with logarithmic returns and assuming that asset movements follow a Brownian motion—with its Wiener increments, which are by definition Gaussian and i.i.d.—the square-root rule is exact.

[7]See also Meucci [9] for an interesting technical treatment of the square-root rule.

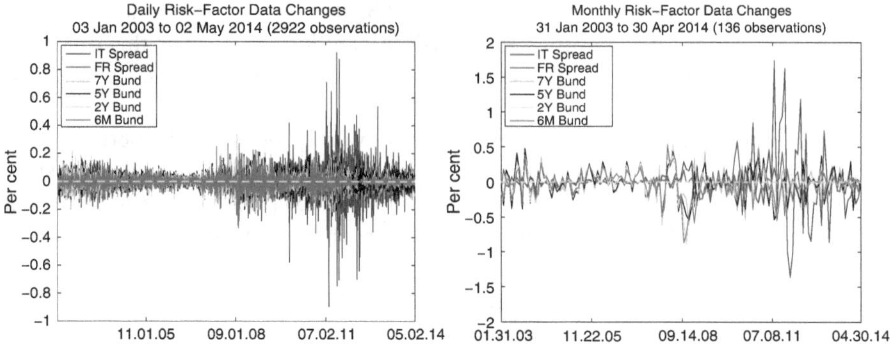


**Fig. 12.2** Are risk factors iid? Perhaps for the treasury yields, but it does not appear to be true for credit spreads. The mean might be constant at zero, but the dispersion is certainly not identically distributed

After this aside, let's return to Table 12.3. Consistent with the previous analysis, the rate volatility estimates appear quite consistent across volatilities. That is, whether computed using daily or monthly data or scaled, the values are similar. This also appears true, although somewhat less so, for the Italian spread. The French spread, however, has a very different result. The daily estimate of the French-Bund spread is substantially higher than the monthly estimate. At a daily frequency, the daily estimate of the French credit spread is 1.75 larger than the scaled monthly equivalent. At a monthly frequency, this rises to 2.3. Clearly, the scaling is responsible for part of the difference, but not all of it. Part of the difference seems to arise from scaling and part simply from differences in the underlying data. In brief, use of daily or monthly data as an input to covariance matrix estimates can matter.

We know that the covariance matrices may vary depending upon our choice of data frequency. Let's see by how much by examining our three portfolios. Recall that the active risk is defined as,

$$\sigma_a = \sqrt{\underbrace{\left(K^T\omega\right)^T}_{\xi_a^T} \Omega_M \underbrace{K^T\omega_a}_{\xi_a}}, \tag{12.5}$$

where $\Omega_M$ is the covariance matrix of the daily or monthly risk-factor changes. Table 12.4 applies Eq. (12.5) to our example and summarizes the equally weighted estimates for each portfolio using our two alternative data frequencies.[8]

The final ex-ante tracking error estimates, employing both daily and monthly frequencies, are not terribly different for Portfolios #1 and #3. Portfolio #2, with the

[8]By convention, these are annualized estimates expressed in basis points. Once again, we need to make use of the square-root rule to obtain annualized estimates—for the daily and monthly estimates, we use the scaling factors $\sqrt{252}$ and $\sqrt{12}$, respectively.

**Table 12.4** Portfolio volatilities

| Portfolio | Daily | Monthly |
|---|---|---|
| Portfolio #1 | 24.6 | 22.8 |
| Portfolio #2 | 66.6 | 39.6 |
| Portfolio #3 | 96.2 | 84.4 |

This table summarizes—for both daily and monthly data frequencies—the annualized volatility estimates for our three sample portfolios. Note the important difference for Portfolio #2.

largest French sovereign position, in contrast, has dramatically different estimates using daily and monthly data. This is not terribly surprising since this is precisely where we observed the largest differences between the daily and monthly volatility estimates.

Sadly, this analysis does not tell us which frequency should be employed. There is something of a trade-off. Since tracking-error is generally reported as an annualized figure, monthly volatility estimates suffer less from scaling problems. Monthly data, however, is only updated every 21 business days. This means that changing market conditions are incorporated less frequently, and less quickly, into our risk estimates. In practice, most risk management systems make use of daily data to ensure a closer link between risk estimates and market conditions. This would suggest that, for most market participants, scaling issues are less important than the need to incorporate market developments in a timely manner.

### 12.1.3 Weighting Scheme

In the previous chapter, we highlighted the notion of Exponential weighting to place more weight on more recent observations and less weight on more distant elements of our dataset.[9] In particular, we indicated that the exponential weighting function is well approximated by,

$$\sum_{t=0}^{T-1} \underbrace{(1-\lambda)\lambda_t}_{\theta_t} X_{T-t}^2, \tag{12.6}$$

where $\theta_t$ is a new weight replacing the standard $\frac{1}{T-1}$ in the variance formula. This is Eq. (11.35) from Chap. 11. As is often the case, there is somewhat more to the story. To be entirely precise, the *exact* exponential-weighting function has the following form,

[9]The following discussion is inspired by the excellent treatment in the original RiskMetrics™[10] document.

$$\sum_{t=0}^{T-1} \underbrace{\left(1/\sum_{s=0}^{T-1} \lambda^s\right) \lambda_t}_{\hat{\theta}_t} X_{T-t}^2, \tag{12.7}$$

where, at the risk of stating the obvious, $\hat{\theta}_t$, is the not the same as $\theta_t$. Fortunately, we may easily retrieve Eq. (12.6) when we consider the limit as $T \to \infty$. In this case, we have that

$$\lim_{T\to\infty} \left(1/\sum_{s=0}^{T-1} \lambda^s\right) = 1 - \lambda. \tag{12.8}$$

Thus, the first term in Eq. (12.7) simply reduces—by virtue of it being a geometric sum—to $1 - \lambda$. In plain language, Eqs. (12.6) and (12.7) are equivalent for large $T$. Equation (12.6) is thus a good approximation and, quite frankly, easier to interpret and explain.

Nevertheless, we do not live in the limit. For finite $T$, it is not entirely true. It can be written more precisely by using the properties of a non-infinite geometric series,

$$\left(1/\sum_{s=0}^{T-1} \lambda_s\right) = \frac{1-\lambda}{1-\lambda^T}. \tag{12.9}$$

Consequently, the *precise* exponential weighting formula is given as,

$$\sum_{t=0}^{T-1} \underbrace{\frac{1-\lambda}{1-\lambda^T} \lambda_t}_{\hat{\theta}_t} X_{T-t}^2. \tag{12.10}$$

For reasonably sample sizes (i.e., $T > 100$) and typical values of $\lambda$ (i.e., $\lambda < 0.98$), the denominator in $\theta$ is close to unity. That is, $\theta_t \approx \hat{\theta}_t$. For smaller samples and larger values of $\lambda$, it can make a difference. When using smaller sample sizes, it is probably safer to use Eq. (12.10).[10]

This is perhaps a useful precision about the computation of exponential weights, but it does not help us to determine the appropriate choice of $\lambda$. One potentially

---

[10]Moreover, this formula will also ensure that weights always sum to unity. Summing $\hat{\theta}_t$ over $t$, we have

$$\sum_{t=0}^{T-1} \hat{\theta}_t = \frac{1-\lambda}{1-\lambda^T} \underbrace{\sum_{t=0}^{T-1} \lambda_t}_{\text{Geometric series}} == \frac{\cancel{1-\lambda}}{\cancel{1-\lambda_T}} \frac{\cancel{1-\lambda^T}}{\cancel{1-\lambda}} = 1, \tag{12.11}$$

which is reassuring.

useful way to make this decision is to examine how quickly the weight is applied to the historical observations. Simply put, how much time does it take for the exponential weighting scheme to consume a given proportion of the overall weight? This is clearly a function of one's choice of $\lambda$. Alternatively, we can think of this as a measure of the decay of the weighting function.

It is also something that we can determine mathematically. First, we set $\alpha$ as the proportion of overall weight consumed and $\tau(\alpha)$ is the amount of time required to consume $\alpha$ weight. To find a general expression for $\tau(\alpha)$, we need only resolve the following expression,

$$\begin{aligned}\sum_{t=0}^{\tau(\alpha)-1} \theta_t &= \alpha, \qquad (12.12)\\ \frac{1-\lambda}{1-\lambda^T}\sum_{t=0}^{\tau(\alpha)-1} \lambda_t &= \alpha,\\ \frac{\cancel{1-\lambda}}{1-\lambda^T}\frac{1-\lambda^{\tau(\alpha)}}{\cancel{1-\lambda}} &= \alpha,\\ 1-\lambda^{\tau(\alpha)} &= \alpha(1-\lambda^T),\\ \tau(\alpha) &= \frac{\ln\left((1-\alpha)+\alpha\lambda^T\right)}{\ln(\lambda)}.\end{aligned}$$

This tells us, for a given choice of $\alpha$, how much time is required to consume this proportion of the total exponential weight. Does this expression make sense? If we set $\alpha = 1$, which implies all of the weight is consumed, then $\tau(\alpha)$ reduces to $T$, which is sensible. If $\alpha = 0$, then $\tau(\alpha)$ is also zero. This seems like a reasonable expression.

When $\alpha = 0.5$, this is termed the half-life for the exponential decay of our weighting function. To see what this means, let's now consider a more interesting example. Imagine that you have 250 daily observations and a $\lambda$-parameter of 0.94. You want to know how many days are required to consume 50 % of the weight? With equal weights, the answer is simply 125 days. With exponential weighting, the result is dramatically different. Half of the weight is applied after about 11 days.[11] In other

[11]We simply employ Eq. (12.12) with $\alpha = 0.5$, $\lambda = 0.94$, and $T = 250$ as

$$\tau(0.5) = \frac{\ln\left((1-0.5)+0.5\cdot 0.94^{250}\right)}{\ln(0.94)} \approx 11 \text{ days}. \qquad (12.13)$$

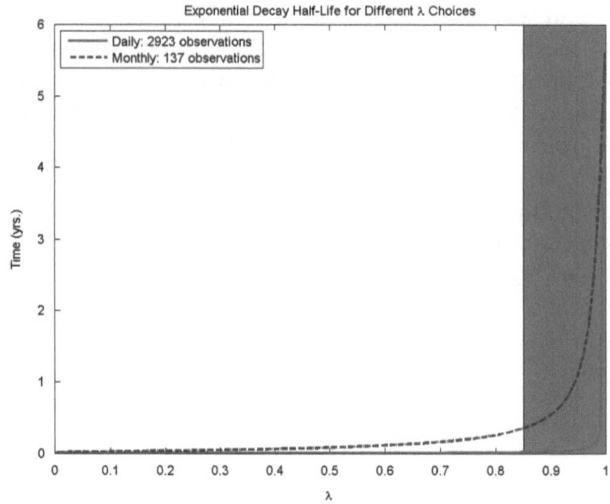


**Fig. 12.3** Exponential half-life. The following figure shows, for each data frequency, the half-life (in years) for each choice of $\lambda$. Note that the daily observations decrease much more quickly—this is simply a function of the larger number of observations

words, the $\lambda$ choice places half the weight on the last 2 weeks of data. Although exponential weighting employs—with extremely small weights—all observations, the practical number of observations employed (i.e., $\alpha = 0.99$) is about 74 days or roughly 15 weeks. This is only about 30 % of the total dataset.

To complete this analysis, Fig. 12.3 demonstrates, for each data frequency, the half-life (in years) for each choice of $\lambda \in (0, 1)$. Recall that the full period is roughly 11.5 years—consequently, the longest possible half-life, with equal weights, is a bit less than 6 years. The shortest half life depends on the data frequency—12 h for daily observations and about 2 weeks for the monthly frequency.[12] The shaded range of $\lambda \in (0.85, 1)$ represents a reasonable range of $\lambda$ choices. Note that the daily observations decrease much more quickly—this is simply a function of the larger number of observations. This implies that one's choice of $\lambda$ parameter is *not* independent of our the decision regarding data frequency.

In a nutshell, therefore, the smaller the choice of $\lambda$, the more weight on more recent observations. Does this make a difference in our ex-ante tracking error estimates? To the extent that volatilities and correlations change over time, this will matter. Let's have a closer look. Figure 12.4 demonstrates in the context of our three sample portfolios, for both daily and monthly risk-factor data, the impact of changing the $\lambda$ parameter in the exponential-weighting function.[13]

What can we conclude? Risk estimates for Portfolio #1 and #2 appear to be relatively invariant to the choice of $\lambda$. To be fair, this is not completely true with monthly data, but with daily data the estimates look to float around the 20 and

[12]This is the obvious extreme, and nonsensical, case of placing all of the weight on the most recent observation.

[13]When $\lambda = 1$, since the exponential weights are undefined, we substitute the equally weighted estimates.

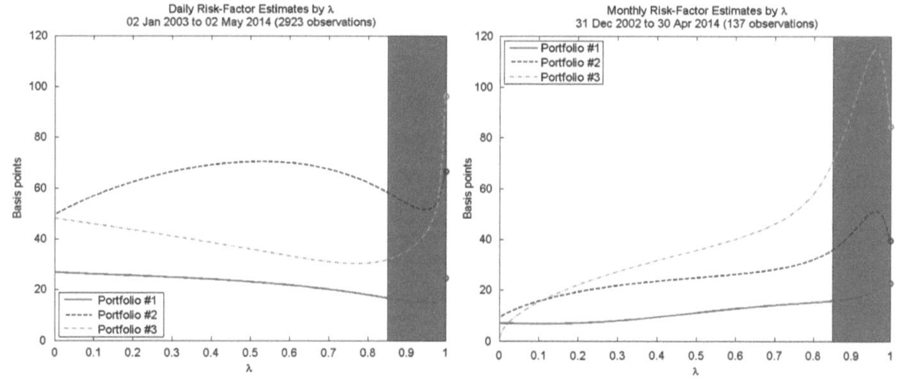


**Fig. 12.4** Impact of $\lambda$. Here we see, for both daily and monthly risk-factor data, the impact of changing the $\lambda$ parameter in the exponential-weighting function

60 basis-point levels, respectively. Portfolio #3 is a different story. The equally weighted estimate—at the daily frequency—is almost 100 basis points, but the estimate when $\lambda$ takes the not unreasonable value of about 0.90, it is less than half as large. This is an enormous difference and shows just how much the choice of weighting parameter can influence one's risk estimates.

Again, the question arises "what should one do?" This is not an easy question to answer. The first observation is that the choice of $\lambda$ can, depending on the risk factor, have an important impact on one's risk estimates. This would suggest that a careful eye must be kept on this parameter. It also seems to suggest that, given its importance, a single estimate may be insufficient. This is not to imply that one should compute 50 separate risk measures for various choices of $\lambda$. Something a bit more balanced would be advisable. It might make sense, for example, to employ two estimates. A slow-moving equally weighted risk estimate and a faster-moving risk estimate with a $\lambda$ value in the 0.94 to 0.98 range. The specific choice, however, has to depend on how quickly you want your volatility and correlation estimates to react to market movements. Two such estimates would bookend the range of possible values and simultaneously provide insight into how much, at each point in time, the choice of $\lambda$ matters.

We've seen that exponential weighting determines—albeit in an indirect fashion—the number of observations employed in covariance computations. You could have more than 1,000 observations, but for a choice of $\lambda < 1$, you will ignore a sizeable portion of them. This implies that one need not be overly concerned about one's choice of sample size. Practically speaking, assuming the use of daily data, it probably makes sense to have somewhere between 1 and 2 years of data—this implies somewhere between 250 and 500

(continued)

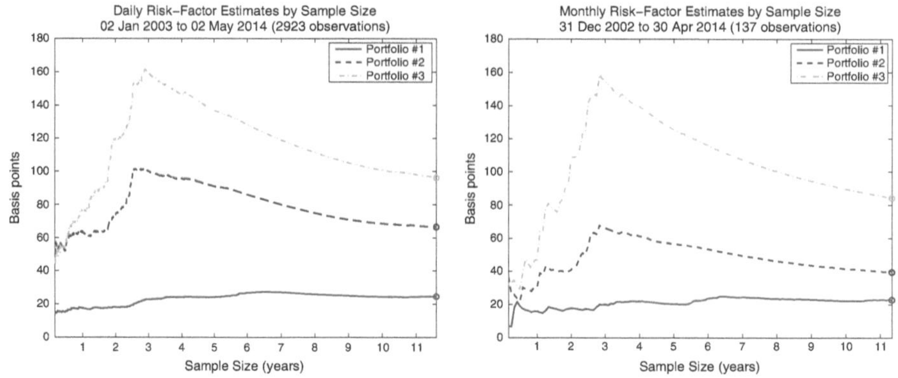


**Fig. 12.5** Impact of sample size. Here we examine—for daily and monthly frequencies—the impact on TE computations of increasing the number of observations from 2 months to the full dataset

observations. You may, however, not wish to employ exponential weighting. There may be a variety of reasons for this, but one common concern is that exponential weighting leads to a high degree of volatility in one's risk estimates.

Whatever the reason, if you do decide to forego exponential weighting, then you need to decide on a sample size. Should you employ 100 or 1,000 days of risk-factor changes? Moreover, what are the implications of changing the sample size? The sensitivity to one's sample size is not so easy to quantify. We can gain some insight by systematically adjusting the number of observations employed in the computation of the covariance matrix.

We do precisely this in Fig. 12.5—for daily and monthly frequencies—to examine the impact on TE computations of increasing the number of observations from 2 months to the full dataset.

The results make logical sense. When the dataset is small, each additional observation has a relatively sizeable impact on the final estimation. This explains the noisiness of the estimates for small sample sizes. As the sample size grows, the estimates become more stable. The estimates still vary for different sample sizes, but the transition is relatively smooth.

This does not answer the question of the appropriate sample size. It merely suggests that one could, absent exponential weighting, accomplish something similar by varying the sample size. A rolling 50-day sample size would lead to fast-moving variance and covariance estimates. The key difference with exponential-weighting is that extreme observations join or exit the estimation in a relatively abrupt manner. In the end, an equally weighted approach probably makes sense when one wants a long-term, slow-moving

(continued)

risk estimate. This analysis would correspondingly argue for a relatively sizeable dataset.

### 12.1.4 Role of Dependence

Having examined the data frequency and the weighting scheme, we now turn our attention to dependence. In our risk-measurement framework, dependence is restricted to pairwise correlation between risk factors. Correlation is an imperfect measure of dependence and there exist a number of alternatives—unfortunately, they all require a rather significant leap in complexity.[14] We opt for simplicity. For better or for worse, therefore, dependence is linked to correlation in our setting.

Understanding the sensitivity of our risk measures to our correlation estimates is conceptually different to the previous sensitivity analysis. In some ways, it is closer to *true* sensitivity analysis. Changing the frequency or the weighting scheme can have an important impact on the risk estimates, but changing correlation gets to the heart of risk. In this section, we are arbitrarily changing a key element of our risk model. We may, for example, have an empirically estimated correlation coefficient between two risk factors of 0.40. We might then proceed to inquire: what if the correlation coefficient was really 0.6? Or 0.9? This is the spirit of sensitivity analysis. We ask what would happen to our estimate if our inputs are dramatically wrong.

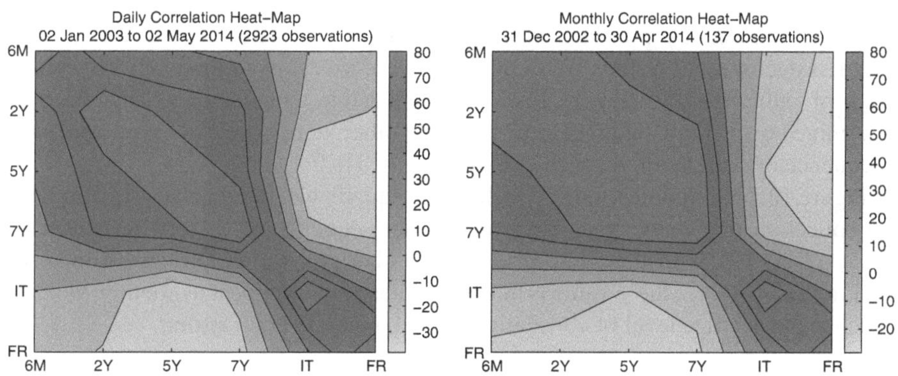


**Fig. 12.6** Heatmaps. Here we observe correlation heat maps for the equally weighted daily and monthly frequency risk-factor data. Negative spread-rate correlations appear more strongly negative with daily data

14The interested reader is referred to Embrechts et al. [7] as a starting point.

Let's begin by examining the correlation structure of the *six* risk factors in our simple example. Figure 12.6 shows the correlation heat maps for the equally weighted daily and monthly frequency risk-factor data. The story is essentially the same for both the daily and monthly correlation estimates.[15] The Bund rates exhibit, as rates typically do, relatively high levels of correlation. The credit spread and Bund rates, however, look to be weakly or even negatively correlated. This is a relatively standard result and it suggests, at least in principle, a substantial diversification effect in our portfolio risk estimates.

Given that one typically has a large number of risk factors—somewhere in the range of 10 to 100—it is very difficult to perturb individual correlation coefficients to see what happens.[16] There are a number of complex techniques that one might employ, but as usual, we will opt for something simpler.

We have seen, in the previous chapter, that we can write the risk-factor covariance matrix as $\Omega_M = VCV$, where $V$ is a diagonal volatility matrix and $C$ is the correlation matrix. We are not interested, in this analysis, in the volatility matrix, $V$. We will hold it fixed. What we seek is a standardized way to examine different alternatives for $C$. We will do this in a very simple and naive manner. Let's write $C \in \mathbb{R}^{m \times m}$ as,

$$C = \begin{bmatrix} 1 & \gamma & \gamma & \gamma & \gamma & \gamma \\ \gamma & 1 & \gamma & \gamma & \gamma & \gamma \\ \gamma & \gamma & 1 & \gamma & \gamma & \gamma \\ \gamma & \gamma & \gamma & 1 & \gamma & \gamma \\ \gamma & \gamma & \gamma & \gamma & 1 & \gamma \\ \gamma & \gamma & \gamma & \gamma & \gamma & 1 \end{bmatrix} \tag{12.14}$$

This specification of $\gamma$ provides us with a single correlation coefficient for all pairwise risk-factor relationships. With this approach, we can systematically consider a range of values for $\gamma \in [-1, 1]$. Each $C$ is indexed to a choice of $\gamma$—that is, $C$ is determined by a single parameter, $C \equiv C_\gamma$. In other words, we can easily construct a sequence of correlation matrices, $\{C_\gamma, \gamma \in [-1, 1]\}$.[17]

Figure 12.7 recomputes, using constant and equally weighted volatility estimates, the collection of ex-ante tracking error for our sequence of correlation matrices, $\{C_\gamma, \gamma \in [-1, 1]\}$. In the same graphic, we also plot the empirical estimate of tracking error—since the $\gamma$ value is unknown, these points appear at about $\gamma \approx 0.18$, which is the average level of correlation between our six risk factors.

[15]Negative spread-rate correlations do appear, however, to be more strongly negative with daily data.

[16]With only 50 risk factors, for example, there are more than 1,200 separate correlation coefficients.

[17]At extreme values of $\gamma$, it is likely that the corresponding covariance matrix is not, in fact, positive definite. This is a technical problem and may even lead, in extreme cases, to negative variance estimates. This is intended, however, to illustrative sensitivity analysis and we should not take these technical challenges too seriously.

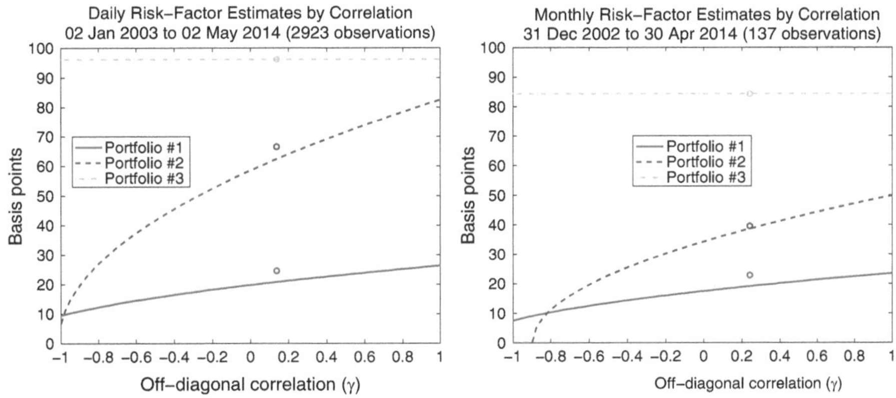


**Fig. 12.7** Impact of $\gamma$. Consider the impact on the TE, by data frequency, for varying choices of $\gamma$. Observe that Portfolio #3 is invariant to $\gamma$. Why?

The first, and perhaps most striking result, is that Portfolio #3 is invariant to the correlation structure. Why is this the case? The answer lies in Table 12.2, where we showed that Portfolio #3 is solely exposed to a single risk factor. With a single risk factor, the correlation with the other five factors is irrelevant.

The second observation is that Portfolio #2 is dramatically affected by the correlation structure. Again Table 12.2 can help us understand why. Portfolio #2, with exposure to all six risk factors, has the most to gain from diversification effects. As the correlation structure changes, its risk estimates—each a function of overall volatility—are strongly affected. Portfolio #1 lies somewhere in between. With a relatively substantial cash position and exposure to only two risk factors, the sensitivity to correlation, while still present, is not as dramatic.

Some caution is nonetheless recommended. This is a very simple way to vary the dependence structure. Our $\gamma$ parameter can be interpreted as something like an average correlation coefficient. It is the same for all risk factors within a given risk estimate. In reality, as evidenced by Fig. 12.6, the level of correlation between different risk factors can vary substantially. Our assumption that correlation coefficients are equal is certainly missing something.[18]

In our defence, Eq. (12.14) is not a model for correlation. Our objective with its use in this analysis is merely to gain an appreciation for the sensitivity of our TE estimates to correlation coefficients. Despite the simplicity of our approach, the results cannot be contested. Correlation matters. Moreover, the larger the number of the risk factors and the greater their heterogeneity, the more it matters. The naivety of our approach does not weaken these conclusions.

[18] There are more sophisticated and realistic ways to understand dependence—copula theory is, in fact, the correct framework to examine dependence, between random variables.

Where does this leave us? As correlations increase, so does risk. Quite simply, the benefits of diversification disappear. What does this imply? The danger lies with underestimates, and not overestimates, of the correlation coefficients. It suggests that one should routinely test the implications of stronger than expected correlations on one's risk measures. An extreme case would involve comparing one's risk estimates with the empirical correlation matrix and the perfectly correlated correlation matrix.[19] This simple test would provide a regular insight into just how much diversification benefits your portfolio. Diversification benefits that, under an extreme set of market conditions, might very well vanish.

### 12.1.5 Summing Up

The previous analysis has involved a large number of moving parts: data frequency, sample sizes, smoothing parameters, and correlation coefficients. Table 12.5 attempts to summarize some of the key results. It illustrates—using daily data—three different choices of weighting parameter, sample size, and correlation coefficient on our three portfolios.[20] The key conclusion is that all of these elements have the potential to importantly influence our risk estimates.

Unfortunately, we cannot rate the various assumptions in terms of their sensitivity on the risk estimates. Depending on the underlying characteristics of your portfolio, the impact can be quite different. What is clear is that one ignores these effects at one's own peril. A prudent approach would probably involve, at the very least, fast- and slow-moving incorporation of market conditions into one's risk measures and a routine examination of stronger-than-expected correlations.

**Table 12.5** Summarizing sensitivity

| | Weighting: $\lambda$ | | | Sample size: $T$ | | | Dependence: $\gamma$ | | |
|---|---|---|---|---|---|---|---|---|---|
| Portfolio | 1.00 | 0.98 | 0.94 | **0.25** | **5.00** | **10.00** | 0.0 | 0.5 | 1.0 |
| #1 | 24.6 | 15.5 | 14.9 | **15.4** | **24.1** | **24.3** | 19.8 | 23.2 | 26.3 |
| #2 | 66.6 | 54.9 | 51.6 | **54.5** | **91.0** | **68.6** | 58.9 | 71.5 | 82.6 |
| #3 | 96.2 | 54.9 | 40.6 | **50.5** | **137.0** | **100.8** | 96.2 | 96.2 | 96.2 |

This table attempts to provide some insight into the range of possible ex-ante TE estimates that might be achieved for different assumptions on weighting, choice of time horizon, and dependence between the risk factors The differences can be quite important.

---

[19]This is merely a matrix of ones in $\mathbb{R}^{m \times m}$ where $m$ is the number of risk factors. While perhaps not very sophisticated, it has the strong advantage of being trivial to compute and implies a lack of possible diversification benefit.

[20]Monthly data offers a similar range of estimates, so we have generously spared the reader the extra numbers. One could also set the $\gamma$ parameter to something more reasonable, such as 0.95 or 0.98. The results would be similar and the covariance matrix would be positive definite.

## 12.2 Backtesting

Sensitivity analysis tells what matters for our risk estimates. It tries to answer the "what-if" questions. Backtesting takes a different perspective. It takes our best risk estimates as given. It then seeks to compare these estimates to realized outcomes. From this perspective, our risk estimates can be viewed as forecasts. They are not, however, forecasts of some specific point in the future—what, in the statistical literature, is often termed a point estimate. Instead, risk estimates are typically a forecast of some dimension of the portfolio's return distribution (i.e., a moment or percentile). The real criterion, therefore, for judging one's parameter settings—and effectiveness of one's risk estimates—is to examine one's forecast accuracy. This is essentially like asking: "are the estimates consistent with the observed reality?" Backtesting essentially seeks to provide a report card on our capacity to assess risk. Indeed, backtesting analysis is analogous to the analysis, in the performance setting, of the residual term in our daily performance attributions. We are essentially verifying that our approximations and assumptions are, in practice, effective and realistic.

Backtesting does, however, present a number of challenges. Unlike examination of the risk term in performance attribution, we cannot assess the accuracy of our risk estimates on a daily basis. If we estimate the average dispersion between portfolio and active returns (i.e., the ex-ante tracking error), we are computing a moment of the active-return distribution. A single outcome of the true distribution will not provide much insight into our success in this venture. Estimating a mean or variance with a single observation will not be a very successful venture. VaR measures, to an even greater extent, face the same challenges. If you expect an estimate to be breached only one out of every 100 realizations, you may need to watch it for some time.[21] Backtesting, therefore, requires data.

We plan to explore the notion of backtesting from *two* alternative perspectives: heuristic and formal. The heuristic perspective is essentially motivational. We will try, without formality, to describe what one would generally expect to observe from a good collection of risk estimates. This amounts to sitting in front of a group of risk estimates and realized returns and trying to judge if the two are consistent with one another.

We will then turn to examine how one might more formally—through the use of statistical tests—assess the accuracy of your risk forecasts. Formal tests, however, are by construction somewhat technical. The heuristic analysis will serve us well in this regard and, hopefully, avoid losing the forest for the trees. In the final section of this chapter, we will try to close the circle with sensitivity analysis in the previous discussion. That is, we will try to select model parameters that are the most consistent with our observed data. In other words, we will reverse engineer

[21]Of course, if you observe ten consecutive daily breaches of a VaR estimate with a 99 % confidence interval, then you probably do not require much more information to assess your model's accuracy.

a collection of risk estimates that, by construction, perform well in a backtesting exercise.

### 12.2.1 A Heuristic Perspective

Comparing risk estimates to return realizations requires a bit of structure. In particular, before examining any data, we need to form an idea of what we would expect to observe, if the risk estimates were sensible. To get started, therefore, imagine that you have $T$ active-return realizations and an *equal* number of active-risk estimates. An obvious place to start is to consider the number of times that the observed active-risk outcomes *exceed* the risk measure. These extreme observations should be consistent with the definition of the risk measures—it will depend, however, if it is a volatility estimate or a percentile measure.

We can probably do a bit better. Risk is a two-sided street. Given that returns are reasonably symmetric, both extreme *positive* and negative return outcomes should be considered. A two-sided perspective, therefore, involves the construction of an interval. We can then pose the following question: how often do you expect the active return to fall into, or outside of, the following illustrative interval?

$$r_{a,t} = \left( - \begin{array}{c}\text{Active-Risk}\\ \text{Measure}\end{array}{}_{t-1}, \begin{array}{c}\text{Active-Risk}\\ \text{Measure}\end{array}{}_{t-1} \right) \text{ for } t = 1, \ldots, T. \tag{12.15}$$

In words, for each point in time we construct an interval $\pm$ our risk measure. To build some terminology, let's call this the risk-measure interval. We may then examine an observed outcome and determine whether it falls inside or outside of the interval. Doing this once or twice will probably not tell us very much. If we do this for many different points in time, however, we might obtain some interesting information.

What do we do with this information? It depends. In fact, it depends importantly on *two* separate, although related, elements:

1. the aspect of the active-return distribution targeted by the risk measure; and
2. the distributional assumption made, if any, by the risk measure.

The first point is rather obvious. If your risk estimate is volatility, your expectation regarding the number of observations falling in the risk-measure interval would be quite different than if you estimated the 99 % VaR. The former measure seeks to describe average dispersion, whereas the latter focuses on the tails of the return distribution. Naturally, we would expect more observations outside the risk-measure interval for a volatility estimate than a VaR-measure with a high level of confidence.

The second point is a bit more subtle. The actual expected number of observations that fall inside and outside the risk-measure interval depend on the distribution of the underlying returns. Consider a volatility measure. Estimating the volatility of a portfolio return is equivalent to estimating the second moment of the underlying return distribution. Simply estimating the second moment does *not* require any distributional assumption. Most distributions possess a second moment. The amount of probability mass that lies between $\pm$ one standard deviation of the mean, however, can be quite different depending on the underlying distribution. We make the fairly uncontroversial suggestion that some thought should be given to understanding the true underlying return distribution.

The same logic applies to percentile (i.e., VaR) risk measures, but it typically impacts both the risk measures and the size of the risk-measure interval. Parametric VaR estimates, as considered in this text, are merely a linear multiple of the volatility measure. The magnitude of the multiple is simply a function of the assumed underlying return distribution—often one assumes that returns follow a normal distribution. Other choices are naturally also possible. It seems logical that the risk-measure interval should be constructed using the same distributional assumption as used in constructing the risk measure. To do otherwise, would be logically inconsistent.

If we have an idea of how many returns should fall in the interval, we may compare it to the actual results. This can be quite informative. Let's look at a practical example. To do this we will examine *two* possible risk measures: one familiar and one invented, although also quite familiar, risk measure. The first measure is the ex-ante TE. The risk-measure interval has the following form,

$$r_{a,t} = (-\sigma_{a,t-1}, \sigma_{a,t-1}) \text{ for } t = 1, \ldots, T. \tag{12.16}$$

What do we expect? It depends, of course, on the underlying return distribution. If $r_{a,t}$ were Gaussian, then you would heuristically assume that about 68 % of the return observations should fall in this interval. If the result is 67 % or 69 %, then we probably would not be overly concerned. Values of 10 % or 90 %, however, would likely be grounds for concern about our risk measure.

For the second measure, we define the following

$$\text{TEVaR}(95\,\%)_{t-1} = \underbrace{\mathcal{N}^{-1}(0.95)}_{\approx 1.64} \sigma_{a,t-1} \tag{12.17}$$

This is a 95 % VaR measure for the active returns, computed under the assumption that the underlying returns are normally distributed. We have taken the liberty of calling this TEVaR. The situation is clear, we expect about 95 % of the active-return outcomes to fall in the following risk-measure interval,

$$r_{a,t} = (-\text{TEVaR}(95\,\%)_{t-1}, \text{TEVaR}(95\,\%)_{t-1}) \text{ for } t = 1, \ldots, T. \tag{12.18}$$

Armed with our two risk measures and our expectations regarding the proportion of observations that should fall in the risk-measure interval, we may proceed to a practical example. We have roughly 5 years of data for a sample UST portfolio—

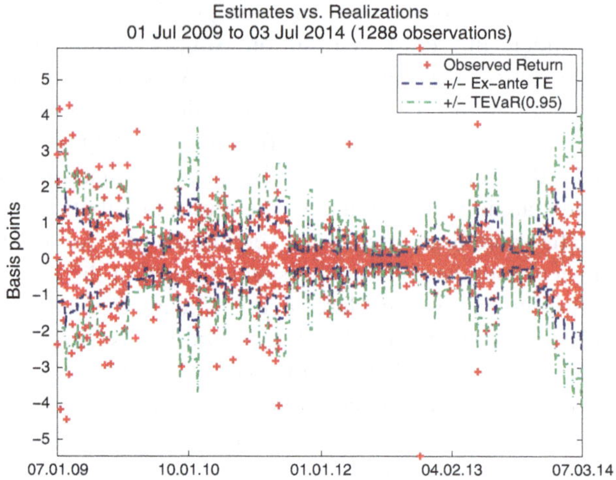


**Fig. 12.8** An example. Here are *5* years of daily return observations along with the two risk-measure intervals for the ex-ante TE and the TEVaR(95 %)

**Table 12.6** A heuristic comparison

| Measure | Observed | Expected |
|---|---|---|
| $T$ | 1,288 | 1,288 |
| $r_{a,t} \notin \left(-\sigma_{a,t-1}, \sigma_{a,t-1}\right)$ | 325 | 409 |
| $r_{a,t} \notin \left(-\mathcal{N}^{-1}(0.95)\sigma_{a,t-1}, \mathcal{N}^{-1}(0.95)\sigma_{a,t-1}\right)$ | 144 | 64 |

Here we compare the number of observations that fall outside of the *expected* interval computed assuming that portfolios returns are normally distributed. TE appears to overestimate risk whereas our VaR estimate seems to underestimate it.

this amounts to almost 1,300 daily return and active-risk estimate observations. Figure 12.8 provides a visual perspective on the active-return observations along with the two risk-measure intervals for the ex-ante TE and the TEVaR(95 %).

It is a bit difficult to make any definitive conclusions from Fig. 12.8. On the one hand, we do not observe anything shocking, such as seeing all of the observations outside or inside the two risk-measure intervals. On the other hand, we see a number of observations outside the risk-measure intervals, but cannot easily judge if this is consistent with our expectations. To say something a bit more definitive, we need to look at some numerical results.

Table 12.6 comes to the rescue with the number of observations, for both risk measures, that fall outside of the risk-measure interval. With 1,288 observations, there are 325 active-return observations outside the tracking-error interval and 144 outside the TEVaR interval. Under the assumption of Gaussianity, we would have expected to observe 409 and 64, respectively, for our two risk measures. It appears, therefore, that *fewer* observations than expected fall outside the TE interval. Simultaneously, it appears that *more* than expected observations fall outside the TEVaR interval.

It is tempting to conclude that we have overestimated TE risk, but underestimated TEVaR risk. Is this fair? In this heuristic setting, it is difficult to say with any precision. Intuitively, the TEVaR seems quite poor. More than twice as many observations fall outside the risk-measure interval relative to our expectation. It does not seem unreasonable to conclude that it is underestimating risk. The TE measure, conversely, is more difficult to judge. Fewer observations fall outside the risk-measure interval, but the difference is smaller. Moreover, it is based on an assumption of normality of the underlying active returns. All we can really say is that there is evidence of overestimating risk, but that it is not definitive.

This is the drawback associated with heuristic analysis. We can easily, and probably even visually, make a decision on very good or very bad risk estimates. When things are close, however, it is difficult to take a decision or make a clear statement about our risk measures without some more formal tools. In the subsequent section, we will introduce a bit more formality to help us resolve this dilemma.

### 12.2.2 A More Formal Perspective

With 1,288 observations, we expect about 400 observations to fall outside the TE interval. We observed about 325, which is less than expected. Is this still reasonable, or are we overestimating risk? It is difficult to answer this question, because we expect some natural variability in this estimate. Indeed, it would be quite surprising if we observed precisely, as suggested by Table 12.6, 409 observations outside of our risk-measure interval. We need some help.

With some assumptions and some statistical machinery, it is possible to statistically test if a given result (i.e., 325) is actually consistent with a robust forecast. There is, in fact, an entire literature dedicated to this topic. There are numerous tests and the treatment can become quite complex. In this text, we will restrict our attention to the original, and still quite popular backtest proposed by Kupiec [8]. This result is intuitive, relatively easily to implement, and represents a natural extension of our previous heuristic analysis.

The general approach, rather unsurprisingly, is statistical hypothesis testing. The basic idea is that we construct a null hypothesis that our model is consistent with the data and then, using the results, look to see if we can reject this hypothesis. If we reject the null hypothesis, at a given level of confidence, this means that we conclude our risk model is either over- or underestimating risk. This is precisely the structure that was missing in our previous heuristic analysis.

Although it sounds simple, some mathematical structure is required.[22] We begin with some important definitions. An observation outside the risk-measure interval is termed a *failure*. This can be easily written as an indicator variable indexed to time,

$$\mathscr{I}_t = \begin{cases} 0 : r_t \text{ \textit{inside} the risk-measure interval} \\ 1 : r_t \text{ \textit{outside} the risk-measure} \\ \quad \text{interval (i.e., a failure)} \end{cases} \quad \text{for } t = 1, \ldots, T. \tag{12.19}$$

The sum of this indicator variable across the entire horizon of analysis is the total number of failures. We call the total number of failures relative to the entire population as the *observed* proportion of failures. This is denoted as $\hat{p}$ and simply defined as follows,

$$\hat{p} = \frac{\sum_{t=1}^{T} \mathscr{I}_t}{T}. \tag{12.20}$$

The theoretical proportion of failures is simply denoted $p^*$. This comes from our model or the features of the assumed underlying return distribution. Table 12.7 restates the results from the previous heuristic analysis in terms of these two variables, $p^*$ and $\hat{p}$.

The backtesting methodology seeks to test the null hypothesis that $p^* = \hat{p}$. That is, the null hypothesis holds that the observed proportion of failures is statistically equivalent to the theoretical proportion of failures. Clearly, in our example, we have overestimated the TE and underestimated the TEVaR(95 %). Is this just normal dispersion of failures? To answer this question, we need to build a test-statistic under the null hypothesis that $\hat{p} = p^*$. This requires some effort.

To start, we wish to characterize the probability of given number of failures. The backtest suggested by Kupiec [8] starts from a simple, but clever, observation. He observed that probability of any single failure can be described by a Bernoulli

**Table 12.7** $p^*$ and $\hat{p}$

| Measure | $p^*$ | $\hat{p}$ |
|---|---|---|
| Ex-Ante TE | 0.32 | 0.25 |
| TEVaR(95 %) | 0.05 | 0.11 |

This table highlights, for our example, the observed proportion of failures ($\hat{p}$) to the theoretical proportion of failures ($p^*$). The next step is to determine if these differences are consistent with normal variation in the data.

[22] This following discussion follows from Kupiec [8].

trial with probability, $p$. That is, each $\mathscr{I}_t$ is a Bernoulli trial with parameter $p$. The Bernoulli distribution has the following form for $\mathscr{I}_t$,

$$\mathbb{P}\left(\mathscr{I}_t = x \,|\, p\right) = p^x(1-p)^{1-x}, \tag{12.21}$$

for $x \in (0,1)$ and $t = 1,\ldots,T$.[23] This can be evaluated for any arbitrary choice of $p$. Two interesting choices, of course, are $p^*$ and $\hat{p}$.

Having an understanding of each single realization of $\mathscr{I}_t$, we need to understand the full collection of realizations. It turns out, quite happily, that $n$ failures in $T$ observations of $\mathscr{I}_t$—ignoring the order of the failures—follows a binomial distribution:

$$\mathbb{P}\left(\sum_{t=1}^{T}\mathscr{I}_t = n \,\middle|\, p\right) = \binom{T}{n} p^n(1-p)^{T-n}, \tag{12.22}$$

for $n = 0,\ldots,T$ and where,

$$\binom{T}{n} = \frac{T!}{n!(T-n)!}. \tag{12.23}$$

Again, this distribution has a parameter, $p$. An important observation is that the binomial failure distribution holds true independent of the assumptions, or lack thereof, made in the computation of the underlying risk measure. This test is thus independent of the underlying risk measure. This is an important advantage.

If you know the distribution of a particular variable, then a popular, quite generally applicable, and useful approach is to use a likelihood ratio test.[24] The idea quite simply is to look at the ratio of the likelihood functions, with parameters selected through maximum-likelihood estimation, under the null and alternative hypotheses.

The likelihood ratio test for the null hypothesis, $H_0 : p^* = \hat{p}$ is

$$\begin{aligned}\Lambda &= -2\ln\left(\frac{\text{Likelihood under the null hypothesis}}{\text{Likelihood under the alternative hypothesis}}\right), \\ &= -2\ln\left(\frac{\mathbb{P}\left(\sum_{t=1}^{T}\mathscr{I}_t = n \,\middle|\, p = p^*\right)}{\mathbb{P}\left(\sum_{t=1}^{T}\mathscr{I}_t = n \,\middle|\, p = \hat{p}\right)}\right),\end{aligned} \tag{12.24}$$

[23]Of course, since $p$ is essentially a probability, we must have that $p \in (0,1)$.

[24]For a thorough and detailed discussion of the likelihood ratio test, please see the excellent reference (Casella and Berger [4]).

$$= -2\ln\left(\frac{\binom{T}{n}p^{*n}(1-p^*)^{T-n}}{\binom{T}{n}\hat{p}^n(1-\hat{p})^{T-n}}\right),$$

$$= -2\ln\left(\left(\frac{p^*}{\hat{p}}\right)^n\left(\frac{1-p^*}{1-\hat{p}}\right)^{T-n}\right).$$

This is termed a proportion of failures (or POF) test. When $p^* = \hat{p}$, this test statistic, $\Lambda$, collapses to zero. The greater the distance between $p^*$ and $\hat{p}$, the larger the value of the test statistic. When it is sufficiently large then, we can reject the null hypothesis. It turns out that, under the null hypothesis, $H_0 : p^* = \hat{p}$, the likelihood ratio follows a chi-squared distribution with one degree of freedom,

$$\Lambda \sim \chi^2(1). \tag{12.25}$$

Thus, at a 95 % confidence level, the critical value is about 3.84. Values of $\Lambda$ greater than this critical value allow us to reject the null hypothesis with a 95 % level of confidence. We now have all that is required for us to perform some hypothesis testing on our example.

Table 12.8 summarizes the test-statistics from Eq. (12.24) for our previous example. We reject the null hypothesis in both cases, although the margin of rejection is rather higher for our TEVaR(95 %) than for the TE. Nevertheless, we can state rather definitively that our TE estimates do indeed appear to overestimate risk, whereas our TEVaR(95 %) computations underestimate risk. For the TEVaR measure this is not much of a surprise, but it is particularly useful for our TE measure.

One might reasonably ask, which is worse? Is it better for a risk measure to be overly conservative or not conservative enough? Neither is particularly good, but an overly conservative estimate is probably preferable. It implies that the estimate is more conservative than reality and this is probably, to the extent that it is not excessive, a not undesirable characteristic for a prudent risk estimate.[25]

**Table 12.8** A formal test

| Measure | $p$ | $\hat{p}$ | $\Lambda$ | Critical value |
|---|---|---|---|---|
| Ex-Ante TE | 0.32 | 0.25 | 26.17 | 3.84 |
| TEVaR(95 %) | 0.05 | 0.11 | 77.85 | 3.84 |

Here we apply the Kupiec-test to the results from Table 12.7 We may, for both risk measures, reject the null hypothesis that $p = \hat{p}$. Statistically speaking, therefore, it appears that there is both under- and over-estimation of risk.

[25]All is not lost for our TE estimate. As will be discussed in the next section, the $p^*$-value depends importantly on the distributional assumption for the asset returns. In this analysis we have assumed

This has been an introduction to formal backtesting, but far from the last word. Indeed, we have only lightly covered the topic. Kupiec [8] developed a useful measure, but it is not perfect. In particular, it only considers the number of *failures*, but not the order in which they occur. This can be quite important. Imagine that your model fails whenever there is a large spike in volatility. Moreover, it tends to fail for a number of consecutive days. This is probably not a desirable feature in a risk measure. Even worse, if these episodes are sufficiently infrequent, it may still pass the Kupiec test, which does not consider the order of the failures. Kupiec's test is, as a consequence of this feature, termed an unconditional test. Considering the ordering of the failures is termed conditionality.

More complex tests are possible to test beyond the unconditional setting—these are imaginatively called conditional tests. There are even ways to jointly test the conditional and unconditional aspects of the model failures. Christofferson [5] and Campbell [2] provide many more details on alternative, and more comprehensive, tests.

### 12.2.3 Thinking Optimally

Almost as important as being able to effectively backtest one's risk measures, is understanding how this information might be employed to generate better risk estimates. Let's look at an alternative example and focus on a portfolio—without strategic benchmark—comprised of the universe of UST notes and bonds from Jan 2008 to May 2014. This is a period of about 6.5 years. Figure 12.9 shows the return observations and the associated risk-measure intervals—computed with equal weights—for portfolio volatility and VaR measures at the 95 % and 99 % levels. Note that, at least initially, our VaR measures are computed parametrically from the portfolio volatility under the assumption of normality.

By virtue of the equal weights with a lengthy sample size, each of the risk measures is generally quite slow moving. Visually, there appears to be some problems with the spike in volatility during late 2008 and early 2009. Other than this episode, it is difficult to conclude more from Fig. 12.9. It should nonetheless be noted that we are asking quite a bit from our risk model. We are asking our model to successfully describe both typical dispersion and simultaneously more extreme tail outcomes, within the same model structure, over an extended period of time. This is not to say that it is a bad idea, but we should recognize that it is challenging. Indeed, jointly testing different aspects of risk is a good idea, and a recommended practice, in the backtesting literature—see again, Campbell [2].

---

normality. If we use a t-distribution with 5.4 degrees of freedom—which is the empirical estimate from the observed data—then your $p^*$ takes the value of 0.26 relative to our $\hat{p}$ estimate of 0.25. This yields a $\Lambda$ value of 0.47, which is well below the critical value. The distributional assumption can make the difference between accepting and rejecting a risk measure.

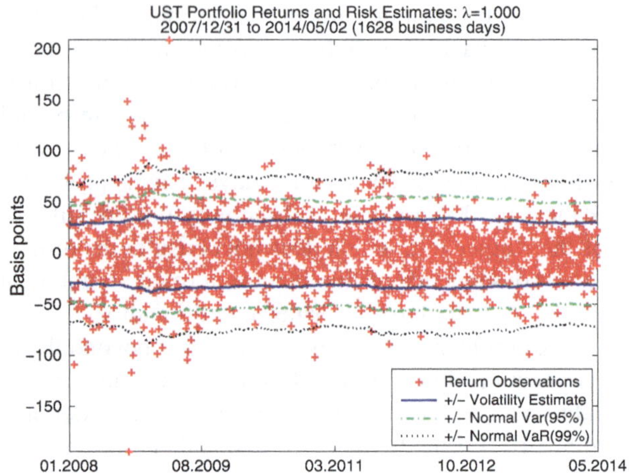


**Fig. 12.9** A more complex example. Here is an example of a portfolio—without strategic benchmark—comprised of UST bonds and notes from Jan 2008 to May 2014. This is a period of about 6.5 years. The risk figures are computed with equal weights

**Table 12.9** Normality and equal weighting

| Measure | $p^*$ | $\hat{p}$ | $\Lambda$ | Critical value |
|---|---|---|---|---|
| Portfolio volatility: $\sigma_p$ | 0.32 | 0.25 | 35.37 | 3.84 |
| VaR(95 %): $\mathcal{N}^{-1}(0.95)\sigma_p$ | 0.05 | 0.09 | 50.68 | 3.84 |
| VaR(99 %): $\mathcal{N}^{-1}(0.99)\sigma_p$ | 0.01 | 0.03 | 34.62 | 3.84 |

Under the assumption of normality and equal-weighting, here are the formal test results for our UST portfolio Again, the null hypothesis is rejected in both cases. Risk appears, under these assumptions, to be incorrectly estimated.

Using our previously developed backtest, we can formally compare our three risk measures to the observed outcomes. Table 12.9 displays the theoretical and observed proportion of failures and computes the Kupiec test statistic. For all three measures the null hypothesis, $p = \hat{p}$, is formally rejected. The portfolio volatility is, once again, overestimated. The two VaR measures, however, are, as before, underestimated. This is a bit discouraging. Can we do better?

Hopefully we can, but to improve the situation, we first need to exert some degree of control over our risk measures. More specifically, we need to identify what might profitably be adjusted to improve the performance of our risk measures. Based on our previous discussion, we can identify *two* important variables that are directly under our control:

1. the $\lambda$ parameter in our weighting function; and
2. the distributional assumption regarding our underlying returns.

We saw how the choice of $\lambda$ can have quite an important influence on the results. We might even postulate that there exists an optimal level, say $\lambda^*$, that maximizes our estimated performance on the back-testing exercise. The weighting parameter is thus an obvious candidate for adjustment and, based on the difficulty of the equally

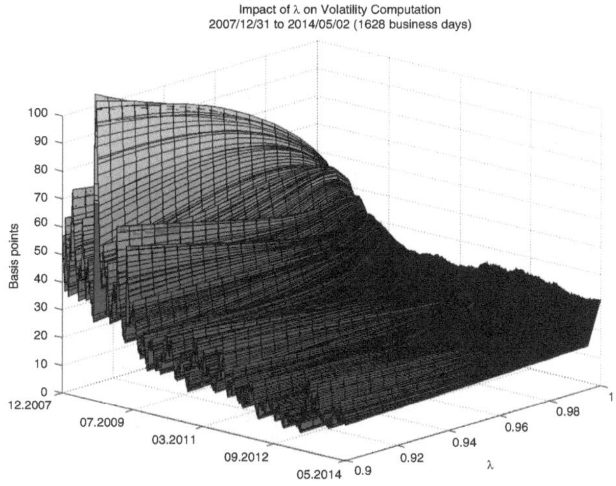


**Fig. 12.10** Choosing $\lambda$. This graphic shows how the volatility estimates vary—for each date over our analysis horizon—for values of $\lambda \in [0.90, 1]$. Note that, in this context, $\lambda = 1$ implies equal weights

weighted risk measures with episodes of volatility, it has the potential to make an important difference.

The second point is equally important. The values of $p^*$ and $\hat{p}$ as well as our parametric VaR estimators also depend importantly on our underlying distributional assumption for the portfolio returns. In fact, they matter in two ways. First in determining the correct choice of $p^*$ and second in the computation of the risk measure itself.[26] Until now, we have assumed an underlying multivariate normal distribution for market risk-factors changes and asset returns. It is time to relax this assumption.

Let's begin with the $\lambda$ parameter from our exponential weighting function. Figure 12.10 illustrates the variation of volatility estimates—for each date over our analysis horizon—for values of $\lambda \in [0.90, 1]$. In this context, as previously, we assume that $\lambda = 1$ implies equal weights.

Although this was already established in previous sections, the impact of the weighting scheme on the volatility estimates is nonetheless remarkable. It is particularly obvious during the high-volatility regime from late 2008 to early 2009. Figure 12.11 provides another perspective on the importance of $\lambda$. Observe how as $\lambda$ tends to unity, the estimates become smoother. The further from unity, the higher the impact of more recent observations, but also the noisier, or alternatively faster moving, the volatility estimates.

Understanding the impact of the weighting parameter on our results, we can now turn to consider the distributional assumption. The simple question is: are the asset return observations consistent with the normal distribution? The simple answer is no. One of the strongest results in empirical finance is the non-Gaussianity of asset

[26]This does not apply to the volatility estimate—it requires no distributional assumption.

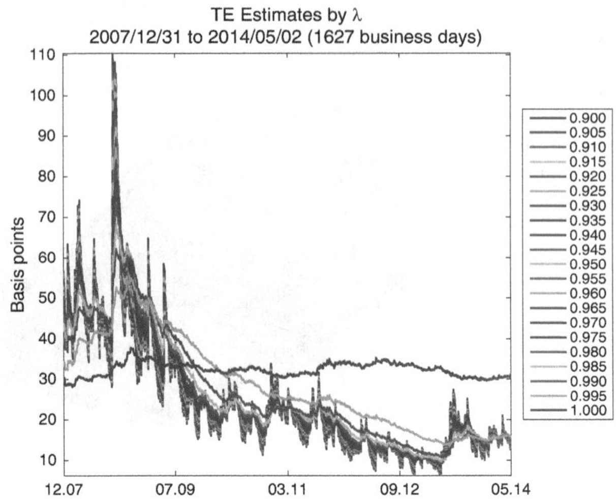


**Fig. 12.11** Another perspective on $\lambda$. Observe how as $\lambda$ tends to unity, the estimates become smoother. The further from unity, the higher the impact of more recent observations, but also the noisier the volatility estimates

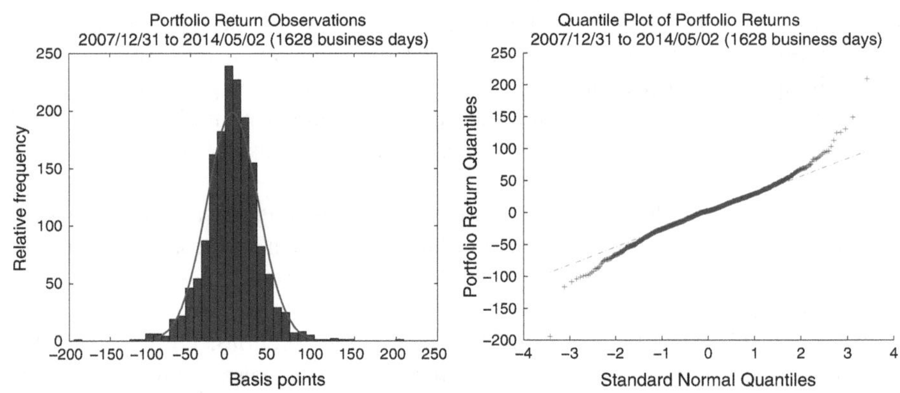


**Fig. 12.12** Judging normality. One of the strongest results in empirical finance is the non-Gaussianity of asset returns. Judge for yourself

returns.[27] We may, however, judge for ourselves. Figure 12.12 outlines a histogram, overlaid with a normal distribution, and a quantile plot of the observed returns from our sample. Both measures strongly suggest non-normality.[28]

[27]The literature on the characteristics of asset returns is enormous. A good starting point is the first few chapters of Campbell et al. [3].

[28]More formal tests are positive. One possibility is the Kolmogorov–Smirnov test.

If returns are not normally distributed, then what are the alternatives? There are other possible symmetric distributions that we might use, which demonstrate a better fit the observed returns. There are many possible choices, but to keep things relatively simple, we opt for the Student-t distribution. This is a family of distributions indexed by the parameter, $\nu$. More information on this distribution can be found in the underlying box or in Casella and Berger [4].

The standard probability density function of the Student-t, or simply t, distribution has the following form,

$$f(x|\nu) = \frac{\Gamma\left(\frac{\nu+1}{2}\right)}{\sqrt{\nu\pi}\Gamma\left(\frac{\nu}{2}\right)}\left(1+\left(\frac{x^2}{\nu}\right)^{-\frac{\nu+1}{2}}\right), \tag{12.26}$$

where $\Gamma(\alpha) = \int_0^\infty x_{\alpha-1}e^{-x}dx$ is a gamma function. Like the Gaussian distribution, the t-distribution has infinite support, $x \in (-\infty, \infty)$. As evidenced by Eq. (12.26), it is not particularly tractable. In the grand scheme of distributions, it is neither extremely important nor very famous. It does, to be fair, play an important role in hypothesis testing in small samples. It nonetheless has a symmetric form and, as the degrees-of-freedom parameter attains about 30, it converges to the normal distribution. For modest levels of $\nu$, it exhibits rather more probability weight in the tails of the distribution. In brief, its symmetry, heavy tails, and easy adjustment through a single parameter make it an attractive choice.

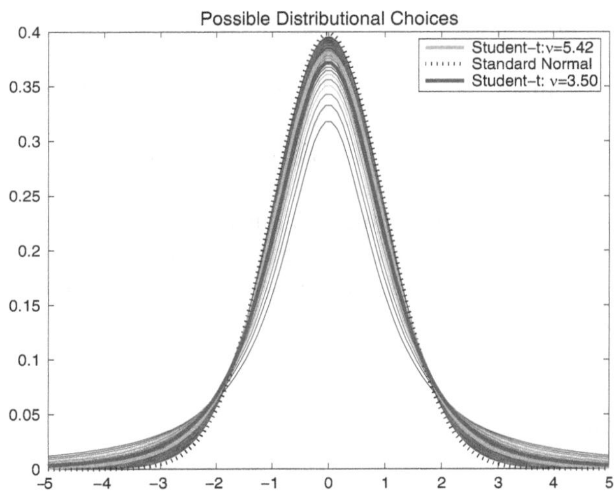


**Fig. 12.13** The student-t distribution. There are other possible symmetric distributions one might use that better fit the observed returns. The Student-t distribution is a family of distributions indexed by the parameter, $\nu$

Figure 12.13 compares the standard normal density function along with a number of alternative Student-t density functions associated with different choices of $\nu$. Note how the standard normal distribution has more probability mass in the centre of the distribution and correspondingly less in the tails. For values of $\nu$ less than about 30, the Student-t distribution has more probability mass in the tails of the distribution and, naturally, somewhat less in the distribution's centre.[29] Under the assumption that our return data follows a Student-t distribution, the empirical $\nu$ parameter is roughly 5.4.

Calibrating our risk estimates to our $6\frac{1}{2}$-year sample of return observations basically amounts to finding sensible choices for the *two* parameters $\nu$ and $\lambda$. We seek the values of these two parameters that maximize our backtesting performance. To some readers, this might feel a bit like cheating. It is probably better to think of it as reverse engineering. If we want robust and reliable risk estimates then we need to think hard about how they are best constructed. Nevertheless, this should not be considered a one-time exercise—analysis like this should be performed regularly.[30]

Our two parameters impact the risk measures in different ways. In particular,

- changing $\nu$ will have an impact on the $p^*$ parameter in our likelihood ratio test;
- changing $\nu$ will also have an influence on the VaR measures—although it has no impact on the volatility; and
- changing $\lambda$ will impact the volatility measure and thus the VaR estimates—it has no influence on the $p^*$ values.

In short, $\nu$ and $\lambda$ do *not* independently affect the backtest results.

We could perform a formal optimization, but that is probably a bit excessive. The objective is *not* to solve for two optimal $\nu$ and $\lambda$ parameters to five significant digits. Rather, we seek sensible values that make logical sense and are broadly consistent with the data.[31] To accomplish this, we examined a broad range of $\nu$ and $\lambda$ values. Although no optimization was involved, it did require a reasonable amount of computing and a few iterations. Ultimately, we obtained the most sensible results by setting $\nu^* = 3.5$ and $\lambda^* = 0.97$. Figure 12.14 illustrates our risk estimates—for these choices of $\nu$ and $\lambda$—along with the true daily return realizations.

As usual, it is difficult to make any strong conclusions from the visual evidence in Fig. 12.14. It is nonetheless clear that the performance is superior to the equally weighted Gaussian setting outlined in Fig. 12.9 on page 410. These risk estimates react more quickly to changing market conditions and failures and, for the most part, appear to be evenly spread out over the time horizon.

---

[29]For $\nu = 1$, the Student-t distribution is actually equivalent to the Cauchy distribution, which is famous for having neither mean, variance, nor indeed any other higher moments.

[30]Daily backtesting probably does not make much practical sense. Every quarter or twice a year, however, would not be unreasonable.

[31]In principle, one wants to select parameters that will work across many periods and datasets—a hyper-specialized estimate to a single dataset is unlikely to be very robust.

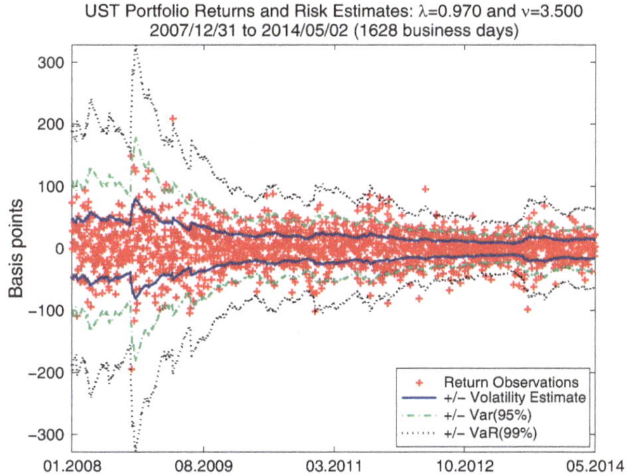


**Fig. 12.14** Getting the right $\nu$ and $\lambda$. Here are the results for our selected choices of $\nu$ and $\lambda$

**Table 12.10** Judicious $\nu$ and $\lambda$ values

| Measure | $p^*$ | $\hat{p}$ | $\Lambda$ | Critical value |
|---|---|---|---|---|
| Portfolio volatility: $\sigma_p$ | 0.21 | 0.20 | 1.04 | 3.84 |
| VaR(95 %): $\mathscr{T}^{-1}(0.95, 3.5)\sigma_p$ | 0.05 | 0.06 | 1.98 | 3.84 |
| VaR(99 %): $\mathscr{T}^{-1}(0.99, 3.5)\sigma_p$ | 0.01 | 0.01 | 0.34 | 3.84 |

Employing the same data sample, but changing the assumptions from Table 12.9, we demonstrate a sensible set of risk estimates.

The real test, however, comes from formally testing the three risk measures. Table 12.10 summarizes the results. The null hypothesis, $p^* = \hat{p}$, cannot be rejected for any of the three risk measures. Given that the estimates were constructed specifically to pass the backtest, we should not be overly surprised. A more important question is: are these sensible parameter values? The weighting parameter seems reasonable. A $\lambda$ value of 0.97 appears to be a good trade-off between incorporating market data and being over-reactive. It falls comfortably into the typical range. For a sample of 250 days, it has a weighting half-life of about 22 days with 99 % of the weight applied within about 150 days. The choice of $\nu$ is perhaps less easy to defend. A value of 3.5 compares reasonably to the empirical value of 5.4, but it still remains slightly arbitrary.

As usual, the previous numerical analysis is only suggestive. The calibration exercise cannot tell you precisely how to construct your risk estimates, but it does provide you with valuable information. In the same way that residual analysis of performance attributions is meaningful, backtesting tells you if your risk estimates are consistent with the true return observations. In both cases, this information can be used to circle back and formulate better risk or performance analysis.

## 12.3 Concluding Thoughts

Risk measurement is *not* an exact science. Instead, it is a forecasting exercise. More specifically, it is an exercise in forecasting the future joint distribution of the returns in your portfolios. Moreover, this is based upon a second-order Taylor series expansion of the bond-price function combined with an estimation of the covariance matrix of the market risk factors impacting your portfolio. Much can go wrong. As a consequence, it is advisable to keep a close eye on your risk estimates. As a first step, we need to think carefully about what factors exhibit the most influence on these estimates. This is termed sensitivity analysis and, when done well, it becomes part of your overall framework. Examples would include multiple risk measures—slow and fast-moving—and regular tests on the correlation coefficients.[32] As a second step, it is important to periodically compare one's risk estimates to the true returns outcomes they purport to forecast. This process, commonly known as backtesting, provides helpful information on the general success of your risk estimates. In fact, understanding how your forecasts perform—and more particularly, how they fail—can be instrumental in improving their overall accuracy.

## References


1. D.J. Bolder, Y. Romanyuk, Combining canadian interest-rate forecasts. Bank of Canada: Working Paper 2008–34, Ottawa, 2008
2. S.D. Campbell, A review of backtesting and backtesting procedures. Federal Reserve Board, Finance and Economics Discussion Series 2005–21, Apr 2005
3. J.Y. Campbell, A.W. Lo, A.C. MacKinlay, *The Econometrics of Financial Markets* (Princeton University Press, Princeton, 1997)
4. G. Casella, R.L. Berger, *Statistical Inference* (Duxbury Press, Belmont, 1990)
5. P.F. Christoffersen, Evaluating interval forecasts. Int. Econ. Rev. **39**(4), 841–862 (1998)
6. F.X. Diebold, A. Hickman, A. Inoue, T. Schuermann, Converting 1-day volatility to h-day volatility: scaling by $\sqrt{h}$ is worse than you think. Wharton Financial Institutions Center Working Paper, pp. 97–34, July 2007
7. P. Embrechts, A. McNeil, D. Straumann, Correlation and dependence in risk management: properties and pitfalls. ETH Working Paper, 1999
8. P.H. Kupiec, Techniques for verifying the accuracy of risk measurement models. J. Deri. **3**, 73–84 (Winter 1995)
9. A. Meucci, Visualizing the propagation of risk: square-root rule, covariances, and ellipsoids. GARP Risk Professional - "The Quant Classroom", Feb 2010
10. J.P. Morgan/Reuters, *RiskMetrics™— Technical Document* (Morgan Guaranty Trust Company, New York, 1996)


[32]This is not as crazy as it sounds. There is evidence in the forecasting literature—see, for example, Bolder and Romanyuk [1]—that multiple models with different assumptions perform better jointly than individually.

# Part V
# Risk and Performance

Risk and performance are often examined separately. Given that return can only be achieved, at least in expectation, by exposing oneself to risk, it makes good logical sense to consider them jointly. It is nevertheless not a completely trivial task to combine risk and return. The next two chapters examine, from different temporal perspectives, different possible approaches to combining these two important dimensions.

# Combining Risk and Return 13

*Combine the extremes, and you will have the true centre.*

Karl Wilhelm Friedrich von Schlegel

It is generally difficult to read any finance article without stumbling, at some point, across the notions of risk and return. Modern portfolio theory has been built on the underlying premise that increased expected return cannot be attained without exposing oneself to greater risk. Strategic asset allocation (SAA)—the process of determining one's medium- to long-term portfolio given constraints, objectives, and assumptions on the distributional characteristics of asset returns—concerns itself principally with finding an optimal trade-off between risk and return. A broad range of ratios—some well-known examples including the Sharpe, information, Treynor, Calmar and the appraisal ratios—combine historical return observations in such a manner as to approximate the amount of risk earned per unit of risk assumed.[1]

Risk and return also dominate the practitioner world. Risk management groups across the globe compute a wide range of ex-post and ex-ante notions of risk for their portfolios in an effort to monitor and manage their exposures. Portfolio returns are also computed on a daily basis and are linked together across time for comparison to their respective benchmarks. Many institutions make extensive use of return attributions to determine the source of returns over the last few days, weeks, or months. Often ex-post performance and risk are combined to compute new ratios to track the ongoing performance of internal and external investment mandates. These combined risk-return computations, however, are generally only performed with monthly data and typically only attain a reasonable level of reliability once one has accumulated between 24 and 36 months of data observations.

[1]For more information on these ideas, please see Chap. 14.

 
D.J. Bolder, *Fixed-Income Portfolio Analytics*,
DOI 10.1007/978-3-319-12667-8_13

In the finance literature, therefore, we see an extensive examination of risk and return in a combined manner. In the practitioner world, however, risk and return are rigorously independently computed on daily basis. Indeed, in the preceding chapters, we invested significant effort to highlight how one might compute such risk and return measures for a broad range of portfolios. These daily notions of risk and return are generally only combined at lower frequencies and examined on a relatively infrequent basis.

The question posed by this chapter is: can one sensibly combine high-frequency (i.e., daily) measures of risk and return? There are reasons for scepticism. One could, for example, postulate that daily risk and return measures are essentially noise and that a clear signal about the risk and return characteristics of a given portfolio can only be ascertained with monthly, quarterly, or even annual data. One could also argue that daily risk and return are fundamentally different objects. Daily risk measurement uses information in the recent past to generate an ex-ante estimate for the range of possible outcomes that may occur in the future. The daily return is merely the realization of one of these outcomes over the course of a single day. Interpretation of a combination of these daily measures may be challenging.

There are nonetheless reasons for taking an interest in this question. A well-known branch of asset-management research suggests that upwards of 90 % of the variance of the return of one's portfolio is determined by the choice of its strategic benchmark: see Brinson et al. [4] and Hood [8].[2] This literature is occasionally used to suggest that, given its relative lack of importance, excessive time and effort is spent upon tactical portfolio management. The corollary is that, although the SAA process is of paramount importance for the overall risk of every institution involved in portfolio management, only a relatively small amount of time and effort is allocated to this task. High frequency examination of risk-adjusted returns may provide more information on the relative success of a given SAA implementation and thus help somewhat to address this shortcoming.

Another reason to take interest in the combination of daily risk and return measures is that it may assist in tactically positioning one's portfolio or making dynamic adjustments to one's benchmark. Levels of risk-adjusted return and associated trends on a portfolio level or with respect to specific risk factors could prove quite illuminating in tactical and dynamic decisions.

This chapter poses a simple question about whether it makes sense to combine daily measures of risk and return. The principal application would be to enhance one's daily or weekly reporting and find a sensible way to jointly interpret the usual high-frequency risk and return measures. One may either immediately dismiss the question or consider it worthy of future examination. To pursue it at all, however, one must accept the premise that there may be information in daily portfolio data above and beyond what is embedded in monthly data.

[2]Although subsequent work, see Ibbotson and Kaplan [9] for example, debate exactly what percentage of return variance is explained, there is general agreement that it is sizeable.

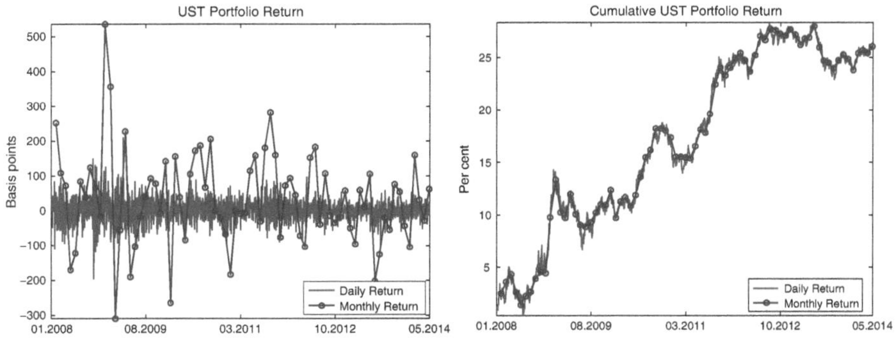


**Fig. 13.1** Daily vs. monthly returns. This underlying table compare and contrast raw and cumulative daily and monthly returns for the Merrill Lynch (MEL) US Treasury tradeable security universe from late December 2007 until May 2014

Figure 13.1 outlines raw and cumulative daily and monthly returns for the Merrill Lynch (MEL) US Treasury tradeable security universe from the beginning of 2008 until early May 2014. While Fig. 13.1 neither confirms nor negates the premise that daily returns hold useful information not found in monthly returns, it is nonetheless suggestive. There appear to be events occurring at a daily frequency that are not entirely reflected at the monthly level.

This discussion does not claim to have a definitive answer to this question. A few points have become nonetheless clear over the course of our analysis. First, there are a few challenges associated with addressing this question. In particular:

- daily returns are extremely noisy; and
- daily ex-ante risk is quite persistent.

Direct combination of daily risk and return is *not* going to be very helpful. Second, in this chapter, we sadly do *not* find a simple and meaningful daily measure of risk and return. We do, however, take some first steps towards usefully combining daily risk and return that nevertheless require some careful transformation of the underlying data.

The remaining discussion proceeds in three main parts. The first section concerns itself with the introduction and examination of our data. The majority of the development is performed within the context of a simple example: 76 months of daily risk and return data for the US Treasury bond and note universe. The second section explores some alternative techniques for dampening the amount of noise in the daily returns and the associated implications of such an adjustment. The third section then proceeds to use the adjusted daily returns in conjunction with the daily risk figures to construct a few alternative measures of risk-adjusted returns.

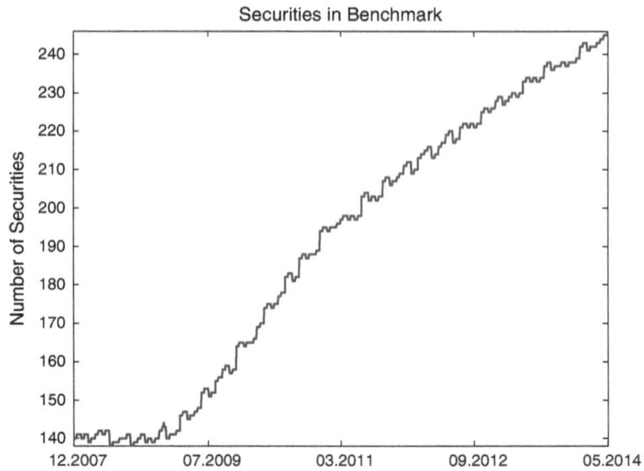


**Fig. 13.2** Number of securities. This figure tracks the number of securities in the MEL US Treasury tradeable security universe since July 2009

## 13.1 The Data

To examine what might be done to address our straightforward, but challenging, question, we will work with a practical, simple, and well-known example: the tradeable universe of US Treasury note and bond securities as encapsulated by the 1–30 US Government benchmark. Figure 13.2 tracks the number of securities in our example benchmark since January 2008. Over the 6+ year period under examination, there are an average of about 190 bonds in the benchmark.

Our choice of data is relatively simple because it is virtually devoid of:

- currency risk;
- liquidity risk; and
- credit-spread risk.

By selecting a single issuer of liquid securities with a very high level of credit quality, we have dramatically reduced the number of possible risk factors driving portfolio returns. The principal risk factors are movements in the underlying US Treasury yield curve. This was done by construction based on the logic that with a smaller number of moving parts, we can better focus on the important ideas.[3]

[3]Although we are not specifically looking at active management of portfolios, the following ideas could apply equally to active return and risk. In this case, one's dataset would be comprised of the risk and return differences between a given portfolio and its benchmark.

### 13.1.1 Understanding Our data

Our detailed sample provides us with a highly detailed time series of both risk and return organized both on the aggregate level and by specific risk factor. We could, of course, start combining them immediately. As we will see in a moment, this will turn out to be a bad idea. Before proceeding further, we should instead begin by gaining an understanding of the characteristics of our data. Specifically, we wish to address the following questions regarding both the time series of risk and return:

- How do they evolve over time?
- Are they slow-moving, fast-moving, or noisy?
- Are there obvious trends?
- How does a given date's values depend on previous values?
- Are the risk and return time series related to one another?

Once we have answers to these questions, we will be in a better position to consider ways of combining them. In the subsequent sections, we will progressively increase the depth of our analysis starting from a preliminary graphical analysis and then move steadily towards the distributional properties of our risk and return data.

Let us begin with an examination of the raw daily return data. Figure 13.3 outlines the roughly 1,600 daily returns over the sample period both as a scatter-plot and a time series. Observe that while the vast majority of returns fall within a range of about $\pm 100$ basis points, they appear to be roughly symmetrically centred around zero. Moreover, it is difficult to ascertain any trends in the data as returns appear to oscillate strongly from 1 day to another.

This result should not be an enormous surprise given that daily returns are essentially a mathematical transformation of daily movements of the underlying risk factors—in this case, the risk factors are US Treasury yields. The behaviour of

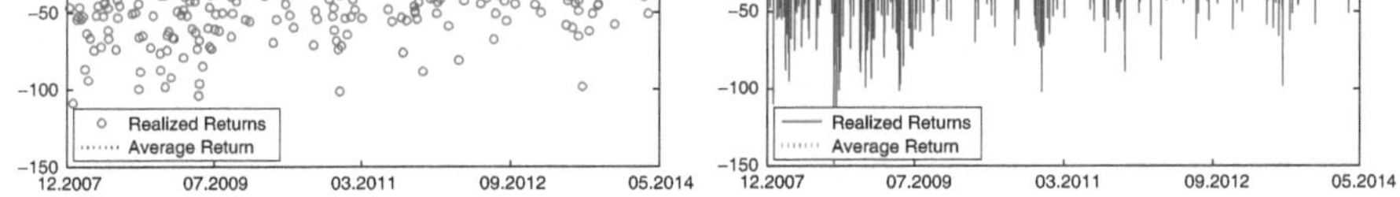


**Fig. 13.3** Daily returns. This figure traces the daily evolution of our US Treasury security benchmark example

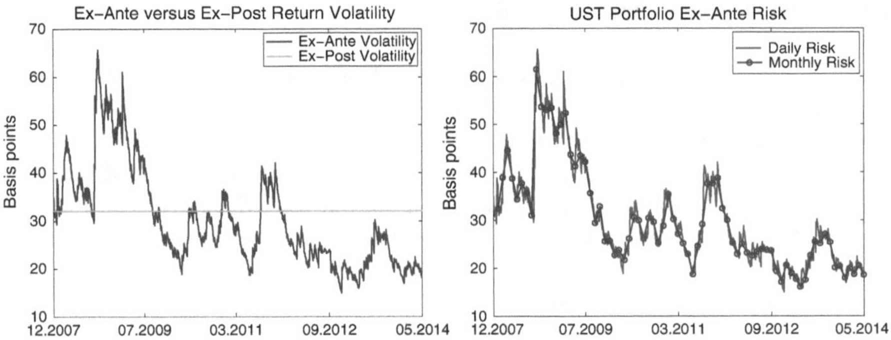


**Fig. 13.4** Daily ex-ante volatility. This figure describes the daily ex-ante volatility of our US Treasury portfolio's returns. It also compares the daily and monthly estimations of ex-ante volatility

the daily return series does not differ dramatically from the daily first differences of say, 10-year US Treasury yields.[4]

Let us now turn our attention to our principle measure of risk: ex-ante volatility.[5] Figure 13.4 highlights the daily evolution of this risk measure over our data horizon. Ex-ante daily portfolio volatility over the period ranges from roughly 20–65 basis points. While it at times demonstrates an ability to vary quite quickly over short periods of time, it appears overall to be a relatively persistent time series. In contrast to the return series, for example, it appears to be remarkably stable.

Figure 13.5 introduces the attribution of the daily risk and return figures to three principal risk factors: the short-end of the US Treasury curve (i.e., 6 months to 5 years), the intermediate part of the US Treasury curve (i.e., 5–10 years), and the long end of the US Treasury curve (i.e., 10–30 years). There are, of course, other ways to decompose one's daily risk and return figures, but we find this to be a reasonable choice.[6]

The smallest contributor to both daily return and risk is the short-end of the curve. Given the shorter modified durations of securities in this sector, and the correspondingly lower levels of exposure to yield movements, this seems quite reasonable. The amount of risk and return stemming from the intermediate and long sectors of the US Treasury curve, by contrast, seem to be quite similar although their relative magnitudes appear to vary significantly over time. That is, at times the

[4]The corollary is that an ability to predict daily portfolio returns would be tantamount to being able to accurately predict daily movements in US Treasury yields. Were this the case, it would imply a violation of the efficient markets hypothesis. While violations of this theory abound in recent years, it is probably not tremendously far from the truth in the deep and liquid US Treasury market. See Brealey et al. [3] for a more detailed description of the efficient markets hypothesis.

[5]The exponentially weighted covariance matrix used in this computation is estimated with daily data and a decay factor of 0.96. Portfolio returns are approximated using the standard additive risk-factor decomposition described in previous chapters.

[6]One could, for example, use model-based or *ad hoc* defined notions of level, slope, and curvature.

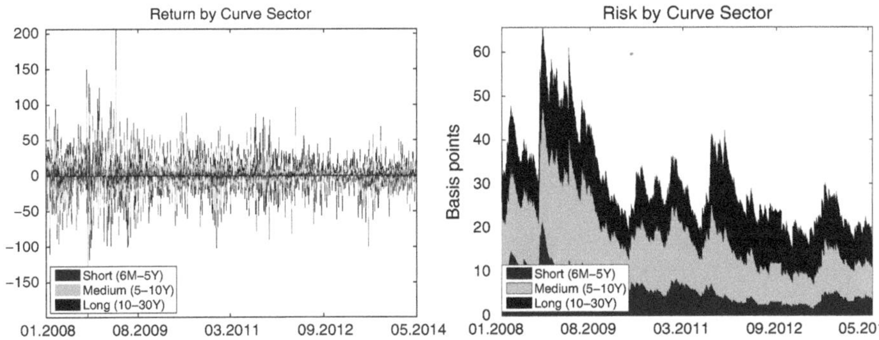


**Fig. 13.5** Risk and return attribution. Here we see the decomposition of risk and return by exposure to three curve sectors: the short- (6 months to 5 years), intermediate- (5–10 years) and long-end (10+ years)

risk (and associated return) arising from the long end appear to dominate. During other periods, intermediate tenors look to generate more risk (and, at times, more associated return).

Even with three rather simply defined risk factors, we are capable of observing rather complex behaviour and thereby gaining substantial insight into our portfolio. Furthermore, despite the fact that daily returns are (by far) the noisiest of our two time series, there is a non-trivial amount of variation in the daily risk of our US Treasury portfolio.[7]

Numerical analysis can also provide some insight into our risk and return data. Table 13.1 summarizes some basic summary statistics surrounding the daily risk and return figures outlined graphically in the previous section. We can make a number of quick observations:

- neither risk nor return appear to be particularly normally distributed;
- there appears to be virtually no correlation between the two measures; and
- the autocorrelation figures highlight the persistence of risk and the virtually memoryless nature of daily returns.

One strong implication of this analysis is that, in their daily form, the individual return observations appear to be essentially independently distributed.

The risk figures are particularly non-Gaussian with a truncated lower end and something of a positive skew. The daily returns, while relatively symmetric, exhibit rather more probability weight in the distribution's tails than one would expect from a normal distribution *Skewness*.

[7]The period in question also coincided with rapid increases in the overall amount of US Treasury debt outstanding and a marked decrease in the overall level of yields. Part of the relative changes of risk between the various sectors, therefore, describes the issuance practices of the US Treasury over this period.

**Table 13.1** Summary statistics

| Statistic | Return | Risk |
|---|---|---|
| Mean | 1.6 | 30.3 |
| Median | 2.1 | 27.2 |
| Maximum | 209.9 | 65.6 |
| Minimum | −195.3 | 15.0 |
| Volatility | 32.2 | 10.2 |
| IQR | 37.5 | 13.1 |
| Skewness | 0.0 | 1.0 |
| Kurtosis | 5.8 | 3.5 |
| Signal-to-noise | 0.1 | 3.0 |
| CV | 19.7 | 0.3 |
| Correlation | 0.0 | |
| Autocorrelation | | |
| 1-Day | −0.0 | 1.0 |
| 1-Week | −0.0 | 1.0 |
| 1-Month | −0.0 | 0.9 |

Below are some useful summary statistics that permit rapid comparison of the distributional features of the daily risk and return observations.

**Table 13.2** Curve risk-factor correlations

| Tenor | Short | Medium | Long |
|---|---|---|---|
| Return | | | |
| Short | 1.0 | 0.9 | 0.7 |
| Medium | 0.9 | 1.0 | 0.9 |
| Long | 0.7 | 0.9 | 1.0 |
| Risk | | | |
| Short | 1.0 | 0.8 | 0.3 |
| Medium | 0.8 | 1.0 | 0.6 |
| Long | 0.3 | 0.6 | 1.0 |

The underlying table provides some summary statistics regarding the curve-related risk and return for short-, medium-, and long tenors.

The contemporaneous correlation between risk and return is quite small. Within the risk and return measures, however, there appears to be rather more correlation. In Table 13.2 we compute the contemporaneous correlation of daily risk and return figures across the various elements of the US Treasury curve.

The correlation among daily returns across the three curve sectors are quite high, but the correlation of daily risk between short and medium tenors and the long end of the US Treasury curve is surprisingly weak.

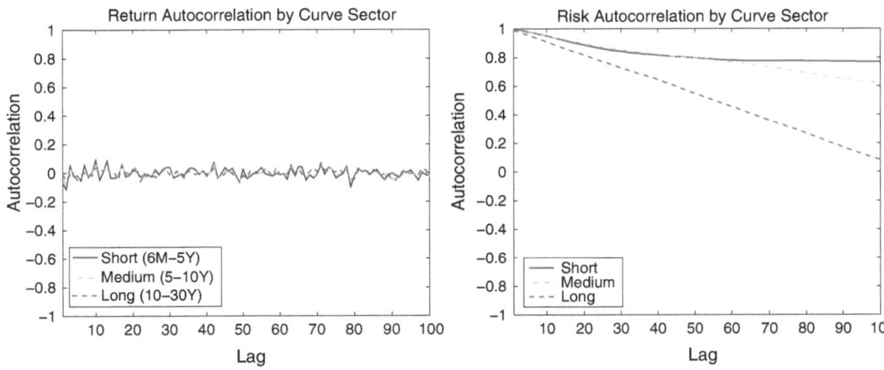


**Fig. 13.6** Autocorrelation. This figure describes respective autocorrelation functions for our daily risk and return observations

**Table 13.3** High-level overview

| | | Daily | | | |
|---|---|---|---|---|---|
| Statistic | Monthly | Total | Short | Medium | Long |
| Return | 34.3 | **1.6** | 0.4 | 0.6 | 0.5 |
| Positive (%) | 56.6 | **54.8** | 54.4 | 54.6 | 54.6 |
| Ex-ante risk | 137.2 | **30.3** | 6.3 | 12.9 | 11.1 |
| Return per unit of risk | 0.2 | **0.1** | 0.1 | 0.0 | 0.0 |
| Risk per unit of return | 4.0 | **18.5** | 14.4 | 21.6 | 23.2 |

The underlying table provides some summary statistics regarding monthly, daily, and curve-related risk and return for short-, medium-, and long tenors.

The persistence and memory of a process is closely linked to its autocorrelation.[8] Figure 13.6 highlights the respective autocorrelation functions for risk and return observations out to 100 daily lags (i.e., almost 5 months).

The forms of these two functions are dramatically different. The daily returns have almost no relationship to their previous values, while the risk figures, across all sectors, exhibit a strong persistence that falls off gradually over time. Again, this is more evidence of the independent nature of raw daily returns.

Before we move any further, let us attempt, on a preliminary basis to organize the unadjusted relationship between risk and return. Table 13.3 provides a high-level description of the average return and risk figures over the range of our data observations. It includes monthly and daily returns and their associated ex-ante volatility.[9] For the daily observations, risk and return figures are also provided

[8] Autocorrelation describes a given value's relationship with previous observations.

[9] Daily risk figures are transformed into a monthly risk approximation by scaling them with $\sqrt{21}$. This is based on the assumption, with roughly 250–255 working days per year, there are approximately 21 working days in each month.

for our three main risk factors: short, medium, and long US Treasury yield-curve movements.

Armed with these summary statistics, we can proceed to perform a few simple computations. With the daily data, the ratio of the average daily return to average ex-ante risk is 0.1 units. This figure can be interpreted as the amount of daily return, measured in basis points, of return earned per unit of risk taken. The inverse of this figure is also interesting: the average amount of ex-ante risk normalized by the average daily return. This ratio, also termed the coefficient of variation (cv), can be interpreted as the amount of risk taken for each unit of return. On a daily basis, this value is almost 19 basis points. On average, therefore, one has to undertake substantial risk to generate a unit of return.

When we examine the corresponding risk factor, less return is generated per unit of risk as we move further out along the US Treasury curve. Shorter tenor yield movements appear, on average over the period, to generate almost three times the amount of return per unit of risk—or equivalently require almost one third of the risk to create one basis point of return—relative to long-tenor yield changes.

The monthly and daily figures tell rather different stories. With an average monthly return of 34 basis points and an associated average ex-ante return volatility of slightly more than 135 basis points, the average monthly return per unit of risk is 0.2 basis points.[10] This is roughly five times more than the daily approximation. Similarly, the risk assumed for each basis point of return is only about 20 % of the value estimated with daily data. Part of the answer stems from the fact that while return grows linearly over time, risk only grows at the square root of time. Specifically, while the monthly return is roughly the equivalent of 21 working days times the average daily figure, the risk increases by a factor of $\sqrt{21}$. The result is more return per unit of risk.

The proportion of positive monthly returns is actually almost 2 % points greater than that observed in the daily return observations. Furthermore, there are differences between the overall daily return and the returns associated with our three different yield-curve sectors. It seems, therefore, that some information is lost when aggregating by risk factors over time.

What can we conclude about daily return and ex-ante risk from this analysis? Daily returns are extremely noisy, exhibit weak autocorrelation, strong correlation among the underlying risk factors, virtually no contemporaneous correlation with risk, and appear that they can be loosely approximated by a Gaussian distribution. Ex-ante risk, conversely, is much more persistent over time, exhibits strong

[10]Assuming that daily returns are independent and approximately normally distributed, then we can easily compute the average monthly return from the daily statistics. If $X_1, \ldots, X_n$ are independent and identically distributed random variables with mean $\mu$ and standard deviation $\sigma^2$, then $\sum_{t=1}^{n} X_t \sim \mathcal{N}\left(n\mu, n\sigma^2\right)$. For a 21-day month, our daily estimates translate into an expected monthly return and volatility of 34 and 139 basis points, respectively. This is quite close to the figures presented in Table 13.3.

autocorrelation, has modest correlation among the underlying risk factors, shows virtually no contemporaneous correlation with return, and is strongly non-Gaussian.

When we examine return per unit of risk, we find that on average over our data sample, one has to assume roughly 19 units of risk to generate one unit of return. Although this is an interesting result when computed as an average over the entire data sample, it is essentially worthless when computed on a day-to-day basis. The reason is that any measure of raw daily returns and daily ex-ante risk will inherit the noisiness and weak autocorrelation of the daily-return series. To make such a daily measure meaningful, we need to find a technique to reduce the noise in daily returns and, if possible and it actually exists, extract a trend.

## 13.2 Dampening Return Noise

In their raw form, daily returns are virtually impossible to use. Raw daily returns, therefore, require some form of adjustment to dampen their noise and highlight the recent trend. Ideally, we could decompose the return into these two categories. To accomplish this, we have identified *four* different techniques that we might employ to perform such an adjustment. These include:

- a moving average;
- a non-parametric moving-average technique (i.e., kernel regressions);
- a filtering technique; or
- a noise reduction technique.

Each of these approaches attempts to find a trend in the underlying daily return series. The objective is to determine a return trend at both the overall daily-return level and for each risk factor that can be sensibly compared to its associated ex-ante risk exposure.

### 13.2.1 The Moving Average

The moving average is the most obvious and simple approach. We merely define a window, say 5 or 10 days, and compute a rolling average return using this window. As we move forward each day, we include the most recent observation and exclude the furthest observation while always conserving a fixed number for the computation of our average. Figure 13.7 computes precisely such a measure using a number of different frequencies: 5, 10, and 30 days.[11]

[11]This basically amounts to a rolling window of 1, 2, and 6 weeks of daily return observations.

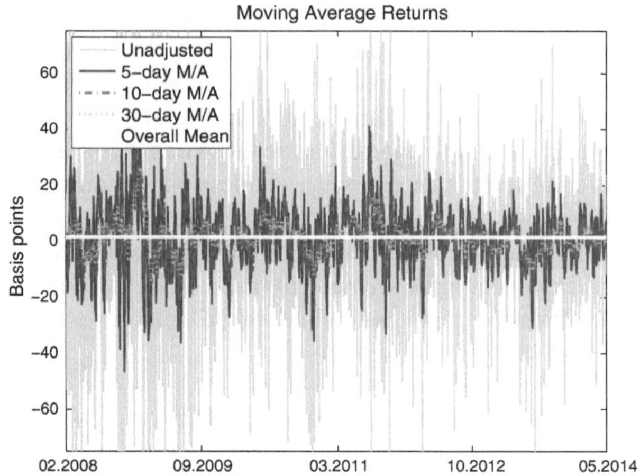


**Fig. 13.7** The obvious approach. Here we compare upon a number of different frequencies, different moving average returns

A moving average, therefore, is basically a local average of the recent values in a time series. Merely by averaging the most recent five daily-return observations we reduce the volatility of the daily returns by a factor of three. Moreover, the application of a moving average to the daily returns extracts a discernible trend that can be easily interpreted.

### 13.2.2 The Hodrick–Prescott Filter

The Hodrick–Prescott (HP) filter is commonly used in economics to extract a cyclical trend from a raw time series with noise. Originally proposed by Hodrick and Prescott [7], it linearly breaks down a time series $\{x_t, t = 1, \ldots, T\}$ into:

$$x_t = g_t - c_t, \tag{13.1}$$

where $g_t$ and $c_t$ are referred to as the growth and cyclical components, respectively.[12] They construct the following optimization problem in $g$:

$$\min_{\{g_t, t=1,\ldots,T\}} \left( \sum_{t=1}^{T} c_t^2 + \lambda \sum_{t=1}^{T} \left( \underbrace{(g_t - g_{t-1})}_{\Delta g_t} - \underbrace{(g_{t-1} - g_{t-2})}_{\Delta g_{t-1}} \right)^2 \right), \tag{13.2}$$

where, once solved, $c_t$ is easily determined from Eq. (13.1). If $\lambda$ is equal to zero, then the choice of $g$ does not matter and the cyclical component is the same as

[12] See also Baxter and King [1] and Guay and St. Amant [6] for more discussion of filtering techniques.

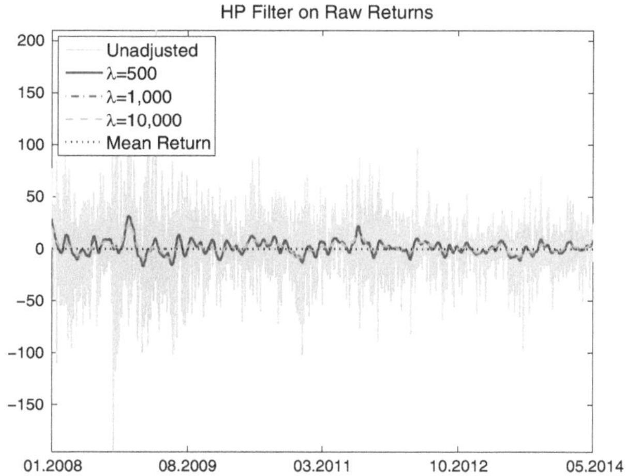


**Fig. 13.8** The Hodrick–Prescott filter. This figure outlines the application of the Hodrick–Prescott filter to US Treasury portfolio's returns. The HP filter is commonly used in economics to extract a trend from a noisy time series

the actual time series. As the $\lambda$ parameter tends to infinity, the sum of the second differences takes on enormous importance and the best fit is a linear model. For moderate choices of $\lambda$, however, the second term in the optimization problem penalizes variability in $g_t$ and one obtains a smoothed cyclical trend from the raw data.

Figure 13.8 provides an illustration of the application of the HP filter to the raw daily return data for a range of alternative choices of the smoothing parameter, $\lambda$. Observe that the cyclical trend from the HP filter demonstrates dramatically less noise than the raw daily return series. It remains somewhat oscillatory, but one is capable of discerning return behaviour that last for several days and even weeks.

### 13.2.3 The Kernel Regression

Another possible noise-reduction or smoothing technique that may be employed stems from non-parametric statistics: the kernel regression.[13] The simplest kernel regressions for a time series $\{x_t, t = 1, \ldots, T\}$ have the form,

$$\hat{x}_t = \text{mean}\left(x_i \mid x_i \in N_h(x_t)\right), \tag{13.3}$$

for $t = 1, \ldots, T$ where $N_h(x_t)$ is a collection of points defined in a neighbourhood of points around $x_t$. For this reason, this simple version of the kernel regression is often called the *nearest-neighbour* regression. The size of the neighbourhood is determined by parameter, $h$, which is typically termed the bandwidth. All of the

[13] A good treatment of the kernel regression can be found in a wide variety of flavours in many statistic textbooks. The original references are Nadaraya [10] and Watson [12].

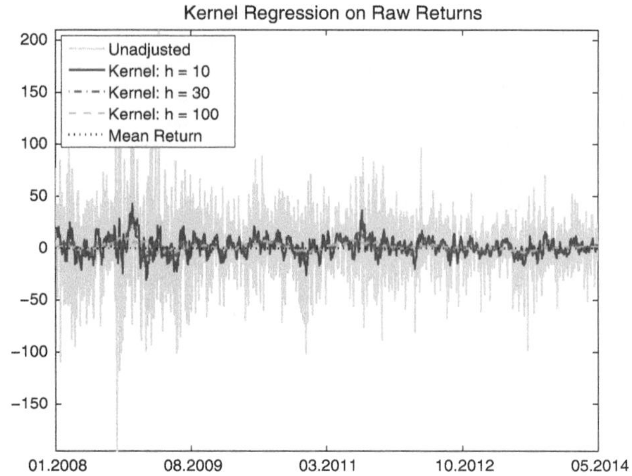


**Fig. 13.9** Kernel regression. Again, we attempt to reduce the noise of our daily US Treasury portfolio returns. Here we employ a kernel regression, which essentially amounts to a weighted average within a pre-defined neighbourhood

work involved in applying this algorithm resides with finding a reasonable value for $h$.[14]

Figure 13.9 summarizes a smoothed set of daily returns for our US Treasury portfolio employing the basic simple kernel regression described in Eq. (13.3). Three alternative choices of the neighbourhood size parameter, $h$, were selected: 10, 30, and 100 days. As the size of the neighbourhood increases, the estimate becomes smoother.

### 13.2.4 An Engineering Approach

The final noise reduction technique that we will consider has its origins in the engineering literature and is described in Shin et al. [11]. This approach, termed Quotient Singular Value Decomposition (QSVD), has been used to isolate a signal from noise in numerous engineering applications, particularly in signal processing.

Conceptually, it is not dissimilar to the notion of principal component analysis where one approximates a variance-covariance matrix with a subset of the eigenvalues and eigenvectors from its spectral decomposition. With QSVD one uses a generalized notion of the eigenvalue decomposition and applies it to the so-called Hankel (or trajectory) matrix of our time series $\{x_t, t = 1, \ldots, T\}$,

[14]It is always useful to consider the limits. For extremely small values of $h$, one will recover the original time series. As $h$ approaches infinity, the kernel regression will merely produce the unconditional mean of the time series for each $x_t$. A reasonable value obviously lies somewhere between these two extremes.

$$X = \begin{bmatrix} x_1 & x_2 & x_3 & \cdots & x_n \\ x_2 & x_3 & x_4 & \cdots & x_{n+1} \\ x_3 & x_4 & x_5 & \cdots & x_{n+2} \\ \vdots & \vdots & \vdots & \ddots & \vdots \\ x_{T-n+1} & x_{T-n+2} & x_{T-n+3} & \cdots & x_T \end{bmatrix}, \tag{13.4}$$

which is essentially collection of $n$-lagged values of $x_t$. If we set $N = T - n + 1$, then the matrix has the following cleanly defined dimensions: $X \in \mathbb{R}^{N \times n}$. Generally speaking, $n$ is selected such that $N \gg n$. This specification implies that $X$ is not a square matrix—the consequence is that an eigenvalue decomposition is impossible.[15]

The basic idea is that one additively separates the trajectory matrix, $X$, into a deterministic trend component, $\bar{X}$, and a noise component, $\bar{N}$:

$$X = \bar{X} + \bar{N}. \tag{13.5}$$

The question is how to perform this decomposition. The QSVD technique addresses this questions by recalling that the matrix, $X$, can always be decomposed as follows,

$$X = S\Sigma C^T, \tag{13.6}$$

where $S \in \mathbb{R}^{N \times n}$, $C \in \mathbb{R}^{n \times n}$, and $\Sigma \in \mathbb{R}^{n \times n}$ is diagonal. This is termed the singular value decomposition of the matrix, $X$. Combining Eqs. (13.5) and (13.6), we may write:

$$X = \underbrace{S_1\Sigma_1 C_1^T}_{\bar{X}} + \underbrace{S_2\Sigma_2 C_2^T}_{N}, \tag{13.7}$$

where $S_1 \in \mathbb{R}^{N \times k}$, $C_1 \in \mathbb{R}^{k \times k}$, and $\Sigma_1 \in \mathbb{R}^{k \times k}$ are selected by one's choice of $k$. In this context, $k$ boils down to selecting the most important singular values from the decomposition of $X$. Having made this choice, one merely computes $S_1\Sigma_1 C_1^T$ and then backs out the deterministic trend for the time series, $\{x_t, t = 1, \ldots, T\}$. There are alternative, typically rather more complex, approaches to determining the optimal choice of $k$—and thus one's separation of the trend and noise elements—but the algorithm is already sufficiently complex for our purposes.

Figure 13.10 illustrates the result of the use of 250 data lags (i.e., $n = 250$) and 10 singular values (i.e., $k = 10$). The technique does not, unfortunately, perform very well. Much of the reason stems from the relative lack of memory in the daily-return series. A relatively small number of singular values do *not* dominate

[15]Very roughly, a singular value decomposition is a generalization of the eigenvalue decomposition for non-square and, even, non-singular matrices. See Appendix A for more information on the eigenvalue decomposition.

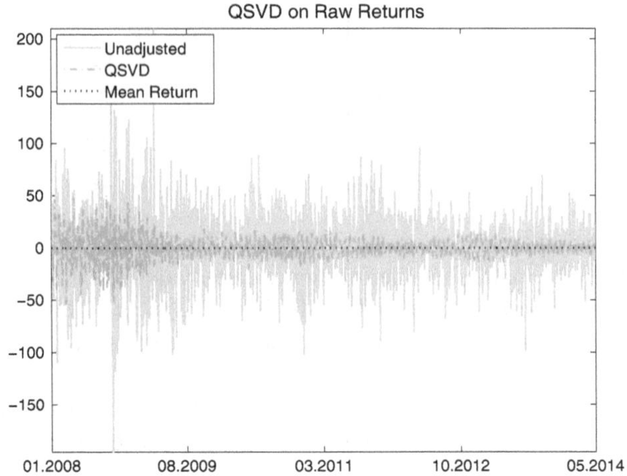


**Fig. 13.10** An engineering approach. This approach, termed quotient singular value decomposition or QSVD, is used to isolate a signal from noise in numerous engineering applications particularly signal processing

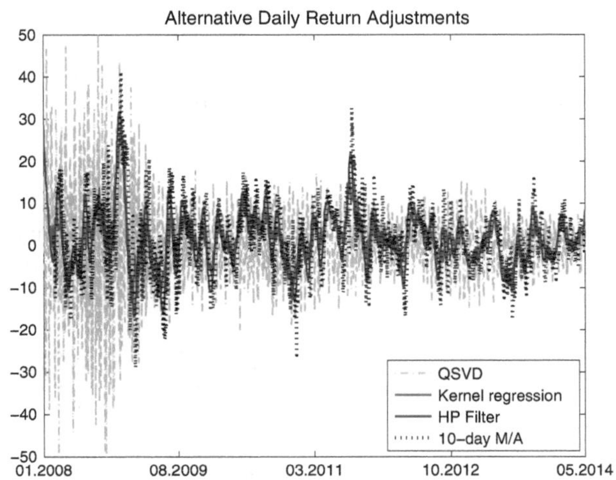


**Fig. 13.11** Model selection. This figure directly compares the four previously described adjustment approaches

the decomposition described in Eq. (13.6), making it difficult to separate the trend from the noise. Similarly, given the independence of the raw daily returns, we also observe that the covariance matrix of lagged daily returns also contains relatively little information.

### 13.2.5 Model Comparison

Having selected and introduced four different possible approaches for the reduction of daily return volatility, we are now in a position to select the most appropriate choice. Figure 13.11 summarizes the noise-reduced estimates of the daily return for our four approaches: a 30-day moving average, the HP filter with $\lambda = 500$, a kernel regression with $h = 30$, and the previously described QSVD.

**Table 13.4** Model correlation

| Model | QSVD | K/R | H/P | M/A |
|---|---|---|---|---|
| QSVD | 1.0 | 0.1 | 0.0 | 0.1 |
| K/R | 0.1 | 1.0 | 0.8 | 0.6 |
| H/P | 0.0 | 0.8 | 1.0 | 0.7 |
| M/A | 0.1 | 0.6 | 0.7 | 1.0 |

The underlying table underscores the contemporaneous correlation between the adjusted daily returns stemming from each of our return-adjustment models.

All of the approaches, with the exception of the QSVD yield visually similar estimates. Table 13.4 underscores this point by showing the contemporaneous correlation between the adjusted daily returns stemming from each of our noise-dampening models.

The kernel regression and the Hodrick–Prescott filter are strongly similar with a correlation of slightly more than 0.8. The moving average is also similar to the HP filter and the kernel regression with correlation coefficients between 0.6 and 0.7. The QSVD method neither performs particularly well nor does it correlate strongly with the other techniques. We can, at this point, reject this as a possible model for the dampening of return volatility.

Given the relatively high degree of similarity between the approaches, therefore, how should we choose between the remaining three models? The answer is that, because of their strong similarities, we need to consider other factors. The moving-average model has the disadvantage that one essentially loses data during its computation. For a 30-day moving average, one basically loses the first 6 weeks of one's data sample. The HP-filter and the kernel regression do not share this drawback. On this basis, we can exclude the moving-average approach.

It now reduces to a choice between the HP filter and the kernel regression. The principal distinction between these two approaches is what type of data is used in the computation. The kernel filter estimates a particular value using values that lie both in the past *and* in the future. The HP filter, during its optimization, makes use of all of the observations simultaneously, but each step in the actual filter computation looks backwards from the current point in time. This offers something of a conceptual advantage for the daily reduction of the noise in returns. On this basis, we will adopt the HP filter as our technique of choice in this chapter.

### 13.2.6 Implications of Filtering

Having selected the HP filter to dampen the noise in daily returns, it is important to examine the implications of the application of this filter on our data. Specifically, we wish to understand the differences between the filtered and raw daily return series. The noise-dampening process essentially seeks to extract a cyclical trend from the raw daily return series. If this cyclical trend differs too much, however, from the

**Table 13.5** Implications of return filtering

| Statistic | Raw | Filtered |
|---|---|---|
| Mean | 1.6 | 1.6 |
| Median | 2.1 | 1.2 |
| Maximum | 209.9 | 31.5 |
| Minimum | −195.3 | −16.7 |
| Positive (%) | 54.8 | 57.6 |
| Volatility | 32.2 | 7.0 |
| IQR | 37.5 | 9.1 |
| Skewness | 0.0 | 0.7 |
| Kurtosis | 5.8 | 5.1 |
| Signal-to-noise | 0.1 | 0.2 |
| CV | 19.7 | 4.2 |
| Correlation | 0.3 | |
| Autocorrelation | | |
| 1-Day | −0.0 | 1.0 |
| 1-Week | −0.0 | 0.8 |
| 1-Month | −0.0 | −0.0 |

The underlying table provides some insight into what happens to our daily portfolio return time series when it is filtered.

underlying raw data, then it may lead to misleading results. Table 13.5 provides some basic summary statistics to facilitate this comparison.

What happens to the time series when it is filtered? Table 13.5 suggests that the mean daily return is preserved, but that all measures of volatility are reduced by roughly a factor of five. The skew and the kurtosis of the two data series do not appear to exhibit any important differences, although there is some evidence of a slight positive skew in the filtered returns. The two series also exhibit weakly positive correlation and the autocorrelation of the filtered returns are dramatically higher over short time horizons.

This latter observation is supported by Fig. 13.12, which provides the autocorrelation function for the raw and filtered returns for lags of up to 100 days. We note that the autocorrelations start out strong and decay rather quickly and are essentially absent beyond about 20 days or 4 weeks.[16] Not surprisingly, therefore, our cyclical trend demonstrates significantly greater memory from 1 day to the next as compared to the raw return data.

Figure 13.13 displays the empirical distributions for the filtered and raw daily returns. This illustrates quite dramatically the impact of the filtering algorithm—it essentially squeezes the observations around the same central point while preserving

[16]Roughly speaking, therefore, the choice of $\lambda = 500$ appears to be roughly equivalent to a moving average window of 4 weeks.

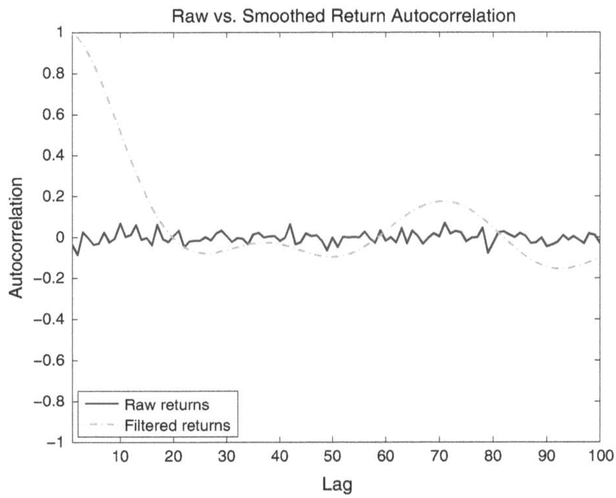


**Fig. 13.12** Autocorrelation: raw vs. filtered returns. This figure demonstrates the impact of dampening the noise in return observations on the autocorrelation function

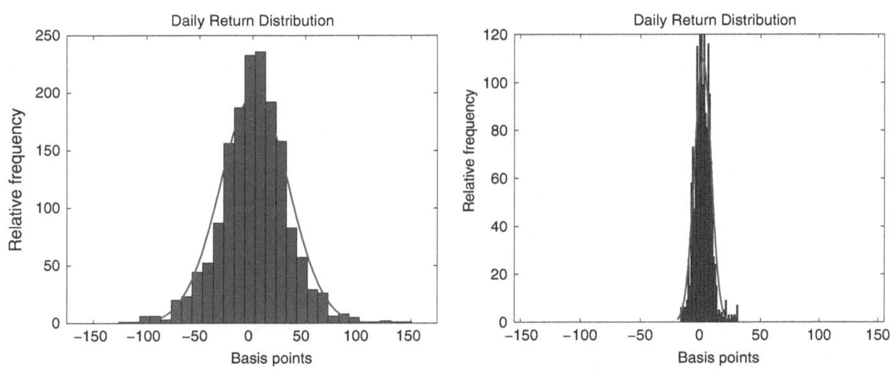


**Fig. 13.13** Distributional impact. This figure contrasts the unfiltered and filtered empirical return distributions

their basic shape and form. Clearly, the dampening of return noise adjusts the raw daily returns, but the extracted cyclical trend appears to contain much of the same information embedded in the noisier raw returns. This should hopefully make it a more useful object when combined with risk data.

## 13.3 Combining Risk and Return

Our objective is to construct a measure of return per unit of risk—or, conversely, the amount of risk required by basis point of return—that can be monitored on a daily or weekly basis. In our previous discussion, we determined the average amount of

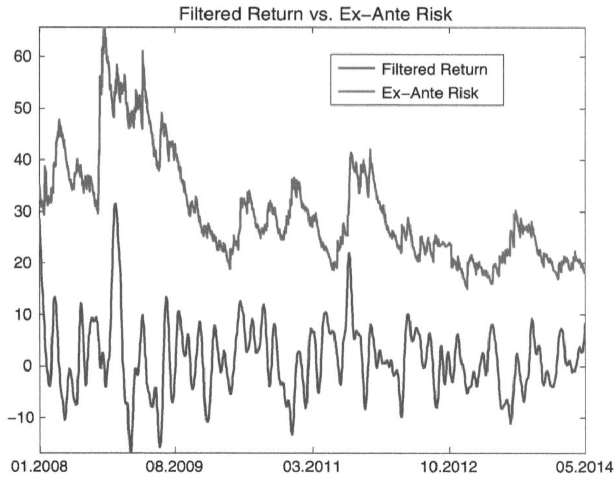


**Fig. 13.14** Risk and (filtered) return. We may now jointly examine our daily ex-ante risk and *adjusted* return time series. The reduction in return noise permits a more sensible comparison

return per unit of risk over the entire period. Having now found a technique to extract a cyclical trend from our daily return series, we can now return to the notion of combining risk and return on a daily basis. Figure 13.14 plots out the joint evolution of ex-ante volatility and the filtered daily returns. The noise in the cyclical daily return trend thankfully no longer overwhelms the associated ex-ante risk measure.

We are finally in a position to be somewhat more precise about how we might combine daily risk and return. The first step is a bit of notation. Let us denote the realized raw return over the interval, $[t-1, t]$, as $r_t$. The filtered return over this same period is denoted $\tilde{r}_t$, making the set of filtered daily returns $\{\tilde{r}_t, t = 1, \ldots, T\}$.

The estimated ex-ante daily return volatility also applies over the interval, $[t - 1, t]$, but in a slightly different fashion. The estimated ex-ante volatility computed at time $t$ applies over the period, $[t, t + 1]$. More formally, we define it as,

$$\hat{\sigma}_t \approx \sqrt{\mathbb{E}\left(\text{var}(r_t) \middle| \mathscr{F}_t\right)}. \tag{13.8}$$

In other words, the ex-ante volatility is the conditional daily return volatility incorporating all of the information available up until time $t$.[17] The consequence of this difference is that the return over the interval $[t - 1, t]$ must be compared to the ex-ante risk figure computed on the previous day, $t - 1$. This avoids peeking into the future.

[17]As encapsulated by the information set (or rather, $\sigma$-algebra) $\mathscr{F}_t$. See Durrett [5] or Billingsley [2] for more detail on this concept.

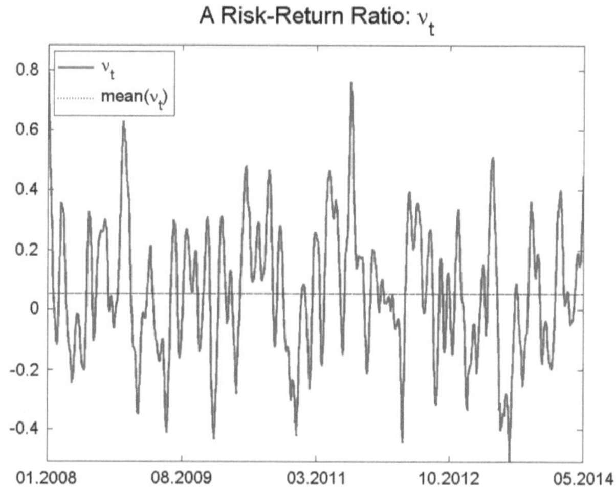


**Fig. 13.15** A simple ratio. This figure provides a description of our simple ratio over time

Using these concepts, perhaps the simplest and most straightforward combination of risk and return would take the following form,

$$\nu_t = \frac{\tilde{r}_t}{\hat{\sigma}_{t-1}}, \tag{13.9}$$

for $t = 1, \ldots, T$. This simple measure is the filtered daily return normalized by its expected volatility. In the engineering literature, this is often referred to as the signal-to-noise ratio. Figure 13.15 describes the evolution of this simple ratio over our time horizon. It also includes a comparison of the average risk-normalized return over the period amounting to 0.06 units of daily return per unit of risk—this coincides precisely to the value computed using raw data and displayed in Table 13.3 on page 427. Our simple measure appears to fall roughly symmetrically on both sides of this overall average. Although it remains reasonably oscillatory, it does appear to demonstrate substantial periods where it remains positive or negative. In other words, there appears to be a reasonable amount of persistence in this simple measure.

The inverse of our simple ratio—which we have described as the risk required to obtain a basis point of return—cannot be employed in this context. The reason is simple; when our filtered return approaches zero, the ratio of risk to filtered return approaches infinity. The inverse of $\nu_t$, also known as the coefficient of variation, is known to fail when values are equal to or approach zero.

As a next step, we'd like to see if we can identify trends in our simple ratio. To accomplish this, we propose two basic normalizations our risk-adjusted-return ratio, $\nu_t$. The first looks at the sign of $\nu_t$,

$$\mu_t = \begin{cases} 1 : \nu_t = \dfrac{\tilde{r}_t}{\hat{\sigma}_{t-1}} > 0 \\ -1 : \text{Otherwise} \end{cases}, \tag{13.10}$$

This measure provides some insight into the consistency of the ratio over time. For example, once it moves into positive territory, does it stay there for a while? Is there also some persistence when it becomes negative?

The second transformation normalizes $\nu_t$ by subtracting its mean and dividing by its standard deviation. This will ensure that it has a zero mean and a variance of unity,

$$\gamma_t = \frac{\nu_t - \frac{1}{T}\sum_{k=1}^{T}\nu_k}{\sqrt{\frac{1}{T-1}\sum_{j=1}^{T}\left(\nu_j - \frac{1}{T}\sum_{k=1}^{T}\nu_k\right)^2}} \tag{13.11}$$

If one is willing to assume that $\nu_t$ is approximately normally distributed, then this transformation can help to identify situations where the cyclical component of the daily return is either statistically quite large or small. For such an application, we can think of the transformation, $\gamma_t$, as essentially being a *z-score* for $\nu_t$.[18]

Figure 13.16 illustrates the evolution of these two normalizations of our risk-adjusted return measure, $\nu_t$. We can make two observations. First, the transformation, $\mu_t$, demonstrates a significant amount of persistence. It appears to spend, on average, multiple days or even weeks in a positive or negative state. The second observation is that the $\gamma_t$-transformation typically lies between $\pm 1$ standard deviation of the mean. It attains 1.5–2 standard deviations roughly 10–12 times over our 6+ year sample period.[19]

In addition to the graphical examination of our base measure, $\nu_t$, and its two transformations, $\mu_t$ and $\gamma_t$, it is also interesting to examine some summary statistics. Table 13.6 therefore examines a range of different summary statistics for our three suggested risk-return ratios.

Perhaps the most interesting statistic in Table 13.6 is the average daily persistence. Naturally, the $\nu_t$ and $\gamma_t$ measures take a slightly different value every single day. The $\mu_t$-transformation, however, has only two states: positive and negative. Over the 1,600-odd sample of daily observations, it has only changed value 62 times overall leading to an average persistence of 26 days or almost 5 weeks. For the monthly observations, however, it changed 36 times out of 76 months, implying an average persistence of about 2 months.

We also observe that the daily sign-based transformation, $\mu_t$, appears to have essentially the same strength as its monthly equivalent. The daily signal exhibits a similar proportion of positive observations and consistent average returns. Not much

---

[18]A distributional assumption is not strictly necessary, but it does allow us to probabilistically interpret the outcomes.

[19]For a standard normal distribution and sample of 1,600 independent observations, you would expect to observe slightly more than three observations $\pm 3$ standard deviations from the mean.

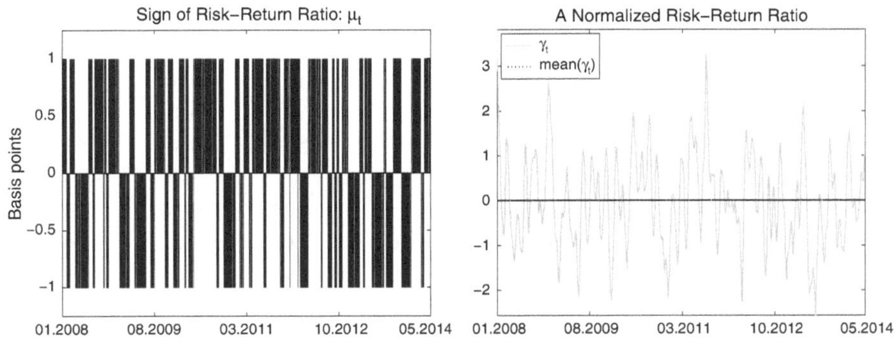


**Fig. 13.16** Transformations of our simple ratio. This figure outlines evolution over our sample period of the two transformations of our simple ratio described in Eqs. (13.10) and (13.11)

**Table 13.6** Numerical ratio comparison

| | Monthly | | | Daily | | |
|---|---|---|---|---|---|---|
| Statistic | $\nu_t$ | $\mu_t$ | $\gamma_t$ | $\nu_t$ | $\mu_t$ | $\gamma_t$ |
| Mean | 0.2 | 0.1 | 0.0 | 0.1 | 0.2 | −0.0 |
| Median | 0.2 | 1.0 | 0.0 | 0.0 | 1.0 | −0.0 |
| Maximum | 2.2 | 1.0 | 2.1 | 0.9 | 1.0 | 3.8 |
| Minimum | −0.5 | −1.0 | −2.6 | −0.5 | −1.0 | −2.6 |
| Positive (%) | 56.6 | 56.6 | 51.3 | 57.6 | 57.6 | 48.2 |
| Volatility | 0.9 | 1.0 | 1.0 | 0.2 | 1.0 | 1.0 |
| IQR | 1.3 | 2.0 | 1.4 | 0.3 | 2.0 | 1.4 |
| Skewness | −0.2 | −0.3 | −0.2 | 0.2 | −0.3 | 0.2 |
| Kurtosis | 2.7 | 1.1 | 2.7 | 2.9 | 1.1 | 2.9 |
| Persistence (days) | | | | | | |
| Observations | 76.0 | 76.0 | 76.0 | 1627.0 | 1627.0 | 1627.0 |
| # Of value changes | 75.0 | 36.0 | 75.0 | 1626.0 | 62.0 | 1626.0 |
| Persistence | 1.0 | 2.1 | 1.0 | 1.0 | 25.9 | 1.0 |

The underlying table examines a range of summary statistics for our three suggested risk-return ratios. It also includes a comparison of these ratios to both monthly and daily data.

information is being lost as we aggregate to monthly returns using the cyclical trend in returns extracted by the HP filter. Daily returns are at the shortest possible frequency one can reasonably employ and it appears that they can provide some insight into returns computed at lower frequencies.

### 13.3.1 Moving to the Risk-Factor Level

It is natural to ask how does the risk-adjusted return, as captured by our measure $\nu_t$, compares across our curve risk factors. Table 13.7 displays some selected summary statistics for the filtered returns for each of our three yield-curve sectors.

**Table 13.7** Sector-based ratios by the numbers

| Statistic | Short | Medium | Long |
|---|---|---|---|
| $\nu_t$ | | | |
| Mean | 0.05 | 0.04 | 0.05 |
| Volatility | 0.26 | 0.22 | 0.23 |
| Positive (%) | 62.4 | 56.8 | 53.8 |
| $\mu_t$ | | | |
| Mean | 0.25 | 0.14 | 0.08 |
| # Of value changes | 58.0 | 60.0 | 66.0 |
| Positive (%) | 62.4 | 56.8 | 53.8 |
| Mean persistence | 27.2 | 26.8 | 24.0 |
| $\gamma_t$ | | | |
| Minimum | −3.1 | −3.1 | −2.4 |
| Maximum | 4.6 | 3.7 | 3.8 |
| Positive (%) | 54.5 | 50.3 | 46.4 |

Here we examine a few summary statistics for our two suggested risk-return ratios.

The filtered returns and associated risk-adjusted return ratios are strongly correlated among the three yield-curve sectors. Given the high degree of correlation between US Treasury yield movements, this should come as no surprise.[20] Nevertheless, there are also periods when the behaviour, in terms of risk and return, of our three risk factors visibly differ.

Table 13.7 focuses on the key elements by providing different summary statistics for each of the three measures.[21] We can see that the short end of the curve has provided, over the period of analysis, the best risk-adjusted return. The daily return for each unit of risk is approximately constant at 0.5 for each of the three sectors of the US Treasury curve. The short end does demonstrate the highest proportion of positive returns; this is again, amounting to 6 % points more than that observed in the long end. We also note that the average persistence—at 26 days or roughly 5 weeks—does *not* appear to vary much between the three sectors.

## 13.4 So What?

After all this effort, the reader might be asking: this is all quite interesting, but what should I do with this analysis? It is, to be frank, not entirely obvious. A useful start, however, might be to incorporate daily risk-return measures, or some simple

[20]For portfolios that also have foreign-exchange and credit-spread risk, such a high degree of correlation between the individual risk factors will be smaller.

[21]The mean and variance, for example, are not terribly interesting for the $\gamma_t$-measure since they are always by construction zero and one, respectively. Moreover, the average persistence for the continuous measures $\nu_t$ and $\mu_t$ are equally uninteresting.

**Table 13.8** A possible daily report

| | A date in late November | | | | | |
|---|---|---|---|---|---|---|
| Statistic | '08 | '09 | '10 | '11 | '12 | '13 |
| Latest return ($r_t$) | **113.9** | −4.4 | **4.6** | 22.9 | **−21.1** | −27.1 |
| 21-day mean return | **15.8** | 4.1 | **−6.5** | 6.8 | **4.2** | −3.7 |
| Mean return since inception | **4.0** | 2.4 | **2.3** | 2.5 | **2.2** | 1.6 |
| Latest risk ($\tilde{\sigma}_{t+1}$) | **49.2** | 27.9 | **28.0** | 35.7 | **18.9** | 21.4 |
| 21-day mean risk | **53.4** | 30.2 | **26.2** | 37.5 | **19.9** | 21.1 |
| Mean risk since inception | **40.5** | 42.2 | **36.9** | 35.2 | **33.0** | 31.1 |
| $\nu_t$ | **0.6** | 0.3 | **−0.2** | 0.1 | **0.1** | −0.0 |
| $\gamma_t$ | **2.6** | 0.9 | **−1.3** | 0.2 | **0.0** | −0.4 |
| Days in $\mu_t$ | **23.0** | 18.0 | **29.0** | 21.0 | **21.0** | 17.0 |

This table demonstrates how a daily report might look for a collection of different dates across our sample period.

variations of them, into one's daily reporting. Figure 13.8 takes a first step in this direction by showing a collection of daily risk and return measures for six separate dates in our data sample. Each date, falling in late November, is separated by roughly 1 year.

Table 13.8 begins with the latest return and ex-ante risk observations. From the preceding analysis, we know that a naive combination of these two pieces of information is of questionable use—the raw daily returns are simply too noisy. As a consequence, it is sensible to supplement the latest observations with the mean values over the last month (i.e., 21 days) and since the inception of our data-set.

These additional statistics provide some perspective. Taking the date in November 2009 as an example, the most recent return is −4.4 basis points and the ex-ante volatility is about 28 basis points. These two figures alone are not enough. The average return over the last 21 days is 4 basis points, however, while the average risk over this period is around 30 basis points. Interestingly, the average 21-day return is almost twice the mean return since inception. The average risk over the last 21 days is, however, significantly lower than the 42 basis-point average since the start of the data sample. We may, therefore, conclude that November 2009 was characterized by higher returns with slightly lower risk.

Without any context these figures may seem to be of limited usefulness. Adding the expected risk and return arising from one's SAA process into the report, however, might permit a useful comparison to the most recent, the 21-day average, and the longer-term average values. In principle, this could potentially help to bring the strategic perspective more explicitly into one's daily-reporting discussions.

The final three columns of Fig. 13.8 include the current levels of our simple risk-return ratio ($\nu_t$), the normalized risk-return ratio ($\mu_t$), and the number of periods that the risk-return ratio has had its current sign.[22] Again, we observe that our risk-return ratio has a reasonably high level of persistence. In November 2010, for example, the ratio took a negative value of −0.4—this amounted to a normalized signal of −1.4. As of late November 2010, however, it had been negative for 29 consecutive working days or almost 6 weeks.

These final rows of Fig. 13.8 potentially provide a tactical perspective to the strategic viewpoint in the previous rows. They basically indicate the sign and strength of the return signal and its persistence. One might imagine that this might be useful information for portfolio managers and investment committees.[23]

The bottom line is that the preceding analysis is neither intended to provide a revolutionary trading measure nor a path-breaking strategic tool. Instead, it represents an attempt to combine daily risk and return measures in a sensible way that may provide some additional strategic and tactical perspective in our daily reporting. Significant effort is required for the computation of these risk and return figures. They also include important and useful information about our portfolios. As such, it makes logical sense to find ways to make more targeted use of them.

## 13.5 Concluding Thoughts

What have we accomplished in this chapter? We posed a simple question: *can one sensibly combine high-frequency (i.e., daily) measures of risk and return?* Although no definitive solution was provided, a proposed, albeit preliminary, approach attempts to extract a signal from daily returns and compares this signal to the contemporaneous ex-ante risk estimate. The ratio of these two values is intended to provide some insight—at a daily frequency—of the amount of return being generated per unit of risk embedded in the portfolio. The results suggest that this measure demonstrates a reasonable amount of persistence, can be transformed in a few useful ways, and may also be applied at the underlying risk-factor level. This measure, and its transformations, appear to provide additional information beyond what one observes in monthly returns. We would argue that this presents a sensible supplement to daily reports including solely risk and return information.

The principal drawback of this approach is the reliance on a mathematical technique for the extraction of a cyclical return trend. It is necessary to extract such a trend due to the large amount of noise in the raw daily return data. What this approach is basically doing is trying to put the risk and return on the same footing. Daily risk is computed using recent observations to form an estimate of the uncertainty in the return for the next period. Daily returns are a single realization

[22]This is the number of periods that $\mu_t$ has had the same value. In other words, it is essentially the current persistence of the sign of the risk-return ratio, $\nu_t$.

[23]It would be even more interesting if it was broken down by the individual risk factors. In the interests of space—and the reader's patience—we have omitted these details, but they are easily computed and presented.

over a 1-day period. In other words, more information is used to construct a daily risk measure than is employed for the associated daily return observation. The extraction of the cyclical trend in returns makes use of past return observations to provide some indicators of the recent (and hopefully future) direction of daily returns. The combined risk-return measure proposed in this chapter, therefore, attempts to describe a recent trend in the amount of return generated per unit of risk—also computed as a recent trend—from both an aggregate and risk-factor perspective.

This is far from the last word on this topic. Indeed, it is little more than an investigation of what might be possible in this area. Finding better approaches for combining high-frequency risk and return figures is a challenge for the research community. To conclude, although it does not appear that a simple and definitive measure of daily risk-adjusted return can be computed, subject to some assumptions, we have demonstrated that one might nonetheless extract useful information about the local risk and return characteristics of one's portfolio. When compared on an ongoing basis to the assumed return and risk characteristics of the risk-factors embedded in one's strategic benchmark, this could provide, within one's daily reporting, a helpful link to the strategic asset allocation process.

## References


1. M. Baxter, R.G. King, Measuring business cycles: approximate band-pass filters for economic time series. NBER Working Paper No. 5022 (1995)
2. P. Billingsley, *Probability and Measure*, 3rd edn. (Wiley, New York, 1995)
3. R.A. Brealey, S.C. Myers, F. Allen, *Principles of Corporate Finance* (McGraw-Hill, New York, 2011)
4. G.P. Brinson, L.R. Hood, G.L. Beehower, Determinants of portfolio performance. Financ. Anal. J. **42**(4), 39–44 (1986)
5. R. Durrett, *Probability: Theory and Examples*, 2nd edn. (Duxbury, Belmont, 1996)
6. A. Guay, P. St-Amant, Do the Hodrick–Prescott and baxter-king filter provide a good approximation of business cycles. Annales d'économie et de statistique **77**, 133–155 (2005)
7. R.J. Hodrick, E.C. Prescott, Postwar U.S. business cycles: an empirical investigation. J. Money Credit Bank. **29**(1), 1–16 (1997)
8. L.R. Hood, Determinants of portfolio performance: 20 year later. Financ. Anal. J. **61**, 6–8 (1986)
9. R.G. Ibbotson, P.D. Kaplan, Does asset allocation policy explain 40, 90 or 100 % of performance. Financ. Anal. J. **56**, 26–33 (2000)
10. E. Nadaraya, On estimating regression. Theory Probab. Appl. **9**(1), 141–142 (1964)
11. K. Shin, J.K. Hammond, P.R. White, Iterative svd method for noise reduction of low-dimensional chaotic time series. Mech. Syst. Sig. Process. **13**(1), 115–124 (1999)
12. G.S. Watson, Smooth regression analysis. Sankhya **26**(4), 359–372 (1964)

# The Ex-Post World 14

*History never repeats itself, but it often rhymes.*

Mark Twain

The portfolio-analytic framework, developed in this book, examines exposure, risk, and performance from a high-dimensional perspective. Using detailed information about the individual positions in our portfolios and strategic benchmarks, we can gain substantial insight into the risk and return characteristics of our portfolios. This is our central thesis. Such a detailed level of analysis is, unfortunately, not always possible. This chapter, therefore, changes tactics somewhat and considers a different set of portfolio-analytic techniques that may be used to compare the relative risk and performance of portfolios. We collectively refer to these techniques as ex-post analysis.

When is it *not* possible to perform an instrument-level analysis? The answer is whenever high-dimensional data is unavailable. A good example is the outsourcing of a specific asset class to an external manager. This requires selection of a single manager from a set of potential candidates. It is not generally possible to gather the detailed information required for a full-blown performance- and risk attribution from each external manager. Even if an external manager were able to provide sample performance attributions, it would typically only be available for a short periods of time.[1] It would also be complicated—in a multi-manager search—by the fact that each manager is likely use an alternative approach. Moreover, it is possible that none of the performance attributions capture the dimensions that really interest you.

[1]Moreover, they would certainly not provide the underlying data for the computations. Instead, you would only receive the high-level results.



D.J. Bolder, *Fixed-Income Portfolio Analytics*,
DOI 10.1007/978-3-319-12667-8_14

External manager selection is not the only application of these techniques. One may wish to examine *internal* portfolios over several years at a high level without drilling down to the security level. A more global perspective on portfolio risk and performance can be a useful complement to the high-dimensional view.

Ex-post analysis techniques also offer a few advantages relative to their high-dimensional counterparts:

- Less data is needed. While these methods need a much longer data history, they require much less information at each individual point in time.
- The basic computations are straightforward. Most, if not all, of the ex-post analysis can be performed in a spread-sheet.
- Ex-post analysis provides an alternative perspective. These techniques are designed to determine if an internal or external manager is adding value over the long run. Performance and risk attribution, in contrast, is primarily designed to examine portfolios over short- to medium-term horizons.
- Combining high-dimensional risk and performance attributions, as we saw in the previous chapter, is neither easy nor, as a research area, well developed. Ex-post techniques have the advantage of being designed to permit a joint examination of both return and risk.

This does not imply that ex-post analysis is superior to instrument-level risk and performance attribution, but rather that it is different and complementary. Ex-post analysis provides an alternative lens for the examination of portfolio-manager risk and return choices. In a perfect world, of course, one would be in a position to perform and consider *both* types of analysis.

To make our discussion of this perspective more concrete and keep things interesting, we present these ex-post analysis techniques in the context of a single running example. We begin with an empirical analysis and then, along the way, introduce some basic theory. At each point, however, we attempt to keep our example at the forefront.

## 14.1 Basic Statistical Analysis

Imagine your institution has short-listed three managers for a US$100 million investment in a fixed-income mandate for your company's pension fund. You have seen a number of fancy presentations from each of the individual managers where they discussed their investment process, their fantastic performance, and the many virtues of their company. Your objective is to determine which of these managers (if any) that you would recommend as the *best* external manager for your organization. Remember, your pension dollars depend on this decision!

The head of the selection committee has given you a spreadsheet with the return data for each of the three managers and asked you to see if you can make any sense

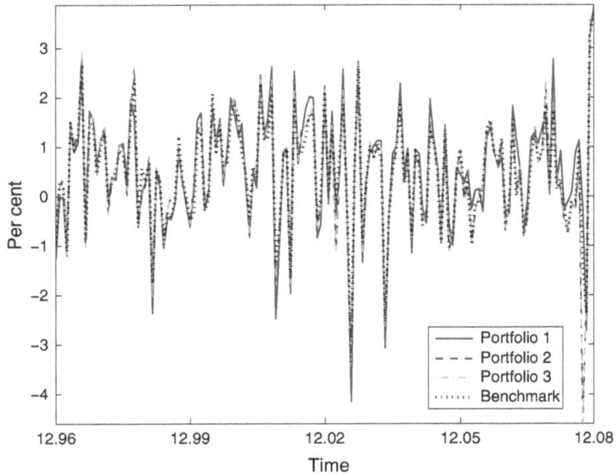


**Fig. 14.1** Absolute returns. This figure outlines the absolute monthly returns, denoted as $R_p$, for our three sample portfolio managers along with the benchmark returns: the U.S. Lehman Aggregate Index

of it. This is to be the foundation of your analysis. You identify a few important facts:

- Each manager invests relative to the same benchmark: the U.S. Lehman Aggregate Index.
- The data covers a 12-year, or 145-month, period spanning December 1996 to December 2008.
- The stated investment style of each investment manager is *fixed-income core*, suggesting a preponderance of relatively high-credit quality, fixed-income assets.

So far so good, but you notice that your spreadsheet has only *five* columns: the return dates, the three portfolio manager return series, and the corresponding benchmark returns. That's all. Staring blankly at the numbers in the spreadsheet or flipping a coin are *not* viable options.

What you need is some structure. Let's start by performing some simple data analysis to better understand the underlying return data. The first thing is to graphically look at the return data.[2] Figure 14.1 summarizes the contents of your spreadsheet. These are the absolute monthly returns for each of the three portfolio managers and the benchmark over the 145-month period ending December 2008.

Returns, ranging between $\pm 3$ %, appear to have been predominately symmetric although a number of fairly sizeable negative monthly returns are evident. There is also significant correlation between the three portfolio managers and their common benchmark. This is disappointing. It is challenging to extract much useful information from Fig. 14.1. Our meagre observations aside, if we want to properly

[2]This should be one's first task with any dataset.

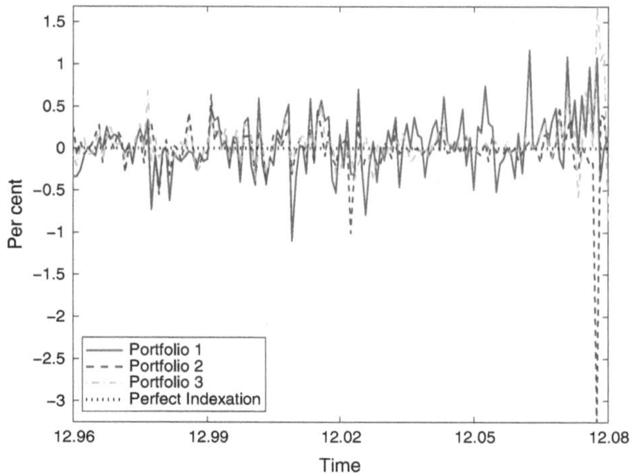


**Fig. 14.2** Active returns. This figure outlines the active monthly returns, computed as $R_p - R_b$, for our three sample portfolio managers. These are essentially the active returns of the portfolio relative to the common benchmark: the U.S. Lehman Aggregate Index

distinguish between our portfolio managers, we definitely need to extend our analysis.

A slight extension involves examining the difference between the monthly returns of each portfolio manager and the common strategic benchmark return. This quantity, which we will refer to as the active return, is defined as,

$$\underbrace{\text{Active Return}}_{R_a} = \underbrace{\text{Portfolio Return}}_{R_p} - \underbrace{\text{Benchmark Return}}_{R_b}. \tag{14.1}$$

Figure 14.2 describes the active returns for each portfolio manager over our 145-month analysis period. While also relatively difficult to interpret, more information can be extracted from it. The black dotted line running through the origin represents the active returns associated with perfect indexation; that is, completely replicating the benchmark. The active returns, for all three managers, appear to be approximately centred around the perfect indexation line. This implies, consistent with our previous observation of symmetric returns, that the probability of outperforming the benchmark is not markedly different than under-performing it.

The typical over- or under-performance in Fig. 14.2 seems to fall in a range of about ±75 basis points. An exception occurs at the end of the period when the third portfolio manager has a return of greater than 150 basis points while the second portfolio manager experienced a loss of approximately 3 %. There is also a trend towards increased active-return volatility as we move through time. In other words, the volatility of active returns, for all portfolio managers, does not appear to be constant over time.

Figure 14.3 examines the active returns from another perspective. It provides *four* separate histograms summarizing the empirical active-return distributions for all portfolios and for each of the three individual portfolio managers. Again, the majority of active returns for all three managers are centred around zero and

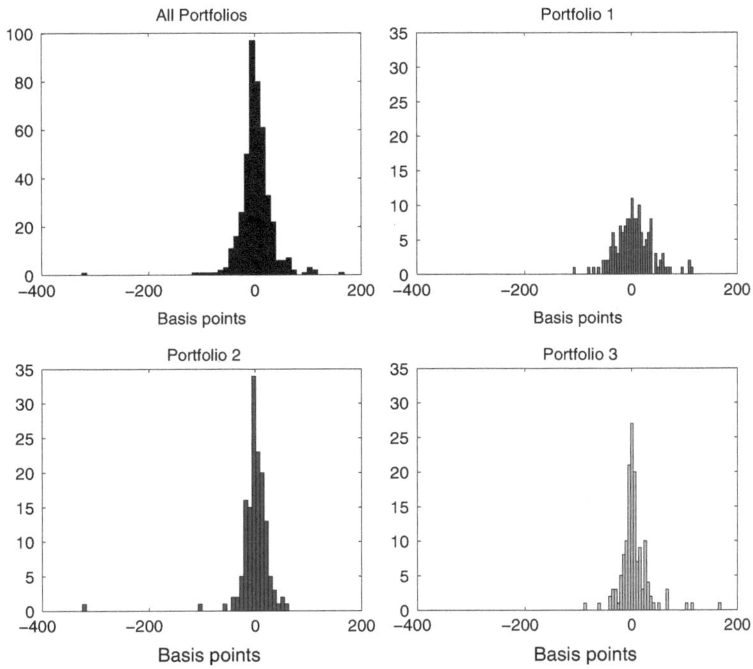


**Fig. 14.3** Active return distributions. This figure outlines the empirical distributions of the active monthly returns, which are displayed in time-series format in Fig. 14.2

generally do not exceed ±75 basis points. The third portfolio manager has a few large out-performances over the period, whereas the second portfolio manager has a number of sizeable negative active returns. The first portfolio manager, in contrast, seems to have more widely spread-out returns with less concentration around perfect indexation.

Having looked at a number of graphical indicators, let us now turn our attention to some numerical summary measures. Table 14.1 provides a number of summary statistics for the absolute and active monthly returns for each of our three portfolio managers and their benchmark.

This numerical analysis supports the conclusions drawn from Figs. 14.1, 14.2, and 14.3. The mean and median absolute monthly returns are positive for all portfolio managers in the area of approximately 50–60 basis points. The range of return outcomes is substantially larger at ± 300–400 basis points. The average dispersion—as measured by the inter-quartile range (IQR) and the standard deviation—is approximately 100–150 basis points.[3]

[3]The inter-quartile range is defined as the difference between the 75th and 25th percentile. It is, roughly speaking, the order-statistic equivalent of the standard deviation.

**Table 14.1** Some return summary statistics

| Statistic | Portfolio 1 | Portfolio 2 | Portfolio 3 | Benchmark |
|---|---|---|---|---|
| **Absolute monthly return** | | | | |
| Mean | 54.7 | 48.7 | 54.5 | 50.0 |
| Median | 57.0 | 55.0 | 54.0 | 56.6 |
| Maximum | 387.0 | 363.0 | 438.0 | 373.1 |
| Minimum | *−415.0* | *−460.0* | *−322.0* | *−336.2* |
| IQR | 143.5 | 151.3 | 134.8 | 129.6 |
| Volatility | 122.5 | 121.3 | 109.0 | 108.1 |
| **Active monthly return** | | | | |
| Mean | 4.8 | *−1.2* | 4.5 | 0.0 |
| Median | 2.6 | *0.0* | 1.0 | 0.0 |
| Maximum | 116.7 | 63.7 | 168.3 | 0.0 |
| Minimum | *−109.1* | *−325.7* | *−90.1* | 0.0 |
| IQR | 42.8 | 23.0 | 19.9 | 0.0 |
| Volatility | 35.5 | 33.7 | 27.9 | 0.0 |

This table highlights a number of summary statistics for both the absolute and active returns of our three portfolio managers. All figure are in basis points.

The active-return numbers also tell an interesting story. Only the first and third portfolio managers succeeded in generating positive mean or median active returns over the period. The second manager not only has a negative active return, but also experienced a staggering minimum active return of about −3.3 %. This is more than three times larger than the other two managers.[4] The average dispersion of active returns, again measured by volatility and IQR, ranges from 20–40 basis points.

Another way to think about return is to consider a cumulative perspective. This is equivalent to posing the question: what if you invested $100 with each manager and the strategic benchmark at the beginning of the data period in December 1996? Figure 14.4 answers this question by showing the path of the cumulative returns for each manager and the benchmark over our 145-month data period.[5]

Although it is difficult to see large differences in cumulative returns, we may nonetheless make a few observations. All of the investments have approximately doubled the original $100 investment over the last 12 years. For much of the period,

[4]This extreme loss, while certainly important, does not appear to be the only reason for the second manager's under-performance. His median active return, less sensitive than the mean return to large outliers, also appears to be lower than his competitors.

[5]Each series in Fig. 14.4 is merely the geometric sum of the monthly returns over the period. Mathematically, the cumulative dollar return on $100 from the first period to the $i$th period is given as,

$$\begin{matrix}\text{Cumulative}\\ \text{Dollar}\\ \text{Return}\end{matrix}_{1,i} = \$100\left(\prod_{k=1}^{i}\left(1+R_{p,k}\right)\right). \tag{14.2}$$

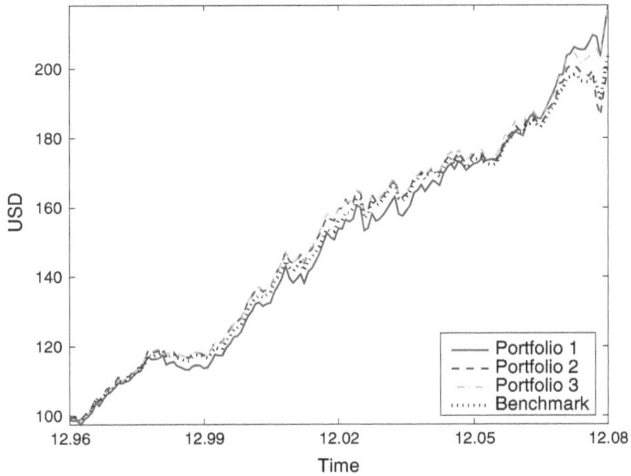


**Fig. 14.4** Cumulative absolute returns. This figure outlines the cumulative absolute monthly dollar returns for our three sample portfolio managers and their benchmark. The cumulative absolute return is the geometric sum of the monthly returns at each point over the analysis horizon starting from December 1996

the first portfolio manager lagged the other two managers as well as the benchmark. In the latter periods, however, the first portfolio manager's strategy appears to outperform. Finally, the third portfolio manager had substantial difficulty towards the end of the period.

We can also pose the previous question in a slightly different way: if you invested $100 with each manager, how much more money will you have relative to a passive investment in the benchmark. This is tantamount to asking if a portfolio manager's investment process has added any active return over and above the benchmark. Over a 1-, 2-, or even 6-month horizon, this may not be a fair question. Active management is difficult and one should expect to observe periods of under-performance. Over a 12-year time horizon, however, it seems like a reasonable and fair question. Figure 14.5 answers it by illustrating the cumulative active return for each of the three portfolios over the investment horizon.[6]

Figure 14.5 provides some interesting insights. Over the first half of the period, a fairly consistent story emerges. The second and third portfolio managers generate small but consistent positive returns, while the first portfolio manager seemed to consistently under-perform. Slightly past halfway through the period, the trend changes. The second and third managers, who had attained a cumulative return of almost $5 on our original $100 investment, appear to go through a few years of negative returns whereas the first manager begins accumulating active return. During the last 18–24 months, this trend has accentuated. The first and third portfolio managers added almost $15 of cumulative return by the end of the period, while the second portfolio manager finds himself with a negative cumulative return of about $5. Moreover, while the last few periods have not been kind to the second

[6]This computation is trickier than one might expect. The interested reader is directed to the shaded section for more detail.

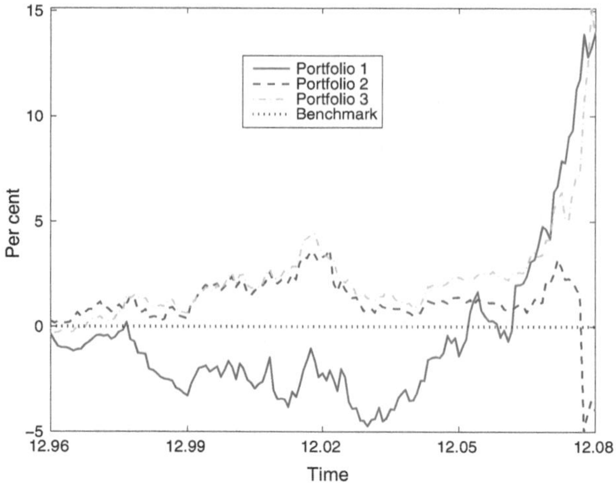


**Fig. 14.5** Cumulative active returns. This figure outlines the cumulative active monthly returns for our three sample portfolio managers relative to their common benchmark. The cumulative active return is the difference between the geometric sum of the portfolio returns and the benchmark return at each point over the analysis horizon starting from December 1996. It is *not* the geometric sum of the active returns

portfolio manager, things do not appear to have been going particularly well since early 2003.

A bit of caution is required with the computation of the cumulative active return. It is tempting to merely compute the geometric sum of the active returns as follows,

$$\begin{matrix}\text{Cumulative}\\ \text{Active}\\ \text{Return}\end{matrix}_{1,t} = \prod_{k=1}^{t}\left(1 + \underbrace{(R_{p,k} - R_{b,k})}_{\substack{\text{Active return}\\ \text{for period } k}}\right) - 1, \tag{14.3}$$

where the cumulative return runs from the first to the $t$th period. The problem is that it doesn't work. Mathematically, the product of the differences is not equal to the difference of the products. The computation in Eq. (14.3) does not correctly account for the compounding effect of the returns within the portfolio and benchmark. The proper computation, which captures the compounding effects, is performed as follows,

$$\begin{matrix}\text{Cumulative}\\ \text{Active}\\ \text{Return}\end{matrix}_{1,t} = \underbrace{\prod_{k=1}^{t}\left(1 + R_{p,k}\right)}_{\substack{\text{Cumulative portfolio}\\ \text{return to } t}} - \underbrace{\prod_{k=1}^{t}\left(1 + R_{p,k}\right)}_{\substack{\text{Cumulative benchmark}\\ \text{return to } t}}. \tag{14.4}$$

Trends in the data must also be explicitly considered. We have tried to highlight trends in our description of the preceding tables and figures, but only in an *ad hoc* manner. This analysis encompasses more than 12 years of data. In all of our computations, an observation occurring 145 months ago receives equal weight to last month's observation. This might be sensible if we believe that the distribution generating the returns is stable. This is a strong and questionable assumption. A better approach involves isolating trends in the data and thereby gaining a better understanding of patterns in our portfolio managers' behaviour.[7]

One useful approach for isolating trends is the use of a moving average. A moving average merely involves a rolling window for the computation of literally any summary statistic. Formally, a moving average for a sequence of observations $\{x_1, \ldots, x_n\}$ at time $t$ and with a window of $\delta$ periods is defined as,

$$\begin{array}{c}\text{Moving}\\ \text{Average}_{\delta,t}\end{array} = \frac{\overbrace{x_{t-\delta+1} + x_{t-\delta+2} + \cdots + x_{t-1} + x_t}^{\substack{\text{The } \delta \text{ most recent} \\ \text{observations of } x}}}{\delta}, \tag{14.5}$$

where $\delta \ll t$. A moving average is identical to a normal average except that the observations used to perform the computation change from one period to the next. The furthest observation falls out and the most recent observation enters into the computation. For this reason, a moving-average is sometimes also termed a rolling average.[8]

Figure 14.6 provides an illustration of the moving-average absolute returns for our three portfolio managers and the benchmark. We have elected to use a 36-month, or 3-year, window for our computations—we will make liberal use of the moving-average concept in this chapter and will always use a 36-month window.[9] Thus, we can see that the time interval in Fig. 14.6 begins in November 1999 instead of the usual December 1996. This is because we require the first 36 observations to compute the first moving-average observation.

It is now much easier to distinguish between the different portfolio-manager return series, because the moving-average dampens much of the noise in the data and permits one to focus on the general trend. Three secular trends in the absolute returns are evident. The period from 1999 to 2003 was characterized by steadily increasing returns. From 2003 to 2006 returns changed course and fell gradually from more than 80 basis points per month to less than 20 basis points. From 2006, the trend is again generally positive, albeit with more dispersion between the individual managers.

---

[7]This is particularly important in this data set as the final 18 months of the dataset—July 2007 to December 2008—coincided with a period of substantial turmoil in financial markets.

[8]This is not entirely unlike the exponential weighting concept that is used for the computation of covariance matrices in previous chapters.

[9]There is nothing particularly special about 36 months. Other choices are, of course, possible.

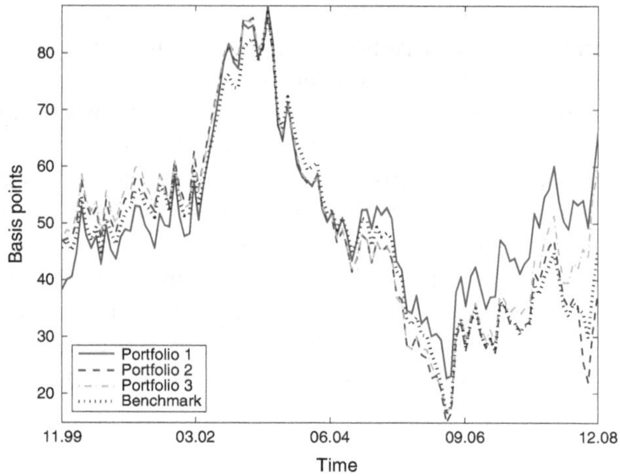


**Fig. 14.6** Rolling absolute returns. This figure outlines a 3-year, or 36-month, moving average of the absolute returns for our three portfolio managers and their common benchmark

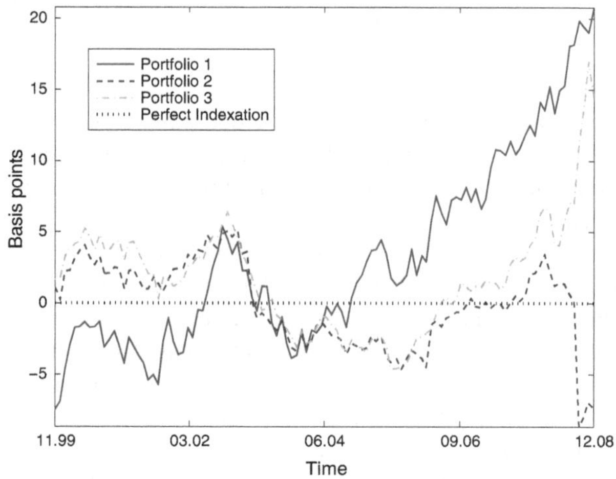


**Fig. 14.7** Rolling active returns. This figure outlines a 3-year, or 36-month, moving average of the active returns for our three portfolio managers relative to their common benchmark

Figure 14.7 examines the moving-average active returns for each of our managers. The under-performance of the first portfolio manager during the earlier part of the analysis horizon and his out-performance in the latter periods are quite evident. The under-performance of the second portfolio manager from around 2003, with a particularly difficult period in 2008, is also relatively easy to identify. Neither of these trends were clearly visible in our original view of the data in Fig. 14.1.

At this point, let us make a brief aside to consider an interesting risk measure. It is not theoretically based, but rather a statistically motivated measure and, as such, we will address it in this section. The idea comes from the observation that it is interesting and useful, when looking at cumulative returns, to think about peaks and valleys. The distance between a peak (i.e., highest cumulative return to a given point) and a valley (i.e., lowest cumulative return to some point) essentially represents your change in wealth over a given interval. The larger these distances, following this logic, the riskier your investment.

Let's make this more precise. We define the drawdown, a statistical measure, as the loss from the peak return to the current value. Mathematically, it is defined as

$$DD(T) = \max_{s \in [t,T]} \left( \hat{R}_s - \hat{R}_T \right). \tag{14.6}$$

where $DD(\cdot)$ denotes drawdown and

$$\hat{R}_s = \prod_{\tau=t}^{s} (1 + R_{p,\tau}) \tag{14.7}$$

is the cumulative return over the interval, $[t, s]$. In words, therefore, drawdown is the distance from the highest level of cumulative return—over some interval $[t, T]$—to the current value at time $T$. Another way to think about this notion is that you have a starting point, $t$, and for each choice of $T$, you will have a different drawdown value.

A drawdown series is useful, but we would probably also like to reduce it to a single statistical measure. A natural choice would be the largest drawdown over a given period. The maximum drawdown (MDD)—interpreted as the biggest peak to trough movement over an interval—is simply defined as

$$\max_{t \in [t,T]} DD(t). \tag{14.8}$$

The maximum drawdown is a perspective on portfolio risk that, from a practical perspective, may be used to complement other risk measures such as return volatility.

Figure 14.8 provides the evolution of the cumulative return and drawdown for the first portfolio manager across our data sample. The values—reported in percentage terms since they are the differences in cumulative returns—range from 0 to 7 %. A value of 0 % for the drawdown measure implies that the portfolio has obtained a new peak cumulative return. The maximum drawdown for this portfolio, slightly above 7 %, is denoted with the square marker.

As a final step, let's examine in Table 14.2, the maximum drawdown for our *three* portfolios. There appears to be a reasonable amount of variation between the managers.

While maximum drawdown is considered a measure of risk, note that it does *not* completely coincide with volatility. The first portfolio manager has, for example, the highest volatility, but not the highest maximum drawdown value. We may conclude that, at least in this case, this measure appears to capture another dimension of risk.

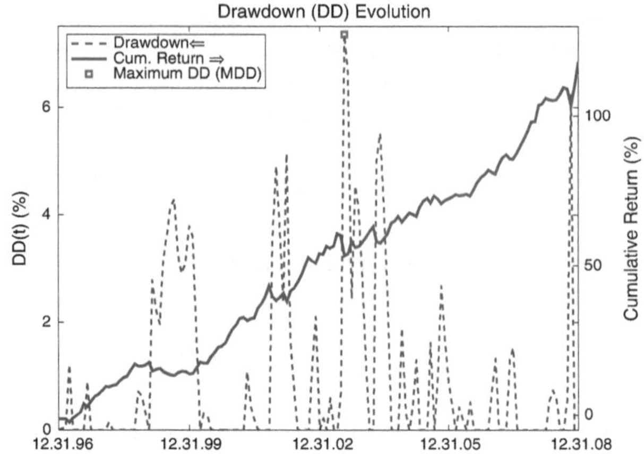


**Fig. 14.8** Drawdown. This figure illustrates, for the first portfolio manager, the evolution of the cumulative returns and associated drawdowns over the data period. The largest drawdown over the interval is termed the maximum drawdown

**Table 14.2** Drawdown statistics

| Statistic | Portfolio 1 | Portfolio 2 | Portfolio 3 | Benchmark |
|---|---|---|---|---|
| **MDD** | **7.3 %** | **14.5 %** | **5.8 %** | **7.6 %** |
| Volatility | 122.5 | 121.3 | 109.0 | 108.1 |

This table outlines the maximum drawdown, in percent, for our three portfolio managers. It also provides, for the purposes of comparison, the absolute return volatility. There appears to be information in drawdowns that is independent of the volatility measure.

The key difference is the importance of the *ordering* of returns to the notion of drawdown. Changing the order of returns will have no impact on a measure like volatility, but it can have a dramatic impact on the drawdown. Consequently, more than many measures, the selected time period has an important impact.[10]

The analysis in this section has been unencumbered by any finance theory. We have merely taken the return history in our spread-sheet and performed some basic statistical analysis. This analysis involved a number of transformations of our raw return data: these included active, cumulative, and moving-average returns. Each transformation help us examine the portfolio managers' performance from a different perspective.

While it will not help us make a definitive judgement on the individual managers, basic statistical analysis is nonetheless an excellent way to gain familiarity with your underlying data. It is a necessary and important first step. In the following sections, we will introduce some basic theory to further guide our analysis.

[10]See also Burghardt et al. [2] and Ismail et al. [8] for more detailed information on this point in particular and the notion of drawdowns in general.

## 14.2 Some Theory

Most of the key ideas in the following sections come from the Capital Asset Pricing Model (CAPM). Some are a direct consequence of this theory, while others are inspired by the basic principles.[11] A (very) brief review of this important finance theory is thus in order.

The main objective behind CAPM is to find the rate of return required for an arbitrary asset assuming that it is added to a diversified portfolio. CAPM relates the expected return on any given asset $(R_i)$ to expected return of the market portfolio $(R_m)$.[12] The key insight of the CAPM is that covariance is risk. Each asset has some component of its return correlated with the market return, called *systematic* risk, and part of its return arising from unique security-related factors—this latter risk is termed *idiosyncratic* risk. Any expected return deviation from the market return for a given asset must stem from systematic risk. Covariance with the market portfolio, therefore, fully determines expected return. Since idiosyncratic risk can be avoided by diversification, it is not compensated.

The basic structure of the CAPM begins with a linear relationship between the return on a given asset, $R_i$ and the return on the market portfolio, $R_m$ as

$$R_i = \beta R_m + \epsilon, \tag{14.9}$$

where $\beta \in \mathbb{R}$ is a coefficient and $\epsilon$ is an error term. This is a linear mapping between of asset $i$ returns and the market portfolio's returns. Instead of direct returns, however, the CAPM typically works with excess returns over the risk-free rate, $R_f$.[13] Thus, we rewrite Eq. (14.9) as,

$$\underbrace{R_i - R_f}_{\text{Excess return}} = \beta \underbrace{\left(R_m - R_f\right)}_{\text{Excess return}} + \epsilon. \tag{14.10}$$

Taking expectations of both sides, we arrive at

$$\mathbb{E}\left(R_i\right) - R_f = \beta\left(\mathbb{E}\left(R_m\right) - R_f\right) + \underbrace{\mathbb{E}\left(\epsilon\right)}_{=0}, \tag{14.11}$$

[11]The CAPM has been much maligned in recent years. It still remains, however, a useful structure for the consideration of portfolio decisions.

[12]The market portfolio is essentially a maximally diversified portfolio exposed to a wide range of underlying risk factors.

[13]Active returns, in this chapter, refer to the difference between a portfolio's return and its underlying benchmark. Excess returns, conversely, refer to the difference between portfolio or benchmark returns and the risk-free rate. This is sadly far from widely accepted terminology. Excess returns can, and often do, refer to both concepts.

$$\mathbb{E}(R_i) = \underbrace{R_f}_{\text{Intercept}} + \underbrace{\beta}_{\text{Slope}} \left(\mathbb{E}(R_m) - R_f\right).$$

This is the fundamental CAPM result: the rate of return of asset $i$ is a linear function of the risk-free rate and the excess return of the market portfolio over the risk-free rate.[14]

The slope parameter $\beta$ has a special significance. It represents the systematic component of the asset's return. This is the risk that cannot be diversified away by holding the maximally diversified market portfolio. You can, as an investor, scale your expected return by taking more ($\beta > 1$) or less ($\beta < 1$) risk than the market portfolio.[15]

That was a brief recap of the theory.[16] While the CAPM model is the natural starting point for much of ex-post analysis, there is more to the story. We make *five* observations regarding the basic CAPM theory:

1. the market return, $R_m$, is vaguely defined;
2. excess instead of absolute returns are used;
3. there is no true intercept term (just the risk-free rate, $R_f$);
4. not much is said about the error term; and
5. there is only one factor driving expected returns–the market portfolio.

Considering each of these points in more detail will help us introduce a set of ex-post analytic measures. While we cannot derive everything directly from the CAPM, it remains the motivation behind most of these ideas. Understanding these links to the theory is, in our view, essential for appropriately using and interpreting these analytic measures.

### 14.2.1 Introducing $\beta$

The slope coefficient $\beta$ describes the systematic risk of asset $i$ in relation to the market portfolio. For our analytic purposes, there are *three* slight changes that we'd like to make to Eq. (14.11):

1. We are more interested in the performance relative to a benchmark portfolio than the market portfolio. We thus replace the return of the market portfolio ($R_m$) in the CAPM equation with the return on the benchmark ($R_b$).

---

[14]This is also referred to as the security market line.

[15]The error term, $\epsilon$, denotes idiosyncratic risk, which has zero expectation ($\mathbb{E}(\epsilon) = 0$) so it should not impact one's expected return.

[16]There are many excellent references for this work. The natural starting point is Treynor (1961, Market value, time, and risk, unpublished), Sharpe [10, 11], and Lintner [7].

2. We are not interested in the return of an arbitrary security, but rather the return of our portfolio. Thus, we will replace $R_i$ with $R_p$.
3. We will ignore the risk-free rate and consider absolute returns.

These changes lead to the following adjustment to the Eq. (14.11):

$$\mathbb{E}\left(R_p\right) = \beta\mathbb{E}\left(R_b\right). \tag{14.12}$$

What have we done? We have written the expected return of our portfolio as a linear function of the expected return on its benchmark.

Is it defensible to replace the market portfolio with your benchmark? If the benchmark, in our case the Lehman Aggregate Index, is a large diversified portfolio, it can be considered the market portfolio for the asset class under consideration.[17] Even if the benchmark is relatively small and specialized, it still represents your target portfolio. The idea is to focus our attention on the set of assets of interest—the market portfolio is simply too broad to be practically useful.

How do we interpret the value of $\beta$? If $\beta$ exceeds one, it implies that for a given change in the benchmark return, either positive or negative, the portfolio is expected to have a larger return. This would imply that the portfolio manager is acting more aggressively than the benchmark.

If $\beta$ is less than one, the opposite is true. A given change in the benchmark return is expected to lead to a smaller return, either positive or negative, compared to the benchmark outcome. A portfolio manager with a $\beta < 1$ is less aggressively positioned than the benchmark. For a fixed-income portfolio one may, loosely speaking, think of $\beta$ as describing one's duration positioning relative to the strategic benchmark—should $\beta$ be greater (less) than unity, then the portfolio manager is systematically long (short) duration.

If $\beta$ is equal to one, then the portfolio returns move one for one, or rather follow perfectly, the benchmark portfolio. In short, therefore, our definition of $\beta$ measures one's systematic risk *relative* to one's benchmark. Passive managers take this idea to heart, by closely replicating the underlying benchmark.

How do we determine the value of $\beta$ for a given portfolio and choice of benchmark? $\beta$ is not something that is observable in the market. You cannot go to a screen to look it up, as you would with a price. We will need to compute it. Fortunately, it is not difficult to compute with the appropriate data. All that we require is a time series—as found in our spread-sheet—of portfolio and benchmark returns. One then performs the following linear regression.

$$R_{p,t} = \beta R_{b,t} + \epsilon_t, \tag{14.13}$$

---

[17] In asset-management circles, the notion of $\beta$ is often used in a different way. Often, when one speaks of obtaining $\beta$ exposure to a particular asset class, one essentially means achieving diversified exposure to that specific asset class and not the degree of systematic risk undertaken.

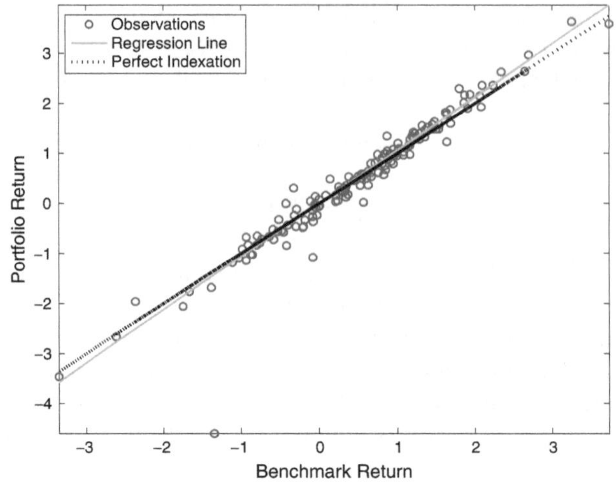


**Fig. 14.9** Computing $\beta$. This figure depicts graphically how one identifies the regression $\beta$ using our benchmark data and the returns from the second portfolio manager—it is merely the slope term from a linear regression of benchmark returns on portfolio returns

for $t \in \{1, \ldots, T\}$. $\beta$ is the slope coefficient that minimizes the error term, $\{\epsilon_t, t = 1, \ldots, T\}$. Recall that from a univariate regression, the $\beta$ coefficient can simply be written as,

$$\beta = \frac{\text{cov}\left(R_p, R_b\right)}{\text{var}\left(R_b\right)}. \tag{14.14}$$

The regression coefficient is thus defined as the covariance between the portfolio and benchmark returns normalized by the variance of the benchmark returns. This is entirely consistent with the fundamental CAPM notion of risk being intimately related to the covariance with the market portfolio (or, in our case, the benchmark portfolio).

Figure 14.9 provides a graphical illustration of how the $\beta$ coefficient is determined. The dotted black line denotes a perfect one-to-one relationship between the second manager's portfolio and strategic benchmark returns—practically speaking, this amounts to passive replication of the benchmark. The solid line, representing the line that best describes the linear relationship between portfolio and benchmark returns, has a slightly steeper slope than that suggested by passive benchmark replication. The value of $\beta$ is, in this case, greater than one implying that the portfolio manager is assuming more risk than the benchmark.

### 14.2.2 Introducing $\alpha$

The classic CAPM model does not have an intercept term.[18] There is nothing, of course, stopping us from adding one to our regression. This may seem like a simple gesture, but in asset-management circles, this intercept term is a highly discussed object. The intercept coefficient, or $\alpha$ as it is often referred to, describes the residual return on the portfolio relative to the benchmark once one has adjusted for systematic risk. This positive active return over the benchmark is exactly what an active manager is trying to accomplish. The intercept term, $\alpha$, is thus a measure of the asset manager's *skill* level. It should come as no surprise, therefore, that it is highly discussed.

Technically, we may compute $\alpha$ by merely adding an intercept term to our regression as follows,

$$R_{p,t} = \alpha + \beta R_{b,t} + \epsilon_t, \tag{14.15}$$

for $t \in \{1, \ldots, T\}$. If we re-arrange Eq. (14.15) slightly and take expectations of both sides,

$$\begin{aligned} \alpha &= R_{p,t} - \beta R_{b,t} + \epsilon_t, \\ \mathbb{E}(\alpha) &= \mathbb{E}\left(R_p\right) - \beta \mathbb{E}(R_b). \end{aligned} \tag{14.16}$$

we see clearly that $\alpha$ is the systematic-risk adjusted active return on the portfolio.

Figure 14.10 provides a geometric illustration of the regression intercept by zooming in on the origin of Fig. 14.9 and simultaneously displaying the vertical and

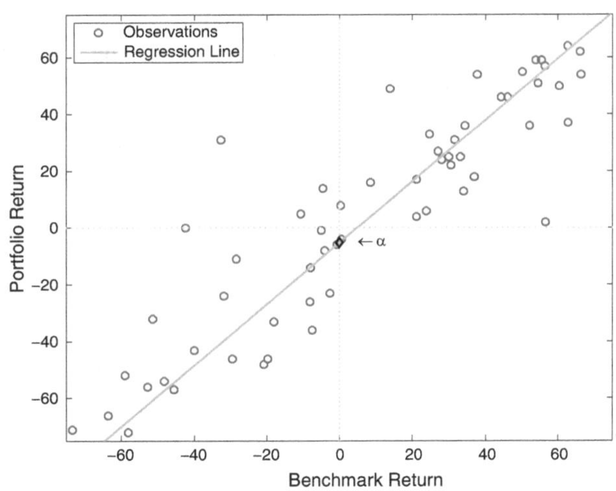


**Fig. 14.10** Computing $\alpha$. This figure depicts graphically how one identifies the regression $\alpha$ using our benchmark data and the returns from the second portfolio manager—it is the intercept term from a linear regression of benchmark returns on portfolio returns

[18]This is not entirely true. The intercept term is, in the formal CAPM setting, the risk-free rate.

horizontal axes. $\alpha$ is, of course, where the linear regression line intersects the vertical axis. In this case, it crosses the vertical axis slightly below the origin implying that the second portfolio manager has, when we have adjusted for systematic risk, a negative active return over the full period.

In the asset-management world, there does not appear to be a consistent definition of $\alpha$. This will be a recurring theme throughout this chapter. The quantity that we have just discussed, and demonstrated in Fig. 14.10, is actually termed the regression $\alpha$.

There are at least *two* other possible definitions. If we slightly adjust the regression equation illustrated in Eq. (14.16) to incorporate excess returns over the risk-free rate, we arrive at another definition,

$$R_{p,t} - R_{f,t} = \tilde{\alpha} + \tilde{\beta}\left(R_{b,t} - R_{f,t}\right) + \epsilon_t, \tag{14.17}$$

for $t \in \{1, \ldots, T\}$. The intercept in this regression is often called Jensen's $\alpha$, which we have denoted $\tilde{\alpha}$.[19] More insight into this measure is found by re-arranging Eq. (14.17) and applying the expectation operator to both sides as,

$$\begin{aligned} \tilde{\alpha} &= R_{p,t} - R_{f,t} - \tilde{\beta}\left(R_{b,t} - R_{f,t}\right) + \epsilon_t, \\ \mathbb{E}(\tilde{\alpha}) &= \mathbb{E}\left(R_p\right) - R_f - \tilde{\beta}\left(\mathbb{E}(R_b) - R_f\right), \\ \tilde{\alpha} &= \mathbb{E}\left(R_p\right) - \left(\underbrace{R_f + \tilde{\beta}\left(\mathbb{E}(R_b) - R_f\right)}_{\substack{\text{The predicted return} \\ \text{from CAPM equation}}}\right). \end{aligned} \tag{14.18}$$

Jensen's $\alpha$ should thus be interpreted as the *excess* return of the portfolio over the risk-free rate adjusted for systematic risk—the term in the brackets on the right-hand-side of Eq. (14.18) should look quite familiar as it is none other than the CAPM equation.

To confuse matters further, there is yet another definition of alpha. Asset managers typically define $\alpha$ as the mean active return computed over a given period of time. Specifically, $\alpha$ is often defined as,

$$\alpha = \frac{1}{T}\sum_{t=1}^{T} \underbrace{R_{p,t} - R_{b,t}}_{\substack{\text{Active return} \\ \text{from period } t}}. \tag{14.19}$$

[19]We use the notation, $\tilde{\alpha}$ and $\tilde{\beta}$, to describe the regression coefficients from Eq. (14.17), because they will not be the same as the regression $\alpha$ and $\beta$ estimated without considering the risk-free rate.

This is a practical definition of alpha that, although it ignores systematic risk, is extremely simple to compute. This is also the definition of active return that one is most likely to observe in practice.[20]

To summarize our discussion, the concept of $\alpha$ can be defined in, at least, the following three ways:

1. regression $\alpha$;
2. Jensen's $\alpha$; or
3. the average active return.

All three statistics attempt to measure the skill of the asset managers in generating active return and all three must be computed from a return history. The first two measures are likely cleaner measures of a portfolio managers' skill insofar as they adjust for amount of systematic risk being taken by the portfolio manager. Despite this theoretical advantage, $\alpha$ is typically computed using the third method. When discussing $\alpha$ with external managers, it is sensible to ensure that you have a common understanding of how $\alpha$ is computed. Better yet, you should ask for the underlying data and compute your own $\alpha$.

### 14.2.3 $\alpha$ and $\beta$

Having introduced the concepts of $\alpha$ and $\beta$, we return to our three portfolio managers. Table 14.3 provides, for the full 145-month sample period, the three alternative $\alpha$ measures for each portfolio manager as well as the associated regression $\beta$.

Not surprisingly, our $\alpha$ measures are not entirely consistent with one another. When looking at the first portfolio manager, the average active return is almost five basis points, while the regression $\alpha$ suggests a value of almost zero; at the same time, the Jensen's $\alpha$ is somewhere in between at about 2.5 basis points. For the second portfolio manager, the figures range from −1 to −5 basis points. Only the third portfolio manager exhibits consistency across the three measures of alpha.

**Table 14.3** $\alpha$ and $\beta$ estimates

| Statistic | Portfolio 1 | Portfolio 2 | Portfolio 3 |
|---|---|---|---|
| $\alpha_1$: Regression $\alpha$ | 0.4 | −5.2 | 5.8 |
| $\alpha_2$: Jensen's $\alpha$ | 2.7 | −2.9 | 5.0 |
| $\alpha_3$: Mean active return | 4.8 | −1.2 | 4.5 |
| Regression $\beta$ | 1.1 | 1.1 | 1.0 |

This table summarizes our estimates of the various $\alpha$ and $\beta$ measures for our three portfolio managers.

[20]It has the important advantage of being easy to compute and requires neither assumptions nor theory.

The first two portfolio managers exhibit regression $\beta$ figures significantly greater than unity. This suggests that these managers are more aggressively exposed to systematic risk, as measured by the benchmark portfolio. This aggressive stance is likely responsible for the inconsistency of the $\alpha$ measures—each measure takes systematic risk exposure relative to the benchmark in a slightly different manner. The fact that the third manager's $\beta$ value is essentially neutral may explain the consistency of its $\alpha$ measures.

It does not immediately follow that a low regression $\alpha$ implies that a manager does not have skill. Systematic deviations from the strategic benchmark (i.e., $\beta \neq 1$) are another way to take active risk. Let's return to our basic CAPM relationship:

$$\mathbb{E}(r_p) = \underbrace{\alpha}_{\text{Skill}} + \underbrace{\beta \mathbb{E}(r_b)}_{\substack{\text{Systematic}\\ \text{return}}} . \tag{14.20}$$

This is the basic logic. Now, we perform a little trick by adding and subtracting one from the $\beta$ coefficient,

$$\begin{aligned} \mathbb{E}(r_p) &= \alpha + (\beta + 1 - 1)\,\mathbb{E}(r_b), \qquad (14.21)\\ &= \mathbb{E}(r_b) + \underbrace{(\beta - 1)\,\mathbb{E}(r_b)}_{\substack{\text{Systematic}\\ \text{skill}}} + \underbrace{\alpha}_{\substack{\text{Pure}\\ \text{skill}}} . \end{aligned}$$

It is subtle, but active risk can be taken from a systematic or an idiosyncratic perspective. It can be a source of skill. A low regression $\alpha$ thus need not immediately imply a poor active manager. Note that the practical active risk measure combines both of these elements.

Given the range of $\alpha$ values provided in Table 14.3, which $\alpha$ measure should one use? The simple answer is: all of them. Each measure provides a slightly different perspective on the value added by the portfolio manager over and above the portfolio return. None of the measures are particularly onerous to compute and each helps to broaden one's understanding of the asset manager's performance. There is no compelling reason to restrict one's attention to a single measure.

The values in Table 14.3, while useful, suffer from the drawback that they focus on the entire 145-month analysis horizon. It is also important, as we've seen, to identify trends in the data. We proceed, therefore, to use the moving-average concept again in our computation of $\alpha$ and $\beta$. Using the first 36 months of data, one simply computes the $\alpha$ and $\beta$ measures. Then one takes a step forward, eliminates the furthest observation, adds the most recent observation, and recomputes the two

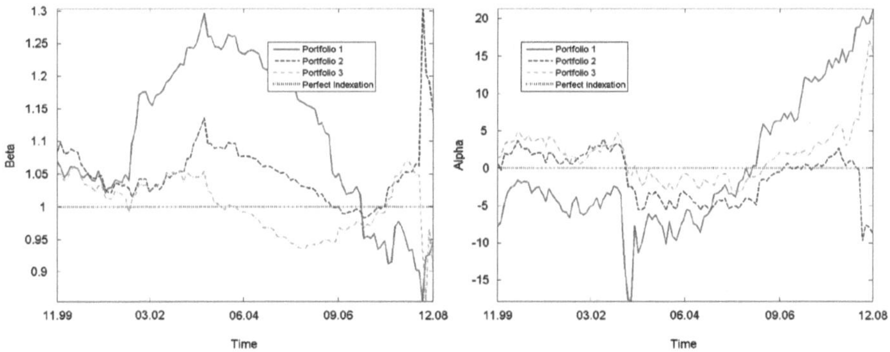


**Fig. 14.11** Rolling $\beta$ and $\alpha$ measures. This figure describes a 3-year, or 36-month, moving average of the systematic beta exposure and Jensen's $\alpha$ for each of our three portfolio managers

regression measures. This is repeated until the end of the sample. The results are provided in Fig. 14.11.

The first portfolio manager started off with a relatively neutral systematic risk stance and then proceeded to position himself quite aggressively relative to the benchmark with a $\beta$ value of almost 1.25 in 2003. This was followed by a long and gradual reduction in systematic risk over the next several years. The second and third portfolio managers have taken, relatively speaking, reasonably consistent systematic risk positions over the period; the second manager has been generally more aggressive than the benchmark while the third manager was typically more defensively positioned. As we enter into more turbulent financial markets at the end of the period, we observe some fairly sharp movements in systematic risk. The general trends nonetheless remain fairly constant.

The rolling $\alpha$ values—Fig. 14.11 shows Jensen's $\alpha$—are consistent with the rolling active return figures illustrated in Fig. 14.7. The first and third portfolio managers have, since about half way through the analysis period, experienced a relatively long and steady positive contribution to active returns. The second portfolio manager has had a recent bout of trouble in recent months, but also had some difficulty through the middle part of the analysis horizon.

Entering into the final part of our 145-month period, reducing one's systematic risk appears to have been a sensible strategy. The first and third portfolio managers did so and have continued to add $\alpha$ during this period. The second manager, with an aggressive $\beta$ position in 2008, has not faired as well.

## 14.3 Relative Risk

The next dimension involves computation of how much risk the portfolio manager is taking relative to the benchmark. This is similar to the concept of $\beta$ and, in fact, we will see that $\beta$ finds its way into the formal definition. Our risk measure is again derived from the original CAPM equation. The derivation requires a bit of patience and, although it is not the typical way to compute relative risk, it is nonetheless quite illuminating.

We begin with our by-now-familiar CAPM equation in regression form, where we ignore the risk-free rate and represent the market portfolio with our benchmark as,

$$R_{p,t} = \alpha + \beta R_{b,t} + \epsilon_t, \tag{14.22}$$

for $t = 1, \ldots, T$. Let's apply the variance operator to both sides of Eq. (14.22) and solve for the variance of the error terms as follows,

$$\begin{aligned}
\text{var}\left(R_{p,t}\right) &= \text{var}\left(\alpha + \beta R_{b,t} + \epsilon_t\right), \qquad (14.23)\\
\text{var}(R_p) &= \underbrace{\text{var}(\alpha)}_{=0} + \underbrace{\text{var}\left(\beta R_b + \epsilon\right)}_{\substack{R_b \text{ and } \epsilon \\ \text{are independent} \\ \text{by definition}}},\\
\text{var}(R_p) &= \beta^2 \text{var}(R_b) + \text{var}(\epsilon),\\
\text{var}(\epsilon) &= \text{var}(R_p) - \beta^2 \text{var}(R_b).
\end{aligned}$$

Following from the linear CAPM relation, the relative uncertainty surrounding the portfolio and the benchmark is described by the variance of portfolio returns less the variance of the $\beta$-adjusted benchmark returns.

Recalling the definition of the regression slope coefficient and covariance, we can further extend Eq. (14.23) as follows,[21]

$$\begin{aligned}
\text{var}(\epsilon) &= \text{var}(R_p) - \left(\underbrace{\frac{\text{cov}(R_p, R_b)}{\text{var}(R_b)}}_{\text{Eq. (14.14)}}\right)^2 \text{var}(R_b), \qquad (14.24)\\
&= \text{var}(R_p) - \left(\frac{\rho_{p,b}\sigma_p \cancel{\sigma_b}}{\sigma_b^{\cancel{2}}}\right)^2 \text{var}(R_b),\\
&= \sigma_p^2 - \left(\frac{\rho_{p,b}^2 \sigma_p^2}{\cancel{\sigma_b^2}}\right) \cancel{\sigma_b^2},\\
&= \sigma_p^2 \left(1 - \rho_{p,b}^2\right),
\end{aligned}$$

[21] In the following manipulation, we use the fairly common notation where $\rho_{X,Y}$ denotes the correlation coefficient between any two random variables $X$ and $Y$ and $\sigma_X = \sqrt{\text{var}(X)}$ represents the standard deviation of the arbitrary random variable $X$.

$$\sqrt{\text{var}(\epsilon)} = \sqrt{\sigma_p^2\left(1-\rho_{p,b}^2\right)},$$
$$\sigma_\epsilon = \sigma_p\sqrt{1-\rho_{p,b}^2}.$$

We have reduced the volatility, or standard deviation, of $\beta$-adjusted active returns into a relatively compact and intuitive expression. It is a function of the volatility of the portfolio returns and the correlation between portfolio and benchmark returns.

This quantity, $\sigma_\epsilon$, is the classical definition of a much-applied relative risk measure: ex-post tracking error. The classical ex-post tracking error is the error term stemming from a slightly altered version of the CAPM equation. This seems quite sensible as it quite naturally adjusts the volatility of active returns for systematic risk.

Once again, we find that this is not how the ex-post tracking error measure is typically measured in the asset-management world. Indeed, we saw a rather different definition in Chap. 10. Instead, it is general practice to merely compute the standard deviation of active returns as,

$$\sqrt{\text{var}\left(\underbrace{R_p - R_b}_{\substack{\text{Active return:}\\ R_a}}\right)} = \sqrt{\frac{1}{T-1}\sum_{t=1}^{T}\left(R_{a,t}-\bar{R}_a\right)^2}, \tag{14.25}$$

where $\bar{R}_a$ denotes the average active return. This definition of the ex-post tracking error is merely the standard deviation of active returns.

Using the definition of the variance of the difference between two random variables and re-arranging the basic expressions, then we may simplify this to,

$$\begin{aligned}
\sqrt{\text{var}(R_a)} &= \sqrt{\text{var}(R_p - R_b)}, \qquad (14.26)\\
&= \sqrt{\text{var}(R_p) + \text{var}(R_b) - 2\text{cov}(R_b, R_p)},\\
&= \sqrt{\text{var}(R_p) + \text{var}(R_b) - 2\underbrace{\frac{\text{cov}(R_b, R_p)}{\text{var}(R_b)}}_{\beta}\text{var}(R_b)},\\
&= \sqrt{\text{var}(R_p) + \text{var}(R_b) - 2\beta\text{var}(R_b)},\\
&= \sqrt{\text{var}(R_p) + \text{var}(R_b)(1-2\beta)}.
\end{aligned}$$

What is interesting in this development is that although it was not explicitly part of the computation in Eq. (14.25), the systematic risk as described by the regression coefficient $\beta$ shows up. While it does not enter into the computation in quite the same way as suggested by the classical computation, it is nonetheless present.

We now have two alternative approaches to compute the tracking error, or active risk, of a portfolio relative to a given benchmark:

$TE_1$ a classical definition based on the CAPM equation; and
$TE_2$ a practical definition based on the standard deviation of active returns.

The classical definition is rarely used in practice where virtually all computations are performed using the second practical definition.

There are clear mathematical differences between these two approaches. What is the magnitude of difference, however, between these two methods in practice? Does it make sense to compute both methods to get a full picture of one's active risk exposure? Generally speaking, the values derived from the two methods are quite close, but as before with the $\alpha$ estimates, multiple measures from different perspectives are practically useful. If a range of measures tell the same story, then one may make more robust conclusions. The underlying shaded box provides, for the interested reader, an analysis of the conditions where the two definitions coincide.

There is an interesting relationship between our two definitions of ex-post tracking error. Let us begin with the definition of $TE_2$,

$$\sqrt{\text{var}(R_p - R_b)} = \sqrt{\text{var}(R_p) + \text{var}(R_b)(1 - 2\beta)}. \tag{14.27}$$

We would like to find out under what conditions, if any, these two definitions of tracking error are equivalent. To this end, we focus on the term $\text{var}(R_p)$—our objective is to try to find another quantity to replace it and simplify. If we assume that the CAPM is true, then with a slight re-arrangement of Eq. (14.22), we have

$$\text{var}(R_p) = \beta^2 \text{var}(R_b) + \text{var}(\epsilon). \tag{14.28}$$

This appears promising. If we put this definition of $\text{var}(R_p)$ into $TE_2$ and simplify, we have

$$\begin{aligned}\sqrt{\text{var}(R_p - R_b)} &= \sqrt{\underbrace{\beta^2 \text{var}(R_b) + \text{var}(\epsilon)}_{\text{Eq. (14.28)}} + \text{var}(R_b)(1 - 2\beta)}. \quad (14.29)\\ &= \sqrt{\text{var}(R_b) \underbrace{(1 - 2\beta + \beta^2)}_{\substack{\text{2nd-order}\\ \text{polynomial}\\ \text{in } \beta}} + \text{var}(\epsilon)},\end{aligned}$$

(continued)

$$= \sqrt{\text{var}(R_b)(1-\beta)^2 + \text{var}(\epsilon)}.$$

If $\beta$ is equal to one, then the first term under the square root on the right-hand-side of Eq. (14.29) vanishes. If $\beta$ is approximately equal to one, which is often the case, then this term is approximately equal to zero. More specifically,

$$\underbrace{\sqrt{\text{var}(R_p - R_b)}}_{\text{TE}_2} = \sqrt{\text{var}(R_b)\underbrace{(1-\beta)^2}_{\beta\approx 1?} + \text{var}(\epsilon)}, \qquad (14.30)$$
$$\approx \underbrace{\sqrt{\text{var}(\epsilon)}}_{\text{TE}_1}.$$

In other words, the two definitions of tracking error converge as the $\beta$ of the portfolio tends to unity.

Table 14.4 illustrates the tracking error using both methods for our three portfolio managers. The differences between the two computations are, in all cases, approximately one basis point in magnitude. This supports our claim that, practically speaking, our two definitions of tracking error are very similar.[22]

Over the full period, the first portfolio manager has taken the most active risk, followed closely by the second manager, and then, with a bit of distance, the third manager. This is consistent with previous findings where we noticed that the third portfolio manager was generally positioned defensively relative to the benchmark.

The figures in Table 14.4 are computed over the entire 12-year period and, as such, do not provide much insight into trends in active risk. Figure 14.12 supplements this information by describing the moving-average tracking error, computed with the practical definition, over our analysis horizon.

There is a spike in active risk for all portfolio managers towards the end of the period. This is a natural consequence of the increased uncertainty in financial markets—as markets become more volatile, so do returns on both the portfolio

**Table 14.4** Ex-post tracking error

| Statistic | Portfolio 1 | Portfolio 2 | Portfolio 3 |
|---|---|---|---|
| Classic TE | 35.5 | 33.7 | 27.9 |
| Practical TE | 34.2 | 32.6 | 27.8 |

This table illustrates the ex-post tracking error for all three portfolio managers using the two alternative approaches described in the text. All values are in basis points.

[22] Should $\beta$ significantly differ from unity, however, then this may no longer be the case.

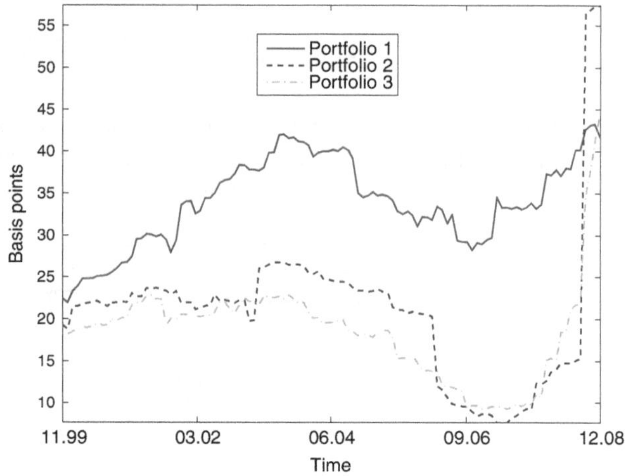


**Fig. 14.12** Rolling ex-post tracking error. This figure describes a 3-year, or 36-month, moving average of the ex-post tracking error for each of our three portfolio managers—it is computed using the practical definition presented in the text

and benchmark and consequently even small active positions tend to become more risky.

Abstracting from the final months in our data sample, we see a fairly consistent trend. The first portfolio manager has, more or less, consistently engaged in more active risk taking than the other managers. The second and third managers have remarkably similar active risk profiles over the entire period with the third manager adopting on average slight less active risk.

Tracking error is a relative measure of risk. This has important implications for its interpretation. Tracking error describes the distance from the benchmark, but it provides no insight into the absolute level of risk adopted in the portfolio. Logically speaking, one can deviate from the benchmark by taking more or less risk than the benchmark. Tracking error will not help differentiate if active risk arises from taking *less* risk than the benchmark or taking *more* risk. It treats both risks in a symmetric manner.

We can see this quite clearly in our practical example. Figure 14.11 shows that the second portfolio manager took consistently greater systematic risk relative to the third portfolio manager. This implies that the second portfolio manager was more aggressive than the benchmark, while the third portfolio manager was positioned more defensively than the benchmark. The tracking error, however, shows similar values for both portfolio managers. The aggressive and defensive positions appear to have consumed similar amounts of active risk.

## 14.4 Risk-Adjusted Ratios

We have examined measures that consider active return or active risk. Can we combine them? The simple answer is, yes. In the financial literature, and in practice, there are a number of ratios that seek to adjust different notions of return by different notions of risk.[23] Each ratio has basically the following form,

$$\begin{array}{c}\text{Risk-Adjusted}\\ \text{Ratio}\end{array} = \frac{\begin{array}{c}\text{Some Notion}\\ \text{of Return}\end{array}}{\begin{array}{c}\text{Some Notion}\\ \text{of Risk}\end{array}}. \tag{14.31}$$

These ratios seek to normalize (or adjust) the return for the amount of risk taken. These ratios are not, in themselves, risk-adjusted measures of return despite the fact they are often referred to as such. The ratios that we will discuss in the coming pages are the units of return per unit of risk. In this respect, they are ratios of return to risk and not returns.[24]

The first ratio that we will consider compares active return to active risk. It has the following general form,

$$\begin{aligned}\text{Information Ratio} &= \frac{\text{Active Return}}{\text{Active Risk}}, \\ &= \frac{\mathbb{E}(R_a)}{\sqrt{\text{var}(R_a)}}.\end{aligned} \tag{14.33}$$

This is perhaps the most common ratio in asset management. It is called the information ratio and is essentially a measure of a portfolio manager's skill. It can be defined in various ways depending on how one wishes to define active risk and active return. We only consider two possibilities, where the *first* is

$$\begin{aligned}\text{Information Ratio } (\text{IR}_1) &= \frac{\text{Mean Active Return}}{\text{Practical TE}}, \\ &= \frac{\frac{1}{T}\sum_{t=1}^{T} R_{a,t}}{\sqrt{\frac{1}{T-1}\sum_{t=1}^{T}\left(R_{a,t}-\bar{R}_a\right)^2}}.\end{aligned} \tag{14.34}$$

---

[23] Bacon [1] provides an excellent and exhaustive description of these ratios.

[24] A risk-adjusted return, which we will not cover in this text, requires the subtraction of some risk adjustment from the original return. It would have the form:

$$\begin{array}{c}\text{Risk-adjusted}\\ \text{return}\end{array} = \begin{array}{c}\text{Unadjusted}\\ \text{return}\end{array} - \begin{array}{c}\text{Risk}\\ \text{adjustment}.\end{array} \tag{14.32}$$

Our *second* definition has the following form,

$$\text{Information Ratio (IR}_2) = \frac{\text{Regression } \alpha}{\text{Practical TE}}, \tag{14.35}$$

$$= \frac{\alpha}{\sqrt{\frac{1}{T-1}\sum_{t=1}^{T}\left(R_{a,t} - \bar{R}_a\right)^2}}.$$

The two definitions for the information ratio are also very similar. The main difference is that the second approach includes an adjustment for systematic risk in active returns through the use of the regression $\alpha$ in the numerator.[25]

Once the computation is complete, how do we interpret the information ratio? Obviously, we hope that it is positive. A negative value indicates that the portfolio manager is, in fact, under-performing the benchmark.[26] Beyond its sign, the larger the information ratio, the better. A large information ratio indicates that the portfolio manager generates more return per unit of risk assumed. There are essentially four distinct cases:

1. low active return with low active risk;
2. high active return with low active risk;
3. low active return with high active risk; and
4. high active return with high active risk.

We expect the highest information ratios where the portfolio manager generates high levels of active return with low levels of active risk (i.e., the second case). This is what one hopes that an internal or external manager can achieve when managing one's funds. It implies that the portfolio manager can skilfully extract significant returns from each unit of risk.

The least desirable outcome, and usually the lowest information ratio values, occur when the portfolio manager achieves low active return with high active risk (i.e, the third case). Here the portfolio manager takes substantial risk to attain modest levels of return; this speaks poorly for his or her skill level.[27]

An information ratio of more than about $\frac{1}{2}$ is considered good and values exceeding $\frac{3}{4}$ would typically put a manager into the top quartile relative to his

[25]We could have used alternative definitions for the denominator, but we previously established that the two definitions provided for the ex-post tracking error do not vary much in our example.

[26]Irrespective of how the information ratio is defined, the information ratio is typically presented as an *annualized* figure. This implies that if you have monthly data, then you need to gross up the average active return and adjust the tracking error in the denominator. Monthly tracking error figures are annualized, using the square-root rule, by multiplying by $\sqrt{12}$.

[27]The first and last cases are more neutral where low (high) returns arise from low (high) levels of active risk.

or her peers in a given asset class. Given that the information ratio is based on statistical estimates of active return and risk, we should be somewhat cautious in its interpretation. It is essential that an information ratio is computed with a reasonable number of data-points. An information ratio computed with less than about 36 months of return data is unlikely to be very reliable.[28]

Another well known risk-adjusted ratio is the so-called Sharpe Ratio. The Sharpe ratio is actually a special case of the information ratio. To show this, let us begin with the most generic definition of the information ratio,

$$\text{Information Ratio} = \frac{\mathbb{E}(R_p - R_b)}{\sqrt{\text{var}(R_p - R_b)}}. \tag{14.36}$$

Imagine that your benchmark over a given time interval is the *deterministic* risk-free rate, $R_f$.[29] Then, the information ratio becomes:

$$\begin{aligned} \text{Information Ratio} &= \frac{\mathbb{E}(R_p - R_f)}{\sqrt{\text{var}(R_p - R_f)}}, \\ &= \frac{\mathbb{E}(R_p - R_f)}{\sqrt{\text{var}(R_p) + \underbrace{\text{var}(R_f)}_{=0} - \underbrace{2\text{cov}(R_p, R_f)}_{=0}}}, \\ &= \frac{\mathbb{E}(R_p) - R_f}{\sigma(R_p)}, \\ &= \text{Sharpe Ratio}. \end{aligned} \tag{14.37}$$

The Sharpe ratio is the equivalent to the information ratio in the special case when one's benchmark is the deterministic risk-free rate. It indicates how much excess return over the risk-free rate is generated for each unit of portfolio risk.[30]

Yet another well-known risk-adjusted ratio has the following form,

$$\text{Treynor Ratio} = \frac{\mathbb{E}(R_p - R_f)}{\beta}. \tag{14.38}$$

This ratio, called the Treynor ratio, places the excess return over the risk-free rate in the numerator and use the regression coefficient, $\beta$, as the risk measure

[28] If one takes the statistical aspect seriously and attempts to test if a given information ratio is statistically different from zero, one is likely to be discouraged. The standard error around our active return and risk estimates are sizeable, leading to correspondingly large confidence intervals for the information ratio.

[29] Also recall that $\text{var}(X + b) = \text{var}(X)$ for scalar $b \in \mathbb{R}$ and random variable, $X$.

[30] The active risk in the denominator collapses to portfolio risk because of the non-randomness of the benchmark, the risk-free rate.

in the denominator. The idea is that only systematic risk relative to the market portfolio matters.[31] The idiosyncratic risk is fully ignored in this case. This is sensible because, given one's capacity to diversify away this idiosyncratic risk, it is not compensated. Nevertheless, portfolio managers continuously take this non-systematic risk in an effort to generate active return and, as such, it is difficult to completely ignore it.

A complement to the Treynor ratio, is the following,

$$\text{Appraisal Ratio} = \frac{\alpha}{\sigma_\epsilon}. \tag{14.39}$$

A bit of reflection reveals that this is nothing other than the information ratio in disguise. Recall from Eq. (14.23) that $\sigma_\epsilon$ was actually the classical definition of tracking error. The appraisal ratio, therefore, is our second definition of the information ratio with the classical definition of the tracking error in the denominator.[32]

A final ratio, termed the Calmar ratio, makes use of the maximum drawdown as its representation of risk. It has the following definition,

$$\begin{aligned} \text{Calmar Ratio} &= \frac{\text{Cumulative Return over } [t, T]}{\text{Maximum Drawdown over } [t, T]}, \\ &= \frac{\left(\prod_{t=1}^{T}(1 + R_{p,t})\right) - 1}{\text{MDD}(t, T)}. \end{aligned} \tag{14.40}$$

In words, the Calmar ratio compares the total return over a period to the biggest peak to trough movement. As a risk measure, the maximum drawdown depends importantly on the sequence, or ordering, of the portfolio's returns. As previously indicated, it also depends importantly on the time horizon. This is probably the principal reason why it is not extensively employed.

We return to our practical example and examine how each of our three portfolio managers performed across this range of risk-adjusted ratios. Table 14.5 summarizes the four alternative risk-adjusted ratios for each portfolio manager computed using the full 145-month sample period.

There is a relatively large difference between the first and second definition of the information ratio.[33] The first portfolio manager, using the first definition, has a quite respectable value of 0.47. As we move to the second definition, however,

[31] In practice, this amounts to one's benchmark.

[32] In this way, it is basically a third possible definition of the information ratio!

[33] The second definition of the information ratio and the appraisal ratio, as expected, are essentially equal. As a consequence, we will ignore it in our discussion.

**Table 14.5** Information ratio

| Statistic | Portfolio 1 | Portfolio 2 | Portfolio 3 |
|---|---|---|---|
| Information ratio$_1$ | 0.5 | *−0.1* | 0.6 |
| Information ratio$_2$ | 0.0 | *−0.5* | 0.7 |
| Appraisal ratio | 0.0 | *−0.5* | 0.7 |
| Sharpe ratio | 0.8 | 0.6 | 0.3 |
| Treynor ratio | 0.03 | 0.02 | 0.03 |
| Calmar Ratio | 16.1 | 6.9 | 20.5 |

This table illustrates the range of risk-adjusted ratios for all three portfolio managers using the definitions provided in the text. Most of these figures can be reproduced using Tables 14.3 and 14.4 although one should not forget to annualize the monthly figures.

we see that the value falls to approximately zero.[34] Recalling from Table 14.3 that the first portfolio manager's average active return was about five basis points relative to a regression $\alpha$ of almost zero, we easily identify the difference between the two measures. Over the entire analysis horizon, only the third portfolio manager provides a consistently respectable information ratio. When all measures, computed from different perspectives, point in the same direction, this is typically an encouraging (or discouraging) sign.

The Sharpe and Treynor ratios use the risk-free rate as the benchmark and attempt to determine the excess return provided per unit of risk—they use, however, alternative definitions of risk in the denominator. From Table 14.1 on page 452, we see that the third portfolio manager has had the lowest absolute monthly return volatility, which likely explains this manager's relative high score on the Sharpe ratio. The second portfolio manager exhibits the lowest Calmar ratio, which is a consequence of having simultaneously lower returns and the largest maximum drawdown.

Since all of the regression $\beta$ figures are relatively close to one, the Treynor ratio is not far from the annualized average excess portfolio return over the risk-free rate. The second portfolio manager, with lower return and a slightly higher systematic risk, exhibits the lowest Treynor ratio. The other two managers have followed different strategies to arrive at a similar Treynor ratio. The first manager has slightly higher returns, but has adopted somewhat more systematic risk. Lower returns and slightly lower systematic risk brought the third portfolio manager to a similar Treynor ratio.

It should be no surprise that we wish to see the trend in each of our risk-adjusted ratios. Figure 14.13 illustrates the moving-average for each of our risk-adjusted ratios with a 36-month window. Despite their alternative definitions, the two information-ratio measures tell very similar stories. In both cases, the second portfolio manager has struggled, beyond the first few years, to generate a reasonable

[34]The difference is due entirely to the choice of numerator—in the first case, we use average active returns while the second employs the regression $\alpha$.

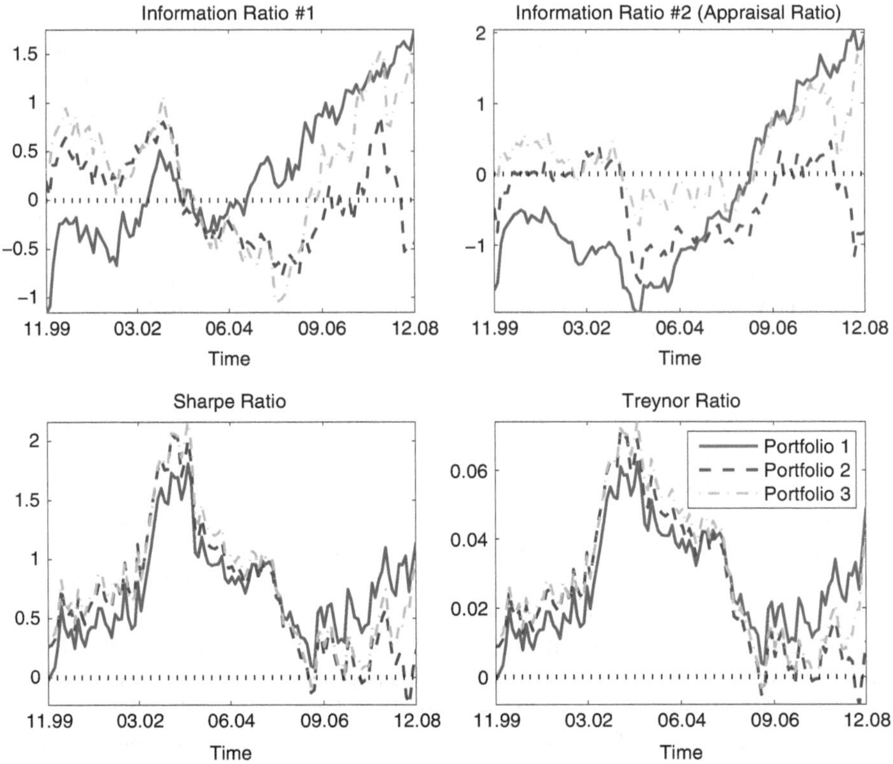


**Fig. 14.13** Rolling risk-adjusted ratios. This figure describes a 3-year, or 36-month, moving average of the four risk-adjusted ratios for each of our three portfolio managers

amount of return per unit of risk. Both measures clearly show a strong positive trend for the first and third portfolio managers starting from approximately mid-2004. Indeed, both of these managers exhibit very strong values in the range of 1–1.5 in the latter parts of the data period.

The Sharpe and Treynor ratio figures, for all portfolio managers, appear to much more strongly correlated. Up until about the last 2 years, each of the managers appears to quite closely track one another on these two dimensions. During the first several years, the third portfolio manager appears to modestly outperform the other two managers with the first portfolio manager generally slightly lagging his counterparts. Performance begins to diverge in the last 2 years. The first portfolio managers actually outperforms the other managers over this period, while the second portfolio manager gives up his middle position through a significant underperformance.

## 14.5 Beyond CAPM

We need to move beyond CAPM to address a challenging aspect of this framework: the identification of the market portfolio.[35] Since it is simply too large for practical use, it is standard practice to proxy the market portfolio with one's benchmark portfolio. This practice can be dangerous.

When one uses the market portfolio, systematic risk covers all possible risk factors. When one uses the benchmark, systematic risk involves only those risk factors included in the benchmark. There is nothing stopping a clever, or perhaps less than scrupulous, portfolio manager from taking systematic risk exposure from other asset classes outside of, and statistically independent from, the benchmark. Our standard techniques would treat this systematic exposure to other asset classes as active return. This is problematic, because such active return is *not* related to skill, but rather to systematic exposure to risk factors outside the strategic benchmark.

A solution to this problem is required. A natural way to identify such behaviour would be to examine the relationship of our portfolio managers' returns with a much broader range of risk factors. We could extend the original CAPM equation as,

$$\mathbb{E}\left(R_p\right) - R_f = \sum_{k=1}^{n} \beta_k \left(\mathbb{E}\left(F_k\right) - R_f\right) + \underbrace{\mathbb{E}\left(\epsilon\right)}_{=0}, \tag{14.41}$$

where $\{F_k, k = 1, \ldots, n\}$ are a collection of risk factors.[36]

What have we done? We have replaced the market portfolio with a collection of different, fundamental risk factors. This amounts to decomposing the market portfolio into its individual sources of systematic risk. One of these risks, of course, could be the benchmark portfolio. In this case, we merely, as before, eliminate the risk-free rate and re-arrange Eq. (14.41) as,

$$\mathbb{E}\left(R_p\right) = \alpha + \underbrace{\beta_1 \mathbb{E}(R_b)}_{\substack{\text{Benchmark}\\ \text{Factor}}} + \underbrace{\sum_{k=2}^{n} \beta_k \mathbb{E}\left(F_k\right)}_{\substack{\text{Other}\\ \text{Factors}}} + \mathbb{E}(\epsilon). \tag{14.42}$$

We have extended our original expression to include other sources of systematic risk.

[35] Recall that the market portfolio is essentially a huge, diversified portfolio incorporating all sources of risk.

[36] This is basically the main idea behind arbitrage pricing theory. See Ross [9] for more details. See also Kryzanowski et al. [6] for an attempt to combine arbitrage pricing theory with performance attribution.

We may cast this result as a multivariate linear regression,

$$R_{p,t} = \alpha + \underbrace{\beta_{0,t} R_{b,t}}_{\substack{\text{Benchmark}\\\text{Factor}}} + \underbrace{\sum_{k=1}^{n} \beta_{k,t} F_{k,t}}_{\substack{\text{Other}\\\text{Factors}}} + \epsilon_t. \tag{14.43}$$

For any given set of systematic risk factors, we may proceed to solve for $\alpha$ and the set of systematic exposures to the risk factors, $\{\beta_k, k = 0, \ldots, n\}$. The key idea is to see, once one controls for the benchmark return, what type of systematic exposure a given internal or external portfolio manager has taken.

Let's make this abstract idea more concrete by returning to our example. We have identified ten potential sources of systematic risk to act as additional regressors in Eq. (14.43). For each month in our 145-month data sample, we have values for each of these risk factors. A number of these factors are very general such as the return on oil and gold prices, inflation rates, and the return on VIX (volatility) index. Other factors relate to specific aspects of the fixed-income market such as changes in the swap spread, changes in the slope and level of the US Treasury curve, and returns on US Municipal, mortgage-backed, and high-yield securities.

One can certainly imagine other possible regressors or even argue that some of our choices should be reconsidered. Should one, for example, use the gold-price level or the monthly return for holding gold? Is it sensible to use a currency factor and should we include commodities other than oil? These are all good questions and in a real-life analysis these issues should be considered.[37] Our analysis, however, is intended to be illustrative and attempts only to demonstrate the general idea.

Table 14.6 illustrates the regression coefficients for each of our three portfolio managers as well as the t-statistic indicating statistical significance. Coefficients that are significant at a 90 % confidence level are highlighted with a single star whereas those statistically significant at the 95 % level are highlighted with two stars. In interpreting the results, we focus principally on the statistical significance of the regression coefficients.[38]

These results are best used as background information for a detailed conversation with each individual portfolio manager about their investment style. A portfolio manager may, for example, speak at length about their categorical avoidance of high-yield markets and yet you may observe a strong, and statistically significant

[37]Performing a multivariate time-series regression requires some caution. One must ensure that the correlation between the state variables is not too high and avoid strong autocorrelation in the individual variables. Failure to avoid these pitfalls can lead to violation of the regression assumptions and generate biased outputs. The regression results presented in Table 14.6 respect these criteria. See Judge [5], Harvey [4] and Hamilton [3] for more information on multivariate linear regression and time-series models, respectively.

[38]It makes little sense to discuss a large regression coefficient on a particular risk factor if we cannot state, with a reasonable level of confidence, that it is statistically different than zero.

**Table 14.6** Multiple risk factors

| | Portfolio 1 | | Portfolio 2 | | Portfolio 3 | |
|---|---|---|---|---|---|---|
| Regressor | Estimate | T-Stat | Estimate | T-Stat | Estimate | T-Stat |
| $\alpha$ | 0.0006 | 1.24 | −0.0003 | −0.70 | 0.0013 | *5.55*** |
| Benchmark return | 0.8535 | *6.25*** | 1.4322 | *11.01*** | 0.6616 | *9.29*** |
| Gold return | 0.0056 | 0.96 | −0.0033 | −0.58 | −0.0016 | −0.51 |
| Oil return | −0.0003 | −0.13 | 0.0031 | *1.43** | −0.0005 | −0.44 |
| Inflation rate | 0.2898 | *4.21*** | 0.0224 | 0.34 | 0.0054 | 0.15 |
| VIX-index return | 0.0026 | *1.64** | −0.0038 | *−2.54*** | 0.0026 | *3.22*** |
| $\Delta$ Swap spread | −0.0032 | *−1.29** | 0.0052 | *2.21*** | −0.0052 | *−3.99*** |
| $\Delta$ UST slope | 0.0033 | 1.20 | −0.0044 | *−1.65*** | 0.0059 | *4.05*** |
| $\Delta$ UST level | −0.0112 | *−3.08*** | 0.0045 | *1.30** | −0.0105 | *−5.53*** |
| Municipal return | 0.0434 | *1.59** | 0.0384 | *1.48** | 0.0143 | 1.01 |
| Mortgage return | −0.0964 | −1.01 | −0.3453 | *−3.81*** | 0.0904 | *1.82*** |
| High-yield return | −0.0066 | −0.40 | −0.0586 | *−3.76*** | −0.0316 | *−3.71*** |

This table provides a summary of the results of a multiple regression of our portfolio manager's returns against a number of market risk factors. Both the regression coefficient and the t-statistic are reported. Values with one star are statistically significant at a 90 % confidence level, whereas values with two stars are statistically significant at a 95 % level.

exposure to this asset class. In such a situation, you would certainly wish to delve a bit deeper into their specific investments. Table 14.6 is nonetheless indicative and hardly definitive evidence. It should be viewed as a form of due diligence on the portfolio manager's investment style and general performance.

What can we say about the results? In all cases, the systematic exposure to the benchmark is strongly statistically significant. At the same time, the magnitude of the benchmark $\beta$'s are dramatically different. The first and third portfolio managers have benchmark $\beta$'s of significantly less than one, while the second portfolio manager, at 1.43, is substantially greater than one. This suggests that all of the portfolio managers are taking systematic exposure outside of the benchmark.

This systematic exposure outside the benchmark does not seem to be providing any active return for the first and second portfolio managers. Only the third portfolio manager generates a statistically significant monthly $\alpha$ value. We can think of this $\alpha$ as being a cleaner version of the base regression $\alpha$, because it includes exposure to a much broader range of systematic risks.

Our portfolio managers appear to be engaging in systematic risk, on a number of dimensions, beyond the benchmark portfolio. In some cases, the systematic risk is very understandable. Statistically significant exposure to swap spreads, inflation rates, and changes in the level and slope of the US Treasury yield curve are fully expected. Positions in spread products, inflation-linked securities, and long-short duration relative to the strategic benchmark are part of a fixed-income manager's tool-kit and are, to a large extent, captured by these four regressors.

The sign of these regression coefficients is nevertheless interesting and might form the basis of a lively conversation with the respective portfolio manager. The third portfolio manager, for example, seems to have had difficulty with duration positioning and spread products, while the first portfolio manager looks to have had success with inflation-linked securities.

There are also a few surprises. The second and third portfolio managers have significant exposures to mortgages, high-yield instruments, and returns on the VIX (volatility) index. These may be acceptable investment classes and a natural part of the portfolio manager's investment universe. If they are not, then systematic exposure to these asset classes may be disguising itself as $\alpha$ and be erroneously attributed to skill. Similarly, one may not wish to have exposure to these asset classes. This analysis can then help to exclude certain managers or help write meaningful investment guidelines to exclude certain asset classes.

Much more can be done in this area. One can experiment with different systematic risk factors. One could expand the set of risk factors. One could look at the evolution of the systematic exposures to see the trend in exposure to various risk factors.[39] Overall, it is an art and a science to distil the information found in these regressions and extract trends in a parsimonious and meaningful way. The effort is, in our view, worthwhile as it provides useful insight into your portfolio manager's activities.

## 14.6 Bringing It All Together

We have examined our three potential portfolio managers from a broad range of perspectives. Which manager do you prefer? This is not an entirely fair question. There is likely *no* best manager and each organization's choice will depend on the relative importance given to each of the individual measures. Having said that, the head of your selection committee is awaiting your recommendation.

To assist you, we have tried, assigning an equal weight to each measure, to compare the three managers. Table 14.7 highlights a number of the most important measures and assigns a ranking from one to three—with one being the best mark—to each of our three portfolio managers.

Table 14.7 separates each ranking into a full-period and a trend component. Most of the measures are obvious: the higher the active return or the risk-adjusted ratio, the better the ranking. For tracking error, for example, we have assigned the highest rank to the lowest tracking error. The ranking for regression $\beta$ and other systematic exposures give preference to those managers who most closely follow the strategic benchmark. One can, of course, disagree with this methodology and use an alternative ranking approach. This is merely our attempt to objectively rank the candidates.

Applying this ranking algorithm over the full period, the third portfolio manager receives the highest marks, while the first portfolio manager exhibits the most

[39]Dimensionality becomes a problem in this case. With ten systematic risk factors, degrees of freedom in a rolling analysis may become a constraint.

**Table 14.7** Bringing it all together

| Statistic | Portfolio 1 Full | Portfolio 1 Trend | Portfolio 2 Full | Portfolio 2 Trend | Portfolio 3 Full | Portfolio 3 Trend |
|---|---|---|---|---|---|---|
| Active return | 2 | **1** | 3 | 3 | **1** | 2 |
| Regression $\alpha$ | 2 | **1** | 3 | 3 | **1** | 2 |
| Jensen's $\alpha$ | 2 | **1** | 3 | 3 | **1** | 2 |
| Benchmark $\beta$ | 3 | 2 | 2 | 3 | **1** | **1** |
| TE | 3 | **1** | 2 | 3 | **1** | 2 |
| Information ratio | 2 | **1** | 3 | 3 | **1** | 2 |
| Sharpe ratio | 2 | **1** | 3 | 3 | **1** | 2 |
| Treynor ratio | **1** | **1** | 3 | 3 | **1** | 2 |
| $\alpha$ significance | 2 | n/a | 3 | n/a | **1** | n/a |
| Other $\beta$ exposure | **1** | n/a | 3 | n/a | 2 | n/a |
| Total | 20 | **9** | 28 | 24 | **11** | 15 |

This table assembles all of the various measures together and attempts to rank the various portfolio managers. The bold figures denote the best ranking for a given category.

positive recent trend. The second portfolio manager is to be avoided. When considering the full period and the recent trend equally, you should probably tell the head of the selection committee that you have a slight preference for the third portfolio manager. Before doing this, however, we recommend a lengthy discussion with the third portfolio manager regarding their range of systematic risk exposures before awarding him the mandate.[40]

The ultimate choice of portfolio manager will depend on how important you deem the various criteria and the relative importance of the recent trend relative to the performance over the entire analysis horizon. Qualitative factors—such as global presence, fees, organizational stability, level of service, and transparency to name a few—also form a critical aspect of one's investment decision. While the quantitative analysis presented in this chapter forms an important part of any decision-making process, it is far from the only dimension to be considered.

## 14.7 Concluding Thoughts

This chapter is a departure from the high-dimensional instrument-level techniques treated in the others chapters. All of the measures and analysis take a backward-looking perspective, while we are all typically more interested in the future. The analysis is restricted to the monthly return series of the portfolio manager. Moreover, we have no information about the range of instruments held in these portfolios nor their exposure to various risk factors. One could easily conclude that these ex-post techniques cannot possibly be useful.

[40] See Table 14.6 on page 481 for more details.

That would be a mistake. Backward-looking analysis, limited data, and a low-dimensional perspective aside, we learned an enormous amount about our three managers. We gained insight into their capacity to generate active return, their exposure to systematic and relative risk, the amount of active risk per unit of risk, and the nature of their systematic exposure to other risk factors beyond the benchmark portfolio. Ultimately, using this analysis, we were able to construct a confident recommendation regarding our preferred manager.

If we broaden portfolio analytics to include techniques that help us take better portfolio-related decisions, then the methodologies discussed in this chapter are easily encompassed in this definition. The techniques described in this chapter should thus be viewed as a useful addition to one's fixed-income portfolio-analytic tool-kit.

## References

1. C.R. Bacon, *Practical Portfolio Performance Measurement and Attribution* (Wiley, New York, 2008)
2. G. Burghardt, R. Duncan, L. Liu, *Understanding Drawdowns.* Carr Futures Research Note, September 2003
3. J.D. Hamilton, *Time Series Analysis* (Princeton University Press, Princeton, 1994). Chapter 22
4. A.C. Harvey, *Forecasting, Structural Time Series Models and the Kalman Filter*, 1st edn. (Cambridge University Press, Cambridge, 1990)
5. G.G. Judge, W.E. Griffiths, R.C. Hill, H. Lütkepohl, T.-C. Lee, *The Theory and Practice of Econometrics*, 2nd edn. (Wiley, New York, 1985)
6. L. Kryzanowski, S. Lalancette, M.C. To, Performance attribution using an apt with prespecified macrofactors and time-varying risk premia and betas. J. Financ. Quant. Anal. **32**(2), 205–224 (1997)
7. J. Lintner, The valuation of risky asset and the selection of risky investments in stock portfolios and capital budgets. Rev. Econ. Stat. **47**(1), 13–37 (1965)
8. M. Magdon-Ismail, A.F. Atiya, A. Pratap, Y.S. Abu-Mostafa, *On the Maximum Drawdown of a Brownian Motion.* Rensaeler Polytechnic Institute, September 2003
9. S. Ross, The arbitrage theory of capital asset pricing. J. Econ. Theory **35**(3), 341–360 (1976)
10. W.F. Sharpe, A simplified model for portfolio analysis. Manag. Sci. **9**(2), 277–293 (1963)
11. W.F. Sharpe, Capital asset prices: A theory of market equilibrium under conditions of risk. J. Financ. **19**(3), 425–442 (1964)

# Some Mathematical Background A

*Only professional mathematicians learn anything from proofs. Other people learn from explanations.*

Ralph Boas

Mathematics is an attempt to express both basic and complex concepts in a concise and precise manner. As an approach, it has proven extremely practical and successful. Indeed, mathematization of the sciences has permitted a common language allowing communication of many important ideas. The field of finance has not escaped this process nor has this book. We have liberally used numerous mathematical ideas in this text and, as such, some background is required.

In the following pages, we consequently review a number of the mathematical concepts necessary for the understanding of this document. This level of detail is provided to keep this book relatively self-contained. The treatment nevertheless remains at a relatively high level. We have deliberately kept this review at a modest level. There will be very few demonstrations, no proofs, no attempts to put mathematical statements into a generalized framework, nor have we taken the time to ensure that all necessary hypotheses are listed for each of the main results. This chapter is, quite simply, a refresher. If you are seeing these concepts for the first time, then it is unlikely that you will gain a deep understanding from our brief and condensed descriptions. It should hopefully act as a quick reference for the main ideas in the preceding text. It is also perfectly acceptable to skip those sections with one already has familiarity or, to be frank, a lack of interest.

This chapter is divided in *four* main sections. We begin with the basic concepts of set theory. We then proceed to review two fundamental areas, which are very important for portfolio analytics: probability and statistics. Finally, we will introduce a range of ideas related to matrices and discuss a few interesting numerical techniques related to matrices. Some ideas addressed in the text, when



D.J. Bolder, *Fixed-Income Portfolio Analytics*,
DOI 10.1007/978-3-319-12667-8

considered too fundamental, are not considered in this appendix. We have, for example, assumed that readers have a basic familiarity with differential and integral calculus.[1]

## A.1 Set Theory

It is often useful to view mathematics as a game. One defines a collection of basic inviolable rules, or axioms, and then one–or rather, the mathematician—can start playing. Playing, in this context, involves analysing the consequences of the chosen rules. Mathematicians have worked extremely hard to define the minimal collection of rules from which all other mathematical results can be constructed. The origins of set theory, in fact, are an attempt to establish these fundamental rules. Sadly, it is something of a *failed* attempt, because the mathematical community has not yet succeeded in establishing a complete set of axioms entirely free of paradoxes.[2] Set theory is, despite this shortcoming, considered the foundation of modern mathematics and proves extremely useful for logical organization of mathematical ideas.

What is a set? A set is a collection of objects, potentially very different, that usually share a common feature. These objects are put in the same abstract container, called a set. The objects in a set are called its elements. If we call $e$ an element of the set $S$, then we say that $e \in S$. The $\in$ operator is our fundamental building block. It is quite usual, whenever possible, to define a set, $S$ by listing its elements the following way:

$$S = \{e_1, e_2, \ldots., e_n\}. \tag{A.1}$$

It took a rather long time before mathematicians made a clear distinction between the element $e$ and the set containing only this element $\{e\}$. These two objects are, however, very different. One may safely write $e \in \{e\}$, but one cannot write $\{e\} \in e$. A set containing only one element is usually called a singleton.

Using the $\in$ operator, it is now possible to define the subset operator $\subset$. A set $A$ is said to be a subset of the set $B$ if every element of $A$ is also an element of $B$. In this case, we write this relationship as $A \subset B$. Note that $\in$ links an element with a set where $\subset$ links two sets. Equality between two sets can now be easily defined. We say that $A$ and $B$ are equal if $A \subset B$ and $B \subset A$.

The union of two sets is the next important operator. It is typically denoted with the symbol, $\cup$. The union of two sets is a set that contains all the elements that are in any of the two sets. It is important to mention that elements that are in each of the

[1]For those missing this background, we recommend Harris and Stocker [5] as a good starting point.

[2]Indeed, it may not even be possible. Set theory is a rather modern endeavour and we recommend Weiss [11] for readers seeking to learn more about this fascinating area of mathematics.

two sets are only counted once. So, if $A = \{a, b\}$ and $B = \{b, c\}$ than the union $A \cup B = \{a, b, c\}$ and *not* $A \cup B = \{a, b, b, c\}$.

After the union comes naturally the intersection operator. The intersection operator is usually denoted with the symbol, $\cap$. The intersection of two sets is a set that contains all the elements that are in both sets. So, using the previous definition for $A$ and $B$, their intersection is given by $A \cap B = \{b\}$. Again, the elements are not double counted, despite the fact that they appear in both $A$ and $B$.

It is now possible to define two additional concepts. First, a set that contains no elements is termed the *null* set. It is quite common to denote the null set with the following symbol, $\emptyset$. We define *disjoint* sets as two, or more, sets which have the null set as their intersection. In other words, if two sets are disjoint, none of the elements in the first set are found in the second, and vice versa.

For completeness, we will define two additional operators. The complementary operator is an unitary operator that creates a set containing all the elements that are not in the input set. It is common to denote the complement of a given set $A$ by $A^c$. And last, but not least, the difference operator of two sets $A$ and $B$ is defined as $A - B = A \setminus B = A \cap B^c$.

It is not always obvious to grasp the meaning of complex set expressions, particularly when one does not have much experience in working with them. The definition of the difference operator, for example, may not be immediately straightforward to the reader.[3] These ideas are, with a bit of practice, quite powerful. We make extensive use of set theory in Chap. 8 when distinguishing between changes in the holding of individual securities—additions, reductions, liquidations, or acquisitions of instruments—for our daily performance attributions.

## A.2 Probability

The fundamental tools in statistics, while often taught on their own, actually originate from probability theory. Consequently, it is useful to begin with a review of the key ideas in probability. Probability theory is an attempt to quantify the concepts of randomness, hazard and uncertainty. Philosophers have, at times, struggled with the notion of randomness and asked whether it really exists or if, instead, it is merely the consequence of our limited computational capacity. As practitioners, however, we simply accept the existence of randomness and seek tools to help us manage it.

A *random event* is an outcome, or set of outcomes, of an experiment that cannot be predicted. Probabilities can be assigned to these events. There are *two* main types of random event. The first concerns experiments where all possible outcomes are known in advance—we just do not know which of these outcomes will actually be observed. The most famous, and perhaps driest, example is coin-flipping. The outcome of a coin-flip is unknown, but the possible outcomes are well known:

[3]One may employ Venn diagrams to help visualise the meaning of more complex set expressions.

it may either be a head or a tail. A second type of random event occurs where the possible outcomes are not known.[4] The second type of random event is quite obviously the most difficult to handle. In the field of finance, the main random events relate to the observation of market data: prices, rates, volatilities, correlations, and so on. For these objects we typically know, or are able to approximate, the range of possible outcomes, even if there are potentially an infinite number of them.[5]

Probabilists use the term *sample space* to describe the set of all possible events and outcomes associated with a given experiment—it is typically denoted as $\Omega$. As previously mentioned, the set $\Omega$ potentially contains an infinite number of elements making it often a bit tricky to handle. The set $\Omega$ is often only of limited interest. We are usually more interested in some subset of $\Omega$. We might, for example, only care about all of the possible *negative* returns of a given portfolio. This is a subset of *all* the possible returns of a given portfolio summarized by the sample space.

The principal objective of probability theory is to define a concept of probability—or, more formally, to assign a consistent measure to different outcomes and events in $\Omega$—that allows us to answer practical questions. This is accomplished through the definition of the following *three* fundamental axioms:

1. The *first* axiom is:

$$\mathbb{P}(A) \geq 0 \tag{A.2}$$

   where $A$ is any subset of $\Omega$. The key idea here is that the probability of any set $A \in \Omega$ is always positive (or zero).
2. The *second* axiom is:

$$\mathbb{P}(\Omega) = 1. \tag{A.3}$$

   In words, this implies that the probability of the set of all possible outcomes and events of a given random experiment is unity.
3. The *third*, and final, axiom is:

$$\mathbb{P}(A \cup B) = \mathbb{P}(A) + \mathbb{P}(B), \tag{A.4}$$

[4]A good example is the evolution of biological species. Although biologists try predict how species will evolve, the set of possible future mutations, however, is not known. Former US Secretary of Defense, Donald Rumsfeld, also discussed this idea in a military setting—calling them *unknown unknowns*—during a February 2002 media conference.

[5]A key thesis of Taleb [10] however, is that the second type of random event presents, in fact, the greatest amount of risk and that markets are not immune to *unknown unknowns*, or in the language of Taleb [10], *black swans*.

if $A$ and $B$ are disjoint (i.e., $A \cap B = \emptyset$). This axiom tells us that we may sum probabilities, in the same way as numbers, when the sets on which those probabilities are defined do not share any common elements (i.e., are disjoint).

There are a number of other features of probability measures that follow from these fundamental axioms. We ask the reader to trust that these three axioms are necessary and sufficient to define a probability measure that is aligned with our intuition.[6]

## A.2.1 Conditional Probability

Information is critical in understanding random phenomena. Often knowledge of the occurrence of one random event provides some insight into the relative probability of another event. One can *condition* upon this information. This idea, in a probabilistic context, is termed conditional probability and it is used occasionally in finance and this book. The conditional probability can be written as,

$$\mathbb{P}(A|B) = \frac{\mathbb{P}(A \cap B)}{\mathbb{P}(B)}. \tag{A.5}$$

One should read the conditional probability as the probability of the event $A$ will occur, if we known for *certain* that the event $B$ has happened.

Let's consider an example. Suppose you live in Basel and you need to fly regularly to New York. Your usual routine is to take the train to Frankfurt airport and then fly onwards directly to your final destination. Imagine that the probability that you miss your flight is 0.02. This is small, but non zero. You want the probability of missing your flight because the train to Frankfurt is late. This is a conditional probability. The lateness of the train is the conditioning information. Suppose further that the probability that your train is late is 0.05 and the probability that your train is late *and* that you miss your flight is 0.01.[7] This is a joint probability, not a conditional probability. A joint probability refers to the likelihood that both events occur—there is no conditioning.

[6]The reader familiar with probability and measure theory will have noted that we have omitted the notion of $\sigma$-algebras, which is critical for the rigorous definition of measures of subsets of $\Omega$. While essential to the proper application of probability theory, we felt that consideration of $\sigma$-algebras was too technical and, therefore, have omitted it. Readers seeking more information on this important topic are referred to Billingsley [1], Durrett [3] and Royden [8].

[7]Note, it has to be smaller than, or equal to, 0.02 as you might miss your flight for another reason–you could, for example, forget to set up your alarm clock properly.

Using Eq. (A.5), we may compute the probability that you miss your flight if your train is late as,

$$\mathbb{P}\left(\begin{matrix}\text{Miss your}\\ \text{Flight}\end{matrix}\middle|\begin{matrix}\text{Train}\\ \text{is Late}\end{matrix}\right) = \frac{\mathbb{P}\left(\begin{matrix}\text{Miss your}\\ \text{Flight}\end{matrix} \bigcap \begin{matrix}\text{Train}\\ \text{is Late}\end{matrix}\right)}{\mathbb{P}\left(\begin{matrix}\text{Train}\\ \text{is Late}\end{matrix}\right)}, \tag{A.6}$$

$$= \frac{0.01}{0.05},$$

$$= 0.2.$$

The probability that you miss your flight if the train is late is 0.20. This is clearly, and naturally, higher than the probability only of missing the flight. Given that your train is late, you may condition on this information and update your probability of missing your flight. In some situations, such a conditional probability may represent a valuable piece of information.

One can swap the two sets in the definition of the conditional probability to arrive at the following expression,

$$\mathbb{P}(B|A) = \frac{\mathbb{P}(B \cap A)}{\mathbb{P}(A)}. \tag{A.7}$$

In our previous example, therefore, this conditional probability amounts to the probability of your train being late assuming that you missed your flight. It might seem an odd question, but it helps us better understand your situation. Numerically, this probability is 0.5.[8] Ultimately, a late train is not the only reason for missing your flight, but it certainly does not help.

Since the probability of the intersection of two sets is commutative, it is possible to obtain the following important relationship,

$$\mathbb{P}(A)\,\mathbb{P}(B|A) = \mathbb{P}(B)\,\mathbb{P}(A|B). \tag{A.9}$$

This identity is called the Bayes' rule.[9]

---

[8] We have all the information required to determine this probability. The computation is merely,

$$\mathbb{P}\left(\begin{matrix}\text{Train}\\ \text{is Late}\end{matrix}\middle|\begin{matrix}\text{Miss your}\\ \text{Flight}\end{matrix}\right) = \frac{\mathbb{P}\left(\begin{matrix}\text{Miss your}\\ \text{Flight}\end{matrix} \bigcap \begin{matrix}\text{Train}\\ \text{is Late}\end{matrix}\right)}{\mathbb{P}\left(\begin{matrix}\text{Miss your}\\ \text{Flight}\end{matrix}\right)} = \frac{0.01}{0.02} = 0.5. \tag{A.8}$$

[9] This is, in fact, the starting point for the study of Bayesian statistics.

### A.2.2 Independence

There are, however, some cases where information has no value. Generally speaking, knowledge of the weather in Montréal, Québec will tell you little about, say, the exchange rate between Swiss francs and the Euro—this is because these two events are somehow independent. They do not influence one another. This important concept is generalized in probability theory. Events $A$ and $B$ are said to be statistically independent if and only if,

$$\mathbb{P}(A \cap B) = \mathbb{P}(A)\,\mathbb{P}(B). \tag{A.10}$$

Equations (A.7) and (A.9) imply, however, that when $A$ and $B$ are independent then

$$\begin{aligned} \mathbb{P}(A \cap B) &= \underbrace{\mathbb{P}(B|A)}_{\mathbb{P}(B)}\mathbb{P}(A), \\ &= \underbrace{\mathbb{P}(A|B)}_{\mathbb{P}(A)}\mathbb{P}(B), \\ &= \mathbb{P}(A)\mathbb{P}(B). \end{aligned} \tag{A.11}$$

This implies that $\mathbb{P}(A|B) = \mathbb{P}(A)$ and $\mathbb{P}(B|A) = \mathbb{P}(B)$. The conditioning set provides no information. In other words, knowledge of set $B$ tells you nothing about set $A$ and vice versa.[10] We will return to this idea in the next section on statistics.

## A.3 Statistics

A key—although unfortunately poorly named—concept in statistics is the *random variable*. It is poorly named, because a random variable is *not* actually a variable, but rather a function. It is a function than assigns a real value to each event in one's sample space. Most of the time, we will denote a random variable by $X$ without writing down the input variable, which simplifies the notation and avoids confusion.[11] Statisticians rarely work with the underlying sample space and thus we will also move it quietly into the background.

Returning to the age-old example of coin-flipping, we could assign the following values to our random variable, $X$,

[10]There is an important distinction between independent and uncorrelated events. Independence implies uncorrelated, while a lack of correlation does not immediately imply independence.

[11]In principle, it can become quite involved. For a real-valued random variable, one would write $X : \Omega \to \mathbb{R}$ or $X(\omega)$ for $\omega \in \Omega$. To do this correctly, however, one has to introduce the notion of a $\sigma$-algebra. Again, the reader is referred to Billingsley [1], Durrett [3], and Royden [8] for more rigour.

$$X = \begin{cases} 1 : \text{Heads} \\ -1 : \text{Tails} \end{cases}. \tag{A.12}$$

We have, with this simple definition, mapped events in our sample space into the space of integers, $\mathbb{Z}$. For an arbitrary random experiment, one may define a random variable taking a large, or indeed even infinite, number of values in the set of real numbers, $\mathbb{R}$. Moving from the strange world of sample spaces to the real numbers permits the use of a much broader and more standard range of mathematical tools—most specifically, the calculus.

### A.3.1 Distributions and Densities

We now have the necessary background required to define arguably the most important concept in statistics: the distribution function. The distribution function is mathematically described as,

$$F(x) = F_X(x) = \mathbb{P}(X \leq x) \tag{A.13}$$

for any value $x \in \mathbb{R}$. The distribution function provides the probability that a random variable takes a value that is *below* the pre-defined threshold, $x$. Equation (A.13) is also often referred to as the cumulative distribution function.

Knowledge of the distribution function for a given random variable provides virtually everything one needs to know about it. It is common to assume, at the outset of any analysis, a specific distribution function for one's key random variables. While practical, this is not without danger. The problem is that, particularly in a financial setting, the true distributions may *not* be known. Moreover, even if a given historical dataset is consistent, at least empirically, with a given distribution, this could also change over time. A good example is the assumption of normality distributed financial random variables—such as prices, rates, and spreads—during the recent financial crisis. Such a choice would have significantly underestimated the probability of extreme outcomes.

Limit cases for the distribution function may—by virtue of its straightforward definition—be easily evaluated. When choosing a very small threshold, $x$, the probability that the random variable falling below this threshold should be correspondingly small. Imagine a simple example: height. We could construct a distribution function for the height of an average woman. If we set the threshold at, say, 1.5 m, then we would expect the probability that a randomly selected woman to be shorter than 1.5 m to be fairly modest. If we set it at 1.2 m, we expect it to be quite small. At 80 cm, we would expect it to be extremely small. In the limiting case, we let the threshold tend towards minus infinity. In this case, the probability must be zero or rather,

$$\lim_{x \to -\infty} F(x) = 0. \tag{A.14}$$

The other extreme is also intuitive. Setting an extraordinarily large threshold—say an upper threshold for women's height to be 2.2 m—implies that the probability our random variable falls under this threshold approaches one. It is thus almost certain the random variable—or the height of our randomly selected woman—will be below the threshold. We express this limiting expression with the following relation,

$$\lim_{x \to \infty} F(x) = 1. \tag{A.15}$$

The cumulative distribution function is a monotonically increasing function. More specifically, this means that if $x < y$ then,

$$\begin{aligned} \mathbb{P}(X \leq x) &\leq \mathbb{P}(X \leq y), \\ F(x) &\leq F(y). \end{aligned} \tag{A.16}$$

This is quite a reasonable property. We cannot, for example, have the probability that a randomly selected woman is shorter than 1.6 m exceeds the probability that she is shorter than 1.7 m.

The introduction of the distribution function permits the use of a broad range of well-known mathematical tools, most importantly the calculus, in statistics. As a direct consequence, it is generally easier to work with continuous variables and functions in statistics.[12] We must nonetheless be aware that the probability a continuous random variable takes a specific value is always equal to zero. To assess the probability associated with continuous variables, therefore, one must consider the probability that the random variable takes a value over a range—for example, between $a$ and $b$, where naturally $a \leq b$. This probability is computed as follows,

$$\begin{aligned} \mathbb{P}(a \leq x \leq b) &= \mathbb{P}(X \leq b) - P(X \leq a), \\ &= F(b) - F(a). \end{aligned} \tag{A.17}$$

This definition is a direct consequence of the fact that a cumulative distribution function is a monotonically increasing function. As a probability cannot be negative, it is clear that the distribution function must be increasing.

Figure A.1 provides a graphical view of *four* fairly well-known cumulative distribution functions. Except for extreme situations, these distribution functions tend to look fairly similar; they take the form of an elongated S. These functions are not, graphically at least, terribly easy to interpret. Consequently, it is often useful to introduce the following transformation of the distribution function,

[12] The opposite is true in probability, where one typically prefers the use of discrete variables.

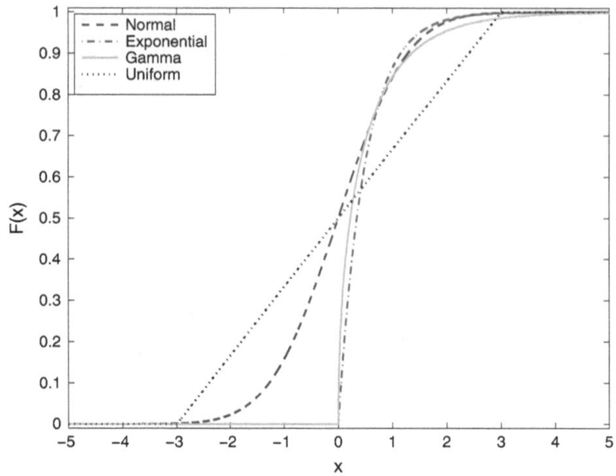


**Fig. A.1** Some cumulative distribution functions. This figure describes the cumulative distribution function for a number of fairly common statistical distributions

$$f(x) = \frac{\partial F(x)}{\partial x}. \tag{A.18}$$

Equation (A.18) is defined as the probability density, or simply density, function.

Applying the fundamental theorem of calculus to Eq. (A.18), it is, of course, also possible to express the probability function in terms of the density function, as follows

$$F(x) = \int_{-\infty}^{x} f(y)\,dy = \mathbb{P}(X \leq x). \tag{A.19}$$

A useful way to think about the cumulative distribution function—as suggested by Eq. (A.19)—is as the sum (i.e., integral) of all the probability mass from $-\infty$ up to the point, $x$.[13] The density function, in contrast, helps us understand where this probability mass sits. Many density functions have a symmetric bell shaped form, but, as we can see from Fig. A.2, a number of other shapes are also possible.

---

[13]For discrete distributions—with mass only at discrete points $x_0, x_1, x_2, \ldots \in S$—we define the cumulative distribution function as,

$$f(x) = \mathbb{P}(X = x), \tag{A.20}$$

and the associated probability density function as,

$$F(x) = \sum_{\{s \in S : s \leq x\}} f(s) = \mathbb{P}(X \leq x). \tag{A.21}$$

Given the continuous nature of financial variables, we will not focus much on the discrete setting. It does, however, provide useful intuition into the basic results.

We can link the density function with our intuition about the probability of an random variable taking a value over the interval, $[a, b]$. Starting with the basic definition of this probability from Eq. (A.17) and keeping in mind the definitions of cumulative distribution and probability density functions, we arrive at the following result,

$$\begin{aligned} \mathbb{P}(a \leq x \leq b) &= F(b) - F(a), \\ &= \int_{-\infty}^{b} f(y)\,dy - \int_{-\infty}^{a} f(y)\,dy, \\ &= \int_{a}^{b} f(y)\,dy. \end{aligned} \tag{A.22}$$

Quite simply, if one integrates the density function over the interval, $[a, b]$, one arrives at the associated probability. This is a useful result.

There are mathematical constraints associated with the density function. Since $\mathbb{P}(a \leq x \leq b)$ is a probability regardless of the value of $a$ and $b$ when $a \leq b$, then $f(x)$ must be a positive function. Moreover, given that a random variable must take values somewhere between minus infinity and plus infinity, then

$$\int_{-\infty}^{\infty} f(x)\,dx = 1. \tag{A.23}$$

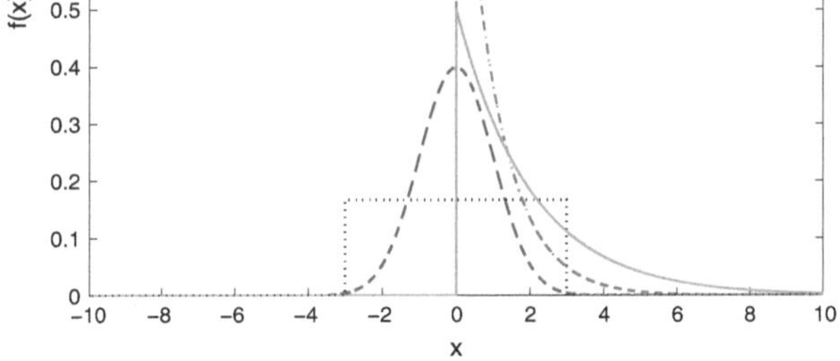


**Fig. A.2** Some probability density functions. This figure describes the probability density function for the fairly common statistical distributions introduced in Fig. A.1

Simply put, the total probability associated with all outcomes of a random variable must, quite reasonably, sum to unity.[14]

## A.3.2 Working with Distribution and Density Functions

There a number of operators that may be applied to the probability distribution function, which are very relevant for financial analysis. The two best known operators are the mean and the variance. The mean or average—more formally termed the expectation—of a random variable $X$ is simply defined as,

$$\underbrace{\mathbb{E}(X)}_{\mu} = \int_{-\infty}^{\infty} xf(x)\, dx. \tag{A.24}$$

This is basically the central value of the distribution.

The variance is defined as,

$$\underbrace{\text{var}(X)}_{\sigma^2} = \int_{-\infty}^{\infty} (x-\mu)^2 f(x)\, dx. \tag{A.25}$$

The square root of the variance, or simply $\sigma$, is termed the standard deviation, or in finance circles, volatility. The variance provides insight into the dispersion of the random variable.

Since it is not always possible to analytically evaluate the integral expression in Eqs. (A.24) and (A.25), it is *not* always possible to obtain a closed-form solution for the mean and the variance of a given probability distribution. It is generally possible to perform a numerical approximation.[15] For one-dimensional random variables, numerical methods to compute these integrals are both extremely fast and robust.[16]

The mean is actually part of the definition of the variance. This may, in some cases, prove problematic or inefficient, particularly for certain applications such as stochastic simulations. Standard integral manipulations, however, permit us to rewrite the variance definition as,

---

[14]Some random variables have a more restricted domain. A number of random variables—stock prices, for example—cannot take negative values. The range over which a random variable, and its associated distribution function, can take values is called its support.

[15]It is still not always possible. The Cauchy distribution, for example, is a random variable where neither the mean, the variance, nor any other higher moments are defined. See Casella and Berger [2] for more information.

[16]This is unfortunately not always true when the number of dimensions increases.

$$\sigma^2 = \int_{-\infty}^{\infty} x^2 f(x)\, dx - \left( \int_{-\infty}^{\infty} x f(x)\, dx \right)^2, \tag{A.26}$$
$$= \int_{-\infty}^{\infty} x^2 f(x)\, dx - \mu^2.$$

The second line of Eq. (A.26) might seem useless as we are actually bringing back into the definition the mean, $\mu$, that we originally sought to exclude. The main difference, however, between the two definitions of variance is that the mean is embedded in Eq. (A.25), whereas it is only a corrective term in Eq. (A.26). One need only apply this correction after having computed the first integral term.

### A.3.3 Some Sample Statistical Distributions

It is quite usual to select a distribution function that well suits one's problem and to work with it. The natural question, therefore, is which distribution function should one select? Unfortunately, the list of possible distributions is sufficiently long that we simply do not have the time to list and describe them all.[17] A bit of colour on different distribution functions, however, is warranted. Consequently, we have selected *three* statistical distributions for closer examination.

Probably the most commonly used distribution function is the *uniform* distribution. People use it constantly without either thinking about it or formalizing it. It is nonetheless worthwhile formalizing. A uniform distribution is defined on a finite interval with endpoints $a$ and $b$ with $a < b$. The probability density function definition is merely,

$$f_X(x) = \frac{1}{b-a} \tag{A.27}$$

The idea is that probability mass is equally spread over the interval, $[a, b]$. In other words, all points have equal, or uniform, probability. All points in the interval have equal probability of occurrence.

The average, or expectation, of a random value variable following a uniform distribution is fairly intuitive. It should lie at the mid-point between the two endpoints, or rather $\frac{b+a}{2}$. To verify our intuition, let's try to recover this expression from the formal definition of the mean provided in Eq. (A.24). If we recall that the distribution is defined only between $a$ and $b$, then the rest is simply calculus,

[17]The interested reader is referred to Johnson et al. [6] for a detailed description of different distribution functions.

$$\begin{aligned}\mu_X &= \int_a^b \frac{x}{b-a} dx, \\ &= \frac{1}{b-a}\int_a^b x dx, \\ &= \left(\frac{1}{b-a}\right)\frac{x^2}{2}\bigg|_a^b, \\ &= \frac{(b-a)^2}{2(b-a)}, \\ &= \frac{(b-a)(b+a)}{2(b-a)}, \\ &= \frac{(b+a)}{2},\end{aligned} \tag{A.28}$$

which, happily, perfectly matches with our intuition.

It is doubtful that many people have a similarly useful intuition for the variance of the uniform distribution. One could probably imagine that the variance should increase when the distance between the endpoints increases. That is, if the a random variable is uniformly distributed over the interval, [0, 1], then we should somehow expect it to have less variance than another uniform random interval defined over the interval, [0, 100]. Applying Eq. (A.25), we arrive at the following result,

$$\begin{aligned}\sigma_X^2 &= \int_a^b \left(x - \frac{(b+a)}{2}\right)^2 \frac{1}{b-a} dx, \\ &= \frac{1}{4(b-a)}\int_a^b (2x-b-a)^2\, dx, \\ &= \left(\frac{1}{4(b-a)}\right)\frac{(2x-b-a)^3}{6}\bigg|_a^b, \\ &= \frac{1}{24(b-a)}\left[(2b-b-a)^3 - (2a-b-a)^3\right], \\ &= \frac{(a-b)^2}{24(b-a)}\left[(b-a)-(a-b)\right], \\ &= \frac{(a-b)^2}{12}.\end{aligned} \tag{A.29}$$

As one can see from this non-trivial example, the variance does indeed increase with the distance between $a$ and $b$ as our intuition was telling us. It increases, in fact, with the square of the distance between the endpoints of the distribution.

Another interesting probability distribution, often found in statistical applications, is the exponential distribution. This distribution has only one parameter, that we will denote as $\lambda$. Its probability density function has the following form,

$$f(x) = \lambda e^{-\lambda x}, \tag{A.30}$$

and it is only defined for positive values of $x$ (i.e., $x \in \mathbb{R}^+$).[18]

It is good form to verify that Eq. (A.30) is indeed a probability density function. The function is always positive, which is already a good start, but we also need to verify that the integral over the positive real numbers is equal to 1. This is easily verified as,

$$\begin{aligned} \int_0^\infty \lambda e^{-\lambda x} dx &= \lambda \int_0^\infty e^{-\lambda x} dx, \\ &= \lambda \frac{e^{-\lambda x}}{-\lambda}\bigg|_0^\infty, \\ &= -e^{-\lambda x}\big|_0^\infty, \\ &= -0 - (-1), \\ &= 1. \end{aligned} \tag{A.32}$$

Equation (A.30) appears to be, in fact, a potential probability density function.

It is also interesting to compute the mean and variance of a random variable, $X$, whose distribution is described by Eq. (A.30). The mean, or expectation, is determined by resolving the following integral equation,

$$\mu = \lambda \int_0^\infty x e^{-\lambda x} dx. \tag{A.33}$$

To solve this equation, we make use of an old trick. The following approach, known as the by-parts integration formula, provided without proof, is occasionally helpful,

$$\int_a^b \frac{df(x)}{dx} g(x)\, dx = \int_a^b f(x) \frac{dg(x)}{dx} dx - \left(f(x)\, g(x)\right)\Big|_a^b. \tag{A.34}$$

If we will apply Eq. (A.34) by setting $g(x) = x$ and $f(x) = e^{-\lambda x}$, it leads to the following result,

[18] The associated cumulative distribution function is defined as,

$$F(x) = 1 - \lambda e^{-\lambda x}, \tag{A.31}$$

for all $x \in \mathbb{R}^+$.

$$\begin{aligned}\mu &= \lambda \int_0^\infty x e^{-\lambda x} dx, \\ &= \lambda \left( \int_0^\infty \frac{e^{-\lambda x}}{-\lambda} dx - \left( x \frac{e^{-\lambda x}}{-\lambda} \right) \Big|_0^\infty \right), \\ &= -\int_0^\infty e^{-\lambda x} dx, \\ &= \frac{e^{-\lambda x}}{-\lambda} \Big|_0^\infty, \\ &= \frac{1}{\lambda}.\end{aligned} \tag{A.35}$$

The mean, or expected, value of an exponentially distributed random variable is the inverse of its parameter, $\lambda$. It is also, of course, possible to analytically compute the variance, which is equal to $\sigma^2 = \frac{1}{\lambda^2}$. We leave this computation as an exercise for the reader.

The exponential distribution has an interesting property: it is memoryless. This implies that the probability of an exponentially distributed random variable taking a given value is independent from your current position. Imagine that you define a threshold at one unit above your current position, whatever that might be. The memorylessness of an exponentially distributed random variable implies that the probability of breaching the threshold never changes—no matter what happens to your current position.

This idea is, admittedly, a bit hard to understand. A few examples are helpful. Imagine that a company has a AAA rating. If the probability of default is described by an exponential distribution, the probability that a company with this rating will default in the next 12 months might be set at 0.01. Memorylessness implies that if, in 6 months time, the company is still AAA rated and has not yet defaulted, the probability of default over the next 12 months remains at 0.01.

Another example relates to a production line. The random variable $X$ is the time to the next work stoppage on your assembly line. The assembly line has survived for $x$, or $\{X > x\}$, units of time and you want to know the probability it will survive another $t$ units of time. Mathematically, you want to know $\mathbb{P}(X > x + t)$ given that the event $\{X > x\}$ has occurred. This is expressed using conditional probability as follows,

$$\mathbb{P}(X > x + t | X > x) = \mathbb{P}(X > t). \tag{A.36}$$

This conditional probability reduces to $\mathbb{P}(X > t)$, which is statistically independent of $x$. The probability that your assembly line survives for another $t$ periods is independent of how long it has already gone without a work stoppage. Conditioning on the current survival time—that is, the set $\{X > x\}$ provides no useful information for the computation of $\mathbb{P}(X > t)$.

**Fig. A.3** A 10 Deutsche-Mark note. This figure illustrates a 10 Deutsche-Mark bank note. In addition to a picture of Carl Friedrich Gauss, the inventor of the normal distribution, we also observe a small picture of the normal density function along with its mathematical representation

The final density distribution that we will consider, and perhaps the most important for finance practitioners, is the Gaussian, or normal, distribution—we will use both terms interchangeably. This is quite likely the most well known statistical distribution. Millions of Germans, and probably hundreds of thousands of foreign tourists visiting Germany, prior to the introduction of the Euro in 2001, carried around the probability-density-function formula for the Gaussian distribution in their pocket or handbag. The reason is simple: it was printed on the bottom left-hand-corner of the 10 Deutsche-Mark banknote—see Fig. A.3 for an illustration.[19]

The density probability function of the Gaussian distribution is given by the following mathematical expression:

$$f(x) = \frac{1}{\sigma\sqrt{2\pi}} e^{\frac{-(x-\mu)^2}{2\sigma^2}} \tag{A.37}$$

The Gaussian distribution has two parameters: $\mu$ and $\sigma$. Consequently, it is often denoted $\mathcal{N}(\mu, \sigma)$. Notationally, it might seem dangerous to use the same naming convention as employed for the mean and the standard deviation in our previous discussion. There is, however, an excellent reason behind this notation: the parameters $\mu$ and $\sigma$ correspond exactly to the mean and the standard deviation of the Gaussian distribution. Indeed, the mean and variance provide all the information required to characterize a given Gaussian distribution.

Our first task is to verify that Eq. (A.37) is indeed a density distribution. This requires, among other things, determining if the total probability sums to unity as

[19]This distribution was discovered by Carl Friedrich Gauss, a legendary German mathematician, explaining the reason it is often called the Gaussian distribution. It is often used to model currency fluctuations. The Deutsche Mark quite famously experienced, during the period of the Weimar republic of the early 1920s; an extreme hyper-inflationary period. The irony is that hyper-inflationary periods are poorly described by the Gaussian distribution.

follows,

$$\int_{-\infty}^{\infty} \frac{1}{\sigma\sqrt{2\pi}} e^{\frac{-(x-\mu)^2}{2\sigma^2}} dx \,?=?\, 1. \tag{A.38}$$

This is, unfortunately, not a trivial problem. The first step is to eliminate the parameters by a suitable change of variables. It is sufficient to define a new variable of integration, $y = \frac{x-\mu}{\sigma\sqrt{2}}$. This has no impact on the area of integration and leads to the following, much nicer, form for the integral,

$$\frac{1}{\sqrt{\pi}} \int_{-\infty}^{\infty} e^{-y^2} dy \,?=?\, 1. \tag{A.39}$$

The main issue with the expression that we are trying to integrate is that it has no analytical solution. Direct methods will not work. We need a trick. It involves bypassing the problem by making it, at first glance, more complicated. We increment the complexity by integrating the expression not only in the $y$-axis but also in an additional axis, $z$. This transforms Eq. (A.39) into the following bi-dimensional problem,

$$\frac{1}{\pi} \int_{-\infty}^{\infty} \int_{-\infty}^{\infty} e^{-y^2} e^{-z^2} dydz. \tag{A.40}$$

This problem is indeed more complicated, but in two dimensions, we may employ an alternative coordinate system. The problem is, in fact, much simpler if we express our problem in polar coordinates. Polar coordinates express the position of a point by the distance to a fixed center (denoted as $r$) and an angle between a reference axis (described as $\theta$). Thus, we define our two variables as,

$$\begin{aligned} y &= r\cos\theta, \\ z &= r\sin\theta. \end{aligned} \tag{A.41}$$

Simple trigonometric manipulations lead to $r^2 = y^2 + z^2$. Trickier is the construction of a surface element. We ask the reader to believe that an infinitesimal square $dydz$ in the $yz$ coordinate system can be expressed as $rdrd\theta$ in the polar coordinate system. The integration is done for all the angles from 0 to $2\pi$ and all of the positive distance $r$. With this in mind, our integral becomes,

$$\begin{aligned} \frac{1}{\pi} \int_{0}^{\infty} \int_{0}^{2\pi} e^{-r^2} rdrd\theta &= \frac{1}{\pi} \int_{0}^{\infty} e^{-r^2} rdr \underbrace{\int_{0}^{2\pi} d\theta}_{2\pi}, \\ &= \frac{2\cancel{\pi}}{\cancel{\pi}} \int_{0}^{\infty} e^{-r^2} rdr, \\ &= 2 \int_{0}^{\infty} e^{-r^2} rdr. \end{aligned} \tag{A.42}$$

The integral can, quite surprisingly, be solved. The absence of $\theta$ variable in the integrand makes it trivial in the angular variable, $\theta$. The remaining integrand is easily resolved through use of substitution with $u = -r^2$, as follows,

$$\begin{aligned} 2\int_0^{\infty} e^{-r^2} r dr &= -\int_0^{\infty} e^u du, \\ &= 1, \end{aligned} \tag{A.43}$$

which, quite satisfyingly, is actually equal to unity.[20]

Haven shown—via geometrical detour—that Eq. (A.37) is a density function, we can now proceed to show that the parameter $\mu$ is, in fact, the mean of the Gaussian distribution as,

$$\begin{aligned} \int_{-\infty}^{\infty} \frac{x}{\sigma\sqrt{2\pi}} e^{\frac{-(x-\mu)^2}{2\sigma^2}} dx &= \frac{1}{\sigma\sqrt{2\pi}} \int_{-\infty}^{\infty} x e^{\frac{-(x-\mu)^2}{2\sigma^2}} dx, \\ &= \frac{1}{\sigma\sqrt{2\pi}} \int_{-\infty}^{\infty} (y+\mu) e^{\frac{-y^2}{2\sigma^2}} dy, \\ &= \frac{\mu}{\sigma\sqrt{2\pi}} \int_{-\infty}^{\infty} e^{\frac{-y^2}{2\sigma^2}} dy + \frac{1}{\sigma\sqrt{2\pi}} \int_{-\infty}^{\infty} y e^{\frac{-y^2}{2\sigma^2}} dy, \\ &= \frac{\mu}{\sigma\sqrt{2\pi}} \int_{-\infty}^{\infty} e^{\frac{-y^2}{2\sigma^2}} dy = \mu, \end{aligned} \tag{A.44}$$

which is the desired result.[21] We leave it as an exercise for the reader to prove that the standard deviation of the Gaussian distribution is actually the second parameter, $\sigma$.[22]

The importance of the Gaussian distribution function is not found in its analytical tractability. There are clearly simpler distributions available in the mathematical literature. The main reason of its success is instead linked to the central limit theorem. This key statistical result describes the distribution of the mean of large numbers of independent random variables. If you compute the mean of large numbers of random variables, which need not be normally distributed, the distribution of this mean value converges to a normal distribution.[23] The central limit theorem is a remarkable result.

---

[20] Indeed, this development explains the reason behind the constant term $\frac{1}{\sqrt{2\pi}}$ found in the Gaussian density function: it is necessary to ensure that the total probability mass sums to unity.

[21] Observe that we use a change of variable, $y = x - \mu$, to move to the second line and employ a symmetry argument to eliminate the second integral on the third line.

[22] As a hint, we recommend the use of integration by parts—the difficult part is finding the appropriate $f$ and $g$ functions.

[23] There is a large number of different versions of the central limit theorem, but one fairly common condition is that the random variables have, at least, finite mean and variance. See Billingsley [1], Durrett [3] for more detail and rigour on this result.

The central limit theorem warrants an example. Consider the price of a liquid stock. At the beginning of each day, it takes a given value. This value, of course, changes during the course of the day. The price changes do *not* follow a Gaussian distribution for a very simple reason: the price can only take discrete values whereas the Gaussian distribution is continuous. We do not, in fact, even know the true probability density distribution of the price changes. We do, however, know that the price changes a large number of times over the course of each day. Assuming that the assumptions of the central limit theorem are respected—which may, or may not, be a strong assumption—then we may conclude that the mean change of the stock price, or rather its average daily price movement, converges to a normal distribution.[24]

### A.3.4 Multivariate Statistics

In finance applications, a significant proportion of quantitative analysis is performed on *not* one, but several random variables. Portfolios are routinely exposed to multiple risk factors, where each of these risk factors is one (or more) random variables. Very often there is a degree of dependence between these random variables—this implies that we need to understand not only the probability of various outcomes of a single random variable, but also the joint outcomes with other random variables. This idea is readily accommodated with the notion of a multivariate distribution function. A joint cumulative distribution function for two random variables, $X$ and $Y$, has the following basic form,

$$F(x, y) = \mathbb{P}(X \leq x, Y \leq y). \tag{A.45}$$

This two-dimensional joint cumulative distribution function has *two* separate thresholds. It provides the joint probability that the first random variable $X$ takes a value below the first threshold, $x$, at the *same* time that the second random variable $Y$ falls below the second threshold, $y$. We can see that this is, conceptually at least, a simple generalization of the one-dimensional cumulative distribution function. Note that Eq. (A.45) can be generalized for an arbitrary number of random variables.[25]

If the joint cumulative distribution function is known, then the individual cumulative distributions are also known. In the statistical literature, the cumulative distribution functions associated with the single random variables that comprise the

---

[24]This is something of a justification for the reliance of the Gaussian distribution—or more formally geometric Brownian motion with Gaussian distributed Wiener increments—in the development of the Black–Scholes option-price formulae and the use of the normal distribution more generally in the finance literature.

[25]For multiple random variables, we call

$$X = (X_1 \cdots X_n) : \Omega \to \mathbb{R}, \tag{A.46}$$

a random vector.

joint cumulative distribution function are known as the *marginal* distributions. It is interesting to think how we can consider situations when an individual random variables ceases to influence the joint distribution function. It turns out that it is enough to set the threshold for a given single random variable to its largest possible value. Doing this essentially renders the random variable irrelevant. In other words, when the upper threshold for a given random variable is sufficiently large, it not longer has an impact on the probability of the other random variables. Mathematically, this is expressed as

$$F(x) = F(x, \infty) = P(X \leq x), \tag{A.47}$$
$$F(y) = F(\infty, y) = P(Y \leq y),$$

where $F(x)$ and $F(y)$ are called the marginal distributions of the two-dimensional cumulative distribution function, $F(x, y)$.

Knowing the joint cumulative distribution function provides you full information about the individual marginal distributions—they can always be extracted using Eq. (A.47). Unfortunately, the reverse is not true. Perfect knowledge of the marginal distributions does *not* give you full information about the joint distribution. It is indeed, in most cases, possible to build an *infinite* set of joint distributions that have the same marginal distributions. Consider the following analogy. If you know precisely the colors used in the creation of an oil painting (i.e., the marginal distributions), you still do not know much about the painting itself (i.e., the joint distribution). Knowing the joint cumulative distribution function is, therefore, basically equivalent to seeing the whole painting. After having seen the painting, in contrast, you may easily identify the individual colours.

This is a bit depressing. It may not seem very useful—if not a complete waste of time—to perform univariate statistical analysis. Happily, this is not true. Univariate analysis remains useful and pertinent thanks to a remarkable result derived by French mathematician Abe Sklar in 1959. Without entering into technical details, Sklar's theorem tells us that if we have precise knowledge of the marginal distributions, then there exists a unique function $C$ that links them with the joint cumulative distribution function,

$$F(x, y) = C(F(x), F(y)). \tag{A.48}$$

The result in Eq. (A.48), known as Sklar's theorem, is the entry point into an active area of statistical research called *copula* theory. A copula is a function that only contains the information related to the dependence between two random variables. While the full copula theory is beyond the scope of this discussion, we still wish to invest a bit of time to redefine a few concepts

(continued)

based on the copula theory. It provides an alternative perspective on the notion of statistical dependence and provides some basic insights into this growingly important areas of statistics. The independent copula is given as,

$$C(x, y) = xy. \tag{A.49}$$

The only random variables that may be declared independent are those for which their copula is the independent copula—this is also termed the product copula. Any other copula has some form of dependence. Thus, copula theory is essentially the study of dependence between multiple random variables. A fascinating starting point is Sklar [9].

In an analogous way to the univariate setting, it is possible to define a *joint* probability density function. Here is the two-dimensional case,

$$F(x, y) = \int_{-\infty}^{x} \int_{-\infty}^{y} f(s, t)\, ds dt. \tag{A.50}$$

The two-dimensional integral on the whole real plane must sum to unity, or rather

$$\int_{-\infty}^{\infty} \int_{-\infty}^{\infty} f(x, y)\, dx dy = 1. \tag{A.51}$$

This implies that the total probability mass associated with the joint outcomes of our *two* random variables may not exceed unity. Naturally, all of these multivariate concepts generalize easily from two to $n$ random variables. It naturally begins with the cumulative distribution function,

$$F(x_1, x_2, \ldots, x_n) = \mathbb{P}(X_1 \leq x_1, X_2 \leq x_2, \ldots, X_n \leq x_n). \tag{A.52}$$

With a collection of random variables, it is *not* enough, unless they are independent, to know only the mean and variance of each of the individual random variables.[26] We also need some notion of their dependence. To this end, there is a another measure which attempts to capture the dependence between two random variables. This measure is termed the *covariance* and is defined as,

$$\mathrm{cov}(x, y) = \int_{-\infty}^{\infty} \int_{-\infty}^{\infty} xy f(x, y)\, dx dy. \tag{A.53}$$

[26]The mean and variance may be determined from the joint distribution function. The simplest approach, however, involves writing down the marginal distributions and computing the required values.

The covariance, quite simply, describes how two random variables move together.

This concept is more clearly understood in the context of an example. One multivariate distribution function that is used extensively in finance applications—some people might say over-used—is the multivariate Gaussian distribution. This book is no exception—we make extensive use of this distribution in the preceding chapters. The mathematical expression of this joint probability density function is, however, far from simple. It is described as,

$$f(x,y) = \frac{1}{2\pi\sigma_x\sigma_y\sqrt{1-\rho^2}} e^{-\frac{k(x,y)}{2}}, \tag{A.54}$$

where the function $k(x,y)$ is itself given by,

$$k(x,y) = \frac{1}{1-\rho^2}\left[\left(\frac{x-\mu_x}{\sigma_x}\right)^2 - 2\rho\left(\frac{x-\mu_x}{\sigma_x}\right)\left(\frac{y-\mu_y}{\sigma_y}\right) + \left(\frac{y-\mu_y}{\sigma_y}\right)^2\right], \tag{A.55}$$

where $\rho$ is an additional parameter associated with the two-dimensional Gaussian distribution that describes the correlation, or co-movement, between the two random variables, $X$ and $Y$.

It is a long, and rather tedious, exercise but it is possible to demonstrate that the covariance of the two-dimensional Gaussian distribution is given by,

$$\text{cov}(x,y) = \rho\sigma_x\sigma_y. \tag{A.56}$$

Equation (A.56) is interesting because it demonstrates the linear relationship between the covariance and the parameter $\rho$ of the Gaussian distribution. For the Gaussian distribution, setting the parameter $\rho$ to zero, leads to a zero covariance and consequently to the independence between the two random variables. The zero covariance is clear from the previous equation but the independence still need to be proven. Rewriting Eq. (A.54) with $\rho = 0$,

$$f(x,y) = \frac{1}{2\pi\sigma_x\sigma_y} e^{-\frac{1}{2}\left[\left(\frac{x-\mu_x}{\sigma_x}\right)^2 + \left(\frac{y-\mu_y}{\sigma_y}\right)^2\right]}. \tag{A.57}$$

Recalling the algebraic properties of the exponential, it is possible to split the argument to arrive at,

$$\begin{aligned} f(x,y) &= \underbrace{\frac{1}{\sigma_x\sqrt{2\pi}} e^{-\frac{1}{2}\left(\frac{x-\mu_x}{\sigma_x}\right)^2}}_{\substack{\text{Marginal density} \\ \text{of } x}} \underbrace{\frac{1}{\sigma_y\sqrt{2\pi}} e^{-\frac{1}{2}\left(\frac{y-\mu_y}{\sigma_y}\right)^2}}_{\substack{\text{Marginal density} \\ \text{of } y}}, \\ &= f(x)f(y), \end{aligned} \tag{A.58}$$

where $f(x)$ and $f(y)$ are the univariate marginal Gaussian distributions. Now that we see the joint density probability function is the product of the two marginal density functions, we can now, and only now, safely state that the two random variables are independent. Equation (A.58) is, in fact, the definition of statistical independence.

It is possible, but a bit cumbersome, to generalise this result for an arbitrary, but finite, number of random variables. It is best performed using matrix algebra. The standard deviation of each random variable and the correlation between each pair of random variables are traditionally aggregated into a mathematical object called the covariance matrix. We have treated this object in substantial detail in the Chaps. 10–12 on risk computation and, consequently, it will not be repeated in this appendix.

## A.4 Matrix Theory

Dealing with multiple variables, whether random or deterministic, can quickly become complex and clumsy. Matrix theory was developed to help ease this burden somewhat by making notation concise and permitting operations on systems of multiple variables. Since we often find ourselves, in the preceding chapters, using systems involving multiple deterministic and random variables, a review of the key results in matrix theory is appropriate.

Quite often, one has to deal with a list of related items. Imagine, for illustrative purposes, you have a list of prices for a given security over three consecutive business days. One may allocate each of these prices to separate variables as follows,

$$\begin{aligned} \text{Price \#1 March 2011} &= 101.45, \\ \text{Price \#2 March 2011} &= 101.53, \\ \text{Price \#3 March 2011} &= 102.12. \end{aligned} \tag{A.59}$$

This does not seem like a very efficient strategy. Instead, it would be vastly better to have a single variable name. Let's call this variable, $P$. While simpler, this creates a bit of a problem. We still need to be able to distinguish between the various numerical values. This is resolved by adding to each variable what is commonly called an index. For example, we could denote an individual price as $P_t$ where $t = 1, 2, 3$ is the index. Now, one can refer to the entire information by using $P$.[27] We may refer to a specific value within the series using its index. The third price, for example, would be denoted $P_3$. $P$ is a mathematical object called a *vector* summarizing $\{P_t : t = 1, 2, 3\}$. In what follows, we require, by standard convention, that vectors are displayed as columns,

[27]In matrix notation, this is sometimes written with an arrow or a bar on top, $\vec{P}$ or $\bar{P}$.

$$P = \begin{pmatrix} 101.45 \\ 101.53 \\ 102.12 \end{pmatrix}. \tag{A.60}$$

Vectors are convenient objects for summarizing properties that have a single index, such as our example where the prices are indexed by time. One can easily imagine more general situations where two indices would be very useful. The time evolution, for example, of prices for a set of different securities within a given portfolio. One would need one index to keep track of time and another index to indicate the security. In this case, instead of a single column, we could conveniently display the prices in a table. This table is called a matrix.[28] Consider a portfolio with two securities for which we have observed the prices 3 days in a row. One can summarize this in a matrix $T$ as,

$$T = \begin{pmatrix} 101.45 & 103.13 \\ 101.53 & 102.98 \\ 102.12 & 103.07 \end{pmatrix}. \tag{A.61}$$

It is naturally of critical importance to agree on a convention for discussing the indices of a matrix: we will always use the first index for the row and the second for the column. Consequently, $T_{2,1}$ refers to the second row and first column of the matrix, $T$, or 101.53.

An important property of a matrix is its size or dimensions. One can say that our matrix $T \in \mathbb{R}^{3\times 2}$, which implies that our matrix is composed of real numbers with three row dimensions and two column dimensions. As a consequence, a vector, $v$, containing $n$ items can be called a matrix, $v \in \mathbb{R}^{n\times 1}$ or simply, $v \in \mathbb{R}^n$.[29] As an extension, a single number, $a$, can also be called a matrix $a \in \mathbb{R}$. In this setting, a matrix with a single element is often called a scalar or a singleton.

Let us now define a second vector, $N$, containing the nominal amounts for each security in the our two-security portfolio,

$$N = \begin{pmatrix} 100 \\ 10 \end{pmatrix}. \tag{A.62}$$

Armed with the nominal amount held for each security and their associated prices, we are now in a position to compute the market value, $V$, of the portfolio. The method for computing the market value of a portfolio is straightforward; one need only multiply the price by the nominal amount held for each security and sum across each of the contributions. When performing such an operation with matrices and

[28]It does not, of course, stop with two indices. There also exist more general objects called tensors, which are essentially multidimensional arrays of numerical values.

[29]If the vector took integer or complex values, it would be denoted by $v \in \mathbb{Z}^n$ or $v \in \mathbb{C}^n$, respectively. This applies, of course, equally to matrices and scalars.

vectors, this is termed a matrix product. The market value of our portfolio, therefore, is merely,

$$V = TN. \tag{A.63}$$

When multiplying two numbers together, there are really no constraints. With matrices, however, it is a bit different. When performing a matrix product, one must respect a dimensionality constraint. Recall that our matrices have dimensionality: $T \in \mathbb{R}^{3\times 2}$ and $N \in \mathbb{R}^{2\times 1}$. The multiplication of the price matrix, $T$, with the nominal-amount matrix, $N$, may only be performed because the number of columns of $T$ corresponds to the number of rows of $N$. Practically, this means that we, quite reasonably, require prices and nominal amounts for each security. The result of a matrix product is also a matrix and the dimension of the resulting matrix is simply the rows and columns of the first and second matrices, respectively. Specifically, $V \in \mathbb{R}^{3\times 1}$ is the dimensionality of the portfolio value matrix. This makes logical sense since we have the prices of the securities over three consecutive days. As a result, we are able, with the given information, to compute the portfolio value for each of these 3 days.

Formalizing this important point about matrix multiplication, we may make the following statement: given matrices $A \in \mathbb{R}^{m\times l}$ and $B \in \mathbb{R}^{l\times n}$, then the product $C = AB \in \mathbb{R}^{m\times n}$ is also a matrix with each element defined by,

$$C_{i,j} = \sum_{k=1}^{l} A_{i,k} B_{k,j}. \tag{A.64}$$

We conclude this brief introduction to matrices with a few brief definitions of important matrix concepts.

- A matrix with the same number of rows and columns is called a square matrix.
- $A^T$ is referred to as the transpose of $A$. The transposition of a matrix consists of swapping the rows and the columns. In particular, each element of a matrix transpose, $A^T$, is defined as

  $$A^T_{j,i} = A_{i,j}. \tag{A.65}$$

  Transposition is often used to transform a column vector $x \in \mathbb{R}^{n\times 1}$ into a row vector $x^T \in \mathbb{R}^{1\times n}$. This simple technique is extremely useful in matching the relevant and correct dimensions for addition and multiplication of matrices.
- A symmetric matrix is a square matrix, where $A^T = A$. This may seem somewhat esoteric, but it is a feature of the covariance matrix used extensively in preceding chapters for risk computations.

Now that we have some basic notation and ideas, we can now focus on a few important problems in matrix theory.[30]

### A.4.1 Solving Linear Systems

A classic and important problem in matrix theory involves finding the vector $x \in \mathbb{R}^{n \times 1}$, given the matrix $A \in \mathbb{R}^{m \times n}$ and the vector $y \in \mathbb{R}^{m \times 1}$, where all three objects are linked by the following mathematical relation,

$$Ax = y. \tag{A.66}$$

The problem looks quite simple. If we forget for a moment that $A$ is a matrix, the answer seems trivial. With very basic algebra, we could solve for $x$ and claim that the answer is simply $x = \frac{1}{A}y$. Abstracting from the technical details of the problem in Eq. (A.66), our intuition is correct. $x$ is indeed conceptually equal to $\frac{1}{A}y$. $A$, however, is not a number, but instead a matrix. The tricky part of our problem, is what does $\frac{1}{A}$ or $A^{-1}$ mean when $A$ is not a number, but instead a matrix? This question requires a rather long answer.

We require two rather strong assumptions in the following development. First, we assume that $A$ is a square matrix. Second, we require that it is actually possible to define a matrix $A^{-1}$ such that the relation $x = A^{-1}y$ holds. In other words, we assume that $A^{-1}$ exists. This need not always be the case. When $A^{-1}$ does not exist, the matrix $A$ is termed *singular*.

We consider this problem using numerical methods, simply because analytical solutions are too cumbersome.[31] Numerical methods are readily implemented on a computer and thus finding a solution is generally quite efficient.[32]

Our first case examine the problem when the matrix $A$ is triangular. A triangular matrix has all elements above (or below) the main diagonal equal to zero—the non-zero entries form a triangle, hence the name. In this special situation, the explicit formulation of our problem is given by the following matrix system,

---

[30]This is a very brief and cursory introduction to the subject. An excellent practical and more detailed source on matrix theory is Golub and Van Loan [4].

[31]There are two main families of numerical methods for solving linear systems: direct and iterative methods. Iterative methods typically have convergence errors since the method must iterate several times on the same process to reach a solution. Each iteration brings the result closer to the exact solution, but upon completion, there will typically be a residual difference called the convergence error. We will not discuss iterative algorithms, although they are used extensively in practice.

[32]The main issues with numerical methods are the inherent existence of rounding errors and the efficiency or the number of elementary operations involved in the solution. The latter provides an estimate of the amount of time a computer requires to solve a problem. Both topics are the subject of much discussion in matrix computations. See Press et al. [7] and Golub and Van Loan [4] for more details.

$$\begin{pmatrix} A_{1,1} & A_{1,2} & \cdots & A_{1,n-1} & A_{1,n} \\ 0 & A_{2,2} & \cdots & A_{2,n-1} & A_{2,n} \\ \vdots & \vdots & \ddots & \vdots & \vdots \\ 0 & 0 & \cdots & A_{n-1,n-1} & A_{n-1,n} \\ 0 & 0 & \cdots & 0 & A_{n,n} \end{pmatrix} \begin{pmatrix} x_1 \\ x_2 \\ \vdots \\ x_{n-1} \\ x_n \end{pmatrix} = \begin{pmatrix} y_1 \\ y_2 \\ \vdots \\ y_{n-1} \\ y_n \end{pmatrix} \tag{A.67}$$

This naturally translates into the following set of equations,

$$\begin{aligned} A_{1,1}x_1 + A_{1,2}x_2 + \cdots + A_{1,n-1}x_{n-1} + A_{1,n}x_n &= y_1, \\ A_{2,2}x_2 + \cdots + A_{2,n-1}x_{n-1} + A_{2,n}x_n &= y_2, \\ &\vdots \\ A_{n-1,n-1}x_{n-1} + A_{n-1,n}x_n &= y_{n-1}, \\ A_{n,n}x_n &= y_n. \end{aligned} \tag{A.68}$$

The solution is actually trivial. One need only work from the final equation backwards. Starting from the last equation, one readily finds the value as

$$x_n = \frac{y_n}{A_{n,n}}. \tag{A.69}$$

Given this value, we may move the second last equation, using the newly acquired $x_n$, and solve for $x_{n-1}$ as,

$$x_{n-1} = \frac{y_{n-1} - A_{n-1,n}x_n}{A_{n-1,n-1}}. \tag{A.70}$$

With values $x_n$ and $x_{n-1}$, we proceed to $x_{n-2}$. One merely moves, in this manner, from the last equation to the first equation, recursively. At each step, we need only to solve one equation in one unknown.

A bit of caution is advised with the previous algorithm. When using division, one must ensure that he or she is not dividing by zero. In our problem, this amounts to verifying that $\{A_{i,i}, i = 1, \ldots, n\}$ are non-zero values—that is, all diagonal elements of the matrix, $A$, must be non-zero. It turns out, although we will not demonstrate this fact, that this is always true when the inverse of $A$, or $A^{-1}$, exists. Since we have already assumed this fact, there should be no problem. To complete the picture, the link between these two facts is termed the *determinant*. More specifically, for an inverse of a given matrix to exist, it must have an non-zero determinant. For a triangular matrix, the determinant can be easily computed as the product of the diagonal elements,

$$\det(A) = \prod_{i=1}^{n} A_{i,i}. \tag{A.71}$$

If any element in the product described in Eq. (A.71) is negative, the determinant will be zero. This leads us to the conclusion, for our triangular matrix at least, that the following three points are equivalent: none of the $A_{i,i}$ are null, the determinant of $A$ is non-zero, and the inverse of $A$ exists. Although this logic applies more generally for non-singular matrices, it requires a rather more involved development.

The Gauss method—also named after Carl Friedrich Gauss, the father of the normal distribution—basically tries to convert a given matrix, $A$, into a triangular matrix.[33] In mathematical terms, we seek a matrix $M$ such that,

$$MAx = My, \tag{A.72}$$

where the product $MA$ is a diagonal matrix. The good news is that we do *not* actually need to explicitly compute the matrix $M$. It is the result of simple operations on the matrix $A$. The underlying shaded box provides a detailed example for the interested reader.

Imagine we are given a system of three equations with three unknowns, $\{x_1, x_2, x_3\}$. The linear system is written as,

$$\begin{aligned} x_1 + x_2 + 2x_3 &= 1, \\ 3x_1 + x_2 + x_3 &= 3, \\ x_1 + 2x_2 + 2x_3 &= 2. \end{aligned} \tag{A.73}$$

Equation (A.73) can, of course, be written in matrix format as,

$$\begin{pmatrix} 1 & 1 & 2 \\ 3 & 1 & 1 \\ 1 & 2 & 2 \end{pmatrix} \begin{pmatrix} x_1 \\ x_2 \\ x_3 \end{pmatrix} = \begin{pmatrix} 1 \\ 3 \\ 2 \end{pmatrix}. \tag{A.74}$$

To build a triangular matrix, we work row by row corresponding to each of the equations in our linear system. We do not need to modify the first row, but in the second row one sees that the element $(2, 1) = 3$ must be set to zero. To do so, we modify the second equation, using the first one, in such a way that the coefficient of $x_1$ is zero. This is possible by removing three times the first equation from the second,

$$3x_1 + 3x_2 + 6x_3 = 3, \tag{A.75}$$

(continued)

[33] Gauss did not actually invent this method, although he did make some improvements to it.

$$\frac{3x_1 + x_2 + x_3 = 3,}{2x_2 + 5x_3 = 0,}$$

to arrive at a new representation for the second equation. It may seem like we have irreparably changed the system. A bit of reflection, however, reveals, that we may increase or decrease the size of any individual equation by a constant factor and not alter the solution. More generally, any linear combination of our original system will still have the same solution. The Gauss method essentially exploits this fact. One can now rewrite the linear system as,

$$\begin{aligned} x_1 + x_2 + 2x_3 &= 1, \\ 2x_2 + 5x_3 &= 0, \\ x_1 + 2x_2 + 2x_3 &= 2. \end{aligned} \tag{A.76}$$

In matrix form, this appears as,

$$\begin{pmatrix} 1 & 1 & 2 \\ 0 & 2 & 5 \\ 1 & 2 & 2 \end{pmatrix} \begin{pmatrix} x_1 \\ x_2 \\ x_3 \end{pmatrix} = \begin{pmatrix} 1 \\ 0 \\ 2 \end{pmatrix}. \tag{A.77}$$

This is already a pleasant improvement. We now move to the third and final row. Observe that we can use the first equation to set the element (3, 1) to zero. We can even directly subtract the first equation from the third one. The system becomes now:

$$\begin{aligned} x_1 + 2x_2 + 2x_3 &= 2, \\ \underline{x_1 + x_2 + 2x_3} &\underline{= 1,} \\ x_2 &= 1, \end{aligned} \tag{A.78}$$

or in matrix form,

$$\begin{pmatrix} 1 & 1 & 2 \\ 0 & 2 & 5 \\ 0 & 1 & 0 \end{pmatrix} \begin{pmatrix} x_1 \\ x_2 \\ x_3 \end{pmatrix} = \begin{pmatrix} 1 \\ 0 \\ 1 \end{pmatrix}. \tag{A.79}$$

We are not quite there yet, but we are close. The labelling of the elements of $x$ is, in fact, arbitrary and swapping $x_2$ and $x_3$ will make our matrix triangular. Once this is performed, then our problem is solved. The final form of our problem, therefore, is

(continued)

$$\begin{pmatrix} 1 & 2 & 1 \\ 0 & 5 & 2 \\ 0 & 0 & 1 \end{pmatrix} \begin{pmatrix} x_1 \\ x_3 \\ x_2 \end{pmatrix} = \begin{pmatrix} 1 \\ 0 \\ 1 \end{pmatrix}. \tag{A.80}$$

If solving by hand, this trick makes sense. For a computer algorithm it is not a great idea—the computer would likely spend too much time seeking these types of opportunities relative effort required to blindly to perform the previously described computations. We will, therefore, continue to solve the problem following the Gauss method. We essentially need to set element $(3,2)$ to zero using the second equation. This can be done by subtracting twice the third equation from the second one to arrive at,

$$\begin{array}{r} 2x_2 = 2, \\ 2x_2 + 5x_3 = 0, \\ \hline -5x_3 = 2, \end{array} \tag{A.81}$$

Consequently, our problem can be written as,

$$\begin{pmatrix} 1 & 1 & 2 \\ 0 & 2 & 5 \\ 0 & 0 & 5 \end{pmatrix} \begin{pmatrix} x_1 \\ x_2 \\ x_3 \end{pmatrix} = \begin{pmatrix} 1 \\ 0 \\ -2 \end{pmatrix} \tag{A.82}$$

Despite the fact that the two triangular matrices in Eqs. (A.80) and (A.82) do not look the same, the solutions for $x$ are identical. We leave it as an exercise for the reader to verify that,

$$\begin{aligned} x_1 &= \frac{4}{5}, \\ x_2 &= 1, \\ x_3 &= -\frac{2}{5}, \end{aligned} \tag{A.83}$$

for both choices of triangular matrix.

Absent a need to swap columns, this method is sometimes called an *LU* factorisation. The name *LU* arises from the remarkable fact that the decomposition can be written as the product of a lower triangular matrix, $L$, and an upper triangular matrix, $U$. A lower (upper) triangular matrix has zeroes *below* (*above*) the main diagonal. Note that a lower (upper) diagonal matrix can be transformed into an upper (lower) diagonal matrix merely by taking its transpose.

(continued)

> It is also possible, with additional effort, to make the matrix diagonal and not triangular. A diagonal matrix is, in fact, the solution. This method, termed the *Gauss-Jordan* method, is unfortunately not more efficient.

Overall, the Gauss method is probably the most commonly used method for solving linear systems absent specific properties that can be exploited. In financial risk applications, however, we often work with a special type of matrix with a property called positive semi-definiteness. For these particular type of matrices, there is a more efficient technique than the Gauss method—this technique is addressed in the next section.

### A.4.2 Cholesky Decomposition

In the preceding chapters, we make extensive use of covariance matrices. A covariance matrix takes real values, is symmetric, and is positive semi-definite. These properties can be exploited to solve linear systems involving covariance matrices and to decompose such matrices into a more convenient form.

Positive definiteness is the matrix equivalent of a positive scalar value. A matrix $A \in \mathbb{R}^{n \times n}$ is termed positive semi-definite if,

$$x^T Ax \geq 0, \tag{A.84}$$

for all $x \in \mathbb{R}^{n \times 1}$. This is roughly the equivalent of a scalar value taking values greater than or equal to zero (i.e., $0 \leq a \in \mathbb{R}$). Positive definiteness implies $>$ and not $\geq$ in Eq. (A.84). To see this more clearly, consider the following example matrix,

$$P = \begin{pmatrix} 1 & 0 \\ 0 & 1 \end{pmatrix}. \tag{A.85}$$

Right- and left-multiplying $P$ by the same vector, $\begin{pmatrix} x \\ y \end{pmatrix}$, yields,

$$(x, y) \begin{pmatrix} 1 & 0 \\ 0 & 1 \end{pmatrix} \begin{pmatrix} x \\ y \end{pmatrix} = x^2 + y^2 \geq 0. \tag{A.86}$$

For any values of $x$ and $y$—presuming, they are both not equal to zero—the result will always be a positive number. Thus, the matrix $P$ is positive definite. This general definition of a positive-definite matrix is not terribly convenient since one is not typically willing to test all possible vectors. According to our knowledge, the best method to assess if a symmetric matrix is positive definite, is to try to perform the decomposition that we will describe in this section. If the algorithm fails, one can conclude that the matrix is not positive definite.

Exploiting these properties brings us to another approach to solving linear systems. If a matrix is symmetric and positive definite, then it is possible to prove that the $LU$ decomposition—this is a Gauss decomposition without swapping columns—is indeed the product of a lower triangular matrix and its transpose, an upper triangular matrix. In mathematical notation,

$$A = LL^T. \tag{A.87}$$

Moreover, if all of the diagonal elements are positive, the matrix $L$ is unique.

To see how the Cholesky decomposition works, a bit of heavy lifting is required. We begin with the $LL^T$ multiplication,

$$A_{i,j} = \sum_{k=1}^{\min(i,j)} L_{i,k} L_{j,k}. \tag{A.88}$$

This is a general representation of each element of $A$ in terms of the product of the elements in the triangular matrix, $L$. This a start, but we need to make this much more explicit. We want an explicit description of each of the elements of $L$ using the known elements of $A$. The first element is trivial,

$$\begin{aligned} A_{1,1} &= L_{1,1}^2, \\ L_{1,1} &= \sqrt{A_{1,1}}. \end{aligned} \tag{A.89}$$

If $A_{1,1} < 0$, then Eq. (A.89) will result in a complex number and the decomposition will fail. This is one, among a number, of places in the algorithm where positive definiteness is required. We may further proceed to easily define each of the $L_{1,k}$ elements as,

$$\begin{aligned} A_{1,k} &= L_{1,1} L_{1,k}, \\ L_{1,k} &= \frac{A_{1,k}}{\sqrt{A_{1,1}}}. \end{aligned} \tag{A.90}$$

This defines each of the elements in the first row of $L$ in terms of known values of $A$. The next step is to compute the subsequent row of $L$ starting from the appropriate diagonal column. The first element of the column $L_{i,i}$, where $i \neq 1$, is defined as,

$$A_{i,i} = \sum_{k=1}^{i} L_{k,1}^2, \tag{A.91}$$

$$L_{i,i} = \sqrt{A_{i,i} - \sum_{k=1}^{i-1} L_{i,k}^2}.$$

The associated off-diagonal elements are,

$$A_{i,j} = \sum_{k=1}^{\min(i,j)} L_{i,k} L_{j,k}, \tag{A.92}$$

$$L_{i,j} = \frac{A_{i,j} - \sum_{k=1}^{j-1} L_{i,k} L_{j,k}}{L_{i,i}}.$$

If none of these equations yields a complex number, then we have fully populated the lower-triangular matrix, $L$, and the algorithm is complete. Admittedly, Eqs. (A.88) to (A.92) are intimidating to look at and fairly ugly to verify. They are nevertheless relatively straightforward to incorporate into a computer algorithm.

More than a method for solving linear systems, we use the Cholesky decomposition extensively in risk computations for the simulation of risk-factor outcomes from a covariance matrix.[34]

### A.4.3 Eigenvalues and Eigenvectors

A final aspect of matrix theory addressed in this appendix concerns what is called the *eigenvalue* problem. While the eigenvalue problem shows up frequently in both pure and applied mathematics, in the finance literature, it is very closely linked with the idea of principal components analysis. Given that this is an important technique used in the description of yield-curve dynamics, it is useful to review the basic ideas in detail.

Eigenvalues, denoted as $\lambda$, and their eigenvectors, denoted as $x$, are the solutions of the following relation,

$$Ax = \lambda x, \tag{A.93}$$

where $A$ is a square matrix. To prevent obvious and uninteresting solutions to Eq. (A.93), the eigenvectors may *not* be null vectors. This basic form is somewhat

[34]This technique was invented by André-Louis Cholesky a French military officer involved in the construction of maps. A change, or recalibration, in the definition of the metre forced the military to adjust their existing map grids. Cholesky developed his eponymous decomposition and greatly simplified these calculations. Killed in battle towards the end of the first world war, Cholesky's algorithm was published posthumously.

inconvenient. As such, the eigenvalue problem is typically written in an alternative fashion. Specifically, eigenvalues are the solution of the following characteristic equation,

$$\det (A - \lambda I) = 0, \tag{A.94}$$

where $I$ is the identity matrix. The explicit formulation of the characteristic equation has a polynomial form. Let us illustrate it in the simplest case of an arbitrary matrix $A \in \mathbb{R}^{2\times 2}$,

$$A = \begin{pmatrix} a_{11} & a_{12} \\ a_{21} & a_{22} \end{pmatrix}. \tag{A.95}$$

To find the eigenvalues of $A$, we must solve for the determinant of the following equation,

$$\begin{aligned} A - \lambda I &= \begin{pmatrix} a_{11} - \lambda & a_{12} \\ a_{21} & a_{22} - \lambda \end{pmatrix}, \\ &= 0. \end{aligned} \tag{A.96}$$

Using the definition of a determinant, we re-write Eq. (A.96) as,

$$\begin{aligned} (a_{11} - \lambda)(a_{22} - \lambda) - a_{12}a_{21} &= 0, \\ \lambda^2 - (a_{11} + a_{22})\lambda + a_{11}a_{22} - a_{12}a_{21} &= 0. \end{aligned} \tag{A.97}$$

Examination of Eq. (A.97) reveals that it is a second-order polynomial in $\lambda$—there is a squared $\lambda$ term, a $\lambda$ term, and a constant. The two solutions to this second-order polynomial, called the characteristic polynomial, which may potentially take complex values, are given by the explicit formulae,

$$\begin{aligned} \lambda_1 &= \frac{(a_{11} + a_{22}) + \sqrt{(a_{11} + a_{22})^2 - 4(a_{11}a_{22} - a_{12}a_{21})}}{2}, \\ \lambda_2 &= \frac{(a_{11} + a_{22}) - \sqrt{(a_{11} + a_{22})^2 - 4(a_{11}a_{22} - a_{12}a_{21})}}{2}. \end{aligned} \tag{A.98}$$

Simply put, the eigenvalues are the roots of the characteristic polynomial. The order of the polynomial is directly linked to the size of the matrix. Unfortunately, mathematicians have been aware for a long time, that explicit general solutions are not available for polynomials with an order of greater than or equal to five.[35] Thus,

[35]This result was established by Evarist Galois in 1832 shortly before he was shot in duel.

as before, we are required to use numerical methods to solve for the eigenvalues of an arbitrary square matrix with dimensionality greater than five.

Instead of working through an involved general numerical approach to the eigenvalue problem, let us instead consider a diagonal matrix. This is an interesting example due to its simplicity and because it clearly demonstrates the fundamental logic behind the general approach. The characteristic equation for a diagonal matrix is given as,

$$D - \lambda I = \begin{pmatrix} d_{11} - \lambda & \dots & 0 \\ \vdots & \ddots & \vdots \\ 0 & \dots & d_{nn} - \lambda \end{pmatrix}. \tag{A.99}$$

The associated determinant is given as,

$$\det(D - \lambda I) = \prod_{i=1}^{n} (d_{ii} - \lambda) = 0. \tag{A.100}$$

We can see from Eq. (A.100) that the characteristic polynomial is already factorised. This directly implies that diagonal elements $\{d_{ii}, i = 1, \dots, n\}$ are, in fact, the eigenvalues of the matrix, $D$. A generic numerical algorithm for finding eigenvalues is provided, for the interested reader, in the underlying shaded box.

This general approach to the eigenvalue problem is termed the *Jacobi* algorithm. It is based on the idea that the eigenvalues of a matrix do not change if the matrix is multiplied by an orthogonal matrix. As a consequence, we may safely apply the following transformation to our original matrix, $A$,

$$A_{k+1} = O^T A_k O. \tag{A.101}$$

The basic idea behind this approach is to make very small, or alternatively dwarf, the off-diagonal elements of $A$. We need to make this a bit more precise. The Jacobi algorithm consists of identifying the largest, in absolute terms, off-diagonal element and bringing it to zero using an orthogonal matrix. To make this more concrete and easier to understand, let's work through a sizeable example. We begin with the following symmetrical matrix $A \in \mathbb{R}^{5\times 5}$,

$$A_0 = \begin{pmatrix} a_{11} & a_{12} & a_{13} & a_{14} & a_{15} \\ a_{12} & a_{22} & a_{23} & a_{24} & a_{25} \\ a_{13} & a_{23} & a_{33} & a_{34} & a_{35} \\ a_{14} & a_{24} & a_{34} & a_{44} & a_{45} \\ a_{15} & a_{25} & a_{35} & a_{45} & a_{55} \end{pmatrix}. \tag{A.102}$$

(continued)

Suppose now that you identify the largest off-diagonal element as $a_{24}$. Given that $A$ is symmetric, we have that $a_{24} = a_{42}$. We will now multiply the matrix $A$ with the following orthogonal matrix,

$$O = \begin{pmatrix} 1 & 0 & 0 & 0 & 0 \\ 0 & \cos\theta & 0 & \sin\theta & 0 \\ 0 & 0 & 1 & 0 & 0 \\ 0 & -\sin\theta & 0 & \cos\theta & 0 \\ 0 & 0 & 0 & 0 & 1 \end{pmatrix}. \tag{A.103}$$

After a fairly tedious and cumbersome matrix multiplication, we find that $O^T A_0 O$ has the following form where the diagonal element, $a_{24}$ is highlighted (in bold),

$$\begin{pmatrix} a_{11} & a_{12}c - a_{14}s & a_{13} & a_{12}s + a_{14}c & a_{15} \\ a_{12}c - a_{14}s & (a_{22}c - a_{24}s)\,c - s\,(a_{24}c - a_{44}s) & a_{23}c - a_{34}s & \mathbf{(a_{22}c - a_{24}s)\,s + c\,(a_{24}c - a_{44}s)} & a_{25}c - a_{45}s \\ a_{13} & a_{23}c - a_{34}s & a_{33} & a_{23}s + a_{34}c & a_{35} \\ a_{12}s + a_{14}c & \mathbf{(a_{22}c - a_{24}s)\,s + c\,(a_{24}c - a_{44}s)} & a_{23}s + a_{34}c & (a_{22}s + a_{24}c)\,s + c\,(a_{24}s + a_{44}c) & a_{25}s + a_{45}c \\ a_{15} & a_{25}c - a_{45}s & a_{35} & a_{25}s + a_{45}c & a_{55} \end{pmatrix}. \tag{A.104}$$

To fit everything on the page, we have shortened the representation of $\sin\theta$ to $s$ and $\cos\theta$ to $c$. At this point, we can make several observations:

- only the elements of the row and the column corresponding to our biggest off-diagonal elements have been modified by our orthogonal transformation;
- the resulting matrix, $O^T A_0 O$ remains symmetric; and
- the sum of the squared elements of $A$ is identical to the sum of squared elements of $O^T A O$. This final fact is less obvious and we leave it as an exercise for any readers who might be suffering from insomnia.

Our objective remains to make $a_{24} = 0$, which may be accomplished by extracting the element of the second row and fourth column of $O^T A_0 O$—shown in bold in Eq. (A.104)—setting it equal to zero and simplifying,

$$\begin{aligned} (a_{22}\cos\theta - a_{24}\sin\theta)\sin\theta + \cos\theta\,(a_{24}\cos\theta - a_{44}\sin\theta) &= 0, \\ \cos\theta\sin\theta\,(a_{22} - a_{44}) + a_{24}\left(\cos^2\theta - \sin^2\theta\right) &= 0, \\ \frac{\sin 2\theta}{2}(a_{22} - a_{44}) + a_{24}\cos 2\theta &= 0. \end{aligned} \tag{A.105}$$

(continued)

The preceding simplification has made use of a number of fundamental trigonometric relationships. The equation can be further simplified as:

$$\cot 2\theta = \frac{a_{44} - a_{22}}{2a_{24}}. \tag{A.106}$$

Using this result, the matrix $A_1$ can then be written as,

$$\begin{pmatrix} a_{11} & a_{12}c - a_{14}s & a_{13} & a_{12}s + a_{14}c & a_{15} \\ a_{12}c - a_{14}s & (a_{22}c - a_{24}s)\,c - s\,(a_{24}c - a_{44}s) & a_{23}c - a_{34}s & 0 & a_{25}c - a_{45}s \\ a_{13} & a_{23}c - a_{34}s & a_{33} & a_{23}s + a_{34}c & a_{35} \\ a_{12}s + a_{14}c & 0 & a_{23}s + a_{34}c & (a_{22}s + a_{24}c)\,s + c\,(a_{24}s + a_{44}c) & a_{25}s + a_{45}c \\ a_{15} & a_{25}c - a_{45}s & a_{35} & a_{25}s + a_{45}c & a_{55} \end{pmatrix}. \tag{A.107}$$

We are now in a position to identify the largest off-diagonal element in $A_1$ and iterate upon this procedure. Say, for example, the largest off-diagonal element is $A_1\,(2, 3)$. Repeated application of the previous steps will also modify the element $(2, 4)$ since it falls on the same row as the new target off-diagonal element. This may seem somewhat discouraging since we just struggled to set this value to zero. It seems, therefore, that our algorithm will continuously undo the hard work from the previous step. Fortunately, this is not the case. Despite the perturbation, each iteration applies to the elements of the same row and column of the selected element, the matrix is indeed converging to the diagonal matrix. While we will not prove this fact, we will nonetheless highlight the logic behind the proof. To do so, we need to distinguish the elements of original matrix $A_k$ from those of the transformed one $A_{k+1}$. Let us, therefore, agree on the notation $a_{ij}^{(k)}$ for the elements of $A_k$, which should then not be confused with the power of the element.

The starting point is the result of the Jacobi transformation along with the definition of $\theta$. This results in the following system of equations,

$$\begin{aligned} a_{pp}^{(k+1)} &= a_{pp}^{(k)} \cos^2\theta + a_{qq}^{(k)} \sin^2\theta - 2a_{pq}^{(k)} \sin\theta\cos\theta \\ a_{qq}^{(k+1)} &= a_{pp}^{(k)} \sin^2\theta + a_{qq}^{(k)} \cos^2\theta + 2a_{pq}^{(k)} \sin\theta\cos\theta \\ a_{pq}^{(k)}\left(\cos^2\theta - \sin^2\theta\right) &= \left(a_{qq}^{(k)} - a_{pp}^{(k)}\right) \sin\theta\cos\theta. \end{aligned} \tag{A.108}$$

After a number of tedious algebraic manipulations, it is possible to show that,

$$\left(a_{pp}^{(k+1)}\right)^2 + \left(a_{qq}^{(k+1)}\right)^2 = \left(a_{pp}^{(k)}\right)^2 + \left(a_{qq}^{(k)}\right)^2 + 2\left(a_{pq}^{(k)}\right)^2. \tag{A.109}$$

(continued)

What is the significance of Eq. (A.109)? Recall that the sum of the squared elements of $A$ is conserved by our orthogonal transformation. Equation (A.109) essentially tells us that the diagonal elements are, as a consequence of our transformation, growing. This implies that the off-diagonal elements are shrinking. If we perform this transformation often enough, therefore, we will eventually converge to a diagonal matrix whose eigenvalues are simply its diagonal elements. As a final point, it is also possible to prove that the product of the orthogonal matrices used to transform our initial matrix into a diagonal form will converge to a matrix whose column vectors are the eigenvectors of the initial matrix.

The eigenvalue decomposition is a bit hard to conceptually understand. A useful description, however, is to think of the eigenvectors of a matrix $A$ as a given set—under a particular linear transformation—of orthogonal, or independent, directions. The eigenvalues are the scaling factors, where each is associated with a given eigenvector, or direction. Those eigenvectors, or again directions, associated with the largest eigenvalues are, in some sense, the most important. They are, in essence, the most important directions for $A$. Often a small subset of eigenvectors, and their associated eigenvalues, describe a large matrix surprisingly well. Knowledge of these key directions can, therefore, be rather useful in solving a broad range of practical problems.

## References

1. P. Billingsley, *Probability and Measure*, 3rd edn. (Wiley, New York, 1995)
2. G. Casella, R.L. Berger, *Statistical Inference* (Duxbury Press, Belmont, 1990)
3. R. Durrett, *Probability: Theory and Examples*, 2nd edn. (Duxbury Press, Belmont, 1996)
4. G.H. Golub, C.F. Van Loan, *Matrix Computations* (The John Hopkins University Press, Baltimore, 2012)
5. J.W. Harris, H. Stocker, *Handbook of Mathematics and Computational Science* (Springer, New York, 1998)
6. N.L. Johnson, S. Kotz, N. Balakrishnan, *Continuous Univariate Distributions*, Wiley Series in Probability and Statistics, vol. 2 (Wiley, New York, 1997)
7. W.H. Press, S.A. Teukolsky, W.T. Vetterling, B.P. Flannery, *Numerical Recipes in C: The Art of Scientific Computing*, 2nd edn. (Cambridge University Press, Cambridge, 1992)
8. H.L. Royden, *Real Analysis* (Prentice-Hall, Englewood Cliffs, 1988)
9. A. Sklar, *Random Variables, Distribution Functions, and Copulas: A Personal Look Backward and Foreward.* Institute of Mathematical Statistics Lecture Notes–Monograph Series, vol. 28 (Institute of Mathematical Statistics, 1996), pp. 1–14
10. N.N. Taleb, *The Black Swan: The Impact of the Highly Improbable* (Random House, New York, 2007)
11. W.A.R. Weiss, *An Introduction to Set Theory* (University of Toronto, Toronto, 2008)

# A Few Thoughts on Optimization

B

*Science is what we understand well enough to explain to a computer. Art is everything else we do.*

Donald Knuth

Portfolio optimization typically implies the selection of a strategic benchmark that—based on some set of pre-defined criteria—is considered optimal. It may, for example, provide the highest expected return while simultaneously restricting one or more dimensions of risk. This is a difficult problem. As previously indicated, this entire book takes the strategic benchmark as given.[1] Portfolio optimization, in our context, instead implies finding a collection of instruments matching the strategic benchmark's risk-factor exposures. In short, when we talk about portfolio optimization, we essentially mean portfolio replication. This is, conceptually at least, an easier challenge. It may, therefore, be tempting to conclude that one may trivially use optimization techniques to ensure that one's portfolio is always neutral—from a risk-factor exposure perspective—to the strategic benchmark. Sadly, this is not entirely true.

To understand why it may not be so easy, let us consider the basic problem. Every month asset managers typically must rebalance—or consciously decide to deviate from—a changing benchmark. A portfolio's composition is, of course, the responsibility of the portfolio manager and is determined by personal views and the investment process. The safest position, from a strategic perspective, is always to be neutral to one's strategic benchmark. Even relatively simple government benchmarks, however, can be comprised of more than 100 bonds. Multi-currency benchmarks may have substantially more instruments. Selection of a relatively small, neutral replicating portfolio, therefore, is not always a trivial task. Understanding one's neutral portfolio would be a useful starting point.

[1] See Meucci [2] for more detail on this important and interesting problem.



D.J. Bolder, *Fixed-Income Portfolio Analytics*,
DOI 10.1007/978-3-319-12667-8

Constructing a neutral portfolio that eliminates a large set of risk-factor exposures is not terribly demanding. Indeed, in the subsequent section, we will highlight how this might be accomplished through the use of a simple linear program that can be solved by any number of different computational softwares. The solution is quite compelling. For a strategic benchmark with over 100 securities, with aggregate exposure to ten separate risk factors, the optimizer would suggest a neutral replicating portfolio with roughly ten securities.

This is very helpful. The difficulty, however, is that optimizers are not particularly intelligent. They seek only to optimize within the pre-defined objectives and constraints. An extremely small numerical change in the objective function—even if the economic difference is negligible—can lead to sizeable changes in the final solution. One could run, for example, an optimization every month on the rebalancing date and the solution will almost certainly eliminate the risk-factor exposures. The portfolios suggested by the optimizer on subsequent dates may nonetheless be dramatically different. That is, from 1 month, or even 1 day, to the next, small changes in market conditions may make it optimal to hold a completely different set of securities. When rebalancing a portfolio, however, it rather defeats the purpose to liquidate numerous positions and acquire new ones. Not only is it expensive, it is also logically inconsistent. We simply do not expect to see drastic changes in a neutral replicating portfolio over relatively short time intervals.

There are possible solutions. One might add additional constraints into one's optimization routine that penalizes the objective function for purchasing and selling existing securities.[2] This will make it less desirable—given changing market conditions—for the optimizer to completely change the portfolio structure from one period to the next. While such a technique can and does help, our experience suggests that such an approach is both fragile and requires careful oversight by a well-trained hand.[3]

Does this imply that this type of portfolio optimization is useless? No, it does not. It means, however, that use of an optimizer to mechanize one's monthly rebalancing process is probably *not* a great idea. Portfolio management, even the relatively passive sort, is as much an art as it is a science. An optimizer will be hard-pressed to reliably replace human judgement. An optimization approach, however, is particularly useful when staring fresh with the funding of a new portfolio. Neither changing portfolios nor transaction costs are an issue in this setting. A straightforward linear optimization technique may thus be relatively easily and fruitfully employed. This would allow the portfolio manager to begin with a neutral portfolio construction, permitting her to gradually add positions after the initial funding. An optimal replicating portfolio is also useful, at any time, as a kind of lighthouse indicating a possible collection of safe and neutral positions. This might provide useful information to the portfolio manager. In the coming pages,

[2]One practical approach, along these lines, is found in Ramaswamy and Scott [4].

[3]In short, it is a bit messy and requires significant judgement.

therefore, we will outline one possible way that such a replicating portfolio might be constructed.

## B.1 A Linear Program

Imagine that we have been given some funds and tasked with investing them in a new fixed-income portfolio. Moreover, our objective is to construct a neutral portfolio relative to strategic benchmark. If one can define what exactly is meant by a *neutral* portfolio, then one can use an optimization procedure to find the *best* neutral portfolio. Construction of an optimization structure will require a bit of thought.

What might we mean by neutral? A portfolio that replicates the risk-factor exposures of the strategic benchmark is a sensible starting point. A good replicating portfolio should, therefore, match the following characteristics of the strategic benchmark:

- the key-rate durations;
- the overall modified duration;
- the overall convexity; and
- the spread duration.

There are probably a few other characteristics, beyond risk-factor exposures, that are required for a good replicating portfolio. In particular, we need something to optimize—we might seek to find the highest possible portfolio yield or credit spread subject to meeting the previously defined constraints. Although it sounds sensible, this may or may not be a desirable criterion for optimization. Bonds with particularly high yields or credit spreads may, depending on the structure of the benchmark, be difficult to purchase in sufficient size. We could alternatively seek the most *liquid* replicating portfolio. The challenge is that liquidity is *not* directly observable—one needs to use a substitute measure, or proxy, for the relative liquidity of a bond. A useful proxy for liquidity could involve minimizing the distance between the yield and coupon of the selected bonds.[4]

We may also wish to only consider bonds with sufficiently large amount outstanding, so they can easily be located and purchased—this is another liquidity based constraint. As a practical matter, we also generally restrict the investment universe such that only bonds from the strategic benchmark may be purchased. This is not strictly necessary, but it substantially simplifies the analysis since, in this case, one only requires knowledge of the strategic benchmark. Finally, a good replicating portfolio has the smallest possible number of bonds consistent with the

[4]The idea is that the longer the lapse of time since the original issuance of a given bond, the lower its liquidity. This is generally correlated with the relative premium or discount of the bond price, which is a function of the distance between the market yield and coupon rate.

aforementioned constraints. This is a good start. We have a much clearer idea of what is required.

### B.1.1 A Simple Case

Before our optimization problem can be given mathematical structure and solved—or, in other words, before we can mathematically construct a sensible replicating portfolio—we need to introduce some notation. Let $n$ denote the number of securities in the strategic benchmark—these are the candidate instruments for our replicating portfolio. For each of the $n$ bonds in the benchmark, let us denote the yield to maturity for each bond as,

$$\{y_i \mid i = 1, \ldots, n\}. \tag{B.1}$$

We set the amount outstanding for each bond as,

$$\{\beta_i \mid i = 1, \ldots, n\}. \tag{B.2}$$

The instrument-level credit spread is correspondingly,

$$\{\theta_i \mid i = 1, \ldots, n\}. \tag{B.3}$$

The spread duration for each bond is written as,

$$\{\phi_i \mid i = 1, \ldots, n\}. \tag{B.4}$$

The $m$ key-rate durations associated with each security are,

$$\left\{\kappa_{i,j} \mid i = 1, \ldots, n; j = 1, \ldots, m\right\}. \tag{B.5}$$

Finally, the coupon rates are described as,

$$\{c_i \mid i = 1, \ldots, n\}. \tag{B.6}$$

The target values for each of these measures in the strategic benchmark, which we seek to replicate, are denoted as $\bar{y}, \bar{\beta}, \bar{\theta}, \bar{\phi}, \{\bar{\kappa}_j, j = 1, \ldots, m\}$, and $\bar{c}$, respectively.

So far, nothing has been mentioned about the modified duration for each security. It is indirectly modelled through the key-rate durations. We need only recall that

$$\sum_{j=1}^{m} \kappa_{i,j} = d_i, \tag{B.7}$$

**Table B.1** Optimization notation

| Description | Instrument level | Portfolio level |
|---|---|---|
| Yield | $\{y_i \mid i = 1, \ldots, n\}$ | $\bar{y}$ |
| Amount outstanding | $\{\beta_i \mid i = 1, \ldots, n\}$ | $\bar{\beta}$ |
| Credit spread | $\{\theta_i \mid i = 1, \ldots, n\}$ | $\bar{\theta}$ |
| Spread duration | $\{\phi_i \mid i = 1, \ldots, n\}$ | $\bar{\phi}$ |
| Key-rate duration | $\{\kappa_{i,j} \mid i = 1, \ldots, n; j = 1, \ldots, m\}$ | $\{\bar{\kappa}_j, j = 1, \ldots, m\}$ |
| Coupon rate | $\{c_i \mid i = 1, \ldots, n\}$ | $\bar{c}$ |
| Modified duration | $\{d_i \mid i = 1, \ldots, n\}$ | $\bar{d}$ |

This table summarizes the key variables employed in the determination of a replicating portfolio using a linear programming approach.

for all $i = 1, \ldots, n$. If we force the replicating portfolio to match each individual key-rate duration of the strategic benchmark, we will, by Eq. (B.7), simultaneously match its modified duration. For completeness, $d_i$ denotes the modified duration of the $i$th bond and $\bar{d}$ is the duration of the strategic benchmark.

This is a large, and overwhelmingly dense, set of symbols. Table B.1 attempts to help digest this information by summarizing our notation in a single place.

We may now proceed to the formal optimization problem. Each of the measures in Table B.1 is a linear function of the portfolio weights.[5] This is an important advantage—linear problems are significantly easier to solve.[6] We intend to exploit this feature of our problem and employ a linear programming approach. Before jumping into the mathematics, let us first describe what we are trying to accomplish in words. Taking our previously stated requirements into account, we seek a set of security weights, $\omega \in \mathbb{R}^n$ that

- maximizes some linear function of the portfolio—yield, credit spread, or a liquidity proxy[7];
- matches the individual strategic key-rate durations (i.e., curve risk of strategic benchmark);
- matches the strategic spread duration (i.e., spread-risk of portfolio);
- ensures no short positions and forces the weights to sum to unity; and

[5]In contrast, most risk measures are functions of higher moments of the joint return distribution and, as such, non-linear.

[6]Non-linear optimization techniques are not necessarily complicated—although they can be—but, absent certain conditions, it is very difficult to know with certainty if one has obtained a global minimum.

[7]The credit spread is also relevant for portfolio of sovereign bonds, since it says something about the relative *richness* or *cheapness* of the securities. Some caution is required, because often bonds are cheap for a reason. One good reason why a particularly bond is inexpensive is that it may be quite difficult to acquire. In short, it is illiquid. This may make it difficult to locate and may make maximizing credit spreads a less desirable choice.

- requires that each bond considered has, at least, one billion outstanding—this is a liquidity constraint.

We may organize all of these objectives into the following linear program.

$$\max_{\omega \in \mathbb{R}^n} \quad \begin{bmatrix} \omega_1 \cdots \omega_n \end{bmatrix} \begin{bmatrix} |y_1 - c_1| \\ \vdots \\ |y_n - c_n| \end{bmatrix}, \tag{B.8}$$

subject to:

$$\underbrace{\begin{bmatrix} \kappa_{1,1} & \cdots & \kappa_{n,1} \\ \vdots & \ddots & \vdots \\ \kappa_{1,m} & \cdots & \kappa_{n,m} \\ \phi_1 & \cdots & \phi_n \\ 1 & \cdots & 1 \end{bmatrix}}_{M} \begin{bmatrix} \omega_1 \\ \vdots \\ \omega_n \end{bmatrix} = \underbrace{\begin{bmatrix} \bar{\kappa}_1 \\ \vdots \\ \bar{\kappa}_m \\ \bar{\phi} \\ 1 \end{bmatrix}}_{b}$$

$$\omega_i \in (0,1) \text{ for } i = 1, \ldots, n$$

$$\beta_i \geq 10^9 \text{ for } i = 1, \ldots, n$$

This problem is highly underdetermined. By underdetermined, we refer to a linear system where the number of constraints (i.e., equations) is less than the number of unknowns. This would suggest that there exist an infinite number of solutions. Which one should be selected? In an optimization setting, this not much of a concern, since we seek the single best solution—from this infinite set of possibilities—that maximizes our objective function.

In our setting, to be more precise, we have 10–12 constraints and (potentially) hundreds of unknowns ($n$). There is, therefore, an infinite number of possible solutions. The optimizer selects the best choice among all possibilities that satisfies the constraints. With nine or ten constraints, it will typically find a similar number of bonds that meet all the constraints and maximize the objective function. Ultimately, the portfolio weights for the neutral replicating portfolio (i.e., $\omega_{\text{neutral}}$) are:

$$\omega_{\text{neutral}} = \{\omega^* \in \mathbb{R}^n \mid \omega_i^* > 0, i = 1, \ldots, n\}. \tag{B.9}$$

where $\omega^* \in \mathbb{R}^n$ is the solution to the linear program.[8] Given its linear structure, solving this optimization problem is both extremely fast and reliable.

---

[8]Practically, the replicating portfolio weights are:

$$\omega_{\text{neutral}} = \{\omega^* \in \mathbb{R}^n \mid \omega_i^* > \varepsilon, i = 1, \ldots, n\}, \tag{B.10}$$

where $\varepsilon$ is a small number, such as 0.001, to avoid unreasonably small allocations.

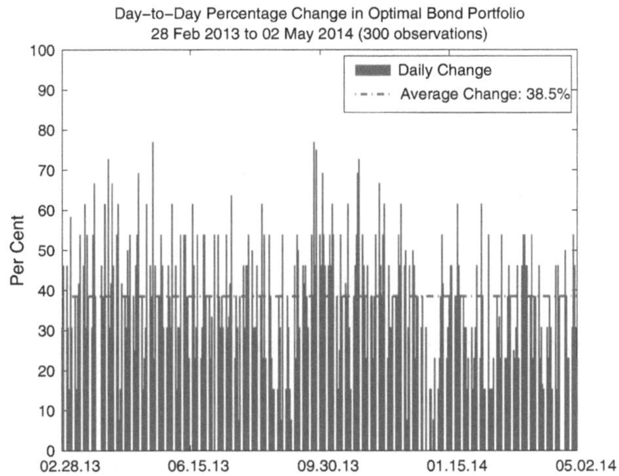


**Fig. B.1** Solution variability. Here we observe 300 consecutive applications of the linear program in Eq. (B.13) to a UST portfolio with an average of 236 bonds and 13 risk factors. The solution, on average, requires 13 bonds to match these risk factors, but a simultaneous turnover of about five bonds per day

The mathematics is not the main issue. The main challenge relates to intertemporal volatility in the solution to the problem in Eq. (B.8). Small changes in the market yield—or some other market variable, depending on the choice of objective function—may force the optimization to jump from one solution to another choice from the infinite pool of possible solutions satisfying the constraints. Exactly, this issue is illustrated in Fig. B.1, where we illustrate 300 consecutive daily applications of the linear program in Eq. (B.13) to a UST portfolio with an average of 236 bonds and 13 risk factors. The solution, on average, requires 13 bonds to match these risk factors. There is, however, an average turnover of about 40 % in the bonds found in the final solution from 1 day to the next—this implies about five transactions per day to maintain the replicating portfolio.[9] In some cases, it even exceeds 70 % or about nine transactions.

Without sensible constraints on the transactions, Fig. B.1 suggests that this is probably not a sensible approach for rebalancing one's portfolio to the strategic benchmark. It is, however, a very powerful technique for portfolio construction or, from time to time, examining how one might quickly return to a neutral stance relative to the strategic benchmark.

[9]This implies roughly 100 transactions per month merely to remain neutral.

### B.1.2 Extending the Simple Case

The approach in the previous section works well with a portfolio of homogeneous bonds—including, for example, only US Treasuries. With multi-currency portfolios or a setting with heterogeneity of credit quality, however, it will not perform as well. The reason is that additional risk-factor exposures need to be considered. In other words, in many cases, we need to ensure that the allocation to different sovereign issuers or credit classes—agency, supranationals, corporates—also match the strategic benchmark. This will require a bit more machinery.

Let's examine this extension within the context of a European sovereign benchmark. We begin by defining the set of $k$ European sovereign issuers as,

$$\mathscr{I} \equiv \{I_1, \ldots, I_k\}. \tag{B.11}$$

Each element in $\mathscr{I}$ represents a separate sovereign issuer. We then proceed to define the following indicator variable,

$$\mathbb{I}_{\omega_i \in I_1} = \begin{cases} 1 : \text{bond } \omega_i \text{ is issued by } I_1 \\ 0 : \text{otherwise} \end{cases}. \tag{B.12}$$

This object will help us determine how the optimizer allocates to each individual issuer. Finally, we define $P_{I_x}$ as the proportion of issuer $x$ held in the benchmark. If the neutral replicating portfolio holds an amount equal to $P_{I_{\text{Germany}}}$—say, for example, it is 0.40—then it will be neutral with respect to Germany. A good replicating portfolio will be neutral to all sovereign issuers in the strategic benchmark.

Mathematically, this is readily incorporated into Eq. (B.8) through the addition of some new constraints,

$$\max_{\omega \in \mathscr{I} \subset \mathbb{R}^n} \quad \begin{bmatrix} \omega_1 \cdots \omega_n \end{bmatrix} \begin{bmatrix} \theta_1 \\ \vdots \\ \theta_n \end{bmatrix}, \tag{B.13}$$

$$\text{subject to:}$$

$$M\omega = b$$

$$\underbrace{\begin{bmatrix} \mathbb{I}_{\omega_1 \in I_1} & \cdots & \mathbb{I}_{\omega_n \in I_1} \\ \vdots & \ddots & \vdots \\ \mathbb{I}_{\omega_1 \in I_k} & \cdots & \mathbb{I}_{\omega_n \in I_k} \end{bmatrix}}_{\hat{M}} \begin{bmatrix} \omega_1 \\ \vdots \\ \omega_n \end{bmatrix} = \underbrace{\begin{bmatrix} P_{I_1} \\ \vdots \\ P_{I_k} \end{bmatrix}}_{\hat{b}}$$

$$\omega_i \in (0, 1) \text{ for } i = 1, \ldots, n$$

$$\beta_i \geq 10^9 \text{ for } i = 1, \ldots, n$$

The solution to this linear program will meet all of the usual risk-factor exposures as well as the sovereign exposures. It can readily be extended, in a similar manner, to incorporate credit categories.

It should be stressed that this is far from an exhaustive treatment of optimization techniques for portfolio construction. Unlike many of the other techniques presented in this book, optimization can go badly wrong. It is not hard, for example, to write down an objective function and a set of constraints for which a solution does not exist. In brief, this short appendix is an illustrative demonstration of what might be done, some ideas on when it might be profitably employed and, more importantly, some of its potential limitations. Use of these techniques can be very helpful, but expertise and caution is advised.

## B.2 Concluding Thoughts

Optimization, even in the linear setting, is not a trivial undertaking. Entire books can, and have, been written on this topic. It would be naive, and probably dishonest, to claim that we can adequately treat this topic in a short technical appendix. The previous discussion is neither complete nor was it intended to be so. The objective of this short appendix, in contrast, is to highlight a complementary quantitative technique for portfolio construction, which may be employed within the context of our fixed-income portfolio analytic framework. Reading the preceding pages will not make you an expert on optimization or linear programming. This would require in-depth study and practical experience.[10] The hope, however, is that it might point one in the right direction when considering the use of these techniques for portfolio construction.

To repeat the principal thesis of this discussion: the use of an optimizer to mechanize one's monthly rebalancing process is probably *not* a great idea. Portfolio management is both art and science. An optimizer cannot, and should not, replace human judgement. An optimization approach, however, can be particularly useful when funding of a new portfolio or trying to get a clear idea of a neutral replicating portfolio. At best, an optimization approach—in a linear setting—can point a general direction towards neutrality with respect to the strategic benchmark and offer ideas to the portfolio manager. For this reason, it should be considered part of our fixed-income portfolio analytic toolbox, even if it is only used occasionally and with caution.

[10]There are, for the interested reader, many excellent books on this topic. A good practical starting point is Press et al. [3, Chapter 10]. The classical reference on linear programming, however, is Gass [1].

## References


1. S.I. Gass, *Linear Progamming: Methods and Applications* (McGraw-Hill, New York, 1985)
2. A. Meucci, *Strategic Asset Allocation*, 1st edn. (Springer, Berlin, 2007)
3. W.H. Press, S.A. Teukolsky, W.T. Vetterling, B.P. Flannery, *Numerical Recipes in C: The Art of Scientific Computing*, 2nd edn. (Cambridge University Press, Cambridge, 1992)
4. S. Ramaswamy, R. Scott, Managing a multi-currency bond portfolio, in *Advanced Bond Portfolio Management: Best Practices in Modeling and Strategies*, ed. by F.J. Fabozzi, L. Martellini, P. Priaulet (Wiley, 2006)

# Index




D.J. Bolder, *Fixed-Income Portfolio Analytics*,
DOI 10.1007/978-3-319-12667-8

# Author Index




D.J. Bolder, *Fixed-Income Portfolio Analytics*,
DOI 10.1007/978-3-319-12667-8